BIOLOGY OF ANIMALS

BIOLOGY OF ANIMALS

Cleveland P. Hickman
Department of Zoology, DePauw University,
Greencastle, Indiana

Cleveland P. Hickman, Jr.
Department of Biology, Washington and Lee University,
Lexington, Virginia

With 852 illustrations, 188 in color

The C. V. Mosby Company
SAINT LOUIS   1972

To

Our wives—Frances and Rae

preface

BIOLOGY OF ANIMALS is intended for two classes of students—those who desire a general knowledge of animal biology as a liberal arts course and those who are taking an introductory course with a view toward advancing further in the field of the life sciences. Although it is based upon the general principles set forth in the senior author's larger work, *Integrated Principles of Zoology,* the present text is really a new work.

Animal biology is a rich and fascinating study. We have attempted to draw together the numerous facts and concepts of zoology into a compact and readable treatment. However, the immense amount of zoological information available now and the ever-increasing complexity of life study place a limit to the bits of information the human brain can handle at one time. (Some think that seven simultaneous information bits is about the maximum for most of us.) Consequently we must try to divide the body of knowledge into segments we can comprehend and then arrange these segments into a logical development. There are at least as many ways to build a zoology text as there are zoology texts to choose from. Our own arrangement has resulted in a text with six parts.

PART ONE explains how matter is the basis of both the nonliving and the living and aims to give the student an understanding of how the living is related to nonliving matter. This part emphasizes the chemical nature of all matter and how life is an outcome of an advanced level of organization of matter.

PART TWO presents the organization of the animal body, with emphasis upon cellular structure and function and the architectural unity of all biological organization.

PART THREE describes the functional systems of animals and explains how problems of life's existence are met. In this part the student is shown the unity that exists between the physicochemical processes of the nonliving and the living.

PART FOUR emphasizes the environmental relations of the animal. Special attention is given to the ecosystems of man and other animals and to the pressing problems facing man in his struggle to understand his responsibility to his environment. Since the causes of pollution and man's efforts to curb them rank high in the priorities of life science studies, much attention is given to their discussion in this part.

PART FIVE considers the continuity of life in the concepts of reproduction, development, hereditary transmission, and the evolutionary processes. Special emphasis is placed upon our present understanding of evolution as a great unifying concept in life science, with special reference to man's position in the animal kingdom.

PART SIX discusses systematically the animal groups showing how these groups fit into the evolutionary blueprint of the living world. We have placed greater emphasis on the physiology, natural history, and behavior of animals, and less emphasis on taxonomy and structure than in the senior author's larger work.

Throughout the text we have tried to incorporate the results of recent investigations, which is admittedly a difficult task in this scientific age of rapid new discoveries and new concepts in the life sciences. References are included at the end of each of the six parts to direct interested students to well-written works that supplement the text material. We regret that space does not allow critical annotations of the references.

Many new illustrations have been drawn especially for this edition. New tables have been added to emphasize significant information, and many illustrated schemes may be of interest to the student. Many of the illustrations are from the larger work, but for the new illustrations we are grateful for the artistic skill of William C. Ober. We are especially indebted to Frances M. Hickman who assumed the burden of condensing the greater portion of Part Six and helped in many other ways. Others who offered advice or checked sections of the manuscript are Rae R. Hickman, Margaret Ober, Michael Pleva, and Henry S. Roberts. We are grateful to them all.

We of course assume full responsibility for mistakes and omissions.

Cleveland P. Hickman
Cleveland P. Hickman, Jr.

contents

CONTENTS

PART FOUR
Environmental relations of the animal

PART FIVE
The continuity of life

PART SIX
The diversity of life

BIOLOGY OF ANIMALS

part one

How life differs from the nonliving

All material things or substances in the world are made up of small units called molecules. Each of these molecules consists of smaller and simpler units known as atoms, or **chemical elements,** of which there are ninety-two basic ones plus about a dozen others that man has created with his cyclotron, nuclear reactors, etc. The atoms of all elements consist of still smaller components, **elementary particles.** No one knows how many elementary particles there are, but the four most common ones that are of greatest concern are the **protons, electrons, neutrons,** and **photons.** This matter or substance may exist in any of three states—solid, liquid, or gaseous. Solid matter exists in a rigid condition with definite form such as crystalline substances. Liquids tend to flow and assume the shape of their containers. Gases have neither definite volume nor shape and will expand to fill any container regardless of its size. It is possible for each of these states to change to another state under certain physical conditions. Matter or substance may thus be defined as anything that occupies space and has mass. Matter and energy are related and interchangeable. This relationship may be expressed in Einstein's equation, $E = mc^2$, in which E represents energy; m, matter; and c, the speed of light. The kinetic molecular theory of matter states that any material body is not a continuous extent of matter, but that the molecules of which it is composed are in constant vibration. Even an apparent solid body is actually made up of tiny particles (molecules) vibrating, colliding, and rebounding with each other. Although the fundamental nature of matter is still unknown to scientists, a great deal is known about its properties and behavior.

The kind of molecules and their arrangement and behavior determine the nature of the various substances, whether of rock, glass, steel, salt, wood, food, or anything else. The whole problem in the study of matter is determining how the different molecules are put together. It is known that some molecules are made up of few and others of many atoms. Some organic molecules are elongated in long threadlike arrangements and others in compact three-dimensional patterns. Chemical changes involve the breaking down of molecules into atoms, and their recombination forms new molecules of different compositions and arrangements.

Present-day biologists generally agree that living things on this planet have evolved from nonliving matter. All matter can be classified as living or nonliving. The science of the living is called **biology,** but as one advances into the nature of matter the distinction between the physical and biological sciences becomes less noticeable. Although **chemistry** is thought of as a science that treats the composition and structure of material substances and **biochemistry** as dealing with living matter and its products, both disciplines are guided and regulated by similar laws. Both living and nonliving matter share the same kind of atoms. The chief differences between the two classes of matter are in different levels of organization and complexity. Organic substances, or those dealing with life, have gradually become more complex by the buildup of the available inorganic compounds. Different levels of organization have different properties, and the properties of protoplasm, or the basic living substances, are those we associate with biologic manifestations.

Basic structure
of matter

To understand and appreciate the nature and origin of life, it is necessary to know the chemical composition of matter and the physical laws that govern the behavior of this matter. The idea that a specific living substance is composed of unique chemical elements is no longer held. Rather, life must be thought of as a special form of activity dependent on interactions between substances. Life involves complex chemical systems. All living things, from the simplest unicellular forms to the highest and most complex organisms, have been derived from the nonliving or inanimate world. Ever since life originated on this planet certain historical processes have affected living organisms and guided them to their present form. So far as we know, most, if not all, the activities of living things can be explained by the general laws of physics and chemistry.

MATTER AS THE SUBSTANCE
OF THE LIVING AND NONLIVING

The fundamental difference between living and nonliving matter lies in the organization of molecules as revealed in different levels of complexity. This hierarchy of organization (Fig. 1-1) affords a broad concept for understanding and appreciating all matter or substance, both living and nonliving. This concept reveals that the whole universe is composed of different levels of organization, from simple units to the most complex units. By realizing that each of these levels has its own characteristic properties, one comes to appreciate how life came about as a natural consequence of the way matter is constructed. The differences between the living and nonliving fall into a gradual and logical interpretation of creation.

On the **first level** of organization so far as is known are the components of the atom known as the **subatomic particles,** or elementary particles. The atoms of all elements are constructed out of the particles—electrons, protons, neutrons, photons, mesons, and many others. No one knows how many elementary particles there are or whether or not some are more elementary than others. Is there a basic stuff of the universe? So far, physicists are unable to answer this question. Although more than ninety subnuclear entities have been defined, the whole of chemistry can be explained in terms of four of these subatomic particles—electrons, protons, photons, and neutrons especially—interacting with each other through electromagnetic forces.

On the **second level** of organization are the **atoms** that consist of the subatomic particles. This level is represented by the familiar elements (an atom may be said to be the smallest complete unit of an element) of oxygen, hydrogen, carbon, nitro-

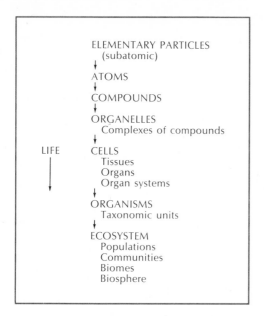

ELEMENTARY PARTICLES
  (subatomic)
  ↓
ATOMS
  ↓
COMPOUNDS
  ↓
ORGANELLES
  Complexes of compounds
  ↓
LIFE   CELLS
  ↓   Tissues
          Organs
          Organ systems
  ↓
ORGANISMS
  Taxonomic units
  ↓
ECOSYSTEM
  Populations
  Communities
  Biomes
  Biosphere

Fig. 1-1. Hierarchy of organization levels. Organization of matter into increasing levels of complexity has made possible the emergence of life from the nonliving, and each level has its own characteristic properties.

gen, etc. The atom may be considered the most basic term in science. Ancient Greeks had a crude concept of the atoms as building blocks, but the first real meaning of the concept of the atoms was given by J. Dalton (1803). He assumed that atoms were indivisible particles of matter, but the discovery of the electron (Sir J. J. Thomson, 1897) showed the atom to be reducible. Lord Rutherford (1911) gave the present concept of the structure of the atom as being made up of a nucleus with electrons revolving around it. Sir J. Chadwick (1932) added the neutron to the nucleus. Other subatomic particles have been discovered since Rutherford's fundamental concept.

The molecular organization is the **third level** and consists of two or more atoms combined to form a **molecule.** Substances differ according to the structure and composition of their molecules. Molecules exist in an almost infinite variety of shapes and sizes — rings, chains, polymers, and other shapes and forms. In chemical action their structure is broken down from their original arrangement and is recombined to form other molecules and different substances. The significance of molecular organization is all the more striking because free

atoms are mostly nonexistent. Perhaps atoms were free in the early origin of the earth, but they later entered into combinations. Biologically, this level also makes possible the formation of the large macromolecules (proteins, DNA, RNA) and the colloidal nature of matter so necessary for life organization.

We are then led to the **fourth level** of **living cells.** The properties of life can be found only in association with the molecular organization of living systems. Life properties do not occur apart from the properties of the separate molecules. Aside from viruses and certain slime molds (whose status is debatable), it may be said that the cell is the smallest portion of protoplasm that is capable of life.

The **fifth level** is represented by **multicellular organisms** that make up the plant and animal kingdoms. All organisms are placed as taxonomic units into schemes of classification. These organisms fit into a still higher organization level in the hierarchy of matter — the **ecosystem** that may be considered the **sixth level,** made up of populations and communities. Each animal is a member of some association or system such as a community or population. Such supraindividual systems exist in many forms in nature, ranging from loose associations to highly integrated associations that may display the characteristics of mutual interdependence and self-regulation. Within each of these levels of hierarchical organizations, subhierarchical levels not only become more apparent and more numerous as one ascends the various levels, but these levels also display a more integrated interaction of specialization and division of labor. Each level can also exist independently of the other levels, and each has its own properties or laws. Each level can be understood not so much by the analysis of its component parts as by the study of its overall organization that enables the parts to function as a whole unit.

It will be seen that the concept of organization levels helps in understanding the broader unity of all matter. Biologic organization may be considered a unique aggregation of chemical structures that fit into a unified pattern of physical constitution. Life substance with its different levels has properties that are unique and give life its own fundamental characteristics. Life resides in organisms

1	2	3	4	5	6	7	8	9	10	11	12	13	14	15	16	17	18
1 H 1.01																	2 He 4.00
3 Li 6.94	4 Be 9.01											5 B 10.8	6 C 12.0	7 N 14.0	8 O 16.0	9 F 19.0	10 Ne 20.2
11 Na 23.0	12 Mg 24.3											13 Al 27.0	14 Si 28.1	15 P 31.0	16 S 32.1	17 Cl 35.5	18 Ar 39.9
19 K 39.1	20 Ca 40.1	21 Sc 45.0	22 Ti 47.9	23 V 50.9	24 Cr 52.0	25 Mn 54.9	26 Fe 55.8	27 Co 58.9	28 Ni 59.7	29 Cu 63.5	30 Zn 65.4	31 Ga 69.7	32 Ge 72.6	33 As 74.9	34 Se 79.0	35 Br 79.9	36 Kr 83.8
37 Rb 85.5	38 Sr 87.6	39 Y 88.9	40 Zr 91.2	41 Nb 92.9	42 Mo 95.9	43 Tc (99)	44 Ru 101.	45 Rh 103.	46 Pd 106.	47 Ag 108.	48 Cd 112.	49 In 115.	50 Sn 119.	51 Sb 122.	52 Te 128.	53 I 127.	54 Xe 131.
55 Cs 133.	56 Ba 137.	57 La 139.	72 Hf 178.	73 Ta 181.	74 W 184.	75 Re 185.	76 Os 190.	77 Ir 192.	78 Pt 195.	79 Au 197.	80 Hg 201.	81 Tl 204.	82 Pb 207.	83 Bi 209.	84 Po (210)	85 At (210)	86 Rn (222)
87 Fr (223)	88 Ra (226)	89 Ac (227)															

58 Ce 140.	59 Pr 141.	60 Nd 144.	61 Pm (147)	62 Sm 150.	63 Eu 152.	64 Gd 157.	65 Tb 159.	66 Dy 162.	67 Ho 165.	68 Er 167.	69 Tm 169.	70 Yb 173.	71 Lu 175.
90 Th 232.	91 Pa (231)	92 U 238.	93 Np (237)	94 Pu (242)	95 Am (243)	96 Cm (247)	97 Bk (247)	98 Cf (249)	99 Es (254)	100 Fm (253)	101 Md (256)	102 No (256)	103 Lr (257)

FIG. 1-2. Periodic table of elements showing those essential to life. Six major elements in living systems are set in bright red; five essential minor elements are set in light red. Eleven trace elements essential to life are set in boldface black. Other elements, notably nickel, aluminum, antimony, mercury, cadmium, lead, silver, and gold, are usually present in living systems in trace amounts but are not dietary essentials. Fluorine and silicon may be essential to some microorganisms. The mass number is shown above, and the atomic weight (approximate) below, each element.

that must undergo a continued change under the impact of environmental changes. Therefore they must be influenced by geographic conditions and the historical element of evolutionary change since life began. These changes and conditions result in a diversity of adaptations, so that each species has a way of life different in some particulars from that of other species.

THE ELEMENTS

Of the ninety-two naturally occurring elements, only eighteen or twenty are considered essential for life. Of these, six have been selected by nature to play especially important roles in living systems. These major elements, shown in red in Fig. 1-2, are carbon (C), hydrogen (H), nitrogen (N), oxygen (O), phosphorus (P), and sulfur (S). Most organic molecules are built with these six elements. Another five essential elements found in less abundance in living systems and shown in Fig. 1-2 are calcium (Ca), potassium (K), sodium (Na), chlorine (Cl), and magnesium (Mg). Several other elements, called trace elements, are found in minute amounts in animals and plants, but are nevertheless indispens-

able for life. These are manganese (Mn), iron (Fe), iodine (I), molybdenum (Mo), cobalt (Co), zinc (Zn), selenium (Se), copper (Cu), chromium (Cr), tin (Sn), and vanadium (V). They are shown in black boldface type in Fig. 1-2. Other elements of the periodic table may be present in living things, often as contaminants, but none have been demonstrated essential for life. Many elements are extremely toxic to life in small amounts.

NATURE OF ATOMS AND MOLECULES

An atom is the smallest structural unit of an element, containing all the chemical properties of that element. The atom consists of a **nucleus** and **elec-**

trons. The nucleus, containing most of the atom's mass, is made up of two kinds of particles, **protons** and **neutrons.** These two particles have about the same weight, each being about 1,800 times heavier than an electron. The protons bear positive charges, and the neutrons have no charges (neutral). Although there is the same number of protons in the nucleus as there are electrons revolving around the nucleus, the number of neutrons may vary. For every positively charged proton in the nucleus, there is a negatively charged electron. The total charge of the atom is thus neutral.

The number of protons in the nucleus is the **atomic number** of the atom. Thus hydrogen, helium, and lithium, containing respectively 1, 2, and 3 protons in their nuclei, have atomic numbers of 1, 2, and 3, respectively. The **mass number** of an atom is the total number of protons and neutrons in its nucleus. The nucleus of an oxygen atom con-

tains 8 protons and 8 neutrons. It therefore has a mass number of 16. The heaviest natural element, uranium, has a nucleus of 92 protons (its atomic number is thus 92) and 146 neutrons, and so its mass number is 238. The elements are designated by convenient symbols that show the atomic number as a subscript, and the total number of protons and neutrons (mass number) as a superscript. Thus oxygen is designated $_8O^{16}$, hydrogen $_1H^1$ (1 proton, no neutrons), helium $_2He^4$ (2 protons, 2 neutrons), and so on (Fig. 1-3).

Atomic weight. The atomic weight of an atom is nearly the same as its mass number. However, a quick examination of a periodic table shows that none of the elements have atomic weights of exact integers. How are atomic weights derived? Obviously an atom weighs far too little to be weighed, or even to serve as a useful index for weight comparison. A hydrogen atom, for example, weighs 1.67×10^{-24} grams. Consequently, physicists have assigned a set of meaningful relative weights to the elements, using carbon as a base for comparison. Carbon, $_6C^{12}$, with 6 protons and 6 neutrons was assigned the integral value 12. By this scale, protons and neutrons have masses of 1.0073 and 1.0087, respectively—masses close to, but not exactly, one. (Electrons have an almost negligible mass of 0.00055.) Thus no element, except carbon 12, has an atomic weight of an exact integer: hydrogen weighs 1.0080; helium, 4.0026; lithium, 6.939; and so on.

Isotopes. It is possible for 2 atoms of the same element to have the same number of protons in their nuclei, but a different number of neutrons. For example hydrogen exists in nature primarily as $_1H^1$,

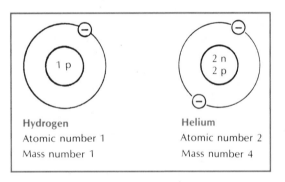

FIG. 1-3. Two lightest atoms. Since first shell closest to atomic nucleus can hold only 2 electrons, helium shell is closed so that helium is chemically inactive.

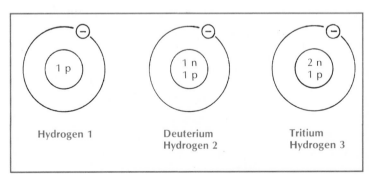

FIG. 1-4. Three isotopes of hydrogen. Of the 3 isotopes, hydrogen 1 makes up about 99.98% of all hydrogen and deuterium makes up about 0.02%. Tritium is found only in traces in water. Numbers indicate approximate atomic weights. Most elements are mixtures of isotopes. Some elements (for example, tin) have as many as 10 isotopes.

that is, it contains 1 proton, but no neutron. However, there are also trace amounts of two other forms of hydrogen: $_1H^2$, which has 1 proton and 1 neutron and is called deuterium, and $_1H^3$, which has 1 proton and 2 neutrons and is called tritium. These three varieties differ only in the number of neutrons in the nucleus (Fig. 1-4). Such forms of an element, having the same charge but different atomic weights, are called **isotopes.**

Radioactive isotopes. Although most of the naturally occurring elements are stable, all elements have at least 1 radioactive isotope. These isotopes undergo spontaneous disintegration with the emission of one or more of three types of particles, or rays—**gamma rays** (a form of electromagnetic radiation), **beta rays** (electrons), and **alpha rays** (positively charged helium nuclei stripped of their electrons). Most of the isotopes of greatest use in biologic tracer studies are beta and gamma emitters. Virtually all are prepared synthetically in nuclear reactors and cyclotrons. Among the commonly used radioisotopes are carbon 14 ($_6C^{14}$), tritium ($_1H^3$), and phosphorus 32 ($_{15}P^{32}$). Using radioisotopes, biologists are able to trace movements of elements and tagged compounds through organisms. Radioisotopes are also used to great advantage in the diagnosis of disease in man, such as cancer of the thyroid gland. Our present understanding of metabolic pathways in animals and plants is due in very large part to the recent use of this powerful analytical tool.

Electron "shells" of atoms. According to Bohr's planetary model of the atom, the electrons revolve around the nucleus of an atom in precise orbits, or shells. This simplified picture of the atom has been greatly modified by recent experimental evidence.

According to the quantum theory the electrons surrounding the nucleus exist at discrete energy levels, called quantum levels. This theory replaces the older idea that electrons revolve around the nucleus in definite shells, or orbit patterns. A quantum level represents a discrete energy value. These energies, and hence the energies of electrons in these quantum levels, increase as the distance from the nucleus increases. Electrons tend to move as close to the nucleus as possible. However, there is a physical maximum to the number of electrons

that can occupy each quantum level. Thus as the inner level becomes filled, additional electrons are forced into more distant quantum levels. These outer electrons are more excited and have a higher energy content.

Although the picture of the atom described by the quantum theory provides a much better basis for understanding the atom, the old planetary model is still useful in interpreting chemical phenomena. The number of concentric "shells," or the paths of the electrons in their orbits, varies with the element. Each shell can hold a maximum number of electrons. The first shell next to the atomic nucleus can hold a maximum of 2 electrons (hydrogen has only 1), and the second shell can hold 8; other shells also have a maximum number, but no atom can have more than 8 electrons in its outermost shell. Inner shells are filled first, and if there are not enough electrons to fill all the shells, the outer shell is left incomplete. Hydrogen has 1 proton and 1 electron in its single orbit but no neutron. Since its shell can hold 2 electrons, it has an incomplete shell. Helium has 2 electrons in its single shell, and its nucleus is made up of 2 protons and 2 neutrons. Since the 2-electron arrangement in helium's shell is the maximum number for this shell, the shell is closed and precludes all chemical activity. There is no known compound of helium. Neon is another inert gas (chemically inactive [Fig. 1-5]) because its outer shell contains 8 electrons, the maximum number. However, stable compounds of xenon (an inert gas) with fluorine and oxygen are formed under special conditions. Oxygen has an atomic number of 8. Its 8 electrons are arranged with 2 in the first shell and 6 in the second shell (Fig. 1-5). It is active chemically, forming compounds with almost all the elements except the inert gases.

The number of electrons in the outer shell varies from 0 to 8. With either 0 or 8 in this shell, the element is chemically inactive. When there are fewer than 8 electrons in the outer shell, the atom will tend to lose or gain electrons to have an outer shell of 8, which will result in a charged ion. Atoms with 1 to 3 electrons in the outer shell tend to lose them to other atoms and to become positively charged ions because of the excess protons in the nucleus. Atoms with 5 to 7 electrons in the outer orbit tend

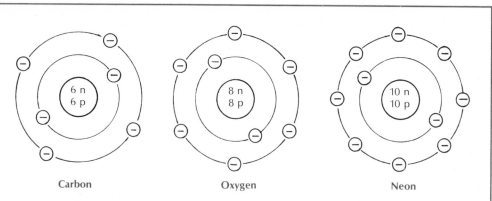

FIG. 1-5. Electron shells of 3 common atoms. Since no atom can have more than 8 electrons in its outermost shell and 2 electrons in its innermost shell, neon is chemically inactive. However, the second shells of carbon and oxygen, with 4 and 6 electrons, respectively, are open so that these elements are electronically unstable and react chemically whenever appropriate atoms come into contact. Chemical properties of atoms are determined by their outermost electron shells.

to gain electrons from other atoms and to become negatively charged ions because of excess electrons over the protons. Positive and negative ions tend to unite.

A combination of two or more elements forms a **compound. A molecule** is the smallest particle of an element or compound that can exist by itself. Compounds always contain different elements, but the term molecule is often applied to single elements. Molecules of most elementary gases are made up of 2 atoms. Thus a molecule of oxygen is O_2 and that of hydrogen is H_2. In such cases the 2 atoms are always found together. Carbon dioxide (CO_2) and methane (CH_4) are true compounds because they consist of different elements, each of which is present in definite proportions.

Gram atomic and gram molecular weights. To appreciate the quantitative expression of compounds, the student should be familiar with the terms **gram atomic weight** and **gram molecular weight.** A gram atomic weight of an element is its atomic weight expressed in grams. The number of atoms in a gram atomic weight is 6.02×10^{23} (Avogadro's number). This means that the gram atomic weight of all elements has the same number of atoms. One gram of hydrogen (its gram atomic weight) has the same number of atoms as does 16

grams of oxygen or 12 grams of carbon, their gram atomic weights. A gram molecular weight, or **mole,** is the sum of the atomic weights of the atoms in a molecule. This also means that the same number of moles of all substances have the same number of molecules. Thus a mole (18 grams) of water (H_2O) has the same number of molecules as does a mole (32 grams) of oxygen (O_2) or a mole (28 grams) of carbon monoxide (CO).

Ions and oxidation states. We have seen that every atom has a tendency to complete its outer shell to increase its stability. Let us examine how 2 atoms with incomplete outer shells, sodium and chloride, can interact to fill their outer shells. Sodium, with 11 electrons, has 2 electrons in its first shell, 8 in its second shell, and only 1 in the third shell. The third shell is highly incomplete; if this third-shell electron were lost, the second shell would be the outermost shell and produce a stable atom. Chlorine, with 17 electrons, has 2 in the first shell, 8 in the second and 7 in the incomplete third shell. Chlorine must gain an electron to fill the outer shell and become a stable atom. Clearly, the transfer of the third-shell sodium electron to the incomplete chlorine third shell would yield simultaneous stability to both atoms. But since electrons bear negative charges, the sodium-to-chlorine transfer

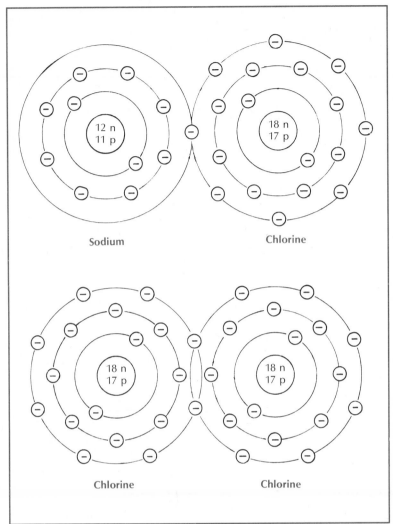

FIG. 1-6. Ionic bond. When an atom of sodium and one of chlorine react to form a molecule, a single electron in outer shell of sodium is transferred to outer shell of chlorine. This causes outer or second shell (third shell is empty) of sodium to have 8 electrons and also chlorine to have 8 electrons in its outer or third shell. The compound thus formed is called sodium chloride (NaCl). By losing 1 electron, sodium becomes a positive ion, and by gaining 1 electron, chlorine (chloride) becomes a negative ion. This ionic bond is held together by a strong electrostatic force.

FIG. 1-7. Covalent bond. Each chlorine atom has 7 electrons in its outer shell, and by sharing one pair of electrons, each atom acquires a complete outer shell of 8 electrons, thus forming a molecule of chlorine (Cl_2). Such a reaction is called a molecular reaction, and such bonds are called covalent bonds.

will create two charged atoms, called **ions.** Both sodium and chlorine become ionized, sodium becoming electropositive (Na^+) and chlorine electronegative (Cl^-). Since unlike charges attract, a chemical bond is formed, called an **ionic bond** (Fig. 1-6). The ionic compound formed, sodium chloride, can be represented in electron dot notation ("fly-speck formulas") as:

$$Na\cdot + \cdot \ddot{\underset{\cdot\cdot}{Cl}}: \longrightarrow Na^+ + (:\ddot{\underset{\cdot\cdot}{Cl}}:)$$

Processes that involve the **loss of electrons** are **oxidation** reactions; those that involve the **gain of electrons** are called **reduction** reactions. Since oxidation and reduction always occur simultane-

ously, each of these processes is really a "half-reaction." The entire reaction is called an **oxidation-reduction** reaction, or simply **redox** reactions. The terminology is confusing because oxidation-reduction reactions involve electron transfers, rather than (necessarily) any reaction with oxygen. However, it is easier to learn the system than to try to change accepted usage.

We now need to introduce the concept of **oxidation number.** This term refers to the charge an atom would have if the bonding electrons were arbitrarily assigned to the more electronegative of two interacting elements. For example, in the sodium chloride reaction, we consider that the bonding electron has been transferrred to the chlorine atom. Consequently sodium, having lost its electron, becomes electropositive and is said to have an oxidation number of +1. Chlorine, with its newly acquired electron, becomes electronegative, and takes an oxidation number of −1. Some elements always exist in compounds in the same oxidation number. For example, oxygen almost always has an oxidation number of −2; sodium, +1; magnesium, +2; potassium, +1; and calcium, +2. However, most metals commonly have two or more oxidation numbers. For example, iron may exist as +2 or +3, chromium as +3 or +6, manganese as +2, +4, or +7. Other elements, such as hydrogen, can take either positive or negative oxidation numbers.

Covalent bonds. Stability can also be achieved when 2 atoms **share** electrons. Let us again consider the chlorine atom, which, as we have seen, has an incomplete 7-electron outer shell. Stability is attained by gaining an electron. One way this can be done is for 2 chlorine atoms to share one pair of electrons (Fig. 1-7). To do this, the 2 chlorine atoms must **overlap** their third shells, so that the electrons in these shells can now spread themselves over both orbits. Many other elements can form covalent (or electron-pair) bonds. For example: hydrogen (H₂)

$$H\cdot + H\cdot \longrightarrow H\!:\!H$$

and oxygen (O₂)

$$\ddot{O}\!: + :\!\ddot{O} \longrightarrow \ddot{O}\!:\!:\!\ddot{O}$$

In this case, oxygen with an oxidation number of −2 must share 2 pairs of electrons to achieve stability. Each atom now has 8 electrons available to its outer shell, the stable number.

Covalent bonds are of great significance to living systems, since the major elements of protoplasm (carbon, oxygen, nitrogen, hydrogen) almost always share electrons. Carbon, which usually has an oxidation number of either +4 or −4 (its outer shell contains 4 electrons), can share its electrons with hydrogen to form methane:

$$\cdot\dot{C}\cdot + 4\,H\cdot \longrightarrow \begin{matrix} & H & \\ H & \!\!\!:\!\!\dot{C}\!:\!\! & H \\ & H & \end{matrix}$$

Carbon now achieves stability with 8 electrons, and each hydrogen atom becomes stable with 2 elec-

trons. Carbon also forms covalent bonds with oxygen:

$$\cdot\dot{C}\cdot + 2\,\ddot{O}\!: \longrightarrow \ddot{O}\!:\!:\!C\!:\!:\!\ddot{O}$$

Carbon can also bond with itself to form, for example, ethane:

$$\begin{matrix} H & H \\ H\!:\!\dot{C}\!:\!\dot{C}\!:\!H \\ H & H \end{matrix} \quad \text{or} \quad \begin{matrix} H & H \\ | & | \\ H\!-\!C\!-\!C\!-\!H \\ | & | \\ H & H \end{matrix}$$

Carbon can join in "double bond" configuration:

$$\begin{matrix} H & \cdot & \cdot & H \\ :\!C\!:\!:\!C\!: \\ H & & & H \end{matrix} \quad \text{or} \quad \begin{matrix} H & & H \\ \diagdown & & \diagup \\ & C\!=\!C & \\ \diagup & & \diagdown \\ H & & H \end{matrix}$$

Or even in "triple bond" configuration:

$$H\!:\!C\!:\!:\!:\!C\!:\!H \quad \text{or} \quad H\!-\!C\!\equiv\!C\!-\!H$$

These examples do not begin to adequately illustrate the amazing versatility of carbon, the element forming the backbone of all life.

ACIDS, BASES, AND SALTS

Every ionic compound is an acid, a base, or a salt. One commonly accepted definition of an **acid** is any compound that dissociates to yield hydrogen ions. An acid is classified as strong or weak, depending on the extent to which an ionic compound is dissociated. Those that dissociate completely in water (H_2SO_4, HNO_3, and HCl) are called strong acids. Weak acids such as acetic acid (CH_3COOH) dissociate only weakly. A solution of acetic acid is mostly undissociated acetic acid molecules with only a small number of acetate and hydrogen ions present.

A **base** contains negative ions called hydroxyl ions and may be defined as a molecule or ion that will accept a proton. Bases are produced when compounds containing them are dissolved in water. NaOH (sodium hydroxide) is a strong base because it will dissociate completely in water into sodium (Na^+) and hydroxyl (OH^-) ions. Among the characteristics of a base is its ability to combine with a hydrogen ion. Like acids, bases vary in the extent to which they dissociate in aqueous solutions into hydroxyl ions.

A **salt** is a compound resulting from the chemical

interaction of an acid and a base. Common salt, sodium chloride (NaCl), is formed by the interaction of hydrochloric acid (HCl) and sodium hydroxide (NaOH). In water the HCl is dissociated into H^+ and Cl^- ions. The hydrogen and hydroxyl ions combine to form water (H_2O), and the sodium and chloride ions combine to form salt:

$$HCl + NaOH \longrightarrow NaCl + H_2O$$
$$\text{ACID} \quad \text{BASE} \qquad\qquad \text{SALT}$$

Organic acids are usually characterized by having in their molecule the carboxyl group ($-COOH$). The carboxyl group contains both a carbonyl group and a hydroxyl group on the same carbon atom:

$$\begin{array}{ccc} R{-}C{=}O & & R{-}C{=}O \\ | & \rightleftharpoons & | \quad + H^+ \\ O{-}H & & O{-} \end{array}$$

R refers to an atomic grouping unique to the molecule. In water, the COO— group will behave as a weak acid. The common organic acids are acetic, citric, formic, lactic, and oxalic. The student will encounter many of these later in discussions of cellular metabolism.

HYDROGEN ION CONCENTRATION (pH)

Solutions are classified as acid, base, or neutral, according to the proportion of hydrogen (H^+) and hydroxyl (OH^-) ions they possess. In acid solutions there is an excess of hydrogen ions; in alkaline, or basic, solutions the hydroxyl ion is more common; and in neutral solutions both hydrogen and hydroxyl ions are present in equal numbers.

To express the acidity or alkalinity (or pH concentration) of a substance, a logarithmic scale, a type of mathematical shorthand, is employed that uses the numbers 1 to 14. In this scale, numbers below 7 indicate an acid range. The number 7 indicates neutrality, that is, the presence of equal numbers of H^+ and OH^- ions. According to this logarithmic scale, a pH of 3 is ten times as acid as one of 4; a pH of 9 is ten times as alkaline as one of 8.

In protoplasmic systems pH plays an important role, for, in general, slight deviations from the normal usually result in severe damage. Most substances and fluids in the body hover closely around the point of neutrality, that is, pH of around 7. Blood, for instance, has a pH of about 7.35, or just

slightly on the alkaline side. Lymph is slightly more alkaline than blood. Saliva has a pH of 6.8, on the acid side. Gastric juice is the most acid substance in the body, about pH 1.6. The regulation of the pH in the body tissue fluids involves many important physiologic mechanisms; one of the most important is the buffer action of certain salts.

BUFFER ACTION

The hydrogen ion concentration in the extracellular fluids must be maintained constantly so that metabolic reactions within the cell will not be adversely affected by a constantly changing hydrogen ion concentration. A change of only 0.2 pH unit from the normal blood pH of 7.35 can cause serious metabolic disturbances. To protect itself from sudden changes in acidity, the body has certain substances that resist any change in the pH when acids or alkalies are added to the body fluids. These are called **buffers.** The hydrogen ion concentration within the cells is probably greater (pH is lower) than the hydrogen ion concentration in the extracellular fluids because of the metabolic production of CO_2, which reacts with the cellular water to form carbonic acid (H_2CO_3). Certain phosphates, sulfates, and organic acid radicals also add to the acidic nature of the intracellular fluids. Within the cells the high content of protein serves as a buffer and thus tends to keep the pH from going too low.

The buffer function of blood is dependent on both plasma and red blood corpuscle buffer mechanisms. The chief buffer of plasma and tissue fluid is sodium bicarbonate ($NaHCO_3$). This salt dissociates into sodium ions (Na^+) and bicarbonate ions (HCO_3^-). When a strong acid (for example, HCl) is added to the fluid, the H^+ ions of the dissociated acid will react with the bicarbonate ion (HCO_3^-) to form a very weak acid, carbonic acid, which dissociates very slightly. Thus the H^+ ions from the HCl are removed and the pH is little altered.

MIXTURES AND THEIR PROPERTIES

Whenever masses of different kinds are thrown together, we have what is called a mixture. All the different states of matter (solids, liquids, gases) may be involved in these mixtures. The mixtures we are

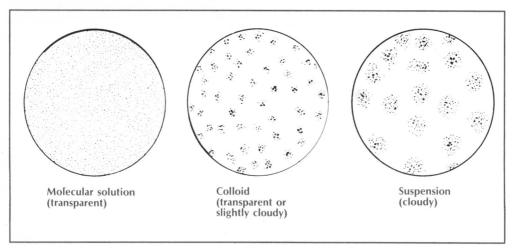

Molecular solution (transparent)   Colloid (transparent or slightly cloudy)   Suspension (cloudy)

FIG. 1-8. Three types of solutions or mixtures.

mainly interested in here are those in which water or other fluid is one of the states of matter. When something is mixed with a liquid, any one of three kinds of mixtures is formed.

Molecular solutions. If crystals of salts or sugars are added to water, the molecules or ions (in the case of salts) are uniformly dispersed through the water, forming a **true solution** (Fig. 1-8). Such solutions are transparent. In such a case the water is the **solvent** and the dissolved salt or sugar the **solute.** The freezing point of solutions is lower and the boiling point is higher than those of pure water.

Suspensions. If solids added to water remain in masses larger than molecules, the mixture is a suspension. Muddy water is a good example. When allowed to stand, the particles in suspension will settle out to the bottom. Suspensions have a turbid appearance and have the same boiling and freezing points as pure water.

Colloids. Whenever the dispersed particles are intermediate in size between the molecular state and the suspension, a third mixture is the result—the colloidal solution. Colloidal particles are rather arbitrarily considered to be between 1 and 500 millimicrons in size. If the particles are smaller, the solution is classified as a true solution; if larger, they are suspensions or emulsions. Colloids consist of two phases—a discontinuous, or dispersed, phase and a continuous, or dispersion, phase. These phases may be represented by the same states of matter

or different ones. Some familiar examples are as follows:

Discontinuous phase	Continuous phase	Example
Solid	Liquid	Ink
Liquid	Liquid	Emulsion
Liquid	Solid	Gel
Solid	Solid	Stained glass
Gas	Liquid	Foam, carbonated water
Liquid	Gas	Fog
Solid	Gas	Smoke

A true colloidal solution is stable (that is, will not settle out), has about the same boiling point and freezing point as pure water, and is either transparent or somewhat cloudy.

Proteins, which are important constituents of protoplasm, form colloidal solutions because their large molecules are well within the size range of colloidal particles and behave like colloids. Since protein molecules also dissolve as molecules in solution, such solutions may also be called molecular.

One special form of colloidal solution is the **emulsion** in which both phases are immiscible liquids (Fig. 1-9). Cream is a good example. Here, droplets of oil, or fat, are dispersed in water. This type of colloidal solution has considerable significance in the makeup of protoplasm.

Colloidal emulsions also illustrate the property of some (but not all) colloids to reverse their phases. When gelatin is poured into hot water, the

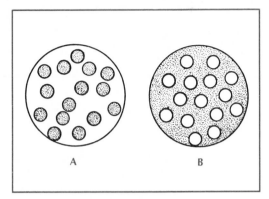

FIG. 1-9. Colloidal solution in which each phase is a liquid. **A,** Oil-in-water emulsion, water being the continuous phase, oil the discontinuous phase. **B,** Water-in-oil emulsion, water being the discontinuous phase, oil the continuous phase. Certain agents can bring about this phase reversal.

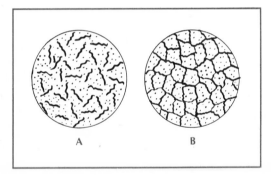

FIG. 1-10. Sol and gel. **A,** Sol condition in which gelatin particles are the discontinuous phase, water the continuous phase. **B,** Gel condition in which gelatin particles form continuous phase (network), enclosing water as discontinuous phase.

gelatin particles (discontinuous phase) are dispersed through the water (continuous phase) in a thin consistency which is freely shakable (Fig. 1-10). Such a condition is called a **sol.** When the solution cools, gelatin now becomes the continuous phase and the water is in the discontinuous phase. Moreover, the solution has stiffened and become semisolid and is called a **gel.** Heating the solution will cause it to become a sol again, and the phases are reversed. Some colloidal emulsions are not reversible. Heating egg white, for example, will change the egg albumin from a sol into an irreversible gel. In such cases the coagulated particles may collect into larger particles and settle out.

Why do colloids play such an important role in the structure of protoplasm? There are several reasons, among which may be mentioned the following:

1. Great surface exposure, which allows for many chemical reactions
2. The property of phase reversal, which helps explain how protoplasm can carry on diverse functions and change its appearance during metabolic activities
3. The property of undergoing gelation or solation, which enables the protoplasm to contract, thus explaining such movements as ameboid movement
4. The inability of colloids to pass through membranes, which promotes the stability and organization of the cellular system, such as cell and nuclear membranes and cytoplasmic inclusions
5. The selective absorption or permeability of the cell membrane, which is largely dependent on the phase reversal of its colloidal structure

2

Structure of living matter

FITNESS OF OUR PLANET FOR LIFE

If the environment is an integral part of the life process, the physical surroundings of living substance must be uniquely suitable for life to exist. Organisms do not merely adapt to a particular environment, however; their own activities produce a change in the environment. It is thus seen that the relation between life and the environment is a reciprocal one. Many years ago (1913) the biochemist L. J. Henderson pointed out many facts about the uniqueness of the physical aspects of our planet for promoting the life process. In his classic book, *The Fitness of the Environment,* he shows how suitable the environment is for the process of organic evolution and the development of life.

The most abundant of all protoplasmic compounds is water, making up about 60% to 90% of most living organisms. All forms, terrestrial or aquatic, maintain an aqueous internal environment. Water therefore plays an important role in all protoplasmic systems. Some properties of this unique compound may be pointed out. It is the most versatile of all solvents. The English physiologist W. M. Bayliss once stated that practically all chemical substances will dissolve in water to some extent. It is evident that this property of water is of the utmost importance for the chemical reactions that must take place in the life process. Water also promotes the ionization of various compounds in the body, a chemical alteration that is essential for bodily processes (promotion of chemical reactions, ionic regulation of cells and organisms, the permeability of cell membranes, active transport of materials, etc.). Water is the only substance that occurs in nature in the three phases of solid, liquid, and vapor within the ordinary range of the earth's temperature. It has a high specific heat. The amount of heat required to raise the temperature of 1 gram of water from 15° to 16° C. is exactly 1 calorie; water thus has a specific heat of 1.0 calorie per gram. Most organic solvents have specific heats of about 0.5 calorie per gram; for iron the specific heat is less than 0.1. Thus a large amount of heat is required to raise the temperature of a given quantity of water. The presence of water has a great effect in moderating environmental temperature changes. Important from a biologic standpoint is the expansion of water before changing to ice instead of contraction (water is heaviest at 4° C.) so that aquatic organisms can live in the bottom of freshwater ponds and lakes in winter without being frozen in. The relatively high surface tension of water is due to cohesion of water molecules at its surface; and the high latent heat of vaporization (requiring many hundred calories to change 1 gram of liquid water into water vapor) is evidence of the strong intramolecular forces in water.

Water possesses its many unique properties in large part because of its dipolar character. Water is a hydride of oxygen. The geometric shape of the molecule, in which the two covalent OH bonds form an angle of 104.5 degrees, produces a separation of electric charge. A water molecule can be represented as:

The dipolar character of water is responsible for the orienting effects water molecules have on each other and on other molecules dissolved in water. Because water molecules are charged, they can participate in **hydrogen bond** formation with the nitrogen and oxygen atoms of large macromolecules, especially proteins. Water can thus be bound to organic molecules and, in turn, serve as atomic bridges that hold numbers of large molecules together.

But water is not the only fitness factor for life on earth. One of these factors is the limited range of temperature extremes on earth. We do not experience on this planet the incredibly hot and cold temperatures that exist on other planets. Especially important in the fitness of earth for life is the abundance and availability of many key elements such as carbon, oxygen, and hydrogen that are crucial constituents in the organic compounds fundamental to life as we know it. Other elements (nitrogen, phosphorus, sulfur, etc.) also contribute to the unique properties of this planet for supporting living substance.

CHEMICAL COMPLEXITY OF LIVING MATTER

The basis of biologic activities is chemical reactions. These reactions involve chemical elements that are found in all organisms as well as in the nonliving world. Life must have had its beginning in combinations and reactions of chemicals. At first, only a small number of combinations and reactions were necessary for the initiation of life. As time went on, more complex substances or compounds with successively higher levels of organization and reactions occurred. Despite the incredible complexity of living

matter, all biologic phenomena operate, so far as is known, according to the physical laws of chemistry and physics.

Analysis of typical living matter reveals that it is composed of about 60% to 90% water (higher animals are usually about 60% to 70% water), 15% protein, 10% to 15% fats and lipids, 1% carbohydrates, and 5% inorganic ions of various kinds (Na^+, K^+, Cl^-, $SO_4^=$, etc.) (Fig. 2-1).

Many different categories of biomolecules are found in every cell of the organism. Most of them are dissolved or suspended in cellular water, either in ionized or nonionized form. Some organic substances form a part of the hard substances (horn, claws, hoofs, keratin, etc.) of the body. Besides being in the cells and tissues of the body, organic components are also found in the environment, especially that of aquatic animals. Surprisingly, lake water may contain a far higher organic content than does ocean water. Some organic substances found in water are organic phosphorus, organic nitrogen, amino acids, carotenoid substances, and vitamins.

Inorganic matter actually outweighs the organic

FIG. 2-1.
Relative composition of major constituents of animal cells.

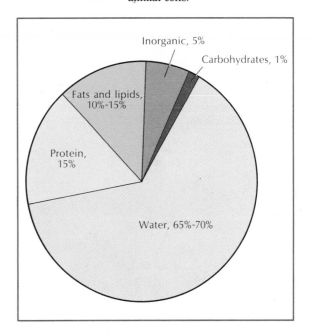

part of living matter. The inorganic portion includes (1) water, (2) mineral solids consisting of crystals, secreted precipitates, bone, and shells, and (3) cellular minerals, either free or combined with organic compounds. Most of the cellular minerals are in the form of ions (H^+, Ca^{++}, Na^+, and Mg^{++}, OH^-, $CO_3^=$, HCO_3^-, $PO_4^\equiv$, Cl^-, and $SO_4^=$ ions). Animals get these constituents from the water in which the minerals are dissolved from rocks and the soil.

ORGANIC MOLECULES

The term organic compounds has been applied to substances derived from plants and animals. All organic compounds contain carbon, but many also contain hydrogen and oxygen as well as nitrogen, sulfur, phosphorus, salts, and other elements. Organic compounds in a specific way are those carbon compounds in which the principal bonds are carbon-to-carbon and carbon-to-hydrogen. Carbon with its four valences has a great ability to bond with other carbon atoms in chains of varying lengths and configurations. More than a million organic compounds have been identified; more are being added daily. Carbon-to-carbon combinations introduce the possibility of enormous complexity and variety into molecular structure. As a chain it can combine with hydrogen to form an **aliphatic** compound:

$$H-\underset{\underset{H}{|}}{\overset{\overset{H}{|}}{C}}-\underset{\underset{H}{|}}{\overset{\overset{H}{|}}{C}}-\underset{\underset{H}{|}}{\overset{\overset{H}{|}}{C}}-\underset{\underset{H}{|}}{\overset{\overset{H}{|}}{C}}-\underset{\underset{H}{|}}{\overset{\overset{H}{|}}{C}}-\underset{\underset{H}{|}}{\overset{\overset{H}{|}}{C}}-\underset{\underset{H}{|}}{\overset{\overset{H}{|}}{C}}-H$$

HEPTANE

or a ring structure (**aromatic** compound)

BENZENE

Other types of configurations include rings and chains joined to each other, multiple branches of chains, helix arrangements, etc. The diversity of carbon molecules has made possible the complex kinds of macromolecules that form the essence of life.

Carbohydrates: nature's most abundant organic substance. Carbohydrates are compounds made of carbon, hydrogen, and oxygen. They are usually present in the ratio CH_2O and are grouped as $H-C-OH$. Familiar examples of carbohydrates are sugars, starches, and cellulose. There is more cellulose, the woody structure of plants, than all other organic materials combined. Carbohydrates are made synthetically from water and carbon dioxide by green plants, with the aid of the sun's energy. This process, called **photosynthesis,** is a reaction on which all life depends, for it is the starting point in the formation of food.

Carbohydrates are usually divided into the following three classes: (1) **monosaccharides,** or simple sugars; (2) **disaccharides,** or double sugars; and (3) **polysaccharides,** or complex sugars. Simple sugars are composed of carbon chains containing 3 carbons (tetroses), 5 carbons (pentoses), or 6 carbons (hexoses). Other simple sugars have up to 10 carbons, but these are not biologically important. Simple sugars, such as glucose, galactose, and fructose, all contain a free sugar group,

$$-\underset{\underset{H}{|}}{\overset{\overset{OH}{|}}{C}}-\overset{\overset{O}{\|}}{C}-$$

in which the double-bonded O may be attached to the terminal C of a chain (an aldehyde) or to a nonterminal C (a ketone). The hexose **glucose** is the most important carbohydrate in the living world. Glucose is often shown as a straight-chain aldehyde,

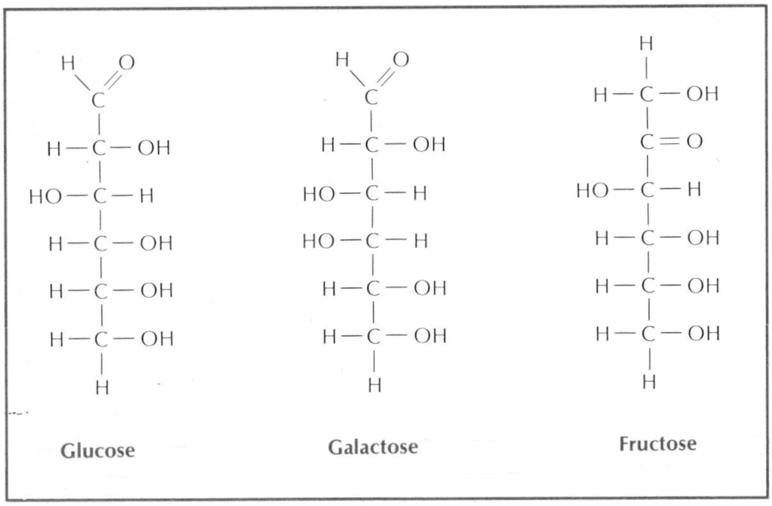

FIG. 2-2. These three hexoses are the most common monosaccharides. Glucose and galactose are aldehyde sugars; fructose is ketone sugar.

but in fact it tends to form a cyclic compound:

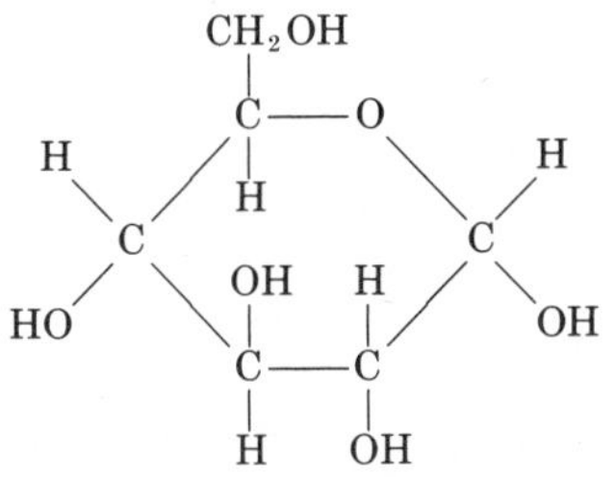

This formula shows the ring structure of glucose, but it is misleading because it obscures the three-dimensional form of the molecule. Rather than lying in a flat plane, the glucose molecule is "puckered," because the 4 atoms bonded to any single carbon molecule occupy the corners of a regular tetrahedron, four-cournered, pyramid-like figure. To better represent the structure of glucose than in the flat-plane ring structure above, organic chemists have devised a "chair" model to show the configuration of glucose and other hexose molecules:

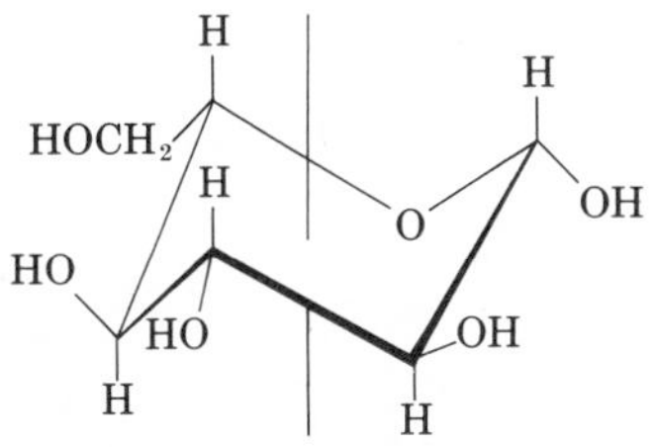

Although the three-dimensional chair conformation is the most accurate way to represent the simple sugars, we must remember that all forms of glucose, however represented, are the same molecule.

Other monosaccharides of biologic significance are galactose and fructose. The straight-chain structure of these hexoses is compared with glucose in Fig. 2-2.

Disaccharides are double sugars formed by the bonding together of two simple sugars. An example is maltose (malt sugar) composed of 2 glucose molecules. As shown in Fig. 2-3 the 2 glucose molecules are condensed together with the removal of a molecule of water. This dehydration reaction, with the sharing of an oxygen atom by the two sugars, characterizes the formation of all disaccharides. Two other common disaccharides are sucrose (ordinary cane, or table, sugar), formed by the linkage of glucose and fructose, and lactose (milk sugar), comprised of glucose and galactose.

Polysaccharides are made up of many molecules of simple sugars (usually glucose) and are referred to by the chemist as polymers. Their empirical formula is usually written $(C_6H_{10}O_5)^n$, where n stands for the unknown number of simple sugar molecules of which they are composed. Starch is common in most plants and is an important food constituent. **Glycogen** (Fig. 2-4), or animal starch, is found mainly in liver and muscle cells. When needed, glycogen is converted into glucose and is delivered by the blood to the tissues. Another polymer is **cellulose,** which is an important part of the cell

17

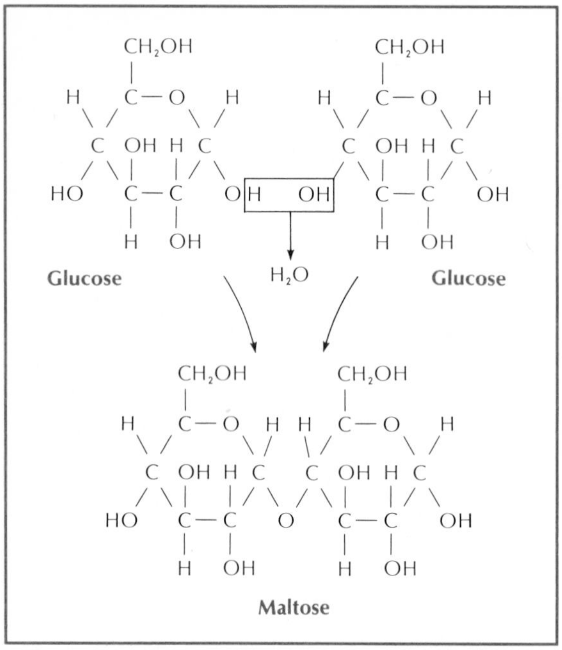

FIG. 2-3. Formation of a double sugar (disaccharide maltose) from 2 glucose molecules with the removal of a molecule of water.

walls of plants. Cellulose cannot be digested by man, but some animals, such as the herbivores, with the aid of bacteria, and termites, with the aid of flagellates, can do so.

The main role of carbohydrates in protoplasm is to serve as a source of chemical energy. Glucose is the most important of these energy carbohydrates. Some carbohydrates become basic components of protoplasmic structure, such as the pentoses that form constituent groups of nucleic acids and of nucleotides.

Proteins: foundation substance of protoplasm. Proteins are large, complex molecules, characterized by a high nitrogen content. Proteins are characteristic for each form of life and for each tissue that composes an organism. The development of an egg into a complex, differentiated animal involves the formation of an enormous number of proteins that are specific for each tissue and organ.

Proteins are composed of twenty commonly occur-

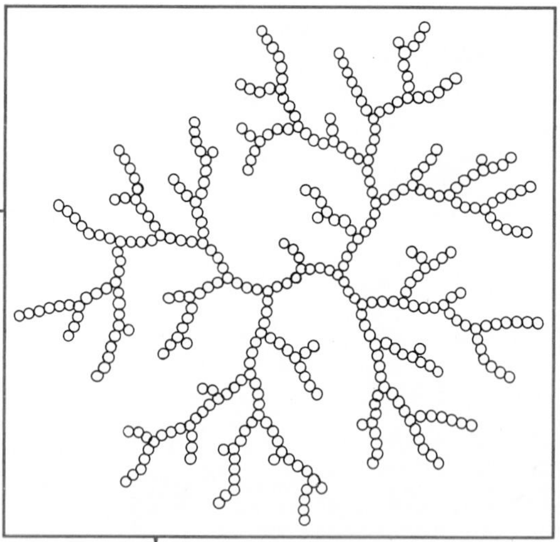

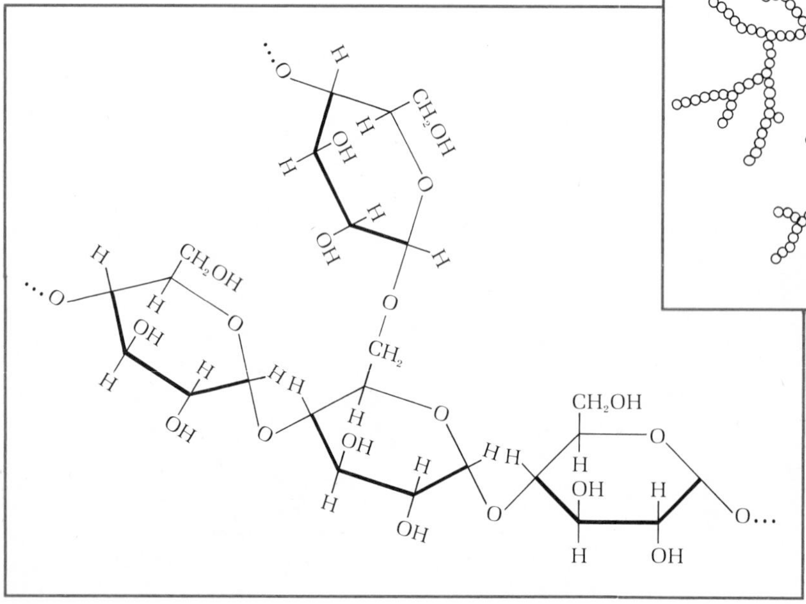

FIG. 2-4. Glycogen is a large, branched polysaccharide with a treelike structure. It is composed of linear chains of linked glucose molecules. A section of a glucose chain is shown enlarged at left.

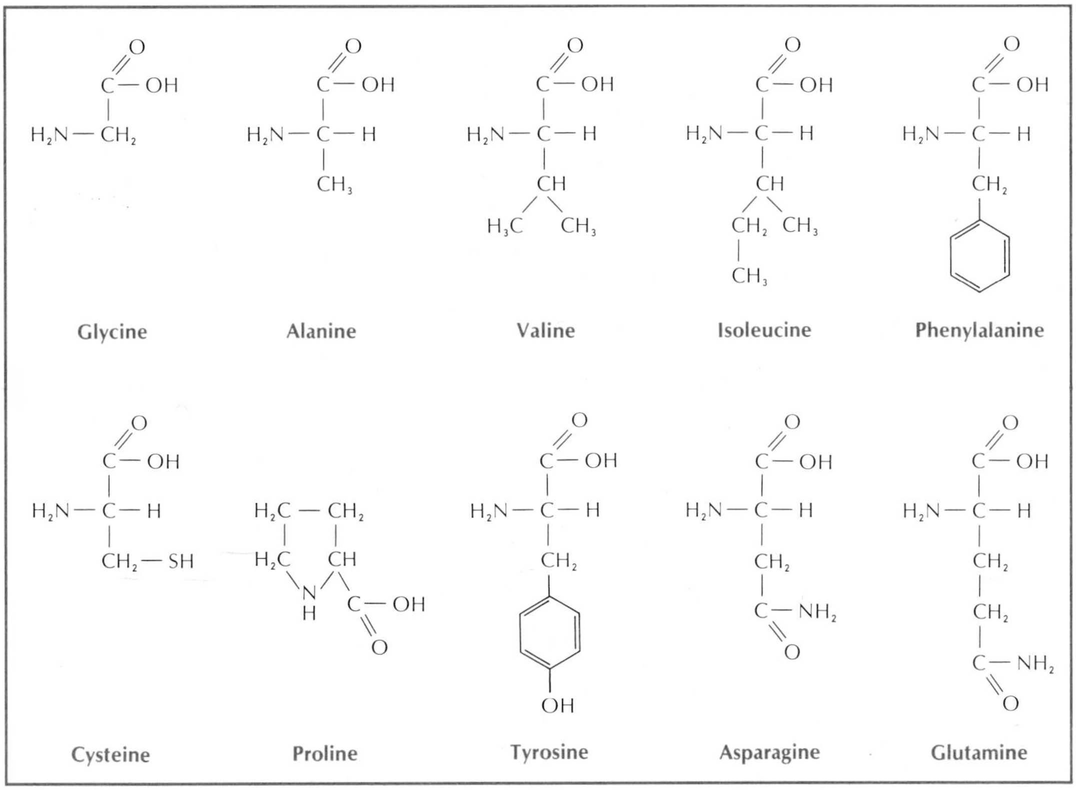

FIG. 2-5. Structural formulas of some important amino acids. Note the sulfur atom in cysteine, which is important in disulfide bond formation in proteins.

ring amino acids (Fig. 2-5). Nearly all of the twenty amino acids are usually present in every protein. Since the amino acids can be, and are, arranged in all possible combinations, it is easy to see how an almost infinite variety of proteins can be produced.

Each of the twenty amino acids contains one amino group ($-NH_2$) and one carboxyl group ($-COOH$) attached to the same carbon atom. The remainder of the molecule is unique for each amino acid. The general formula for an amino acid is as follows:

$$NH-CH-COOH$$
$$|$$
$$R$$

where the symbol R represents an atomic grouping unique for each acid. In forming proteins, amino acids are linked together by a bond between the NH_2 group of one amino acid and the COOH group of another to form a **peptide bond.**

The bonding of two amino acids, as shown below, forms a **dipeptide;** the addition of a third amino acid forms a **tripeptide.** In this manner — sequential addition of amino acids through peptide bonds — long **polypeptide** chains are built.

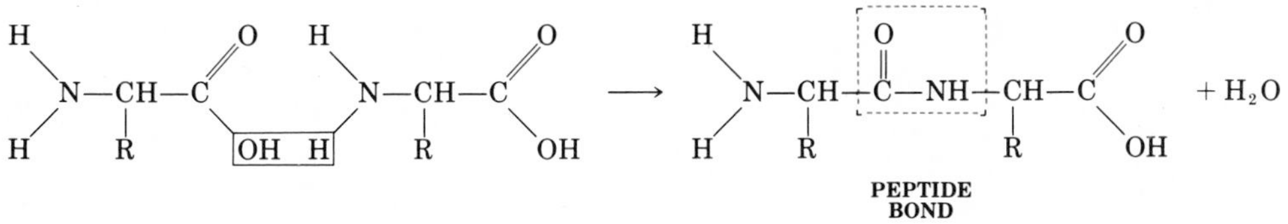

PEPTIDE
BOND

19

$$H_2N-CH-\overset{\overset{\displaystyle O}{\|}}{C}-NH-CH-\overset{\overset{\displaystyle O}{\|}}{C}-NH-CH-\overset{\overset{\displaystyle O}{\|}}{C}-NH-CH-\overset{\overset{\displaystyle O}{\|}}{C}-NH-CH-COOH$$

$$\underset{R_1}{|}\qquad\qquad\underset{R_7}{|}\qquad\qquad\underset{R_{13}}{|}\qquad\qquad\underset{R_4}{|}\qquad\qquad\underset{R_{17}}{|}$$

The polypeptide above consists of five different amino acids linked by peptide bonds. When the chain exceeds about 100 amino acids, the molecule is called a **protein** rather than a polypeptide.

A protein is not just a long string of amino acids. The polypeptide chain tends to spiral into a helical pattern, like the turns of a screw; this conformation is called an **alpha-helix** and is common in natural proteins. As the chain spirals, hydrogen bonds bridge across amino acids in adjacent turns and help to stabilize the pattern. Another factor adding to the complexity of proteins is **disulfide bonds.** Sulfur atoms in pairs of cysteine (sis′tan) molecules (Fig. 2-5) bond together at intervals. These bonds are strong and cause the protein to fold into character-istic globular shapes (Fig. 2-6). Thus polypeptide chains of proteins are not only spiral, the spirals themselves are also bent and folded into complex shapes.

The complete structure of several proteins has now been worked out. This has proved to be a monumental task, for not only must the correct amino acid sequence be determined, but also the complete three-dimensional configuration, that is, the exact way the polypeptide chains are folded and bonded together. Insulin, the pancreatic hormone that governs glucose metabolism, was the first protein to have its amino acid sequence determined. Insulin is a small protein (mol. wt. 5,700), consisting of two polypeptide chains containing fifty-one amino acids.

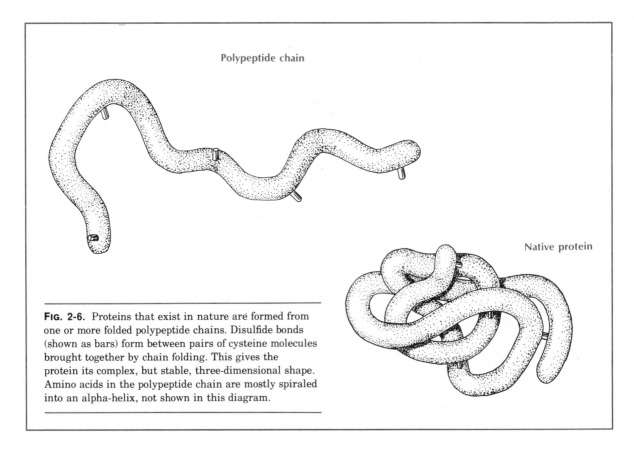

Polypeptide chain

Native protein

FIG. 2-6. Proteins that exist in nature are formed from one or more folded polypeptide chains. Disulfide bonds (shown as bars) form between pairs of cysteine molecules brought together by chain folding. This gives the protein its complex, but stable, three-dimensional shape. Amino acids in the polypeptide chain are mostly spiraled into an alpha-helix, not shown in this diagram.

The amino acid sequence was worked out by Frederick Sanger and his colleagues at Cambridge University in 1953. Sanger could not determine the three-dimensional configuration of insulin by the laborious techniques available at that time. By using x-ray defraction, subsequent workers have constructed what are believed to be fairly accurate pictures of the shape of certain native proteins.

The molecular weights of protein molecules, classed as colloids, vary over a wide range; for example, insulin, 5,700; egg albumin, 40,000; hemoglobin, 68,000; serum globulin, 170,000; thyroglobulin, 660,000; snail blood hemocyanin, 6,600,000; tobacco mosaic virus protein, 60,000,000.

Nucleic acids and nucleoproteins: genetic apparatus of the cell nucleus. These complex substances of high molecular weight represent life at its most fundamental level. Genetic information is deposited in genes, which are built principally of deoxyribonucleic acid (DNA). DNA is composed of structural units called **nucleotides.** Each nucleotide contains three kinds of organic molecules — a **pentose sugar, phosphoric acid, and a nitrogenous base.** The pentose (5-carbon) sugar in DNA is **deoxyribose sugar,** with the structural formula:

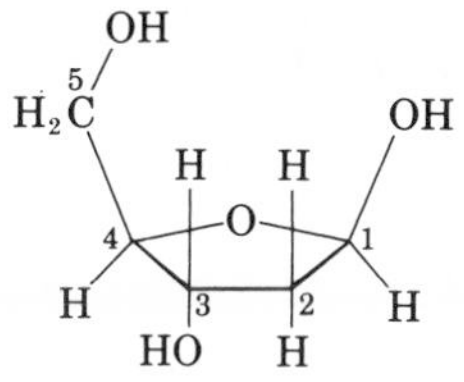

DEOXY-D-RIBOSE

The **phosphoric acid** has the structural formula:

$$HO-P-OH$$

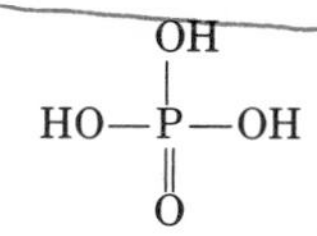

Four nitrogenous bases are found in DNA. Two organically ringed compounds, classed as **purines,** are **adenine** and **guanine:**

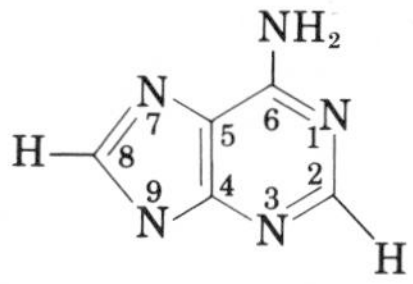

ADENINE 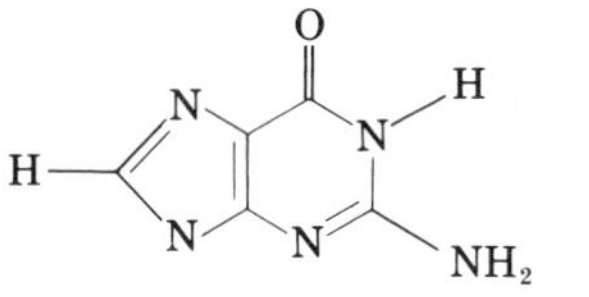   GUANINE

The other two nitrogenous bases belong to a different class of organic compounds called **pyrimidines.** The two pyrimidines found in DNA are **thymine** and **cytosine:**

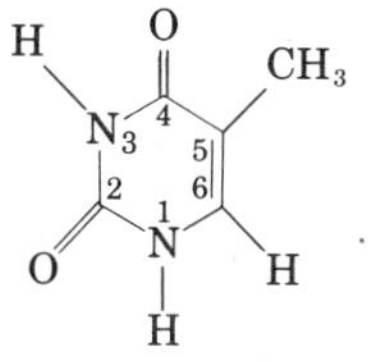

THYMINE 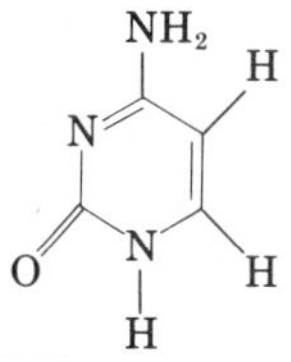   CYTOSINE

In DNA the backbone of the molecule is built of phosphoric acid and deoxyribose sugar; to this backbone are attached the nitrogenous bases (Fig. 2-7). DNA consists of **pairs** of nucleotides (each nucleotide consists of phosphoric acid, sugar, and base) linked together into long chains. The two chains are twisted around one another into a double spiral, or helix. DNA does not exist alone but is conjugated to simple proteins; this combination forms the high–molecular weight compounds called **nucleoproteins.**

Another kind of nucleic acid is **ribonucleic acid (RNA).** RNA is found mainly in the cytoplasm and nucleoli (DNA is restricted chiefly to the nucleus) and is directly concerned in the cellular synthesis of proteins. RNA differs from DNA in several ways. The sugar in the backbone of the RNA is **ribose** instead of deoxyribose:

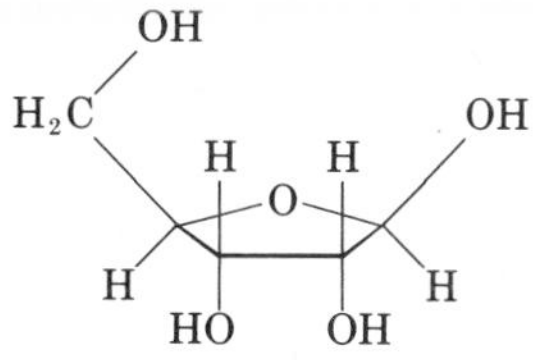

D-RIBOSE

There is also a difference in one of the four bases: the pyrimidine **uracil** is present in place of thymine:

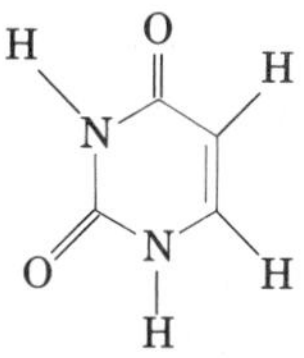

URACIL

RNA also differs from DNA in that its nucleotide

chain is much shorter and consists of only one strand. In other words, RNA is a **single helix** in contrast to DNA, a double helix. The nucleic acids will be discussed in much greater detail in Chapter 20, which deals with the genetic code.

Lipids: fuel storage. Fats and fatlike substances are known as lipids. They include the true fats, oils, compound lipids, and steroids. The true fats, or simple lipids, are sometimes called the neutral fats. They consist of oxygen, carbon, and hydrogen and are formed by the combination of 3 fatty acid molecules and 1 glycerol molecule. True fats are therefore esters, that is, a combination of an alcohol (glycerol [Fig. 2-8]) and an acid. They also bear the term "triglyceride" because the glycerol radical is combined with three radicals from fatty acid groups. A chemically pure fat such as stearin is an ester of glycerol and 3 molecules of a single fatty acid (stearic acid) (Fig. 2-9). Most natural fats, however, such as

lard and butter are mixtures of chemically pure fats, for they usually have two or three different fatty acids attached to the three hydroxyl groups of glycerol. The production of a typical fat by the union of glycerol and stearic acid is shown by the following formula:

$$
\begin{array}{ccc}
C_{17}H_{35}CO\big|OH \quad H\big|O-CH_2 & & C_{17}H_{35}OCO-CH_2 \\
C_{17}H_{35}CO\big|OH + H\big|O-CH & \longrightarrow & C_{17}H_{35}OCO-CH + 3H_2O \\
C_{17}H_{35}CO\big|OH \quad H\big|O-CH_2 & & C_{17}H_{35}OCO-CH_2
\end{array}
$$

STEARIC ACID	GLYCEROL	STEARIN
(3 MOLS.)	(1 MOL.)	(1 MOL.)

In this formula it will be seen that the 3 fatty acid molecules have united with the OH group of the glycerol to form stearin (a neutral fat), with the production of 3 molecules of water. Other common fatty acids in nature are palmitic and oleic acids (Fig. 2-9).

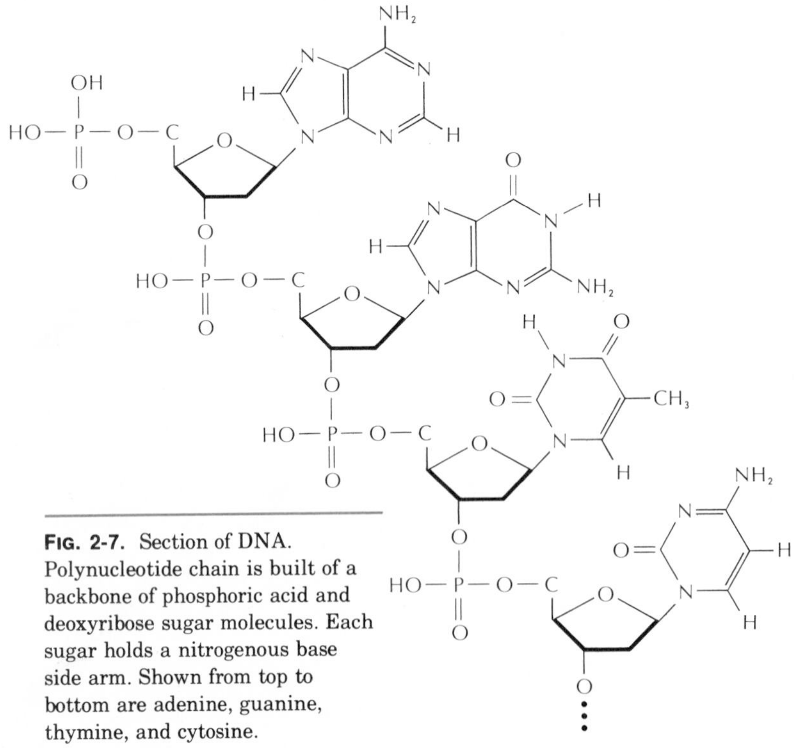

Fig. 2-7. Section of DNA. Polynucleotide chain is built of a backbone of phosphoric acid and deoxyribose sugar molecules. Each sugar holds a nitrogenous base side arm. Shown from top to bottom are adenine, guanine, thymine, and cytosine.

Most true fats are solid at room temperatures, but plant oils (linseed, cottonseed, etc.) and animal oils (fish and whale) are liquid because of the nature of their fatty acids. Waxes such as beeswax differ from true fats in having an alcohol other than glycerol in their molecular structure.

Compound lipids are fatlike substances that, when broken down, will yield glycerol (or some other alcohol), fatty acids, and some other substances, such as a nitrogenous base (for example, choline), phosphoric acid, or a simple sugar. Among these lipids are the phospholipids (lecithin) found in egg yolk and probably every living cell and the cerebrosides (glycolipid) that are common in nervous tissue. The

steroids, or solid alcohols, are not chemically related to fats but are included among the lipids because they have fatlike properties. Cholesterol ($C_{27}H_{45}OH$) is a common example of a steroid. Ergosterol, a plant steroid, becomes the hormone calciferol (commonly known as "vitamin D") when activated by ultraviolet rays. Male and female sex hormones and the adrenal gland hormones are other examples of steroids. Most of these steroids are derived from cholesterol.

Lipids have many functions in protoplasm. The true fats furnish a concentrated fuel of high-energy value and represent an economic form of storage reserves in the body. Excess carbohydrates can be transformed into fat, and to a limited extent fatty acids can be changed into glucose. Some phospholipids form part of the basic protoplasmic structure, such as lecithin, which gives a constant characteristic pattern to all cells. Phospholipids also share with proteins the basic structure of the plasma membrane and of the myelin sheaths or nerve fibers.

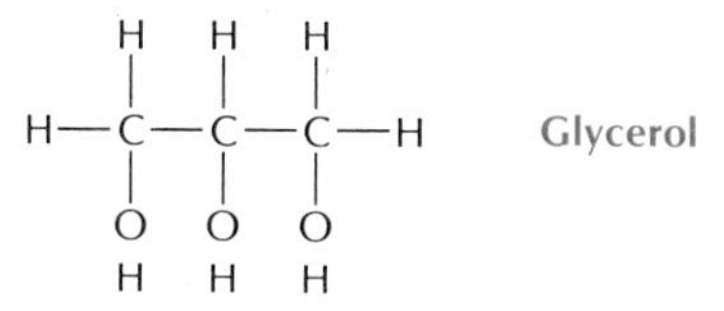

EXTRATERRESTRIAL SOURCES OF LIFE

In this space age, interest in extraterrestrial life has increased and is all the more exciting because scientists seem on the verge of acquiring definite knowledge about such matters, instead of merely speculating about them. Before the development of interplanetary rockets, any significant biologic

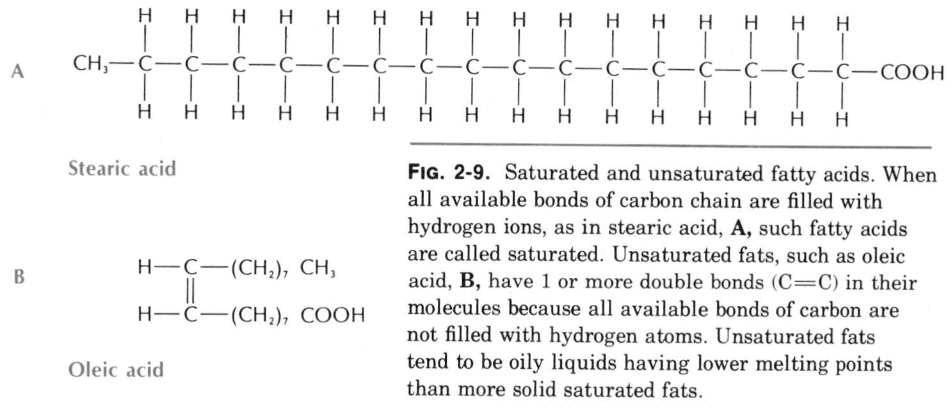

FIG. 2-8. Structural formula of glycerol (glycerin). This substance chemically is a type of alcohol and is obtained from fats and oils. True fats are formed by the combination of 1 molecule of glycerol with 3 fatty acid molecules and with elimination of 3 water molecules. The 3 fatty acid molecules may be the same, or they may be different.

FIG. 2-9. Saturated and unsaturated fatty acids. When all available bonds of carbon chain are filled with hydrogen ions, as in stearic acid, **A,** such fatty acids are called saturated. Unsaturated fats, such as oleic acid, **B,** have 1 or more double bonds (C=C) in their molecules because all available bonds of carbon are not filled with hydrogen atoms. Unsaturated fats tend to be oily liquids having lower melting points than more solid saturated fats.

information about the planets was obtained by astronomic methods. Definite and detailed knowledge of the planets can be acquired only by spacecraft. The presence of carbon seems to be a necessary requirement for life. The carbon on the planet Earth (with the exception of that found in living matter) is chiefly in the form of carbonates from the outgassing of volatile carbon compounds from the hot interior. Some such process must occur on other planets if they are to support a spontaneous origin of life as we know it. Spacecrafts carry automatic instruments that radio information to the earth, and in the extensive space probes planned it may be possible to determine if life exists elsewhere.

In recent years a variety of hydrocarbons have been found in certain meteorites. Some of these hydrocarbons resemble the biogenetic hydrocarbons of the earth, such as amino, carbonyl, and other groups, and have spectral wavelengths comparable to extracts of terrestrial oils. There is even evidence of the presence of nucleic acid, although the possibility of terrestrial contamination of the meteorites can seldom be ruled out. The whole question of extraterrestrial origin of life is still much in doubt, but some scientists think that the nature of stony meteorites might at least throw some light on the primordial material of the solar system and the early origin of the earth.

3

The science of life study

In its development, zoology has followed a certain sequence as man's study broadened and his comprehension became more exact. Zoologic study, of course, is concerned with the myriad forms of animal life and with every aspect of their structure, function, development, evolution, and relationship to their environment. Early biologists were at first primarily interested in finding out the superficial morphologic differences between animals; this resulted in the science of **systematics.** Later, biologists became more interested in learning about the fundamental structure and functioning of animals. Out of this grew anatomy, physiology, histology, embryology, and a host of specialized divisions of study. All this study has emphasized three great zoologic fields—**systematics, morphology,** and **physiology.**

The recent impact of molecular biology has shifted this emphasis to some extent. Molecular biology tries to interpret the structures and functions of organisms in terms of molecular structure. This has resulted in a new method of biologic study in which stress is first laid on an understanding of the chemical background of living matter and the way distinctive macromolecules are formed and function. Structure and physiology still remain important levels of study, but with newer interpretations and viewpoints. Classical fields of zoologic study remain as before important facets of understanding the animal, but there is now a better comprehension of synthesizing and integrating living substance than before.

THE DICHOTOMY OF LIFE (PLANTS AND ANIMALS)

Separate plant and animal traits probably evolved from a common protistan form, or a category of common organisms that included the ancestors of algae, slime molds, fungi, and protozoans. It would be difficult to divide such ancestral, primitive organisms into two distinct categories as we do the modern organisms today. Types of animal and plant traits were intermingled together. Evolution in time led to a rather marked separation into two groups. One line of evolution led to autotrophic organisms that could manufacture their own food from simple inorganic molecules with the aid of photosynthetic processes, and the other line of evolution led to heterotrophic forms that are dependent on outside sources for their complex organic molecules. The first group is the plants and the second group includes the animals. In one sense it may be stated that animals are actually parasites on the plant world.

The earliest animals were without exception unicellular forms and may have been similar to unicellular organisms of today. Since most of them had no hard parts for fossilization, we can only speculate from present forms what they may have been like.

There is a basic unity of life in both plants and animals. No wide gulf separates animals from plants. Plants are usually distinguished from animals in having chlorophyll for photosynthesis, cellulose in their cell walls, and indeterminate growth. Exceptions to some of these characteristics exist in animals. Animals are usually distinguished by the ability to move around because of their muscles and nerves. Plants do move to some extent, but at a much slower rate. The properties of irritability, growth, reproduction, etc. in animals are also found in plants. Biochemically, plants and animals are similar, as are the patterns of structure (cells, nuclei, chromosomes, genes, etc.).

MAN'S ENDEAVOR TO UNDERSTAND LIFE

Scientific developments do not erupt suddenly, but require vast backgrounds of effort for building up facts that can be turned into generalized concepts. It was so with biologic study. Isolated discoveries and developments had been made hundreds and even thousands of years ago, but biologists have begun to "put the pieces" together only in the last few generations. Early in man's existence two opposing points of view must have developed. One view ascribed all the phenomena of life to supernatural causes. The other, which developed slowly and gradually, insisted that natural forces were involved to some degree. Both points of view emphasized a world divided into the living and nonliving. When it was realized that the materials of which organisms are composed could be reduced to inorganic matter in the final analysis, there arose the vitalistic and mechanistic theories of the life process. The vitalistic theory held that there was a "vital force" different from the known laws of physics and chemistry that was responsible for the activating processes. On the other hand, the mechanistic theory held that resultant effects of forces already known actuate the living organism. There are scientists who today support either theory, but the great advancements

in many fields of science tend to support the mechanistic theory in the modern conception of living materials.

The starting point for our modern conceptions of life was early in the nineteenth century when a number of fundamental concepts about the states of matter were propounded. Basic ideas usually arise as the outgrowths of theories, experiments, and logical thinking of many investigators. Such ideas are often brought to a focus by a few individuals who have the ability to see crucial relationships. One of these basic conceptions revolved around the finer structure of the organism, especially the **concept of the cell** as the building block of the organism (1838 to 1839). The theory of the cell was a great simplifying step in understanding the organism and opened up new areas for investigation. This theory naturally led into an investigation of the nature of the cell contents (1839 to 1846). Actually these two fundamental conceptions—the cell as the unit of structure of living substance and the nature of its contents, or protoplasm—were worked out simultaneously. Thus the ideas regarding the building units of the organism and the physical material basic for vital activities were formulated together.

The basic unity of all science was demonstrated when analytic chemists were able to break down different organic compounds into their constituent chemical elements. They found that in living materials only five or six chemical elements made up the bulk of protoplasm. The concepts of the atom and molecule were being clarified as chemistry developed. So the groundwork for our present conception of **molecular biology** was being formed at a relatively early time in modern scientific development. Farseeing investigators of this period had already begun to speculate that structural features below visibility were to be expected in protoplasm. Vague as such ideas were, the substance of living matter was being credited with the structural properties of atoms and molecules.

Another conception of far-reaching consequences was the gross morphologic relationships of **evolution,** which gave a reasonable explanation for the diversity of life (1859). This great concept proved to be a powerful unifying force in man's zoologic thinking, for it paved a way for the logical interpretation of data of embryology, comparative anatomy, plan of body structure, and other aspects of the organism. In time the influence of evolution would be carried down into the basic structural features of atoms and molecules in protoplasm. More than any other concept, evolution showed the common ancestry of animal life and gave impetus to the idea that combinations of known physical forces could account for the origin and behavior of animals.

The study of evolution led naturally into another conception of the utmost importance – **cytogenetics,** the science of the cellular relation to heredity. Although the cell theory had been formulated before the impact of evolution, the development of cytology, the science of cells, made great progress in the comprehension of cellular details during the period when evolution was first being studied. Along with cytologic development, **genetics,** the study of the problem of inheritance, followed a natural line of thinking. Opinions on heredity were vague for a long time, but the idea that something in the cell was responsible for the transmission of inheritance lurked in the minds of cytologists. Precise details about this relationship developed during the last half of the nineteenth century when spectacular advancements were made in our understanding of the role of the nucleus, the continuity of cells and cell divisions, the nature and behavior of chromosomes, and finally the assignment of the mechanisms of heredity to the chromosomes.

The present century has seen additions and extensions to these fundamental concepts. This advancement has been due mainly to improvements in technologic methods and the convergence of the physical sciences of chemistry, geology, and physics to the natural sciences, so that all science may be considered unified. The successes man has accomplished in understanding biologic phenomena within the past few decades may lead to an understanding of the basic features of living substance.

MAJOR SUBDIVISIONS OF ZOOLOGY

Systematic zoology. This group includes taxonomy or classification, ecology, distribution, and evolution of animals.

Morphology. Structural aspects are stressed in this group, which includes comparative anatomy, histology, cytology, embryology, and paleontology.

Physiology. This group has to do with the functional considerations of the organism. It includes general physiology, physiologic chemistry, and animal behavior or psychology.

Experimental zoology. This group is a broad one and includes those subdivisions that are concerned with experimental alterations of the patterns of organisms. It includes genetics, experimental morphology, and embryology.

Molecular biology. This is the study of the ultimate, or ultramicroscopic, structure and function of living matter. At present it emphasizes the four fields of biochemistry, genetics, chemistry of macromolecules, and chemical physics.

• • •

Such groupings cannot be arbitrary, for there is much overlapping and interrelation among the various fields of zoologic investigation. For example, cytogenetics represents the close dependence of two branches of study, cytology and genetics, which were formerly considered more or less separately. As specialization increases, branches of study become more and more restricted in their scope. We thus have protozoology, the study of protozoans; entomology, the study of insects; parasitology, the study of parasites; and many others.

The following are definitions of some of the important areas of zoologic study:

anatomy (Gr. *ana,* up, + *tome,* cutting) The study of animal structures as revealed by gross dissection.

anatomy, comparative The study of various animal types from the lowest to the highest, with the aim of establishing homologies and the origin and modifications of body structures.

biochemistry (Gr. *bios,* life, + *chemos,* fluid) The study of the chemical makeup of animal tissues.

cytology (Gr. *kytos,* hollow vessel) The study of the minute parts and functions of cells.

ecology (Gr. *oikos,* house) The study of animals in relation to their surroundings.

embryology (Gr. *embryon,* embryo) The study of the formation and early development of the organism.

endocrinology (Gr. *endon,* within, + *krinein,* to separate) The science of hormone action in organisms.

entomology (Gr. *entomon,* insect) The study of insects.

genetics (Gr. *genesis,* origin) The study of the laws of inheritance.

helminthology (Gr. *helmins,* worm) The study of worms, with special reference to the parasitic forms.

herpetology (Gr. *herpein,* to creep) The study of reptiles, although the term sometimes includes both reptiles and amphibians.

histology (Gr. *histos,* tissue) The study of structure as revealed by the microscope.

ichthyology (Gr. *ichthys,* fish) The study of fishes.

mammalogy (L. *mamma,* breast) The study of mammals.

morphology (Gr. *morphe,* form) The study of organic form, with special reference to ideal types and their expression in animals.

ornithology (Gr. *ornis,* bird) The study of birds.

paleontology (Gr. *palaios,* ancient, + *onto,* existing) The study of past life as revealed by fossils.

parasitology (Gr. *para,* beside, + *sitos,* foods) The study of parasitic organisms.

physiology (Gr. *physis,* nature) The study of animal functions.

taxonomy (Gr. *taxis,* organization, + *nomos,* law) The study of the classification of animals.

zoogeography (Gr. *zoon,* animal, + *ge,* earth, + *graphein,* to write) The study of the principles of animal distribution.

METHODS AND TOOLS OF THE BIOLOGIST

The higher level of material organization represented by life is made up of properties that are the most difficult to understand in the universe. Protoplasm consists of very large molecules, but the organization of these molecules into particular patterns has given life many of its unique qualities. There are so many variables that reactions cannot always be predicted with certainty. Another difficulty of biologic investigation is that the whole is more than the sum of its parts. The study of one part of a living organism affords a restricted idea of the working of the whole integrated organism.

Biologic investigation was at first purely descriptive, but is now largely experimental. The controlled experiment using a single variable factor is becoming commonplace in investigations in the life sciences. However, in several fields such as evolution and taxonomy little has been done beyond the descriptive phase.

Biologists employ whatever available methods they deem necessary in doing their experimental problems. Some of the problems of greatest interest revolve around cellular biochemistry, energy relationships of metabolic pathways and processes, integrative relationships of physiologic processes, controlling factors of development, the basis of hereditary transmission, neuromuscular and membrane phenomena, and behavior patterns in animals. Most of these problems require complex techniques and apparatus, much of which was drawn originally from the physical and chemical sciences. Biologists modified such borrowed instrumentation to fit their particular purposes and developed many new techniques of their own.

The microscope, with all its types and modifications, has contributed more to biologic investigation than any other instrument developed by man. Its major objectives are magnification, resolution, and definition. The following represents the chief advances in the improvement of the microscope.

1. First compound microscope (Janssen, 1590; Galileo, 1610)
2. Microscope with condenser (1635)
3. Huygenian ocular (Huygens, 1660)
4. Substage mirror (Hertzel, 1712)
5. Achromatic lens (Dolland, 1757; Amici, 1812)
6. Polarizing microscope (Talbot, 1834)
7. Binocular microscope (single objective with double oculars) (Riddell, 1853)
8. Water immersion objective (Amici, 1840)
9. Oil immersion objective (Wenham, 1870)
10. Compensating oculars (1886)
11. Apochromatic objectives (1886)
12. Iris diaphragm (Bausch and Lomb, 1887)
13. Abbé condenser (Abbé, 1888)
14. Double objective binocular microscope (Greenough, 1892)
15. Ultramicroscope (dark-field) (Zsigmondy, 1900)
16. Electron microscope (Knoll and Ruska, 1931)
17. Phase-contrast microscope (Zernicke, 1935)
18. Reflecting microscope (Burch, 1943)
19. Fluorescence microscope (Coons, 1945)

The compound light microscope has been justified in a thousand ways, but the spectacular advances in molecular biology of the past decade or so have been due in large measure to the electron microscope. This microscope makes use of a beam of electrons in a vacuum instead of a beam of light as in an optical system. The electrons produced by a heated filament are focused by a magnetic coil and pass through the prepared object, where they are scattered in accordance with the nature of the object. The electrons have a very short wavelength and their resolving power is much greater than that of the light microscope. After passing through a magnetic objective lens and a projector lens, the transmitted electrons form an image on a photographic plate or fluorescent screen and thus can be made visible to the eye.

The chief limitations of the electron microscope are that the biologic material must be dead and arranged in very thin layers. The resolving power of a microscope is its ability to distinguish adjacent objects. In the light microscope this resolution is about 2,000 Å*; in the electron microscope it is about 5 to 10 Å. Anything smaller will merge and cannot be seen as a discrete particle. The magnifications produced by the electron microscope are enormous compared with those produced by the light microscope.

Other useful methods are (1) isotopic tracers, in which a radioactive isotope of an element can be substituted in a chemical compound and can then be detected by some method (Geiger tube, photographic plate, etc.) so that the chemical pathway or other information can be acquired; (2) chromatography, which involves the separation of organic or inorganic components by allowing a solution of a mixture to flow over or through a porous surface such as filter paper or synthetic gel; (3) centrifugation, which separates materials of different densities from each other and which can be employed for many analytical purposes; and (4) colorimetry and related methods, which measure and identify materials by determining the differential absorption of radiant energy of different concentrations.

In biochemistry the spectrophotometer is widely used to determine different chemical substances. Ultraviolet and visible lights are absorbed at characteristic wavelengths by different chemicals. By determining the wavelength absorbed by an unknown substance and comparing it with known absorbances, identification can be made. Variant forms of this mechanism are in use, but most are based on the selective powers of light absorption of components when traversed by a beam of light.

SOME BASIC CONCEPTS OF ZOOLOGY

Certain broad generalizations based on verified observations may be made in the life sciences. These concepts serve as a framework into which zoologic data can be fitted and serve to integrate biology as a science. These generalizations tend to give mean-

ing to the great mass of facts and the relations of natural phenomena to each other. For convenience, these concepts are grouped according to that division in which they fit best. As zoologic study advances and becomes more integrated, it becomes increasingly difficult to assign a concept to a particular biologic division of study; therefore many concepts overlap divisions of study.

1. The structure of living substance: protoplasm. Living substance contains the same chemical elements found in the nonliving world, but these elements are selective and are arranged in large distinctive macromolecules that have no counterparts in nonliving substance. The reactions of these molecules form the unique characteristics of life. The unique physicochemical organization of protoplasm produces the potentialities of each individual organism and forms the basis of individuality. The most unique and most common of the macromolecules is protein that is typically made up of thousands of atoms. Protein is not only important in the construction of living tissue, but is necessary for many biochemical reactions for sustaining life.

2. Organization within living systems. Protoplasmic systems are organized and differentiated into compartment units. The structural unit of all protoplasm is the cell, which may be considered the lowest unit that can manifest the life process. The cell, however, is a microcosmic life and is endowed with certain properties that by specialization and division of labor can develop into the tissues and organs of higher forms. Specialization and division of labor are correlated with the organization level of the organism. Within each organism there is a tendency of **homeostasis,** or stability of internal conditions (Fig. 3-1). Lower forms have not developed this principle as efficiently as higher forms have. Temperature control, for instance, has been attained only by warm-blooded forms. A diversity of body plans is found among animals, for each group within limits develops a characteristic body form.

3. Functional patterns of animals. Living processes require energy in some form. Animals require food for a source of energy, and they get this energy indirectly from the sun by the photosynthetic processes of the plant. Animals are thus heterotrophic in their food requirements in contrast to the

*An angstrom (Å) = 1/10,000 micron (μ), or 1/100,000,000 cm.; 1 inch = 2.54 cm.

Fig. 3-1. Process of homeostasis.

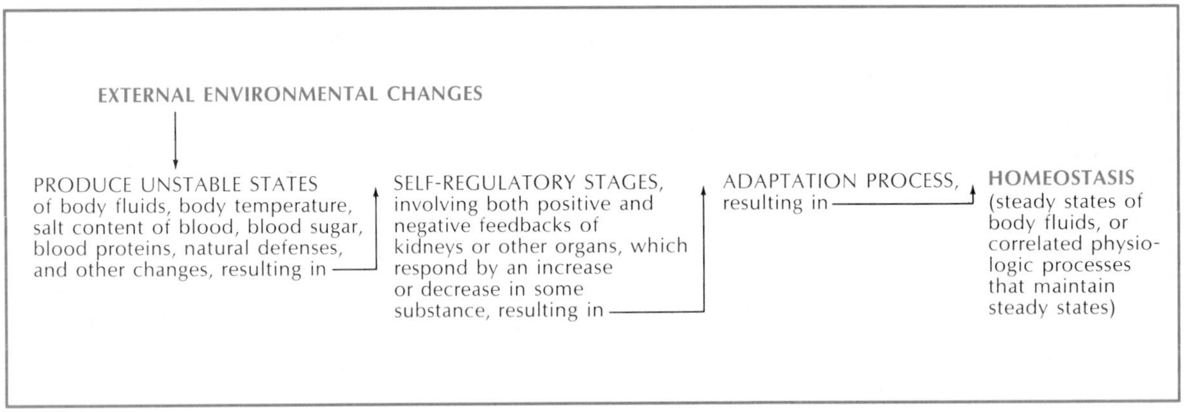

EXTERNAL ENVIRONMENTAL CHANGES

PRODUCE UNSTABLE STATES of body fluids, body temperature, salt content of blood, blood sugar, blood proteins, natural defenses, and other changes, resulting in —

SELF-REGULATORY STAGES, involving both positive and negative feedbacks of kidneys or other organs, which respond by an increase or decrease in some substance, resulting in —

ADAPTATION PROCESS, resulting in —

HOMEOSTASIS (steady states of body fluids, or correlated physiologic processes that maintain steady states)

autotrophic plants. Energy release from ingested foods involves complex chemical processes called metabolism. To carry on multiple chemical reactions, all living organisms require mediation by specific organic catalysts (enzymes). For adjustments and adaptations to their environment, animals must meet certain basic requirements that are fundamentally alike in all organisms. These activities center around digestion, metabolism, respiration, excretion, irritability, reproduction, etc.

4. Transmission of hereditary information. All organisms inherit a certain structural and functional organization from their progenitors. All organisms have the capacity for reproducing their kind and transmitting their characteristic type of organization to their offspring. What is inherited is a type of organization that, under the influence of developmental and environmental factors, gives rise to a certain visible appearance. Heredity is basically the transmission of a sequence code (in DNA molecules) of amino acids for the formation of varied protein patterns, or enzymes, characteristic for each organism. The units of transmission are called genes that are units of DNA and located in the chromosomes. Gene control is involved in every action of the organism. Each biochemical reaction, for instance, is controlled by a specific enzyme, which in turn is controlled by a specific gene (the one gene—one enzyme—one reaction hypophysis). Genetic continuity lies in the chromosomes found in all cells, a fact that may explain

why germ cells in certain instances arise from somatic cells instead of from germ plasm cells.

5. Evolution and phylogeny. The evolutionary doctrine states that existing animals have developed by a process of gradual change from previously existing animals. It is based on the belief that present-day forms have descended with modifications from primitive forms that have been different in structure and behavior.

Evolution explains phylogeny, or racial relationships, and affords a logical interpretation of the classification of animals. The modern synthesis of evolution involves certain major causes that together bring about evolutionary change: (1) the constant reshuffling of genes in the gene pool of a population; (2) mutation, or sudden changes in patterns of organization, induced by alterations in the genes of the genetic code, which in turn produces an alteration or deficiency of the enzyme or enzymes involved in the structure of proteins; (3) natural selection, which favors or eliminates characteristics as they appear; (4) genetic drift, in which small populations have their heredity of genes governed by chance and initially are not oriented by adaptation; and (5) geographic isolation, by which the gene pool of an isolated population cannot interbreed with the gene pool of other populations and thus produces its own evolution.

6. The interrelation of the organism to its environment. To survive an organism must adapt to the conditions imposed by its environment. All

TABLE 3-1. Some life-spans of animals*

Common name, genus, and species	Maximum life-span
Roundworm (*Rhabditis elegans*)	12 days
Fruit fly (*Drosophila melanogaster*)	46 days
Housefly (*Musca domestica*)	76 days
Hydra (*Hydra grisea*)	1-2 years
Mouse (house) (*Mus musculus*)	3 years
Rat (white) (*Rattus norvegicus*)	3-4 years
Earthworm (*Lumbricus terrestris*)	6 years
Leopard frog (*Rana pipiens*)	6 years
Robin (*Turdus migratorius*)	13 years
American crow (*Corvus brachyrhynchos*)	14 years
Bullfrog (*Rana catesbeiana*)	16 years
Domestic cat (*Felis catus*)	21 years
Dog (*Canis familiaris*)	34 years
Horse (*Equus caballus*)	50 years
Asiatic elephant (*Elephas maximus*)	57 years

*Values include animals in captivity and in natural habitat.

animals are influenced by environmental forces and, in turn, influence their environment. These environmental factors may be therefore **biotic** (interrelations between animals or living substances) and **physical,** which is concerned with temperature, climate, moisture, soil, air, etc. Animals and plants live in **communities,** or a naturally occurring group of different organisms interacting with each other, within a common environment, especially through food relations of producer, consumer, and decomposer, and relatively independent of other groups. The term **ecosystem** is often used to express the complete relationships of all the factors (biotic and physical), with the biologic part called a **biocenose.** To understand the complete workings of an ecosystem requires quantitative information about all the interactions between species and their environmental relations, which would include the concepts of food chains, food pyramids, and food webs. The aim of ecologic relations is to obtain a balanced ecosystem that will provide the best economics of the organisms within it (including man who is the major influence in most ecosystems).

7. **The restricted existence of protoplasmic systems.** This refers to the life history, or life cycle, that is characteristic for each species of organisms. Whether the life of an organism is limited to a few hours duration (as in certain insects) or thousands of years (as in the sequoia tree), there is a definite life-span for its existence (Table 3-1). The whole life process has a progressive decline in growth rate from fertilization to death. The biosphere of life at any one time is a transitory phenomenon as generations die and are succeeded by other generations. The life cycle, as J. T. Bonner has so well expressed it, is the unit of evolutionary change for any alteration (innovation or elimination) in the population that occurs at this time. Every organism has the capability of increasing its numbers to fantastic sizes in a relatively short time unless checked by evolutionary processes. All life is a series of chemical reactions that follow one another in sequence, and the reactants involved are the ions, atoms, and molecules that make up the organism. At death these reactants are resolved into the original elements to be used over again in a new generation of organisms. All living beings are decomposable into a chain of closed mechanisms to be later incorporated into the open systems of living organisms.

REFERENCES TO PART I

Borek, E. 1961. The atoms within us. New York, Columbia University Press.

Clark, B. F. C., and K. A. Marcker. 1968. How proteins start. Sci. Amer. **218**:36-42 (Jan.).

Conant, J. B. 1951. Science and common sense. New Haven, Yale University Press.

Gray, P. (editor). 1970. Encyclopedia of the biological sciences, ed. 2. New York, Reinhold Publishing Co.

Handler, P. (editor). 1970. Biology and the future of man. New York, Oxford University Press, Inc.

Kendrew, J. C. 1966. The thread of life. Cambridge, Harvard University Press.

Lanham, U. 1968. The origins of modern biology. New York, Columbia University Press.

Luria, S. E. 1970. Molecular biology: past, present, future. Bioscience. **20**:1289-1293.

Pennak, R. W. 1964. Collegiate dictionary of zoology. New York, The Ronald Press Co.

Ray, J. D., Jr., and G. E. Nelson. 1971. What a piece of work is man. Boston, Little, Brown & Co. (Paperback.)

Reilly, J. G., and A. W. Vander Pyl. 1970. Physical science: an interrelated course. Reading, Mass., Addison-Wesley Publishing Co., Inc.

Von Bertalanffy, L. 1952. Problems of life. New York, John Wiley & Sons, Inc.

Waddington, C. H. 1966. The nature of life: the main problems and trends of thought in modern biology. New York, Harper & Row, Publishers.

part two

The organization of the animal body

The basic living substance (protoplasm) differs from nonliving substance by being organized into definite chemical patterns, some simple and others complex. Chemicals by themselves do not have the properties of life; instead such properties are associated with structural organizations of chemicals arranged in what are known as organic compounds. Many levels of increasing complexity have been pointed out previously, but the fundamental organization unit of life is the **cell.** The cell appears in the hierarchy of the organization of matter at the point where life may be said to appear. It is the lowest structural level that manifests all the properties of life. It is possible for an organism (a living entity) to consist of just one cell (the **unicellular** organisms). Within the cell are found combinations called organelles that have specific functions to perform in the requirements of life. The cell may be considered a microcosm of the life process.

The multicellular animals, or metazoans, have many distinct levels of organization. From an aggregation of a few similar cells (a colony) to complex organisms of many different tissues, organs and systems are found in groups of animals that represent different stages of complexity. At first glance there appear to be numerous morphologic patterns, but closer inspection shows that these can be reduced to a few types of general symmetry. Symmetry refers to the arrangement of body parts into a geometric pattern. Some animal groups may lack definite symmetry (asymmetry), but most animals belong to one of the following types: spherical, radial, biradial, and bilateral. Such types of symmetry have adaptive significance, and each is usually suitable for a certain method of existence.

Differentiation and specialization of body parts is an evolutionary trend that occurs throughout the animal kingdom. There are usually several ways to perform the same function. Differences in organization are probably most marked in internal structures rather than in the more superficial external forms of animals. Among the various external architectural features of interest and obvious to superficial observation (aside from symmetry) are cephalization, or differentiation of the anterior end, and segmentation, or the repetition of body parts.

4

Cellular organization

Life is associated with a heterogeneous substance made up of organic macromolecules arranged according to a specific architecture. Analysis of animal matter indicates that there is a fundamental plan of all living things, whether large or small, complex or simple in structure. No student can appreciate the organization of living substance without a comprehension of the historical record of evolution that explains the great structural plans in the development of living organisms. It is not known, of course, just what the earliest living things were like, although many speculations have been offered by biologists. It is almost firmly established in the minds of zoologists that all the diverse species found today are linked together by genetic connections—in other words, all living things are related. This helps explain the fundamental uniformity present throughout the animal and plant kingdoms.

An analysis of living matter shows many levels of complexity. If we could have laid out before us a complete picture of the transformations organisms have undergone in their long evolutionary history, the task of understanding their present structural patterns would be simpler. But we have only the present forms to study and analyze. So we must speculate about the overall panorama of evolution to a great extent. And yet, putting the pieces to-gether, it is seen that biologic organization is governed by certain principles that give insights into the structural plans of animals, however diverse they appear to be.

Any analysis of an entity involves a description of its components and the way these components are put together. Such a process is used in describing animals. If animals can be resolved into a few units of structure, it is possible to establish schemes of regularity as they are studied at successive levels. This chapter deals with one of the greatest generalizations biologists have formed—the idea that a morphologic entity known as the cell is the basic unit of animal form and function.

THE CELL CONCEPT

Although the idea that the cell represents the basic unit of living forms and functions is still often referred to as the cell theory, it has long passed the status of theory and should be known as the cell concept or doctrine. It is virtually the chief cornerstone for biologic study and understanding.

As is the case with all important concepts, the cell concept has had an extensive background of development. The English scientist R. Hook named the first cells (1665) when he observed the small box-like cavities in the surface of cork and leaves. The classic microscopist A. van Leeuwenhoek, the Dutch lensmaker, described many kinds of cells in addition to his famous protozoan discoveries (1675 to 1680). The French biologist R. Dutrochet gave some basic ideas about cells in 1824. In 1831 R. Brown discovered and described the nuclei of cells. J. Purkinje (1839) not only described cells as being the structural elements of plants and animals but also coined the term **protoplasm** for the living substance of cells. M. J. Schleiden and T. Schwann, German biologists, are often given credit for the cell theory formulation (1838) because of their rather extensive descriptions and diagrams, although they had erroneous ideas about how cells originate. In 1858 R. Virchow stressed the role of the cell in disease, or pathology, and stated that all cells came from preexisting cells. M. Schultze (1864) gave a clear-cut concept of protoplasmic relations to cells and its essential unity in all organisms.

It may be said in summarizing the main features of the cell doctrine that all plants and animals are composed of cells or cell products, that a basic unity exists in their physical construction, and that all cells come from preexisting cells. The total processes and activities of life can be interpreted on the basis of the cellular components of the organism. This statement does not contradict the belief that the whole organism behaves as a unit in its development and in its integration of cell activities or that the action of cells is determined in accordance with the physiologic behavior of the organism at all stages of its existence.

METHODS OF STUDYING CELLS

Cells are small and mostly invisible, so the microscope has been the tool of choice in studying them. But the microscope required the aid of staining methods. Fortunately the discovery and development of the aniline dyes by W. H. Perkin and others gave the investigators of the last half of the nineteenth century the opportunity to work out the details of cellular structures and cell division within the limits of the light microscope.

Among cytologic advances are the careful histologic techniques of fixing the tissues to preserve them as naturally as possible, the art of preparing and slicing tissues with a microtome, and proper staining methods for differential staining of cell constituents, or the selective affinity of the different cell components for the various stains. More precise physiochemical methods for locating specific entities within cells and for identifying them are constantly being sought. Ultraviolet light is employed because different chemical substances absorb rays of characteristic wavelengths. Some of the histochemical techniques for demonstrating inorganic or organic substances in cells and tissues are (1) the periodic acid–Schiff (PAS) reaction for showing carbohydrates, (2) the fluorescent antibody method for determining where the antibodies are localized in the cells, and (3) injecting tagged atoms that have been labeled with a radioactive isotope (tritium [H^3], carbon 14, and many others), and then photographing the desired specimen of tissue on a special photographic emulsion plate that will record the beta or other particles from the radioactive isotope.

ORGANIZATION OF THE CELL

Protoplasm is usually found in small discrete bodies or microscopic droplets known as cells. (An exception is the plasmodium of the noncellular slime molds [myxomycetes]. Plasmodia are slow creeping, jellylike structures with thousands of nuclei and without cell walls.) The cell is the unit of biologic structure and function, and it is the minimum biologic unit capable of maintaining and propagating itself.

Cells vary greatly in both size and form. Some of the smallest animal cells are certain parasites that may be 1 μ (1/25,000 inch) or less in diameter. At the other extreme we have the fertilized eggs of birds, some of which, including the extracellular material, are several inches in diameter. A red blood corpuscle in man has a diameter of about 7.5 μ. The longest cells are the nerve cells because the fibers, which are parts of the cells, may be up to several feet long. Some striped muscle cells or fibers are several inches long.

The various functions of the life process carried on at the unicellular stages tend to be allotted to specialized cells in multicellular organisms. Functional specialization is accompanied by structural

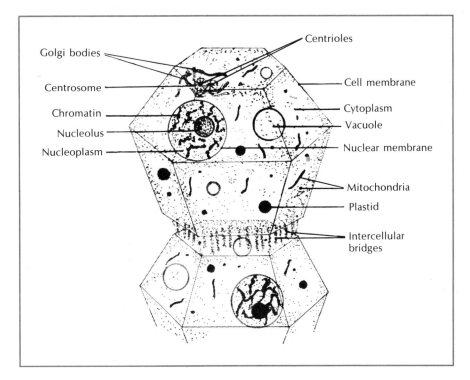

Golgi bodies
Centrioles
Centrosome
Cell membrane
Chromatin
Cytoplasm
Nucleolus
Vacuole
Nucleoplasm
Nuclear membrane
Mitochondria
Plastid
Intercellular bridges

FIG. 4-1. Scheme of generalized cell as revealed by light microscope, showing principal constituents commonly found in most cells. Cell shape is correlated with cell function and with mechanical pressure of adjacent cells. Pressure often produces fourteen-faceted surface that appears hexagonal in cross section. Constituents vary with types of cells and phases of activity. Protoplasmic processes (intercellular bridges) may connect cells in some tissues such as epithelia. Smallest cells, probably bacteria, are less than 1 μ in diameter; ostrich eggs may be several inches around.

specialization or division of labor, and the hierarchy of tissues, organs, and organ systems arise as a consequence in the evolutionary development of life. Each cell retains the capacity to act independently of the others. One cell of a group may divide, secrete, or die, while adjacent cells may be in a different physiologic state.

A cell includes both its outer wall, or membranes, plus its contents (Fig. 4-1). Typically, it is a semi-fluid mass of microscopic dimensions, completely enclosed within a thin, differentially permeable **plasma membrane.** It usually contains two distinct regions—the nucleus and cytoplasm. (A few unicellular organisms, such as bacteria and blue-green algae, do not show a distinctive separation of nuclear and cytoplasmic constituents but have the chromatin material scattered through the cytoplasm.) Cells without nuclei are called **prokaryotic** in contrast to those with nuclei called **eukaryotic.** The **nucleus** is enclosed by a **nuclear membrane** and contains the **chromatin** and one **nucleolus** or more. Within the **cytoplasm** are many **organelles,** such as mitochondria, Golgi complex, centrioles, and endoplasmic reticulum. In addition, plant cells may contain plastids, or chloroplasts.

The different shapes assumed by cells are mostly correlated with their particular function (Fig. 4-2). Many cells will assume a spherical shape when freed from restraining influences; others retain their shape under most conditions because of their characteristic cytoskeleton, or framework.

The electron microscope reveals small cytoplasmic **microtubules** that may serve as cytoskeletal elements in maintaining the shape of cells. These tubules are straight, are of indefinite length, and have a diameter of about 200 to 270 Å. Their walls

37

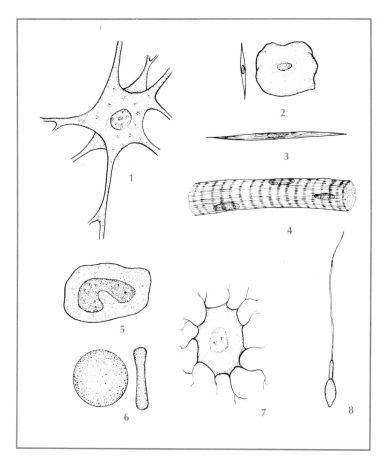

Fig. 4-2. Some common examples of cells. **1,** Nerve cell from spinal cord; **2,** epithelial cell from lining of mouth, with side view on left; **3,** smooth muscle cell from intestinal wall; **4,** striated muscle cell from gross muscle; **5,** white blood corpuscle; **6,** red blood corpuscle, with side view on right; **7,** bone cell; **8,** human spermatozoan. (Not drawn according to scale.)

are made up of ten or more filamentous subunits. They appear to be most common near the cell center and may be closely related to the centriole. They form the spindle apparatus of dividing cells, the caudal sheath of spermatids, the marginal bands of nucleated erythrocytes, and the axoplasm of neurons. Their power to contract may be involved in the movements of the cytoplasm and the alterations in cell shape.

COMPONENTS OF THE CELL
AND THEIR FUNCTION

All structures, or organelles, of the cell have separate, important functions. The **nucleus** (Fig. 4-3) has two important roles: (1) to store and carry hereditary information from generation to generation of cells and individuals and (2) to translate genetic information into the kind of protein characteristic

of a cell and thus determine the cell's specific role in the life process. A component of the nucleus, the **nucleolus,** contains ribonucleic acid (RNA) and may act as an intermediate between the code of the chromosomes and the execution of the code in the cytoplasm. In the **mitochondria** (Figs. 4-3 to 4-6) the energy-yielding oxidations from the breakdown of complex organic compounds are localized. This energy is stored in high-energy phosphate bonds to be used in biologic activities as needed.

The double-layered membranous **endoplasmic reticulum** (ergastoplasm) (Figs. 4-3 to 4-5) is supposed to contain enzymes that synthesize cholesterol and other nonproteins. Often associated with the endoplasmic reticulum are **granules (ribosomes)** (Fig. 4-3) that are the sites of protein synthesis under the influence of RNA, which receives its coded genetic information from deoxyribonucleic acid (DNA) of the chromosomes. The **centri-**

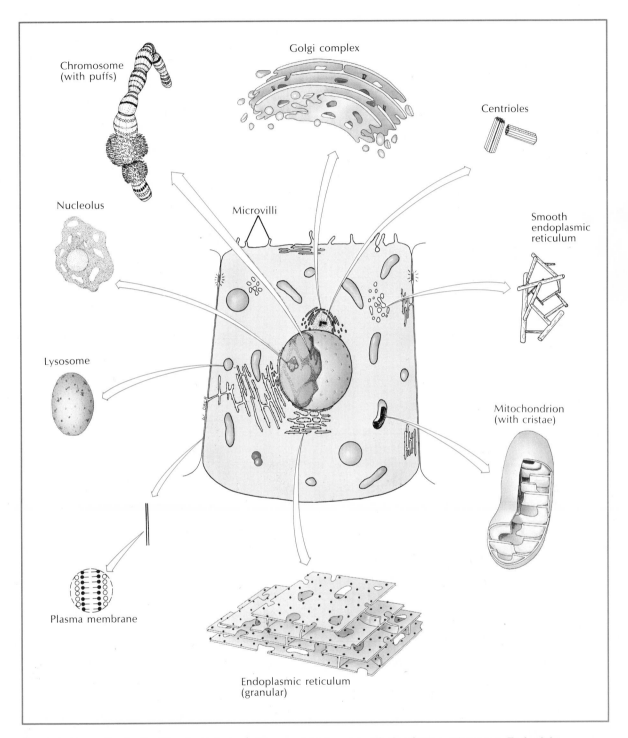

Golgi complex

Chromosome
(with puffs)

Centrioles

Nucleolus

Microvilli

Smooth
endoplasmic
reticulum

Lysosome

Mitochondrion
(with cristae)

Plasma membrane

Endoplasmic reticulum
(granular)

FIG. 4-3. Generalized cell with principal organelles, as might be seen with the electron microscope. Each of the major organelles is shown enlarged. Membranes of organelles are believed to be continuous with, or derived from, the plasma membrane by an infolding process. Structure of other membranes (of nucleus, endoplasmic reticulum, mitochondria, etc.) is probably similar to that of plasma membrane, shown enlarged at lower left.

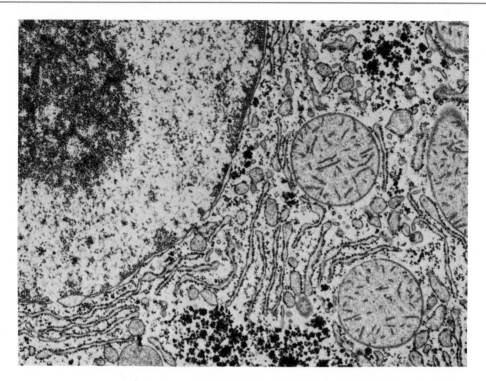

FIG. 4-4. Electron micrograph of part of hepatic cell of rat showing portion of nucleus (left) and surrounding cytoplasm. Endoplasmic reticulum and mitochondria are visible in cytoplasm, and pores are seen in nuclear membrane. (×14,000.) (Courtesy G. E. Palade, The Rockefeller University, New York.)

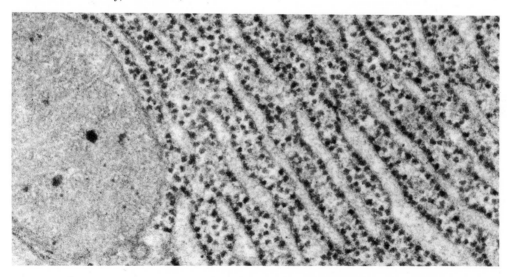

FIG. 4-5. Electron micrograph of portion of pancreatic exocrine cell from guinea pig showing endoplasmic reticulum with ribosomes (small dark granules). Oval body (left) is mitochondrion. (×66,000.) (Courtesy G. E. Palade, The Rockefeller University, New York.)

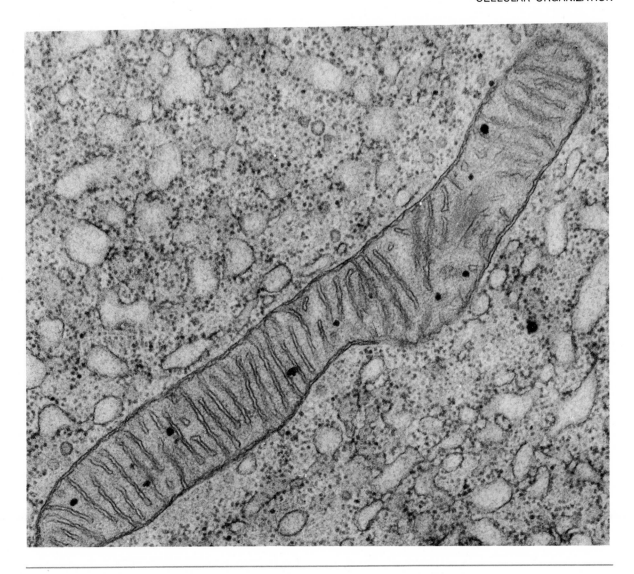

FIG. 4-6. Electron micrograph of enlongated mitochondrion in pancreatic exocrine cell of guinea pig. (×50,000.) (Courtesy G. E. Palade, The Rockefeller University, New York.)

oles (Fig. 4-3) determine the orientation of the plane of cell division and probably supply the **basal granules,** or **kinetosomes,** which are concerned with the formation of motile fibrillar structures, such as cilia and flagella at the surface of cells. Another type of granules, **lysosomes,** are the sites of certain hydrolytic enzymes.

The Golgi complex is the primary site for the packaging of the secretory products that are syn-thesized on the ribosomes and migrate to the sac-cules, or stacks of flattened sacs, making up the Golgi complex. Here also carbohydrate molecules formed by the Golgi complex are added to the pro-tein secretions to form glycoproteins before they are discharged for their various functions.

Although the vast majority of cells contain these basic structures, there are some organisms that lack conformity to this cellular structure.

Whatever function a cell may have depends on the properties of cellular compounds. With the electron microscope, it has been possible to learn a great deal about subcellular organization and to explain the relations between structure and function in terms of the interactions of macromolecules. Among the discoveries are (1) information concerning the submicroscopic membrane systems of the cytoplasm and their roles in the synthesis, storage, and transport of metabolic products and (2) the structure and function of cellular constituents involved in the energy transformations of cells.

Plasma membrane. The plasma membrane encloses the cell and forms the organelles within the cytoplasm. Organelles are specialized parts of a cell that have specific functions, just as organs have in higher forms. The plasma membrane serves as a partition to subdivide the cell space into self-contained compartments in which biochemical reactions may take place for the living process. As seen with the electron microscope, the plasma membrane appears as two black lines with a space between and is about 70 to 100 Å thick (an angstrom is 1/100,000,000 cm.). Its structure suggests a lipid-protein membrane made up of two lipid monolayers between two protein monolayers (Fig. 4-3). Phospholipid molecules are long, with elongated hydrophobic organic compounds and a hydrophilic polar group at one end. Such molecules tend to be absorbed at an air-water or water-oil interface because the lipid end enters the oil or air and the polar end enters the water. The molecules also tend to be tightly packed and oriented in parallel layers at interfaces. The membrane serves to regulate the molecular traffic in and out of the cell.

Plasma membranes have **selective permeability** that is quite necessary to maintain a proper balance of organic and inorganic substances on which life depends.

Mechanisms derived from chemical reactions within the cell (active transport) are often referred to as pumping devices. Thus by **phagocytosis** the membrane forms pockets in which food particles are engulfed. These pockets with the enclosed substance are pinched off and form a vesicle within the cytoplasm. Another pumping device is **pinocytosis,** or cell drinking, which is similar to phagocytosis except that drops of liquid are taken up discontinu-

ously and sucked in to form vesicles within the cell. The concept has been broadened to include the uptake of dissolved substances as well as liquids. By this method it is thought that ions, sugars, and proteins can be pumped into the cell. The average diameter of a pinocytosis vacuole is about 1 to 2 μ but often they coalesce. About 30 minutes is required for the formation of a pinocytosis vacuole.

The formation of cytoplasmic organelles from the infolding of the plasma membranes, invisible with the optical microscope, gives rise to a characteristic cytoskeleton that is revealed to some extent by the electron microscope. Some investigations seem to reveal that the internal membranes of the cell are continuous with the external membrane, so that there is a deeply invaginated surface area for communication with the outside fluid.

Endoplasmic reticulum. A complex membrane system called the endoplasmic reticulum (Figs. 4-3 and 4-5) has been closely associated with the storage and transport of products of cellular mechanisms. The nuclear membrane is formed from parts of this membrane system, and it is so arranged that there is direct continuity between nucleus and cytoplasm by openings in the nuclear membrane (Fig. 4-4). The double-walled endoplasmic reticulum is a highly variable morphologic structure consisting of vesicles and tubules, and it often has the power to fragment and reform its structural features. There are two types—rough- and smooth-surfaced. The rough-surfaced type has on its outer surface small granules called **microsomes** (the dense granules are often called **ribosomes** [Fig. 4-5] because they contain ribonucleic acid). These ribosomes are important sites of protein synthesis. Granules may function without being attached to the membrane. The endoplasmic reticulum is an intracellular transport system for the products synthesized by the ribosomes. Among the substances transported by this system are the zymogen bodies that give rise to digestive enzymes and that are carried to the smooth-surfaced Golgi vesicles. Later these zymogen granules are discharged as enzymes through openings at the surface of the cell.

Mitochondria. Another important cytoplasmic organelle is the mitochrondrion (Figs. 4-3 and 4-6). Mitochondria are found in all cells and can be detected by the light microscope. They show considerable diversity in shape, size, and number. Many of them are rodlike and are about 0.2 to 5 μ in greatest diameter. The electron microscope and the centrifuge microscope reveal a double membrane system that may be formed from detached pockets of the cell membrane. The inner layer of the double membrane is much folded and forms projections (cristae) that extend into the interior fluid or matrix. The mitochondria are the principal chemical sites for cellular respiration. They contain highly integrated systems of enzymes for providing energy in cell metabolism. Among these systems are the important citric acid cycle and the electron transport chain that produce the energy-rich adenosine triphosphate (ATP) so essential for many vital activities.

In addition, mitochondria are known to have a complete apparatus for the synthesis of certain enzymes and proteins and to contain deoxyribonucleic acid (DNA) and ribonucleic acid (RNA). Unlike nuclear DNA, mitochondrial DNA does not associate with the protein histone, and the naked DNA fiber forms a circular molecule of DNA. These conditions are similar to the primitive forms of bacteria and blue-green algae and may indicate that mitochondria were once free-living organisms. There is evidence that other organelles of eucaryotic cells had a similar origin from procaryotic cells. However, mitochondria can synthesize only a few proteins, for most of their enzymes and components are synthesized under the direction of the nuclear DNA genes. Mitochondria are therefore endosymbionts and must depend on the cytoplasm of the cell.

SURFACE OF CELLS AND THEIR RELATIONS

Cell surfaces vary, depending on many factors. Several cell types are free and can move throughout the animal. These free cells have no direct junctional arrangements with other cells and include such types as leukocytes, red blood corpuscles, amebocytes, macrophages, and many others. Interstitial cells are undifferentiated, are located on epithelial structures, and often migrate to injured regions for repair. Pigment cells (chromatophores) have a certain amount of freedom to move about.

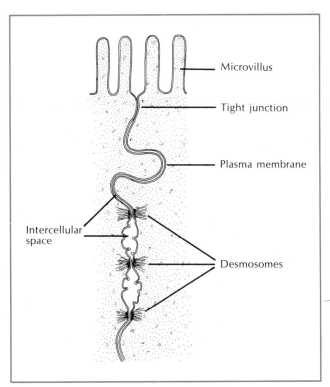

Microvillus

Tight junction

Plasma membrane

Intercellular space

Desmosomes

FIG. 4-7. Two opposing plasma membranes forming the boundary between two epithelial cells. Various kinds of junctional complexes are found. Tight junction is a firm, adhesive band completely encircling the cell. Desmosomes are isolated "spot welds" between cells that serve as sites of intercellular communication. Intercellular space may be greatly expanded in epithelial cells of some tissues.

The surface cells of many types throughout the animal kingdom bear specialized structures of locomotion called cilia and flagella (Fig. 8-2). These are vibratile extensions of the cell surface, and their covering membrane is continuous with the plasma membrane. Internally cilia and flagella have the same structure—nine fibrils surrounding a pair of fibrils. (Exceptions have been noted having but a single central fibril.) At the base where they are anchored in the cell cytoplasm, each flagellum or cilium is attached to a granule, a **kinetosome.** The kinetosome is similar to the cilia and flagella, but has nine triplet filaments and no double central filaments. Flagella and cilia have numerous functions such as propelling individual cells (protozoans), multicellular animals (planarians and ctenophores), or fluids or entities through tubular organs (sponges and gonoducts). Most animal sperm are provided with them for propulsion.

The surfaces of contiguous cells, or cells packed together, have junction complexes between them. There are several types of these specializations. The adjoining surfaces of cells are sealed only in restricted areas. Nearest the free surface, the two opposing membranes appear to fuse to form a **tight junction** (Fig. 4-7). Below this is a slightly widened **intermediary junction.** Next are desmosomes, small ellipsoidal disks scattered between the epithelial cells (Fig. 4-7). Desmosomes act as "spot welds" between apposing plasma membranes. They measure about 250 to 410 millimicrons in their greatest diameter. From the cell cytoplasm tufts of fine filaments converge onto the desmosomes. Between the two apposed plates of a desmosome is a narrow intercellular space (200 to 240 Å wide). All of these special junctional complexes produce the **terminal bar,** found at the distal junctions of adjacent columnar epithelial cells. They form a complete beltlike junction just beyond the luminal or apical portion of the plasma membrane.

Other specializations of the cell surface are the interdigitations of confronted cell surfaces where the plasma membranes of the cells infold and interdigitate very much like a zipper. They are especially common in the epithelium of kidney tubules. The distal or apical boundaries of some epithelial cells, as seen with the electron microscope, show regularly arranged **microvilli.** They are small, fingerlike projections consisting of tubelike evaginations of the plasma membrane, with a core of cytoplasm (Figs. 4-3 and 4-7). They are well seen in the lining of the intestine where they greatly increase the absorptive or digestive surface (Fig. 9-7). Such specializations are seen as striated and **brush borders** by the light microscope. The spaces between the microvilli are continuous with tubules of the endoplasmic reticulum that may facilitate the movement of materials into the cells.

5

Replication and function of cells

CELL DIVISION (MITOSIS)

All cells of the body arise from the division of preexisting cells. All the cells found in most multicellular organisms have originated from the division of a single cell, the **zygote,** formed from the union of an **egg** and **sperm** (fertilization). This division provides the basis for one form of growth, for both sexual and asexual reproduction, and for the transmission of hereditary qualities from one cell generation to another cell generation.

In the formation of **body cells** (somatic cells) the process of cell division is referred to as **mitosis** and is the method of cell division to be described in this section. However, **germ cells** (egg and sperm) have a somewhat different type of cell division known as **meiosis** (to be described later).

There are two distinct phases of mitosis: the division of the nucleus and the division of the cytoplasm. These two phases ordinarily occur at the same time, but there are occasions when the nucleus may divide a number of times without a corresponding division of the cytoplasm. In such a case the resulting mass of protoplasm containing many nuclei is referred to as a **multinucleate** cell. Skeletal muscle is an example. A many-celled mass formed by fusion of cells that have lost their cell membranes is called a **syncytium.**

Mitosis or indirect cell division. The whole process of mitosis has been followed through all stages in living cells. As long ago as 1878 W. Flemming observed it in the skin cells of young salamanders. However, the usual study is made from specially prepared tissues that have been fixed, stained, and mounted on slides (Fig. 5-1). The sequence is observed by finding cells caught in the various stages by the fixatives and arranging them in the proper order.

Chromosomes. The nucleus bears **chromatin** material, which in turn carries the **genes** responsible for hereditary qualities. During cell division this chromatin becomes arranged into definite **chromosomes** of varied shapes that are constant within a species. The metaphase or anaphase chromosomes are most typical in their morphology, for at this time they have reached their maximum contraction. Each animal within a species has the same number of chromosomes in each body cell. Man has 46 in each of his body cells (but not in his germ cells). The number of chromosomes a species possesses has no basic significance nor is there necessarily any relationship between two different species that have the same number. Both the guinea pig and the onion have 16 chromosomes. Since there are thousands of different species of animals, many species must of necessity have the same number.

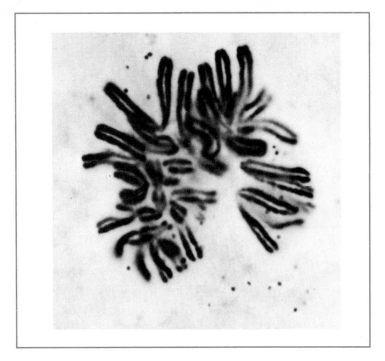

FIG. 5-1. Chromosomes in tail epidermis of salamander *Ambystoma* shown arranged on equatorial plate in metaphase stage of mitosis. (Courtesy General Biological Supply House, Inc., Chicago.)

The range is usually between 8 and 50 chromosomes, although numbers in the hundreds are known (crayfish, for example). Some worms have as few as 2, the least number an organism can have.

Chromosomes are always arranged in pairs, or two of each kind (except sex chromosomes). Of each pair, one has come from one parent and the other from the other parent. Thus in man there are twenty-three pairs. Each pair usually has certain characteristics of shape and form that aid in identification. It will be seen that a biparental organism begins with the union of two gametes, each of which furnishes a **haploid** set of chromosomes (23 in man) to produce a somatic or **diploid** number of chromosomes (46 in man). The chromosomes of a haploid set are also called a **genome.** Thus, a fertilized egg (zygote) consists of a paternal genome and a maternal genome.

The purpose of mitosis is to ensure an equal distribution of each kind of chromosome to each daughter cell. A cell becomes highly abnormal in its reactions if it fails to receive its proper share of chromosomes.

Structure of chromosomes. The shape of the metaphase or anaphase chromosome, although constant within a specific species, differs between species of animals. Some chromosomes have the **centromere** at the midpoint, with equal limbs (metacentric); other chromosomes have the centromere closer to one end, with unequal arms (acrocentric). The ordinary light microscope may reveal only the outline of the metaphase chromosome, with two parallel strands (chromatids) united at one point (centromere) within it. With special techniques and the electron microscope, finer details are found, such as the presence of a coiled multiple filament **(chromonema)** along which are beadlike, dark-staining enlargements called **chromomeres.** The chromomeres, which may represent superimposed coils or nucleoprotein condensations, may contain aggregations of genes, or the genes may be located between them. Many details about the chromosome are still unknown.

Other parts of the chromosome include condensed and variable staining regions called **heterochromatin,** in contrast to the rest of the chromosome or **euchromatin.** The centromere belongs to the heterochromatin. Most chromosomes contain segments of heterochromatin and euchromatin. The ends of the chromosomes, **telomeres,** appear to be functionally different from the rest of the chromosome.

The chemical nature of the chromosome consists of two nucleic acids (deoxyribonucleic acid, or DNA, and ribonucleic acid, or RNA), together with certain proteins (histones, protamines, etc.). How the nucleic acids are arranged in the chromosome is not yet known. (See the discussion on genes, Chapter 20.)

PROCESS OF CELL DIVISION

It is not known what triggers cells to divide, although many biologists believe that division depends on the attainment of a critical mass (for example, when the cell has doubled its mass). However, some cells are known to divide without growth. Specialized cells, such as neurons, lose their power to divide, as do red blood corpuscles that have no nuclei.

The essential plan of cell division is not complicated, but its various steps are known only in part. Briefly, the centrioles double to form two poles, and a mitotic apparatus is established. The chromosomes are exactly duplicated—sister chromosomes migrate to sister poles, the cytoplasm cleaves into two parts, and two cells are formed. In most organisms cell division really consists of two separate processes—nuclear division by mitosis and cytokinesis (cleavage), of the cytoplasm. These two processes usually take place simultaneously. There are, however, variations of this binary fission plan. In some cases nuclear duplication without cytokinesis may occur and form a multinucleated cell, or cytokinesis of a binucleate cell may occur without nuclear duplication.

The process of mitosis is arbitrarily divided for convenience into four successive stages, or phases, although one stage merges into the next without sharp lines of transition. These phases are prophase,

metaphase, anaphase, and telophase. When the cell is not actively dividing, it passes through the "resting" stage, or **interphase.** Before the first visible active phase appears, the chromosome threads and their component genes become chemically duplicated. Genetically, the DNA content of the nucleus is the important constituent that is duplicated between divisions. Thus when the cell begins mitosis, it has a double set of chromosomes.

The mechanism by which the chromosomes of a pair are separated and pushed or pulled to opposite poles ensures that each daughter cell always receives a complete set of hereditary units, or genes. For this separation, a special mechanism or mitotic apparatus is required. In animal cells one of the requirements for mitosis is the presence of centri-

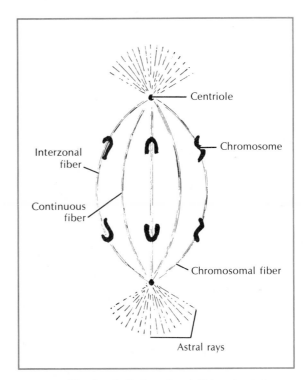

FIG. 5-2. Mitotic spindle (anaphase). It is not yet clear how chromosomes are moved apart. Spindle fibers may act as guides, along which chromosomes move toward poles. Mechanism may involve pushing produced by swelling between two sets of separating chromosomes, or chromosomes may be pulled to poles by folding of protein chains of fibers that produce traction on attached chromosomes. Fibers are tubes, not threads.

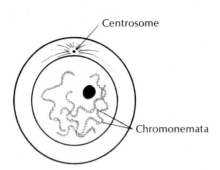

Centrosome

Chromonemata

1. Interphase

Each chromosome reaches its maximum length and minimum thickness; each chromosome composed of coiled thread of genes (chromonema) and centromere; duplication of chromosome occurs at this state

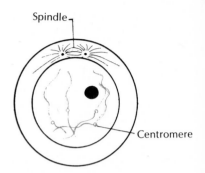

Spindle

Centromere

2. Early prophase

Each elongated chromosome now consists of 2 chromatids attached to single centromere; double nature of chromosome not apparent; centrosome divides and spindle starts development

5. Prometaphase

Nuclear membrane disintegrates or changes structure

6. Metaphase

Chromosomes arranged on equatorial plate; centromeres (not yet divided) anchored to equator of spindle

FIG. 5-3. Mitotic stages.

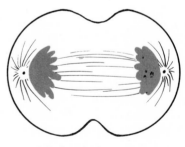

9. Early telophase

Chromosomes lie close together and form a clump; nuclear reorganization begins

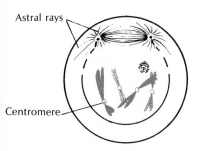

3. Middle prophase

Each chromosome may be visibly double; coils of chromonemata increase and make chromosome shorter and thicker

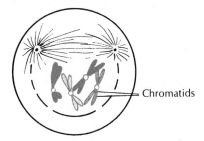

4. Late prophase

Double nature of short, thick chromosome more apparent; each chromosome made up of 2 half-chromosomes or sister chromatids; nucleolus usually disappears

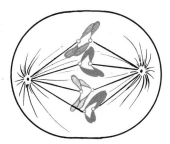

7. Early anaphase

By splitting of centromere each chromatid has its own centromere, which lies on equator and is attached to spindle fiber

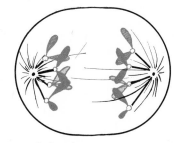

8. Anaphase

Chromatids, now called daughter chromosomes are in 2 distinct groups; daughter centromeres which may be attached at various points on different chromosomes, move apart and drag daughter chromosomes toward respective poles

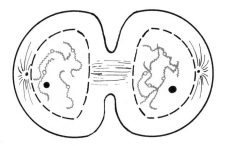

10. Late telophase

Chromonemata lose major coils and chromosomes become longer and thinner; chromosomes may lose identity; nuclear membrane reappears and spindle-astral fibers fade away; cell body divides into 2 daughter cells, each of which now enters interphase

oles that are permanent, self-duplicating, small, dotlike bodies (they have not been found in higher plant cells). Centrioles are generally found in pairs, with the members of a pair lying at right angles to each other. Each cell inherits one set of centrioles and produces another set.

At the start of **prophase** the centrioles, as soon as they have reproduced, migrate toward opposite sides of the nucleus. At the same time portions of the cytoplasm are attracted to the regions of the centrioles and are transformed into fine gel fibrils. Some of these fibrils run between the two centriole complexes to form a **spindle,** and some radiate out from each centriole or pole to form **asters.** The whole structure is called a **mitotic apparatus,** and it increases in size as the centrioles move farther apart. In higher animals the cells at mitosis have two large asters, one at each end of the spindle. At the center of each aster there is a spherical **centrosome** within which is the centriole (Figs. 5-2 and 5-3). During this process the nuclear membrane disappears, and the nucleolus disintegrates or becomes invisible.

After the mitotic apparatus is formed, the chromosomes begin to move under the control of the poles of the spindle. Each chromosome is made up of a double filament and has condensed into visible units. The finer analysis of the chromosome reveals that it is composed of tight coils. Each longitudinal half-chromosome (chromatid) has nucleic acids

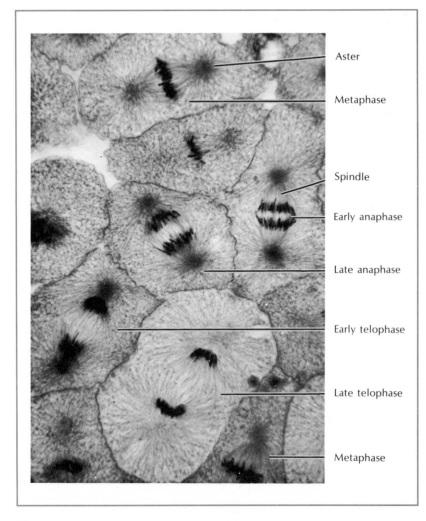

Aster

Metaphase

Spindle

Early anaphase

Late anaphase

Early telophase

Late telophase

Metaphase

Fig. 5-4. Stages of mitosis in whitefish. (Courtesy General Biological Supply House, Inc., Chicago.)

identical to those of the chromosome before duplication. The paired chromatids of a chromosome are joined at a single point by the **centromere**. The position of the centromere varies for different chromosome pairs. From each pole of the spindle a fibril makes connection with the centromere of each double chromosome. The centromere seems to be all important in guiding the sequence of events in mitosis.

During early **metaphase** the chromosomes are somewhat scattered, but they quickly migrate into the equatorial plane set at right angles to the spindle axis. It is actually the centromeres that come to occupy a precise position within the equatorial plane, and the chromosomes to which they are attached trail behind, often in various shapes. When they line up in one plane, the centromeres form the **metaphase plate** (Figs. 5-3 and 5-4). It is thought that the centromeres are pulled into position by the equal tension of the two fibrils that connect each centromere to each pole at this stage.

The two chromatids of each double chromosome now separate. Up to this time each double chromosome has a single centromere that holds the two chromatids together. Now the centromere splits so that two independent chromosomes, each with its centromere, are formed. The same tension of their pole-attached fibrils that pulled the double chromosome into the equatorial plate now pulls the two chromatids (now called chromosomes) of a pair toward opposite poles at a speed of about 1 μ per minute. The chromosomes move in straight lines, and at this **anaphase** stage two sets of chromosomes are plainly visible, one set moving toward one pole and the other set moving toward the other pole. Sometimes the poles move farther apart during this stage and carry the chromosomes with them. At the anaphase, interzonal fibrils connect the separated chromosomes and lie peripheral to the continuous fibrils of the spindle. These interzonal fibrils may be attached to the centromeres of the originally paired chromosomes. They lengthen as the chromosomes move farther apart, while chromosomal fibrils between chromosomes and spindle poles shorten (Fig. 5-2).

When the daughter chromosomes reach their respective poles, the **telophase** has begun. The daughter chromosomes are crowded together and stain intensely. Two other events also occur—the appearance of a **cleavage furrow** encircling the surface of the cell (Fig. 5-3) and a **cell plate** (in plants) that originates from the central portions of the interpolar spindle fibers. Eventually the cleavage furrow deepens and constricts the cell into two daughter cells (Fig. 5-4). Other changes that terminate the telophase period are the disappearance of the spindle fibrils, the gradual assumption of a chromatin network as the chromosomes lose their identity, the formation of a new nuclear membrane, and the manufacture of new nucleoli by the chromosomes.

The duration of the mitotic process varies from 30 minutes to 3 hours for the complete process, prophase and telophase being the longest. The interphase in cells that regularly divide usually lasts from 10 to 20 hours.

The result of cell division is the formation of two cells, each with an identical gene set, so that each daughter cell is potentially the same as the mother cell. Cell division is important for growth and replacement, wound healing, etc. Muscle cells rarely divide, and nerve cells never divide after birth. The more specialized the cell the less frequently it divides. However, some tissues continually divide because the body loses a percentage of its cells daily and these must be replaced. Cell reproduction is faster in the embryonic state and slows down with age, a condition that may be due to metabolic checks brought on by larger cell populations.

THE CELL AS THE BASIC UNIT OF LIFE

A cell is considered to be the minimum unit that manifests the vital phenomena of life. However, the boundary between the living and the nonliving is not as sharply drawn as formerly. Some particles smaller than cells, for example, viruses, are regarded by some as living. All viruses, however, are associated with cells from which they derive their energy. Although viruses have the initial genetic mechanisms and multiply, they do so at the expense of the host's cells.

Many features or aspects of life can be produced in the test tube without cellular organization. Genetic investigations make use of strands of nucleic acid that can synthesize duplications of DNA and proteins. Life must have originated from the non-living world in much simpler units than the complex cell. Some present cells are simpler than others. Bacterial cells as well as those of blue-green algae lack such organelles as mitochondria, lysosomes, and a definite nucleus. Motile bacteria have a flagellum of a single fibrous protein molecule instead of the ninefold symmetry of higher forms. There must have been many precellular forms in the long evolution of the cell because the properties of life did not arise all at once. Many intermediate forms represent a continuity from the nonliving to the living. Whether or not some of these precellular units still exist has never been satisfactorily settled, but the consensus of opinion is that the whole spectrum of the origin of life from inorganic matter could have occurred only once in this planet's history. The cell as representing a combination of all the vital phenomena may logically be considered the basic unit of life.

In the light of these factors a living organism must consist of at least one cell. If one considers all **unicellular** organisms—bacteria, fungi, algae, protozoans, etc.—it is probable that they constitute the majority of all living forms. This unicellular level of complexity, or of hierarchy, may still be considered the favorite one of nature despite the attention we give to multicellular forms.

Cellular differentiation has many advantages over the unicellular condition. For example, it permits the development of large surfaces on which exchange of material occurs. It also promotes functional differentiation and size increase. In all higher organization the principle of cellular construction has been retained because nature has evolved no better method to meet the requirements of the environment.

FLUX OF CELLS

In many organisms certain tissues continually shed their cells because of wear and tear or other causes. The epidermis of the skin, the lining of the alimentary canal, and the blood-forming tissues lose large numbers of cells daily. There must be a constant replacement of the cells that are lost, for there is no net loss or gain in the overall picture. In man it has been estimated that the number of cells shed daily is about 1% to 2% of all the body cells. This amounts to several billion cells each day.

In its contact with the environment the organism is usually subject to a constant attrition of physical and chemical forces. Mechanical rubbing wears away the outer cells of the skin, and emotional stresses destroy many cells. Food in the alimentary canal rubs off lining cells, the restricted life cycle of blood corpuscles must involve a renewal of enormous numbers of replacements, and during active sex life many millions of sperm are produced each day. This loss is made up by a chain reaction of binary fission or mitosis.

At birth the child has about 2,000 billion cells. This immense number has come from a single fertilized egg (zygote). Such a number of cells could be attained by a chain reaction in which the cell generations had divided about forty-two times, with each cell dividing once about every 6 to 7 days. In about five more cell generations by the chain reaction, the cells have increased to 60,000 billion at maturity (in an individual of 170 pounds). However, not all cells divide at the same rate and some cells (nervous and muscular), as we have seen, stop dividing altogether at birth.

The life-span of different cells varies with the tissue, the animal, and the conditions of existence. Nerve cells and muscle cells, to some extent, persist throughout the life of the higher animal. Red blood corpuscles live about 120 days. The normal process of metamorphosis found in many animals involves a great loss of cells. At some point in the life cycle of most cells there is a breakdown of cell substance, the formation of inert material, a slowing down of metabolic processes, and a decrease in the synthetic power of enzymes. These factors lead eventually to the death of the cell. In certain cases, parts of the cell such as scales, feathers, and bony structures may persist after the death of the cell.

PASSAGE OF MATERIALS THROUGH MEMBRANES: EXCHANGE BETWEEN CELL AND ENVIRONMENT

The general metabolism of living cells requires a continuous supply of food materials and oxygen for the energy of the life process. Cells give off, in turn, by-products to the surrounding medium. These substances must all pass through the plasma membrane of the cell, whether they are entering or leaving the cell. Most cells are surrounded by an aqueous solution of some kind. The ameba is surrounded by the fresh water in which it lives. In many-celled animals the cell lives in a medium composed of blood, lymph, or tissue fluid. Before a substance can enter a cell, it must be soluble to some degree in the surrounding medium. The plasma membrane, then, acts as a doorkeeper for the entrance and exit of the substances involved in cell metabolism.

Diffusion, osmosis, and active transport. Since movement of materials within the body fluids is so characteristic of life processes, it is important that we have an understanding of the properties of solutions and the translocation of materials across cell membranes. **Diffusion** may be defined as the redistribution of molecules and ions by random movement. All molecules are in a state of motion because of their heat (kinetic) energy. In solids the molecules are so restricted that they merely vibrate; within a liquid the molecules have more freedom of movement; in gases the freedom to move is restricted only by containing vessels. Molecules move in a straight line until they meet another particle, then they bounce off and take a new direction. The constant agitation of dust particles in a drop of water, called brownian movement, is caused by the constant collision and jostling with other particles and with the molecules of water in which they are suspended.

If some salt is dropped in a beaker of water, salt molecules and ions will spread through the water until the concentration of salt is uniform throughout. The salt particles, as well as the water molecules, are in a continuous state of movement, and through mass movement every component in the solution will diffuse until it reaches equal concentration everywhere in the solution.

Osmosis is essentially diffusion of water through a selectively permeable membrane. If a membrane that is selectively permeable to water, but impermeable to any solutes in the water, is placed between two unequal concentrations of a solution, water will start to flow through the membrane from the more dilute to the more concentrated solution. To understand why this happens we must view the system from the standpoint of the concentration of water on either side. On the side of the stronger solution (higher concentration of solute) the water is present in **lower** concentration than it is on the side of the weaker solution. Water therefore diffuses through the membrane from the side where it is most concentrated (weaker solution) to the side where it is least concentrated (stronger solution). Water, like any other material, tends to diffuse "downhill," that is, from a higher to a lower concentration. Osmosis differs from diffusion in that only the water can diffuse; the solute is restricted by the selectively permeable membrane.

Another difference between osmosis and diffusion is that the movement of water creates a volume change. This can be demonstrated by a familiar experiment in which a selectively permeable membrane, such as collodion membrane, is tied over the end of a funnel. The funnel is filled with a sugar solution and placed in a beaker of pure water so that the water levels inside and outside the funnel are equal. In a short time the water level in the glass tube will be seen to rise, indicating that water is passing through the collodion membrane into the sugar solution (Fig. 5-5). Inside the funnel are sugar molecules as well as water. In the beaker outside the funnel are only water molecules. Thus the concentration of water is greater on the outside because some of the space inside is taken up with sugar molecules. The water therefore will go from the greater concentration (outside) to the lesser (inside).

As the fluid rises in the tube against the force of gravity, it exerts a hydrostatic pressure on the collodion membrane and glass tubing (small arrows in Fig. 5-5). This hydrostatic pressure opposes the movement of water molecules into the funnel. Eventually, the hydrostatic pressure becomes so great that there is no further **net** movement of water from the beaker into the bag, and the fluid level in the glass tube stabilizes. We see, then, that osmosis can perform work. An **osmotic pressure** (large arrows in Fig. 5-5) drives water through the membrane into the solution and creates an opposing **hydrostatic pressure** head. When the hydrostatic pressure (measured by the height of the fluid column) equals the opposing osmotic pressure, no more water enters the osmometer. (Actually, water molecules continue to traverse the membrane, but the

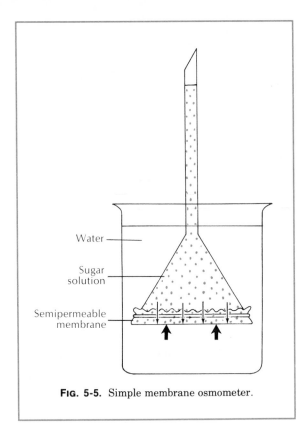

Water

Sugar
solution

Semipermeable
membrane

FIG. 5-5. Simple membrane osmometer.

movement inward is matched by movement outward.) Osmotic pressure can thus be expressed in terms of the height of the fluid column, which in turn depends on the concentration of the sugar solution.

Actually, the direct measurement of osmotic pressure in biologic solutions is seldom done today. This is because the osmotic pressures of most biologic solutions are so great that it would be impractical, if not impossible, to measure them with the simple membrane osmometer described. The osmotic pressure of human blood plasma would lift a fluid column over 250 feet—if one could construct such a long, vertical tube and find a membrane that would not rupture from the pressure. Indirect methods of measuring osmotic pressure are more practical. By far the most widely used measurement is the **freezing point depression.** This is a much faster and more accurate determination than is the direct measurement of osmotic pressure by the collodion membrane osmometer we described above. Pure water freezes at exactly 0° C. As solutes are added,

the freezing point is lowered; the greater the number of solutes, the lower the freezing point. Human blood plasma will freeze at about −0.56° C.; seawater will freeze at about −1.80° C. Although the lowering of the freezing point of water by the presence of solutes is small, great accuracy of measurement is possible because the instruments used by biologists can detect differences of as little as 0.001° C.

Active transport refers to the capacity of a cell membrane to move a substance from a point of lower concentration to a point of higher concentration **against** a diffusion gradient. It is the movement across cell membranes of molecules and ions by forces that cannot be accounted for by the laws of diffusion. Simple diffusion occurs in the same direction as the gradient and tends to equalize the extra- and intracellular concentrations. Active transport requires an expenditure of energy (furnished by ATP), but it may have certain advantages by making possible intracellular concentrations of substances that otherwise might be too low for effective metabolic reactions. Only certain molecules and ions of a given extra- or intracellular fluid can be transported against an electric potential or a concentration gradient. This transport may be facilitated by specific carrier molecules. Without these carriers, the membrane is impermeable to a given compound. The effects of active transport are significant only if the rate of active transport is greater than the rate of diffusion through a membrane in the opposite direction. The rate of active transport through a membrane is largely independent of the concentration gradient. Since active transport is powered by the cells' metabolic machinery, it is very sensitive to temperature change. A rise of 10° C. will about double the rate of active transport. Diffusion, on the other hand, is proportional to the absolute temperature; a rise of 10° C. will increase diffusion rate by only 10/273, or a little less than 4%.

Active transport may occur across all cell membranes, but it has been most studied in nerve, muscle, and kidney cells. Active transport is also involved in the transport of certain nutrients across the epithelium of the digestive tract and of the digestive glands. The excitability of nervous and muscular tissues depends on the relative concentration of ions inside and outside the nerve or muscle fiber. The concentration of K^+ ions is about twenty to fifty times higher inside than outside, whereas Na^+ and Cl^- ions are three to fifty times higher on the outside than on the inside (Fig. 11-2). This distribution of ions is responsible for the potential difference across the membrane known as the resting potential, which depends on the resting membrane being relatively permeable to K^+ and Cl^- ions and relatively impermeable to Na^+ ions. When these fibers or cells are excited by a stimulus, a change occurs in the permeability relationship, so that Na^+ enters the cell and K^+ passes out. This reversal of potential is called the action potential. The resting condition is restored during the recovery period. Na^+ ions that enter the cell during impulse conduction (Chapter 13) are pumped out by the so-called **sodium pump.** Such a pump is supposed to be located in the cell membrane or just beneath it and operates against both a concentration gradient and an electrical gradient.

Many other types of active transport are known. Osmoregulation and ionic regulation involve active transport in the kidney and gills of aquatic animals. Secretions of glands include the synthesis of new matter and the movement of molecules and ions against gradients of electrochemical potential, so that secretory activity is considered to be mostly a matter of active transport. Well-established active transport systems are known not only for ions but also for simple sugars and amino acids. In coupled transport, two substances may be transported in the same direction or they may be carried in opposite directions (as in the case of Na^+ and K^+ ions in muscle and nerve activity).

Facilitated diffusion occurs when certain lipid-insoluble molecules are able to penetrate cell membranes by the facilitation of changed pores so that diffusion is more rapid than can be explained by the concentration gradient of ordinary diffusion. It may be a carrier-mediated, specific transport that requires no metabolic energy and moves in the direction of a downhill concentration gradient. It results in the same equilibrium as that of ordinary diffusion, although its velocity is greater. Sugars such as glucose and mannose are believed to cross the red blood corpuscle membrane by this type of diffusion.

THE CENTRAL ROLE OF ENZYMES IN THE LIVING PROCESS

The whole life process involves numerous chemical reactions occurring within the cells. However, the chemical breakdown of large molecules and the release of energy for cellular activities would not proceed at any meaningful rate without **enzymes.** Enzymes are biologic catalysts that are required for almost every reaction in the body. As every chemist knows, catalysts are chemical substances that accelerate reactions without affecting the end products of the reactions and without being destroyed as a result of the reaction. Enzymes fit this definition, too. Enzymes are involved in every aspect of life phenomena. They control the reactions by which food is digested, absorbed, and metabolized. They promote the synthesis of structural materials to replace the wear and tear on the body. They determine the release of energy used in respiration, growth, muscle contraction, physical and mental activities, and a host of others. Little wonder that enzymes are absolutely fundamental to life.

Nature of enzymes. Enzymes are complex molecules varying in size from small, simple proteins with a molecular weight of 10,000 to highly complex molecules with weights up to 1 million. Some enzymes, such as the gastric enzyme pepsin, are pure proteins—delicately folded and interlinked chains of amino acids. Other enzymes contain special active nonprotein substances in addition to the protein portion of the molecule. Such an active and highly essential component is called a **prosthetic group** (working group). The prosthetic group is usually firmly attached to the protein. In some cases, however, biochemists have been able to detach the prosthetic group, yielding two molecules termed the **coenzyme** (prosthetic group) and the **apoenzyme** (protein part). Apoenzymes are inactive without the all-important coenzyme. Nearly all the vitamins have been shown to be essential parts of enzyme prosthetic groups. Most of the enzymes that have vitamin-containing coenzyme groups play crucial roles in cellular metabolism. Since a vitamin cannot be synthesized by the animal needing it, it is obvious why a dietary vitamin deficiency can be serious.

Naming of enzymes. Enzymes are named for

the reactions they catalyze. Usually the suffix -ase is added to the root word of the substance, or substrate, on which the enzyme works. Thus sucrase acts on sucrose, lipase acts on lipids, and protease acts on proteins. Enzymes may also be named according to the nature of the reaction. For example, dehydrogenases catalyze dehydrogenations.

Action of enzymes. An enzyme functions by combining in a highly specific way with the substance upon which it acts (the **substrate**). According to the lock-and-key concept, each enzyme possesses a unique molecular configuration that will fit only one substrate. By fitting onto the substrate, the enzyme provides a unique chemical environment that converts the relatively inactive substrate into a highly reactive molecule susceptible to chemical change (Fig. 5-6). Enzymes that engage in important main-line sequences, for example, the crucial energy-providing reactions of the cell that go on constantly, seem to operate in enzyme sets rather than in isolation. Main-line enzymes are found in relatively high concentrations in the cell, and they may implement quite complex and highly integrated enzymatic sequences. One enzyme carries out one step, then passes the product to another enzyme that catalyzes another step, and so on.

Specificity of enzymes. Most enzymes are highly specific. Such high specificity is a consequence of the exact molecular fit that is required between enzyme and substrate. However, there is some variation in degree of specificity. Some enzymes, such as succinic dehydrogenase, will catalyze the oxidation (dehydrogenation) of one substrate only, succinic acid. Others, such as proteases (for example, pepsin and trypsin), will act on most any protein. Usually an enzyme will take on one substrate molecule at a time, catalyze its chemical change, release the product, and then repeat the process with another substrate molecule. The enzyme may repeat this process billions of times until it is finally worn out (a few hours to several years) and is broken down by scavenger enzymes in the cell. Some enzymes are able to undergo successive catalytic cycles at dizzying speeds of up to a million cycles per minute; most operate at slower rates. The digestive enzymes, which are secreted into the digestive tract to degrade food materials, are "one-shot" enzymes. After breaking down their substrate,

they are themselves digested and lost to the body. Despite the high specificity of enzymes, they can sometimes be fooled into accepting a molecule that resembles the normal substrate. Malonic acid is such a molecule, and when succinic dehydrogenase combines with it the active site is blocked and the enzyme is inhibited or poisoned (Fig. 5-7). Malonic acid (which undergoes no chemical transformation) thus acts as a competitive inhibitor.

Enzyme-catalyzed reactions. Enzyme-catalyzed reactions are reversible. For example, succinic dehydrogenase catalyzes both the dehydrogenation of succinic acid to fumaric acid and the hydrogenation of fumaric acid to succinic acid.

$$\text{Succinic acid} + H_2O \rightleftharpoons \text{Fumaric acid}$$

Reversibility is signified by the double arrows. Enzymes vastly accelerate reactions in either direction. Many enzymes, however, catalyze reactions almost entirely in one direction. For example, the proteolytic enzyme pepsin can degrade proteins into amino acids, but it cannot accelerate the rebuilding of amino acids into any significant amount of protein. The same is true of most large molecules such as nucleic acids, polysaccharides, lipids, and proteins. There is usually one set of reactions and enzymes that break them down, but they must be resynthesized by a different set of reactions and catalyzed by different enzymes. This apparent irreversibility exists because the chemical equilibrium usually favors the formation of the smaller degradation products. The net **direction** of any chemical reaction is dependent on the relative energy contents of the substances involved, so that other compounds (energy-rich compounds) must participate in the conversion of, for example, amino acids into proteins, thus making the synthetic process different from simply the reverse of the degradation reactions.

Sensitivity of enzymes. Enzyme activity is sensitive to temperature and pH. As a general rule enzymes work faster with increasing temperature, but will do so only within certain limits. Moreover, this increase in velocity is not proportional to the rise in temperature. Usually the rate is doubled with each 10° C. rise, but a change from 20° to 30° C. has a greater effect than one from 30° to 40° C. The optimum temperature for animal enzymes is about

FIG. 5-6. Enzyme action and specificity. Enzymes are thought to have surface configurations that "fit" specific substrates. Here molecules **B** and **C** fit into enzyme surface, but **A** does not. Reactions involving **B** and **C** are speeded up by coming in contact briefly with enzyme. When reaction is complete, the enzyme, still unchanged, can dissociate from the substrate and is free to aid in further reactions. Molecule **A** and others not specific to this enzyme are unaffected by it.

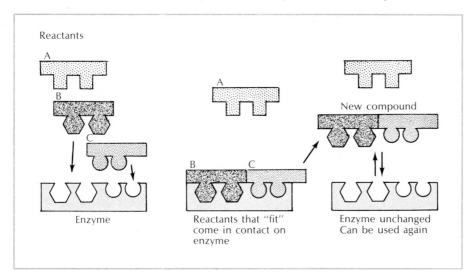

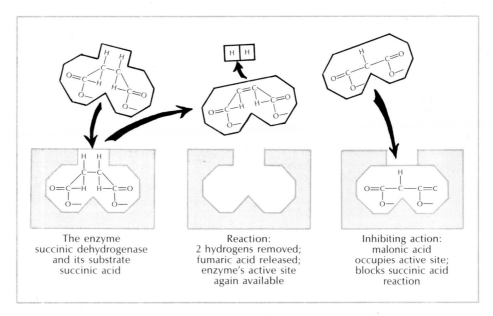

FIG. 5-7. Blocking enzymatic action. Enzyme action is sometimes inhibited by a molecule that resembles the normal specific substrate. Succinic dehydrogenase, for example, catalyzes dehydrogenation of succinic acid to fumaric acid, one of the steps of Krebs cycle. Malonic acid is so similar to succinic acid that, when present, it can temporarily occupy enzyme's active site, thus competing with succinic acid. Some enzyme inhibitors, such as carbon monoxide, nerve gas, or cyanide, remain permanently in the active site, thus putting the enzyme out of commission and poisoning the system.

body temperature. Above 40° to 50° C. most enzymes become unstable and may be inactivated altogether.

Each enzyme usually works best within a certain range of acidity or alkalinity. Pepsin of the acid gastric juice is most active at about pH 1.8; trypsin of the alkaline pancreatic juice is most active at about pH 8.2. Many work best when the pH is around neutrality. In strong acid or alkaline solutions, most enzymes irreversibly lose their catalytic power.

CELLULAR METABOLISM

The term cellular metabolism refers to the sum total of the chemical processes that are necessary for all the phenomena of life, such as the synthesis of new cell materials, the replacement of that which is destroyed during wear and tear, and whatever is needed by the cell to grow, reproduce, move, etc. For these processes cells require a continuous supply of nutrients obtained from the surrounding extracellular fluid. Living cells, like man-made machines, do work and consequently require fuel. This fuel is in the form of organic molecules, which, for animals, must be supplied in the diet. Animals, of course, are totally dependent on plants for ready-made fuels. Animal cells tap the stored energy of organic fuels (for example, simple sugars, fatty

acids, amino acids) through a series of controlled degradative steps. This process makes use of molecular oxygen from the atmosphere. In return the cells give off carbon dioxide as an end product, which is used by plant cells in making glucose and the more complex molecules. In this way the cellular energy cycle of life involves the harnessing of sunlight energy by green plants directly and by animal cells indirectly.

Biologic energy. The combustion, or burning, of fuels is similar in some ways to the cellular respiratory process. In both cases most of the energy is released through the reaction of fuel with molecular oxygen. However, in combustion many chemical bonds are broken simultaneously, liberating a great deal of energy as heat. Respiration, on the other hand, consists of a series of enzymatic reactions that control each step of the process. Energy is released gradually by breaking bonds one after another (Fig. 5-8). Temperatures can thus be kept low. Most of the energy of combustion is dissipated as heat, whereas metabolic energy **creates new chemical bonds** and loses little energy as heat.

There is actually only one way in which the oxidative release of energy is made available for use by cells: it is coupled to the production of ATP (adenosine triphosphate) by addition of inorganic phosphate to ADP (adenosine diphosphate). The

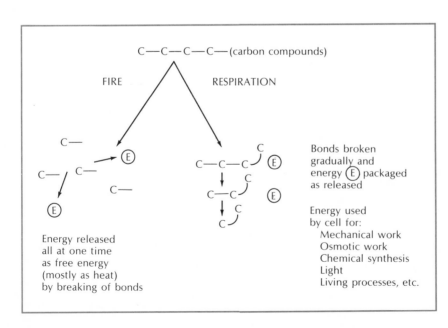

FIG. 5-8. How energy is released gradually in respiration instead of all at once, as in a fire.

structure of ATP is given below:

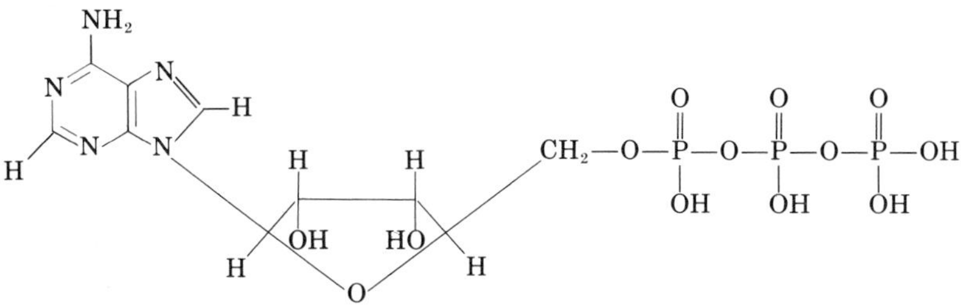

From left to right, the ATP molecule consists of a purine (adenine), a 5-carbon sugar (ribose), and 3 molecules of phosphoric acid linked together by two pyrophosphate bonds to form a triphosphate group. The pyrophosphate bonds are called **high-energy bonds** because they are repositories of a great deal of chemical energy. This energy has been transferred to ATP from other low-energy bonds in the respiratory process. Respiration, by the step-wise oxidation of fuel substrates, redistributes bond energies so that a few high-energy bonds are created and stored in ATP. Obviously this energy is gained at the expense of fuel energy; the end products of cellular respiration, CO_2 and H_2O, contain much less bond energy than the fuel substrates (for example, glucose), that entered the oxidative pathway.

The high-energy pyrophosphate bonds of ADP and ATP are frequently designated by the symbol $\sim$. Thus a low-energy phosphate bond is shown as $-P$, and a high-energy one as $\sim P$. ADP can be represented as $A-P\sim P$ and ATP as $A-P\sim P\sim P$.

The trapping of energy by ATP can be shown as follows:

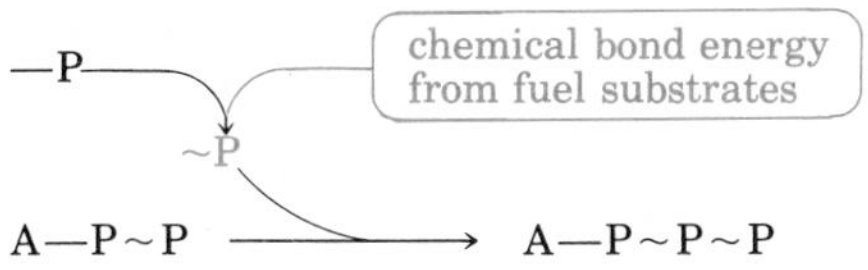

Where does the low-energy phosphate group ($-P$) come from? Ultimately it comes from the diet. But in cellular respiration, ATP itself is the immediate source of the phosphate group necessary to start the oxidation process, donating the $-P$ to the fuel substrate molecule (usually glucose).

The fuel molecule is phosphorylated with a low-energy phosphate group, and the phosphorylated fuel can then be oxidized to yield energy. Quite obviously, the fuel molecule must release more energy — actually it provides much more energy — than is loaned to it by ATP at the start.

The amount of ATP produced in respiration is totally dependent on its rate of utilization. In other words, ATP is produced by one set of reactions and immediately consumed by another. ATP is a great **energy coupling** molecule, used to transfer energy from one reaction to another. For example, ATP formed from the oxidation of glucose is used to synthesize proteins or lipids, or provide power for some other process, such as the contraction of skeletal muscles. The point is, living organisms do not produce and put aside vast amounts of ATP, hoarded against some future energy need. What they do store is the fuel itself, in the form of carbohydrates and lipids especially. ATP is formed as it is needed, primarily by oxidative processes in the mitochondria. Oxygen is not consumed unless ADP and phosphate molecules are available, and these do not become available until ATP is hydrolyzed by some energy-consuming process. **Metabolism is therefore mostly self-regulating.**

There are minor exceptions to the rule that high-energy bonds cannot be stored in cells. Muscle cells contain a type of molecule especially adapted for energy storage. This is phosphocreatine:

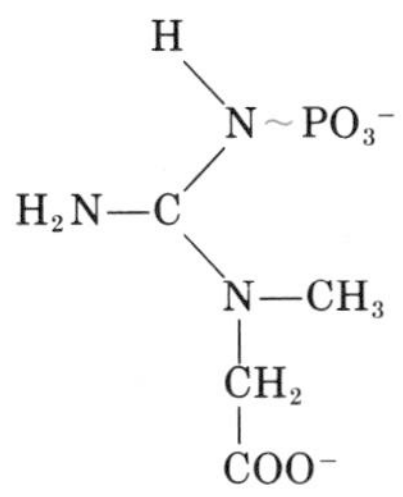

Phosphocreatine contains a high-energy bond that

can provide instant power to the muscle contractile machinery, which often have sudden energy needs. Phosphocreatine is formed from creatine and ATP and is in direct chemical equilibrium with ATP. A short-term burst of activity will rapidly deplete the available ATP, making ADP available. High-energy phosphate is then transferred to ADP from phosphocreatine, providing more ATP for use in muscle contraction.

ATP generated by electron transfer. Having seen that ATP is the one common energy denominator by which all cellular machines are energized, we are in a position to ask how is this energy captured from fuel substrates? The principal means is by the transfer of electrons from fuel substrates to molecular oxygen through a series of enzymes. This electron transfer system is made of a chain of large molecules localized in the inner mitochondrial membranes. The electron carriers are compounds that can be reduced by accepting electrons from the previous carrier, and then be oxidized again by passing electrons to the next carrier. Each successive carrier is a somewhat stronger oxidizing agent than the one before; that is, **successive carriers are increasingly stronger electron acceptors.** Finally, the electrons, as well as the hydrogen protons that accompany them, are transferred to molecular oxygen to form water:

$$\text{Fuel}\,\widehat{(2H)} \quad A \quad B \quad C \quad D \quad \text{Oxygen} \;\to\; H_2O$$

It is conventional to represent electron transfer through the electron carrier system as the transfer of **hydrogen** atoms, although we must emphasize that it is the energized **electrons** and not the **protons** of the hydrogen that are the important energy packets. The proton of each hydrogen atom simply takes a free ride during this electron shuttle until, at the end, it bonds with reduced oxygen and forms water.

The whole function of the electron transport chain is the capture of energy from the original fuel substrate and the transformation of it into a form the cell can use. To do this, the large chemical potential of food molecules is drawn off in small steps (rather than in one explosive burst as in ordinary combustion) by successive electron carriers. At three points along the chain, ATP production takes place by the phosphorylation of ADP. This method of en-

ergy capture is called **oxidative phosphorylation** because the formation of high-energy phosphate is coupled to oxygen consumption, and this depends, as we have seen, on the demand for ATP by other metabolic activities within the cell. The actual **mechanism** of ATP formation by oxidative phosphorylation is not yet known; we can only say that the transfer of electrons does something that is translated into the production of high-energy phosphate bonds.

Nature of electron carriers. Oxidative phosphorylation occurs in the mitochondria; ATP production is the principal metabolic role of these organelles. It will be recalled that mitochondria are composed of two membranes. The outside membrane forms a smooth sac enclosing the inner membrane that is turned into numerous ridges called **cristae** (Fig. 4-3, p. 39). The inner membrane is studded with enormous numbers of minute particles, which some investigators think are actually tiny stalked spheres. These particles, or spheres, bear the electron carriers responsible for oxidative phosphorylation. A section of a mitochondrion, as it might appear under the high-resolution electron microscope, is shown (highly diagrammatically) in Fig. 5-9.

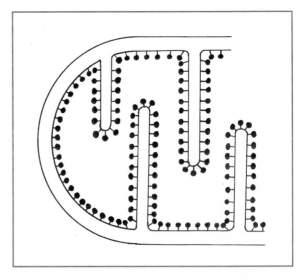

FIG. 5-9. Representation of section of mitochondrion as seen through high-resolution electron microscope.

Pairs of electrons, donated initially from food substrates, flow along the electron carriers in succession (Fig. 5-10). For most food substrates the initial electron acceptor is NAD (nicotinamide-adenine dinucleotide, a derivative of the vitamin niacin). The substrate is oxidized in the process (because it loses electrons) and NAD is reduced (because it gains electrons).

Next FAD (flavine-adenine dinucleotide, a riboflavin derivative) oxidizes the reduced NAD by accepting its electrons. FAD becomes reduced (having gained electrons) and NAD is returned to its original oxidized state. In the same way the pair of electrons is passed sequentially to coenzyme Q, then through the cytochromes, and on to molecular oxygen. Each carrier is successively reduced and then oxidized. The transfer of electrons and the points of high-energy phosphate bond formation are shown in Fig. 5-10.

Thus for every pair of electrons moved along the carriers to oxygen, a total of 3 ATP molecules is formed.

Acetyl–coenzyme A: strategic intermediate in aerobic respiration. At this point the student may be forgiven if he thinks that the metabolic energy fixation process consists entirely of an electron transport chain that is capable of passing bond energy to ATP. Actually, we have only described the last (but especially crucial) step in **aerobic respiration.** The term aerobic respiration describes that kind of respiration requiring atmospheric oxygen, and this, of course, is the familiar sort of respiration practiced by the vast majority of animals. We have seen that ATP production is coupled to electron transfer, which in turn is completely dependent on oxygen, the final hydrogen and electron acceptor. Without oxygen the process stops because the electrons have nowhere to go. The components of the electron transfer chain would quickly become fully reduced and remain so, in the absence of the electron sink, molecular oxygen.

However, organisms do have a backup system enabling them to respire without oxygen. This is called **anaerobic respiration,** or **fermentation.** Although anaerobic respiration does support the lives of bacteria, yeasts, and a few other organisms, it is not nearly as efficient as aerobic respiration and consequently is not capable alone of maintaining life of higher animals. But it is useful, indeed essential, as a supplementary source of energy during rapid and intense muscular contraction. We will be satisfied now with just noting that animal cells

Fig. 5-10. Electron transport system. Electrons are transferred from one carrier to the next, terminating with molecular oxygen to form water. A carrier is reduced by accepting electrons, then is reoxidized by donating its electrons to the next carrier. ATP is generated at three points in the chain. These electron carriers are located on inner membranes of mitochondria.

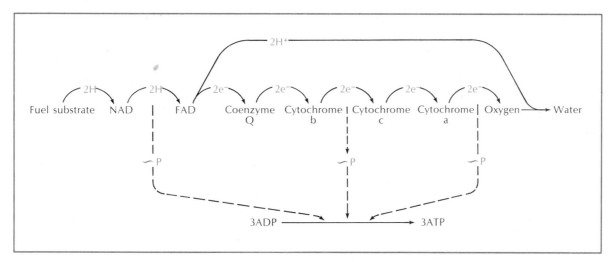

possess the machinery for anaerobic respiration and will treat this subject in more detail later.

In aerobic respiration most fuel molecules are progressively stripped of their carbon atoms until those carbons are finally converted into molecules of CO_2. During this degradation the hydrogens and their electrons are removed and passed into the important energy-yielding electron transport chain we have already described. But to reach this final sequence, most carbon atoms appear in a 2-carbon group, called **acetyl–coenzyme A.** This is a critically important compound. Some two thirds of all the carbon atoms in foods eaten by animals appear as acetyl–coenzyme A at some stage. The strategic metabolic position of acetyl–coenzyme A is illustrated in Fig. 5-11. It is the final oxidation of acetyl–coenzyme A that provides the energized electrons used to generate ATP. Acetyl–coenzyme A is also the source of nearly all the carbon atoms found in the body's fats, as the reverse arrow in Fig. 5-11 indicates. The structure of acetyl–coenzyme A can be shown in abbreviated form as:

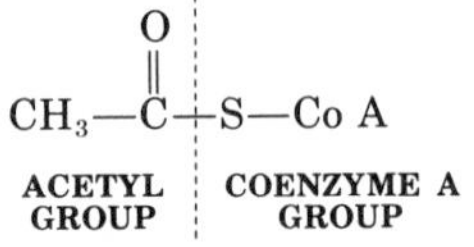

$$\overset{\displaystyle O}{\underset{}{CH_3-\overset{\parallel}{C}}}-S-Co\ A$$

ACETYL   COENZYME A
GROUP   GROUP

The right hand side of the molecule is a coenzyme containing the vitamin **pantothenic acid,** another example of how vitamins play important structural roles in critical cellular functions.

Oxidation of acetyl–coenzyme A. The breakdown (oxidation) of the 2-carbon acetyl group of acetyl–coenzyme A occurs in a sequence called the **citric acid cycle** (or **Krebs cycle,** after its British discoverer Sir Hans A. Krebs). The cycle is composed of a sequence of nine transformations and oxidations. To simplify an otherwise complex story, we have summarized the cycle in Fig. 5-12. The citric acid cycle begins with the condensation of acetyl–coenzyme A with oxaloacetate to form a 6-carbon compound citrate (citric acid). Citrate enters a series of reactions in which 2 molecules of CO_2

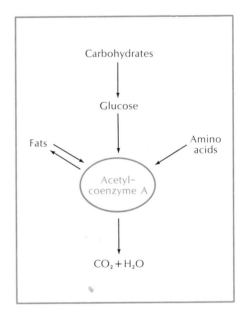

FIG. 5-11. Acetyl–coenzyme A is an important intermediate in oxidation of carbohydrates, proteins (amino acids), and fats.

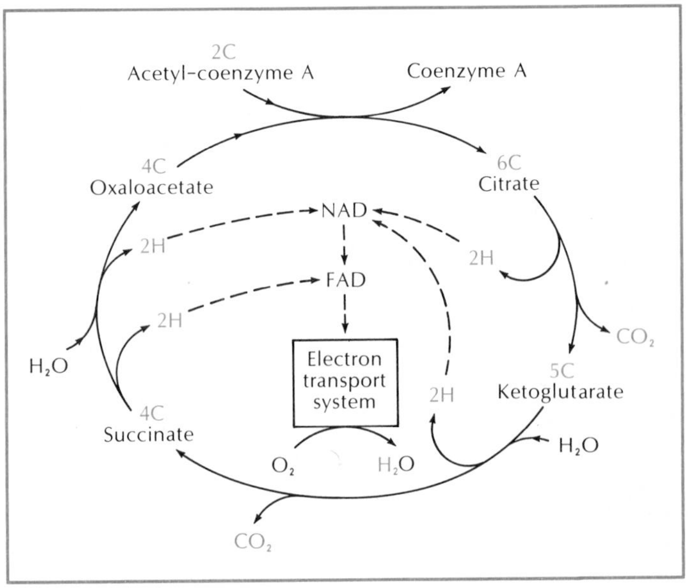

FIG. 5-12. Citric acid cycle. See text for explanation.

are produced from the original acetyl group. When the cycle is complete, the 4-carbon oxaloacetate is returned to its original form, ready to condense with another molecule of acetyl–coenzyme A. Oxaloacetate therefore acts as a carrier for the 2 carbons of the acetyl group; it is not itself used up in the cyclical process. As the acetyl group is oxidized carbon atom by carbon atom, four pairs of electrons and 4 protons (shown as four pairs of hydrogen atoms in Fig. 5-12) are transferred to the electron transfer chain (shown in the center of the cycle in Fig. 5-12). Three pairs of electrons are passed to NAD; the remaining pair is passed directly to FAD. Each pair of electrons then shuttles down the electron transport chain to an atom of oxygen, as already described. Three molecules of ATP are generated

for **each** molecule of NAD receiving electrons; this yields a total of nine ATP per acetyl group. Two more molecules of ATP are generated from the electrons passed directly to FAD. One more high-energy bond is generated at another point in the cycle; it forms a compound called GTP (guanosine-5′-triphosphate), which has the same energy yield as ATP, and for simplicity's sake, we will call it ATP. Thus the net yield is 12 molecules of ATP for the single acetyl group fed into the cycle. We must keep firmly in mind that 11 of these 12 high-energy phosphate bonds are generated by oxidative phosphorylation in the electron transport chain (Fig. 5-10) and not in the citric acid cycle itself. The citric acid cycle simply provides a means for the release of energized electrons during the oxidation

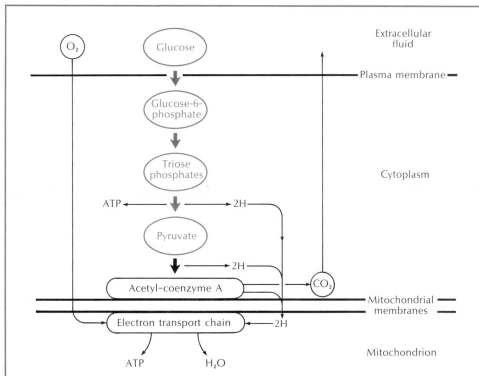

FIG. 5-13. Pathway for oxidation of glucose. Glucose is oxidized to acetyl–coenzyme A via Embden-Meyerhof pathway (in red). Acetyl–coenzyme A then enters citric acid cycle (not shown) on outer mitochondrial membrane. Hydrogens removed in cycle are transferred to electron transport chain on inner mitochondrial membrane. See text for details.

GLUCOSE   **GLUCOSE-6-PHOSPHATE**

of the acetyl group. All of these reactions occur in mitochondria. But electron release through the citric acid cycle is thought to occur in the **outer** membrane, whereas the electron carriers and the coupling to oxidative phosphorylation occurs in the **inner** mitochondrial membrane. Thus there is a spatial as well as functional separation of these processes.

Glucose: major source of acetyl–coenzyme A. All the major fuels (glucose, fats, amino acids) serve as sources of acetyl–coenzyme A. Glucose, however, is a particularly important fuel for most tissues, especially the brain. Glucose is first converted to a 3-carbon compound called **pyruvate** (pyruvic acid) through a series of reactions that are called the Embden-Meyerhof pathway. Pyruvic acid is then enzymatically stripped of a carbon atom to form acetyl–coenzyme A. The general outline for this sequence is shown in Fig. 5-13. Again, we shall simplify a rather complex biochemical story by condensing this glucose metabolism pathway, which actually consists of ten consecutive enzymic reactions, into four major steps.

The metabolism of glucose begins with its phosphorylation by ATP to form **glucose-6-phosphate** (see above). Glucose-6-phosphate is an important intermediate because it is a "stem" compound that can lead into any of several metabolic pathways. However, the predominant metabolic fate for glucose-6-phosphate is entry into the Embden-Meyerhof sequence. Following still another phosphorylation, the 6-carbon glucose molecule is split into two 3-carbon sugars, called **triose phosphates.** Each triose phosphate is oxidized and rearranged to form **pyruvate** resulting in a yield of high-energy phosphate as ATP. A pair of hydrogen atoms and a molecule of CO_2 are then removed from pyruvate, forming acetyl–coenzyme A.

Let us now summarize the entire oxidation of glucose. Glucose first enters the cells of tissues, passing through the plasma membrane by a transport process that requires the presence of the pancreatic hormone **insulin.** Glucose is then phosphorylated and enters the Embden-Meyerhof pathway in the cytoplasm. This is shown in red in Fig. 5-13. Through this sequence it is split in the middle to form two 3-carbon sugars (triose phosphates) that are converted to pyruvate. Pyruvate is decarboxylated to form acetyl–coenzyme A. This sets the stage for entry into the citric acid cycle located on the outer mitochondrial membrane. After condensing with oxaloacetic acid to form citric acid, the 2-carbon acetyl fragment is oxidized—yielding 4 pairs of electrons and 4 protons that are passed along the electron transport chain located on the inner mitochondrial membrane. The electrons finally arrive at oxygen, the ultimate electron acceptor.

What has been accomplished? A molecule of glucose has been completely oxidized to CO_2 and H_2O. ATP has been generated at several points along the way. Let us now balance up the yield. First of all, 2 molecules of ATP were consumed in the initial phosphorylation of glucose. Then 14 molecules of ATP are generated by the transformations leading to the formation of acetyl–coenzyme A. (Remember that each glucose molecule is split into **two** 3-carbon sugars, each of which produces ATP when oxidized to acetyl–coenzyme A. Some ATP is produced directly, the rest results from oxidation followed by oxidative phosphorylation.) Our balance is now 12 ATP. Then we add the 12 ATP generated in the complete oxidation of **each** of the acetyl–coenzyme A molecules. This gives a total yield of 36 ATP. The whole sequence can be summarized as follows:

$$\text{Glucose } (C_6H_{12}O_6) + 6O_2 + 36ADP + 36 \text{ —P} \longrightarrow 6CO_2 + 6H_2O + 36ATP$$

Efficiency of oxidative phosphorylation. It is probably obvious that no energy transforming process can be 100% efficient, not even the remarkable cellular oxidative machinery produced by organic evolution. If we burn a mole of glucose (180 grams) in a bomb calorimeter, it releases about 686,000 calories. This is the **potential** energy for forming ATP. It has been determined that 8,000 to 12,000 calories are required to synthesize 1 mole of ATP. Consequently glucose theoretically **could** provide enough energy to generate 50 to 85 moles of ATP from ADP. It actually turns out 38 moles

(36 + 2) of ATP. If we assume that each ATP mole represents an average energy equivalent of 10,000 calories, then 38 moles of ATP represents 380,000 calories. Thus the efficiency of glucose oxidation is 380:686, or about 55%. Engineers would be delighted if they could build machines that could do as well.

Metabolism of lipids. Animal fats are triglycerides, molecules composed of glycerol and 3 molecules of fatty acids. These fuels are important sources of energy for many metabolic processes in all animals, not just obese victims of misplaced

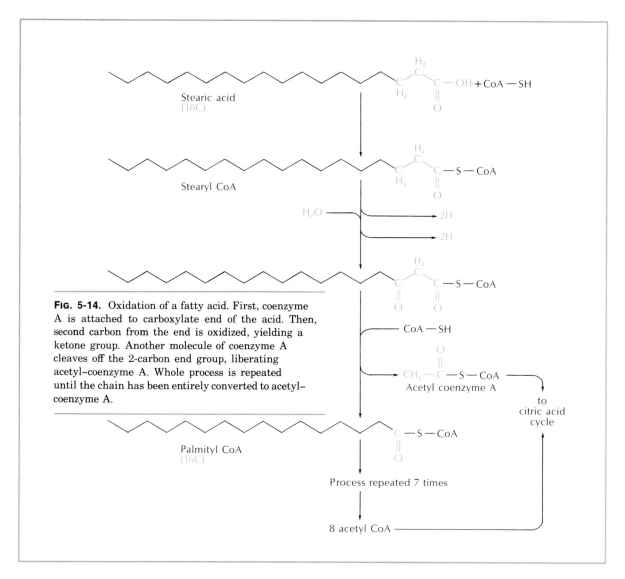

Fig. 5-14. Oxidation of a fatty acid. First, coenzyme A is attached to carboxylate end of the acid. Then, second carbon from the end is oxidized, yielding a ketone group. Another molecule of coenzyme A cleaves off the 2-carbon end group, liberating acetyl–coenzyme A. Whole process is repeated until the chain has been entirely converted to acetyl–coenzyme A.

appetites. Most of fat is fatty acids, carboxylic acids with long hydrocarbon chains (Fig. 2-9, p. 23). We know that fats enter the mitochondrial metabolic processes through acetyl–coenzyme A (Fig. 5-11). What happens in brief is that the long hydrocarbon chain of a fatty acid is sliced up by oxidation, two carbons at a time; these are released from the end of the molecule as acetyl–coenzyme A. The process is repeated until the entire chain has been reduced to several 2-carbon acetyl units. The oxidation of a fatty acid is diagrammed in Fig. 5-14, using a shorthand representation in which each jog in the chain symbolizes a saturated carbon $(—CH_2—)$ of the fatty acid **stearic acid,** one of the abundant naturally occurring fatty acids. First, the fatty acid is combined with coenzyme A. Then in a three-step process the third carbon from the end is oxidized (stripped of its hydrogens). Next, a molecule of acetyl–coenzyme A is sliced off the end, by **another** molecule of coenzyme A that then adds itself to the chain. Thus the hydrocarbon chain, now 2 carbons shorter, is left with coenzyme A on its end. The process is repeated until the whole chain has been chopped up into acetyl–coenzyme A. This material then enters the citric acid cycle to yield ATP in the manner described earlier.

How much ATP is gained from fatty acid oxidation? Five high-energy phosphate bonds are gen-

erated for each acetyl–coenzyme A unit split off. Then oxidation of each 2-carbon fragment produces $12 \sim P$. Allowing for the ATP expended to attach the first coenzyme A, it has been calculated that the complete oxidation of 18-carbon stearic acid will yield 147 ATP molecules. By comparison, 3 molecules of glucose (also totalling 18 carbons) yields 108 ATPs. Little wonder that fat is considered the king of animal fuels! Fats are more concentrated fuels than carbohydrates because fats are almost pure hydrocarbons; they contain more hydrogen per carbon atom than sugars do, and it is the energized electrons of hydrogen that generate high-energy bonds, when they are carried through the mitochondrial electron transport system.

Glycolysis: generating ATP without oxygen. Up to this point, we have been describing **oxidative,** or **aerobic,** metabolism, the kind of respiration that predominates in the vast majority of animals. It hardly needs emphasizing that the availability of oxygen is an obvious basic necessity of animal life. Nevertheless there are microorganisms, notably the yeasts and certain bacteria, that multiply happily with no oxygen at all. These organisms are called **anaerobes.** They occupy important ecologic niches, some of the niches created by man. For example, oxygen depleted streams are becoming regretably common appendages to our industrial-

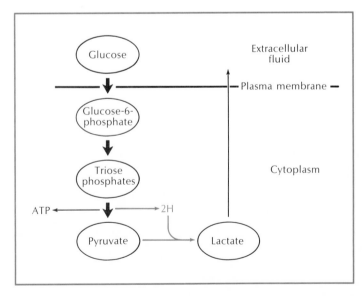

FIG. 5-15. Glycolysis and formation of lactate. See text for details.

ized society. Anaerobic organisms use carbon compounds as fuel, breaking them down by a process commonly called **fermentation.** This term, meaning "cause to rise," was originally used to describe the action of yeasts that break down glucose into alcohol (ethanol) and CO_2. It is now applied to any microorganism that metabolizes foodstuffs without oxygen. The end products, which vary with the nature of the fermentive process, include butanol, acetone, lactic acid, and hydrogen gas.

Most higher organisms also have the capacity to ferment glucose, that is, break it down to produce high-energy phosphate in the absence of oxygen. The process is called **glycolysis.** It is used as a backup system for aerobic metabolism, providing a means for short-term generation of ATP during brief periods of heavy energy expenditure, when the slow rate of O_2 diffusion would be a limiting factor.

In glycolysis, glucose is split eventually into two molecules of **lactate** (lactic acid), yielding 2 molecules of ATP in the process. The glycolytic pathway is shown in Fig. 5-15. It will be seen that glycolysis utilizes the same Embden-Meyerhof pathway that, in oxidative metabolism, directs glucose into the citric acid cycle via acetyl–coenzyme A (compare Figs. 5-13 and 5-15). But in the absence of oxygen, both pyruvate and hydrogen accumulate in the cytoplasm because neither can proceed into their oxida-

tive channels without oxygen. The problem is neatly solved by forming lactate. Pyruvate is converted into lactate that accepts the hydrogen, as shown below:

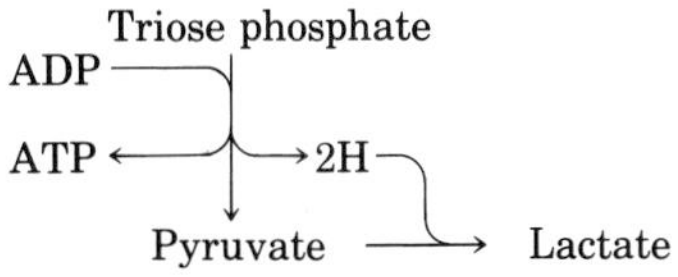

Lactate then diffuses out into the blood, where it is later disposed of in the liver. Thus lactate formation prevents the cytoplasm from being swamped with pyruvate and allows **some** ATP formation. Of course, glycolysis is not an efficient producer of ATP; only 2 moles of ATP per mole glucose are generated by glycolysis as against 36 moles by oxidative phosphorylation. Nevertheless, the capacity to produce a little extra ATP during an emergency may mean the difference between life and death for an animal. Some animals rely heavily on glycolysis during normal activities. For example, diving birds and mammals fall back on glycolysis almost entirely to give them the needed energy to sustain a long dive. And salmon would never reach their spawning grounds were it not for glycolysis that provides almost all the ATP used in the powerful muscular bursts needed to carry them over falls and up rapids.

6

Architectural patterns of animals

The architecture of an animal refers to the pattern in which the body components are arranged. Most animals fall into a few well-defined plans of structure, for there is a uniformity of biologic organization. This uniformity can be explained on the basis of evolution and the genetic kinship of animals. Common ancestry has kept animals in a more or less common plan, with such deviations as adaptive radiation has made. Of course, the level of organization has a lot to do with the architectural pattern of an animal. For instance, certain architectural variations are not available at the unicellular state. The form of an animal determines how it meets its environment. Basic body plans are modified in countless ways to suit the varied environments into which life has moved.

GRADES OF STRUCTURE AMONG ANIMALS

The first grade of structure is that between the acellular (protozoan) and the cellular (metazoan) animals. The **acellular,** or single-celled, forms are complete organisms and carry on all the functions of higher forms. Within the confines of their cell, they often show complicated organization and division of labor, such as skeletal elements, locomotor devices, fibrils, beginnings of sense organs, and many others.

On the other hand, the **metazoan,** or multicellular animal, has cells differentiated into tissues and organs that are specialized for different functions. The metazoan cell is not the equivalent of a protozoan cell; it is only a specialized part of the whole organism and usually cannot exist by itself.

How has complexity arisen in the animal kingdom? In general it is a matter of difference in organization, but certain principles are involved. One of these is size, which will be discussed later in this chapter. Another is specialization and division of labor. An ameba can move without muscles, digest food without an alimentary canal, and breathe without gills or lungs. But higher forms have specialized organs for these functions. The more complicated a device becomes, the more necessary it is to have accessory organs to help out. An alimentary canal is not a mere epithelial tube for secretion and absorption but has muscles to manipulate and nerves to control it. Specialization and division of labor have many advantages for adjustments to specific functions, but they require complicated machinery and more energy.

Does this mean that life is progressing toward higher and higher types, such as man? In the evolutionary picture the first animals were small and relatively simple, but there is no reason to believe that more recent animals are better adjusted to their environments than were their ancient ancestors. Nor is there any evidence that evolution has led in man's direction, for many lines definitely have not.

ORGANIZATION OF THE BODY

The body consists of three different elements—body cells, extracellular structural elements, and body fluids. The body fluids include the blood, tissue or interstitial fluid between the cells, and lymph within the lymphatics. An endothelial barrier separates the tissue fluid outside the lymphatics and the lymph within them. There is an exchange of materials between the blood vascular system and the interstitial (tissue) fluid between the cells. Body fluids fill continuous spaces and are responsible for diffusion and convection. The organization and composition of the body fluids are described in more detail in Chapter 11.

Interstitial or extracellular substance is the material that lies between the cells. It affords mechanical stability, protection, storage, and exchange agents. It is mainly responsible for the firmness of tissues and gives support to the cells. Two types of intercellular tissue are recognized—formed and amorphous. The formed type includes collagen (white fibrous tissue). This is the most abundant protein in the animal kingdom and makes up the major part of the fibrous constituent of the skin, tendon, ligament, cartilage, and bone. Elastin, which gives elasticity to the tissues, also belongs to the formed type. Amorphous interstitial substance (ground substance) is composed of mucopolysaccharides arranged in long chain polymers.

The above description applies especially to vertebrates and higher invertebrates. Modifications of this basic plan may be found in the animal kingdom. Unity of pattern is characteristic of the structure of animals, but structural variations do occur in comparative morphology throughout the gamut of animal life.

GRADES OF ORGANIZATION

An animal is an organization of units differentiated and integrated for carrying on the life processes, but this organization goes from one level to another as we ascend the evolutionary path.

Protoplasmic grade of organization. This type is found in protozoans and other acellular forms. All activities of this level are confined to the one mass called the cell. Here the protoplasm is differentiated into specialized organelles that are capable of carrying on definite functions.

Cellular grade of organization. Here aggregations of cells are differentiated, involving division of labor. Some cells are concerned with reproduction and others with nutrition. The cells have little tendency to become organized into tissues. Some protozoan colonial forms having somatic and reproductive cells might be placed in this category. Many authorities also place the sponges at this level.

Cell-tissue grade of organization. A step beyond the preceding is the aggregation of similar cells into definite patterns of layers, thus becoming a tissue. Sponges are considered by some authorities to belong to this grade, although the jellyfish are usually referred to as the beginning of the tissue plan. Both groups are still largely of the cellular grade of organization because most of the cells are scattered and not organized into tissues. An excellent example of a tissue in coelenterates is the **nerve net,** in which the nerve cells and their processes form a definite tissue structure, with the function of coordination.

Tissue-organ grade of organization. The aggregation of tissues into organs is a further step in advancement. Organs are usually made up of more than one kind of tissue and have a more specialized function than tissues. The first appearance of this level is in the flatworms (Platyhelminthes), in which there are a number of well-defined organs such as eyespots, proboscis, and reproductive organs. In fact, the reproductive organs are well organized into a reproductive system.

Organ-system grade of organization. When organs work together to perform some function, we have the highest level of organization—the organ system. The systems are associated with the basic bodily functions—circulation, respiration, digestion, etc. Typical of all the higher forms, this type of organization is first seen in the nemertean worms in which a complete digestive system, separate and distinct from the circulatory system, is present.

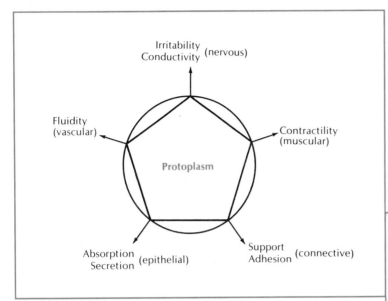

Irritability
Conductivity (nervous)

Fluidity
(vascular)

Contractility
(muscular)

Protoplasm

Absorption
Secretion (epithelial)

Support
Adhesion (connective)

FIG. 6-1. Origin of tissues. Five basic tissues of body are manifestations of properties found in all protoplasm. Tissues are variously specialized to perform functions, as listed in diagram.

MAJOR ANIMAL TISSUES

The different types of tissues originate from the basic properties of protoplasm (Fig. 6-1). These are irritability and conductivity (nervous), contractility (muscle), supportive and adhesive (connective tissue), absorptive and secretion (epithelial), and fluidity and transportation (vascular). These five tissues are specializations of the various protoplasmic properties, and with their varied structures and functions they are able to meet the basic requirements of morphologic patterns, as in organogenesis.

A **tissue** is a group of similar cells (together with associated cell products) specialized for the performance of a common function. The study of tissues is called **histology.** All cells in metazoan animals take part in the formation of tissues.

Epithelial tissue. An **epithelium** is a tissue that covers an external or internal surface. It is made up of closely associated cells, with some intercellular material between them. Most epithelial cells have one surface free and the other surface lying on vascular connective tissue. A noncellular **basement membrane** is often attached to the basal cells. Epithelial cells are often modified to produce secretory glands that may be unicellular or multicellular. Some free surfaces (joint cavities, bursae, brain cavity) are not lined with typical epithelium.

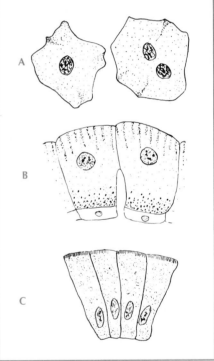

FIG. 6-2. A, Simple squamous epithelial cells from lining of mouth. **B,** Simple cuboidal epithelium. **C,** Simple columnar epithelium.

Epithelia are classified on the basis of cell form and number of cell layers. **Simple epithelium** is one layer thick (Fig. 6-2), and its cells may be flat or **squamous** (endothelium of blood vessels), short prisms or **cuboidal** (glands and ducts), and tall or **columnar** (stomach and intestine). Any of these three forms of cells may occur in several layers as a

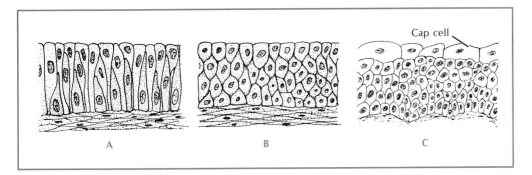

Fig. 6-3. Types of epithelial tissue. **A,** Pseudostratified epithelium. **B,** Stratified columnar epithelium. **C,** Transitional epithelium.

stratified epithelium (skin, sweat glands, urethra) (Fig. 6-3). In some stratified epithelia the number of cell layers can be changed by movement (**transitional**—bladder). Some epithelia have cells of different heights that give the appearance of stratified epithelia (**pseudostratified**—trachea). Many epithelia may be **ciliated** at their free surfaces (oviduct). Epithelia serve to protect, secrete, excrete, and lubricate.

Connective tissue (supporting). Connective tissues bind together and support other structures. They are so common that the removal of all other bodily components would still leave the gross outlines of the body distinguishable. Connective tissue is made up of scattered cells and a great deal of formed materials, such as **fibers** and ground substance (**matrix**). Fibers probably originate from ground substances secreted by the cells. Three types of fibers are as follows: white or collagenous, yellow or elastic, and branching or reticular. Connective tissue may be classified in various ways, but all the types fall under either **loose connective tissue** (reticular, areolar, adipose) or **dense connective tissue** (sheaths, ligaments, tendons, cartilage, bone). The distinction between these two groups is mainly one of emphasis upon fibers, ground substance, or cells (Fig. 6-4). For example, adipose tissue stresses cells, ligaments stress fibers, and cartilage stresses ground substance (matrix).

Muscular tissue. Muscle is the most common tissue in the body of most animals. It is made up of elongated cells or fibers specialized for contraction. It makes possible the movements of the body and its

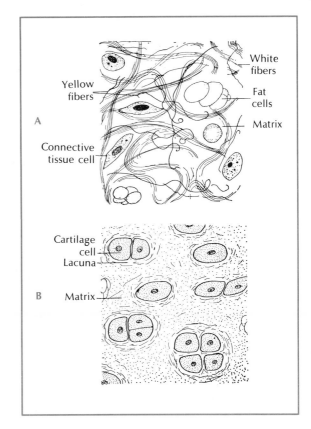

Fig. 6-4. A, Areolar connective tissue. **B,** Hyaline cartilage, most common form of cartilage in body.

parts. It originates (with few exceptions) from the mesoderm, and its unit is the cell or **muscle fiber.** The unspecialized cytoplasm of muscles is called **sarcoplasm,** and the contractile elements within the fiber (cell) are the **myofibrils.** Functionally, muscles are either **voluntary** (under control of will) or **involuntary.** Structurally, they are either **smooth**

71

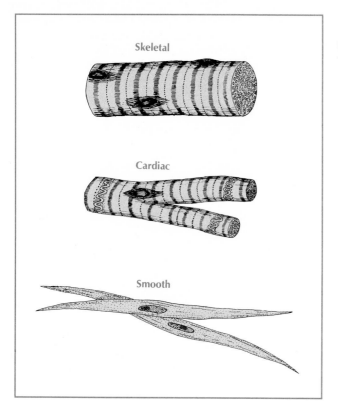

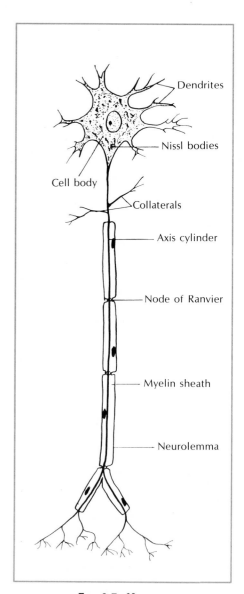

FIG. 6-5. Three kinds of vertebrate muscle fibers, as they appear when viewed with the light microscope.

FIG. 6-7. Neuron.

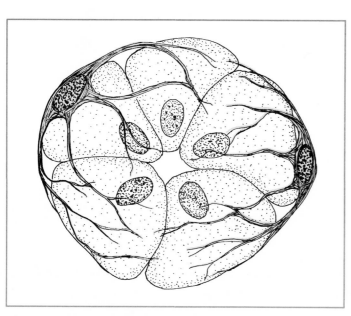

FIG. 6-6. Two myoepithelial basket cells surrounding salivary secretory cells. Each myoepithelial cell is made up of a central body with long cytoplasmic processes and may be considered a fourth type of muscular tissue. Myoepithelial cells resemble smooth muscle cells.

(fibers unstriped) or **striated** (fibers cross-striped). The three kinds of muscular tissue are **smooth involuntary** (walls of viscera and of blood vessels), **striated involuntary** or cardiac (heart), and **striated voluntary** or skeletal (limb and trunk) (Fig. 6-5). Another type of muscular tissue is made up of the **myoepithelial cell** (Fig. 6-6), which is found in sweat, salivary, and mammary glands between the epithelium and connective tissues. The branchlike processes of myoepithelial cells function by squeezing secretions through the ducts of these glands.

Nervous tissue. The structural and functional unit of the nervous system is the **neuron** (Fig. 6-7). This is a nerve cell made up of a body containing the nucleus and its processes or fibers. It originates from an embryonic ectodermal cell called a **neuroblast.** In most animals the bodies of nerve cells are restricted to the central nervous system and ganglia, but the fibers may be very long and ramify through the body. Neurons are arranged in chains, and the point of contact between neurons is a **synapse.** Some of the fibers bear a sheath (medullated or myelin); in others the sheath is absent (nonmedullated).

Sensory neurons are concerned with picking up impulses from sensory **receptors** in the skin or from sense organs and transmitting them to nerve centers (brain or spinal cord). **Motor neurons** carry impulses from the nerve centers to muscles or glands **(effectors),** which are thus stimulated to act. **Association neurons** may form various connections between other neurons.

Vascular tissue. Vascular tissue is a fluid tissue comprised of **white blood cells, red blood cells, platelets,** and a liquid, **plasma.** Traveling through blood vessels, the blood carries to the tissue cells the materials necessary for their life processes. **Lymph** and tissue fluids, which arise from blood by filtration and serve in the exchange between cells and blood, are also vascular tissues.

BODY CAVITY, OR COELOM

The coelom is the true body cavity. It is the space between the digestive tube and the outer body wall; it contains the visceral organs. Although not all animals have a coelom, for example, the jellyfish and flatworms, those that do have a "tube-within-a-tube" arrangement. The outer tube is the body wall; the inner tube is the digestive tract, and the space between is the coelom. A true coelom develops between two layers of mesoderm—an outer somatic layer and an inner visceral layer—and is lined with mesodermal epithelium called the **peritoneum.**

The coelom is of great significance in animal evolution because it provides space for visceral organs, permits greater size and complexity by exposing more cells to surface exchange, and contributes directly to the development of certain systems such as excretory, reproductive, and muscular. This fluid-filled cavity also serves as a hydrostatic fluid skeleton in some primitive forms, aiding them in movement, rapid change of shape, or burrowing. One has only to compare the locomotion of a planarian worm (without a coelom) with that of an annelid worm (with a coelom) to see this advantage. In roundworms and some others the body cavity is not lined with mesoderm and so is given the name of **pseudocoel** (Fig. 6-8).

BODY PLAN AND SYMMETRY

Convenient terms for locating regions of an animal body are **anterior** for the head end, **posterior** for the opposite or tail end, **dorsal** for the back side, and **ventral** for the front or belly side. **Medial** refers to the midline of the body, **lateral** to the sides. **Distal** parts are farther from a point of reference; **proximal** parts are nearer. **Pectoral** refers to the chest region or the area supporting the forelegs, and **pelvic** refers to the hip region or the area supporting the hind legs.

Symmetry refers to balanced proportions, or the correspondence in size and shape of parts on opposite sides of a median plane (Fig. 6-9). A symmetrical body can be cut into two equivalent or mirrored halves. **Spherical symmetry** is found in a few organisms, chiefly protozoans. Such a form, like a ball, could be divided equally by any plane that passed through the center. These are usually floating or rolling forms. **Radial symmetry** applies to a few sponges, most coelenterates, and adult echinoderms that, like bottles or wheels, can be divided into similar halves by any plane passing through the longitudinal axis. In such forms one end of the longitudinal axis is usually the mouth or oral end and the other is the aboral end. A variant form is

73

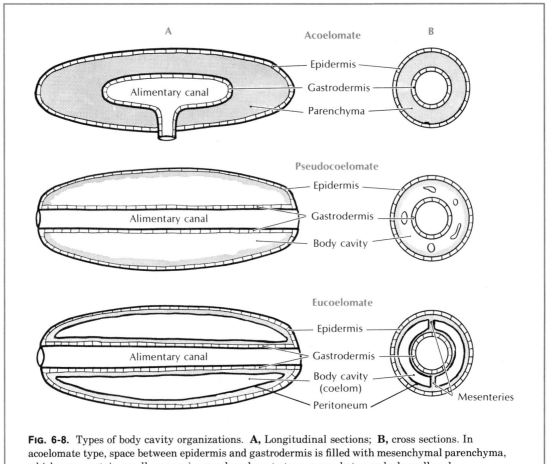

FIG. 6-8. Types of body cavity organizations. **A,** Longitudinal sections; **B,** cross sections. In acoelomate type, space between epidermis and gastrodermis is filled with mesenchymal parenchyma, which may contain small spaces; in pseudocoelomate type, space between body wall and digestive tract is remnant of blastocoel and is not lined with mesodermal peritoneum; and in eucoelomate, or true coelomate, type, body cavity is lined with mesodermal peritoneum, which also covers digestive tract. Mesenteries are made up of two peritoneal layers.

biradial symmetry in which, because of the presence of some part that is single or paired rather than radial, only one or two planes through the longitudinal axis divides the form into mirrored halves. Examples are the comb jellies, which are basically radial forms but have a pair of arms or tentacles, and sea stars, which have a sieve plate located between two of their radially arranged arms. Radial animals are usually well suited to a sessile existence, for they react equally well on all sides.

Most other animals have **bilaterally symmetric** bodies (Fig. 6-10). In these types only a **sagittal**

plane divides the animal into equivalent right and left halves. A sagittal plane passes through the anteroposterior axis and through the dorsoventral axis. A **frontal plane** divides a bilateral body into dorsal and ventral halves by running through the anterior-posterior axis and the right-left axis at right angles to the sagittal plane. A **transverse plane** would cut through a dorsoventral and a right-left axis at right angles to both the sagittal and frontal planes and would result in anterior and posterior portions. Along with bilateral symmetry, we find differentiation of a head end, which has

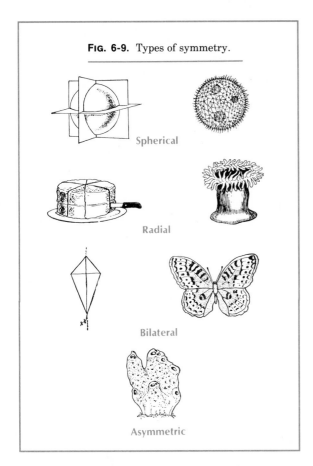

FIG. 6-9. Types of symmetry.

Spherical

Radial

Bilateral

Asymmetric

greater perception than the tail end. These forms are suited for forward movement, which is an asset in the search for food and protection.

Most of the sponges lack any symmetry at all and are called **asymmetric**.

CEPHALIZATION AND POLARITY

The differentiation of a definite head end is called **cephalization** and is found chiefly in bilaterally symmetric animals. Cephalization involves the concentration of nervous tissue (brain and sense organs) in the head. This arrangement in an actively moving animal makes possible the most efficient reaction with the environment. Cephalization is always accompanied by a differentiation along an anteroposterior axis (**polarity**). Polarity usually involves gradients of activities between limits, such as between anterior and posterior ends.

FIG. 6-10. Bilaterally symmetric animal showing three planes of symmetry — frontal, transverse, and sagittal.

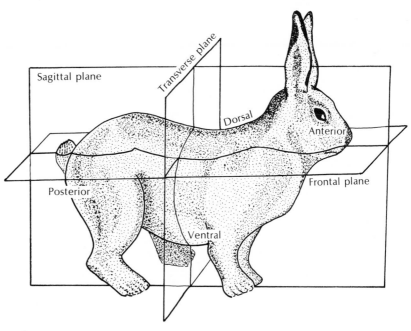

METAMERISM

Metamerism is the linear repetition of similar body segments. Each segment is called a **metamere,** or **somite.** In forms such as the metameric earthworm and other annelids, the segmental arrangement includes both external and internal structures of several systems. There is repetition of muscles, blood vessels, nerves, and the setae of locomotion. Some other organs, such as those of sex, are repeated in only a few somites.

When the somites are similar, as in the earthworm, the condition is called **homonomous metamerism;** if they are dissimilar, as in the lobster and insect, it is called **heteronomous metamerism.** True metamerism is found in only three phyla: Annelida, Arthropoda, and Chordata (Fig. 6-11), although superficial segmentation of the ectoderm and the body wall may be found among many diverse groups of animals.

In higher animals much of the segmental arrangement has become obscure. Segmentation often shows up in embryonic stages in forms in which metamerism is not so evident in the adult. Muscles of vertebrate animals show a marked metamerism in the embryo but little in the adult. In adult vertebrates the arrangement of the vertebrae is metameric.

Tagmatization (tagmosis). In connection with

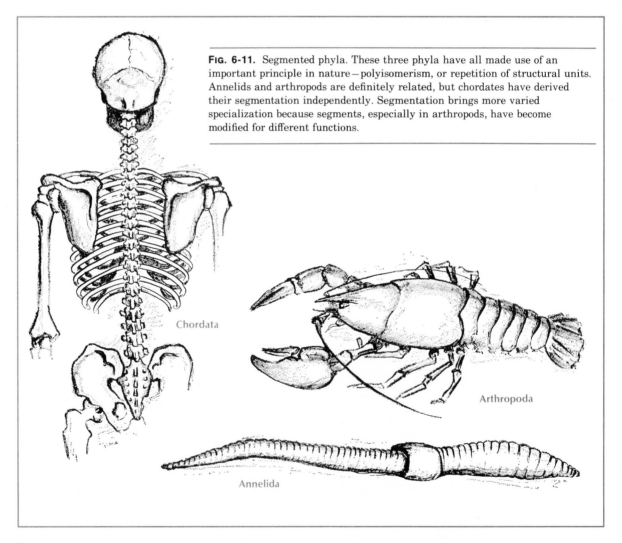

FIG. 6-11. Segmented phyla. These three phyla have all made use of an important principle in nature—polyisomerism, or repetition of structural units. Annelids and arthropods are definitely related, but chordates have derived their segmentation independently. Segmentation brings more varied specialization because segments, especially in arthropods, have become modified for different functions.

Chordata

Arthropoda

Annelida

segmentation as found in the annelids and arthropods, the segments may be united into functional groups of two or more somites, each group being structurally separated from other groups and specialized to perform a certain function for the whole animal. In the primitive forms of both annelids and arthropods the general pattern for the entire organism is a serial succession of identical somites. Tagmosis is more common among the arthropods than in the annelids. A good example of tagmosis is the familiar insect with three tagmata: head, thorax, and abdomen.

HOMOLOGY AND ANALOGY

In comparative studies of animals the concepts of **homology** (similarity in origin) and **analogy** (similarity in function) are frequently used to express relationships between animals, the basic patterns of morphology, and the way these patterns have varied. Homology refers to correspondence between parts of different animals such as the arm of a man, the wing of a bird, the foreleg of a dog, and the pectoral fin of a fish.

However, homology is a relative term and an absolute meaning has to be guarded against. If one always insists on criteria of similarity of structure and development for strict homology, then many structures that are called homologous do not meet these requirements. For instance, the wing of a bird and the wing of a bat are homologous only in so far as both are derived from pentadactyl forelimbs. They are analogous, however, for they have the same function. In examining homologous structures it is best to keep in mind the question: to what extent or degree are they homologous? Convergent evolution, or the independent origin of two similar structures, is caused by independent mutations that are favored under similar environments. Horns have appeared independently many times in mammals, but they should they be considered homologous? Perhaps the best criterion for homology would be the presence of homologous genes, but this is not practical at our present state of knowledge. Embryologists deal with phenotypes, or the visible expressions, which is not the same thing as the genotypes, or hereditary constitution.

Analogy denotes similarity of function. The wing of a bird is analogous to the wing of a butterfly because they have the same function, but they are not homologous because they are formed from different sources. The term analogy in its usage may or may not denote similarity of origin.

The principle of homology, despite the shortcomings mentioned above, has a wide application in zoology. It is used as an argument for evolution because it is based on the idea of inheritance from common ancestors. Homology is also important in classifying animals.

SIZE OF ORGANISM

Different species of animals show an enormous range in size, from the tiny protozoan weighing a fraction of a milligram to the whale weighing more than 100 tons (Table 6-1). Many animals such as birds and mammals have definite age limits to growth, but in reptiles and fishes growth may continue throughout life, although at a reduced pace. Mammals vary from forms as small as shrews to those as large as whales. One may get the idea that the evolutionary trend has been from the small to the large, but this has not always been the case.

It is a general principle that every organism is distinguished by a characteristic size. Of course, many factors may modify the dimensions of an individual animal, such as nutrition and hormone imbalance.

TABLE 6-1. Average length of certain organisms

Common name, genus, and species	Average length
Colon bacterium (*Escherichia coli*)	3.5 μ
Euglena (*Euglena gracilis*)	40 μ
Slipper animalcule (*Paramecium caudatum*)	200 μ
Fruit fly (*Drosophila melanogaster*)	3 mm.
Housefly (*Musca domestica*)	7 mm.
Honeybee (*Apis mellifica*) (worker)	15 mm.
Leopard frog (*Rana pipiens*)	60 mm.
Box turtle (*Terrapene carolina*)	100 mm.
Red fox (*Vulpes vulpes*)	590 mm.
Man (*Homo sapiens*)	1,700 mm.
African elephant (*Loxodonta africana*)	3,500 mm.

When unicellular animals reach a certain size they divide, for a size limitation is imposed upon the animal by its surface-volume ratio. Every cell depends on its surface membrane for the exchange of materials with the surrounding environment. The volume increases as the cube of the radius; the surface, as the square of the radius. Since the volume increases much faster than the surface, as the size of the cell increases, the rate of exchange of food and waste decreases. Protozoans can solve the problem by simply dividing when they reach a certain size. Large animals are made up of cells. By keeping their cells small there is a morphologic increase in surface.

The ratio between the nucleus and cytoplasm may be a controlling factor in determining the size of cells, for these two units must be in a certain balance for the efficient functioning of the cell. The importance of cell size as a regulative force in metabolism is strikingly demonstrated by the fact that the volume is much the same for any cell type and is independent of the animal's size. The cells found in a mouse are about the same size as those found in the largest mammals.

REFERENCES TO PART II

Bonner, J. T. 1965. Size and cycle. Princeton, Princeton University Press.

Bourne, G. H. 1970. Division of labor in cells, ed. 2. New York, Academic Press, Inc.

Clark, R. B. 1964. Dynamics in metazoan evolution: the origin of the coelom and metamerism. New York, Oxford University Press.

DeRobertis, E. D. P., W. W. Nowinski, and F. A. Saez. 1970. Cell biology; ed. 5. Philadelphia, W. B. Saunders Co.

Freeman, J. A. 1964. Cellular fine structure. New York, McGraw-Hill Co.

Gerard, R. W. 1940. Unresting cells. New York, Harper & Row, Publishers.

Ham, A. W. 1970. Histology. Philadelphia, J. B. Lippincott Co.

Kimball, J. W. 1970. Cell biology. Reading, Mass., Addison-Wesley Publishing Co., Inc. (Paperback.)

Levine, L. 1963. The cell in mitosis. New York, Academic Press, Inc.

Loewy, A. G., and P. Siekevitz. 1969. Cell structure and function, ed. 2. New York, Holt, Rinehart & Winston, Inc.

Romer, A. S. 1971. The vertebrate body, ed. 4. Philadelphia, W. B. Saunders Co.

part three

Physiology of animals

All animals have the same basic problems of existence, but they have different ways of solving these problems. Common tasks of life must be performed at all levels of animal structure. All must utilize metabolic processes to extract energy from the food. All have respiration, excretion, coordination and irritability, and muscular movement. These attributes of life are inherent in the simplest animals as well as the most complex. In seeking an understanding of life processes, biologists have broken down traditional barriers between the physical and life sciences. Interdisciplinary studies such as biochemistry and biophysics are now being used effectively to provide a better understanding of the composition and functioning of the organism's tissues.

At present there seems to be no evidence that the laws governing biologic phenomena are different from those that explain physicochemical processes of the nonliving. Structurally, we have seen that both the living and nonliving are made up of the same elements. There is no vital force that distinguishes the animate from the inanimate. Cells cannot synthesize molecules that the chemist cannot synthesize experimentally now or in the near future. It may be true that unique principles operate at the biologic level, but so far such principles have not been revealed. This certainly does not mean that the living process is fully understood, for many problems still baffle biologic investigation.

The functional activities of an animal center around **motility, nutrition, respiration, transport, excretion, hormonal and nervous integration, reproduction,** and **integument.** Animals have evolved various ways to carry out these necessary activities of life. Not all animals have specialized systems for carrying out specific functions. For instance, many groups have no specializations for the exchange of gases in respiration; the necessary gases simply diffuse directly into and out of the body tissues. As animals increased in structural complexity, more complex control mechanisms and greater integration of the varied activities were required. Thus the processes of nutrition, respiration, transport, and excretion have come to be coordinated by a system of physiologic controls that operate to maintain relatively constant conditions within the body. This maintenance of internal constancy is called **homeostasis.** Although variability is greater in lower than in higher forms, physiologic processes of all animals tend to operate within definite limits.

7

Body covering and support

INTEGUMENT AMONG VARIOUS GROUPS OF ANIMALS

The integument is the outer covering of the body; this includes the skin and all structures that are derived from, or associated with, the skin, such as hair, setae, scales, feathers, and horns. The integument is mainly a protective wrapping. In most animals it is rough and pliable, providing mechanical protection against abrasion and puncture and forming an effective barrier against the invasion of bacteria. It provides moisture-proofing against fluid loss or gain. The skin protects the underlying cells against the damaging action of the ultraviolet rays of the sun. But in addition to being a protective cover, the skin serves a variety of other important functions. For example, in warm-blooded (**homeothermic) animals**, it is vitally concerned with temperature regulation, since most of the body's heat is lost through the skin. The skin contains receptors of many senses that provide essential information about the immediate environment. It has excretory functions and, in some forms, respiratory functions as well. Through skin pigmentation the organism can make itself more or less conspicuous. Skin secretions can make the animal attractive or repugnant or provide olfactory cues which influence behavioral interactions between individuals.

INVERTEBRATE INTEGUMENT

Many protozoans have only the delicate cell or plasma membranes for external coverings; others, such as *Paramecium,* have developed a protective pellicle. Most multicellular invertebrates, however, have more complex tissue coverings. The principal covering is a single-layered **epidermis.** Some invertebrates have added a secreted noncellular **cuticle** over the epidermis for additional protection; some groups, such as many parasitic worms, have only a thick resistant cuticle and lack a cellular epidermis.

Arthropods have the most complex of invertebrate integuments, providing not only protection but also skeletal support. The development of a firm exoskeleton and jointed appendages suitable for the attachment of muscles has been a key feature in the great evolutionary success of this largest of animal groups. The arthropod integument consists of a single-layered **epidermis** (also called **hypodermis**) that secretes a complex cuticle of two zones (Fig. 7-1, *A*). The inner zone, the **procuticle,** is composed of protein and chitin. Chitin is a polysaccharide that resembles plant cellulose. The chitin is laid down in molecular sheets like the veneers of plywood, thus providing great strength to the pro-

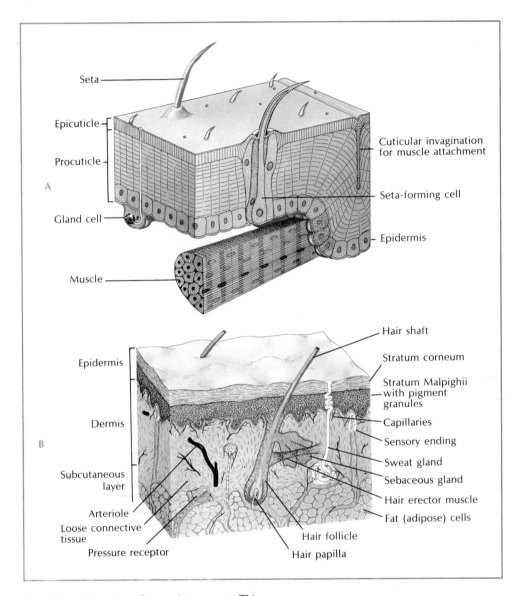

FIG. 7-1. A, Structure of insect integument. This reconstruction shows a block of integument drawn at a point where the cuticle invaginates to provide an exoskeletal muscle attachment. **B,** Structure of human skin.

cuticle layer. The outer zone of cuticle, lying above the procuticle, is the thin **epicuticle.** The epicuticle is nonchitinous, consisting instead of a complex of proteins and lipids. The epicuticle is significant in providing a protective moisture-proofing barrier to the integument.

The arthropod cuticle may remain as a tough but soft and flexible layer, or it may be hardened by one of two ways. In the decapod crustaceans, for example, crabs and lobsters, the cuticle is stiffened by **calcification,** the deposition of calcium carbonate. In insects hardening is achieved by a process called **sclerotization,** in which the protein molecules of the chitin form stabilizing cross-linkages. Arthropod chitin is one of the toughest materials synthesized by animals; it is strongly resistant to pressure and tearing and can withstand boiling in concentrated alkali, yet it is light, having a specific weight of only 1.3.

When arthropods molt, the epidermal cells first divide by mitosis. Enzymes secreted by the epidermis dissolve most of the procuticle; the digested materials are then absorbed and consequently not lost to the body. Then, in the space beneath the old cuticle, a new epicuticle and procuticle are formed. After the old cuticle is shed, the new cuticle is thickened and calcified or sclerotized.

VERTEBRATE INTEGUMENT

The basic plan of the vertebrate integument, as exemplified by human skin (Fig. 7-1, *B*), includes a thin, outer stratified epithelial layer, the **epidermis,** and an inner, thicker layer, the **dermis.** The epidermis is usually comprised of several cell layers. The basal part consists of columnar cells that continually divide by mitosis to form new cells that are pushed toward the surface. As the cells move upward, they become **keratinized** into scaly plates by the deposition of a fibrous protein, **keratin,** which makes the surface resistant to abrasion, chemical change, and water transfer.

The dermis is comprised of connective tissue, fat cells, muscle, blood vessels, and nerves as well as epidermally derived structures such as hair follicles, sweat glands, and sebaceous glands. The latter produce a fatty secretion called **sebum** that lubricates the hair and skin.

Hair, scales, and horns. Although the human species is not very hairy, hair is a distinct mammalian characteristic. The hair follicle from which a hair grows is an epidermal structure even though it lies mostly in the dermis and subdermal (subcutaneous) tissues. The hair grows continuously by rapid proliferation of cells in the base of the follicle. As the hair shaft is pushed upward, new cells are carried away from their source of nourishment and turn into the dense type of keratin (**hard keratin**) that constitutes nails, claws, hooves, and feathers as well as hair. On a weight basis, hair is by far the strongest material in the body. It has a tensile strength comparable to rolled aluminum, which is nearly twice as strong, weight for weight, as the strongest bone. A small erector muscle is attached to each hair follicle and slants upward

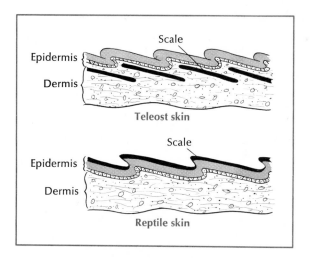

Fig. 7-2. Integument of bony fish and reptiles. Bony (teleost) fish have bony scales from dermis, and reptiles have horny scales from epidermis. Dermal scales of fish are retained throughout life. Since a new growth ring is added to each scale each year, fishery biologists use scales to tell age of fish. Epidermal scales of reptiles are shed periodically.

toward the epidermis. Contraction of this muscle causes the hair to stand up and the skin to dimple above the muscle attachment. The result is the "gooseflesh" characteristic of certain emotional states (fear and excitement) and of cold stimulation. The color of hair is determined by the amount and quality of pigment in the outer shell, or cortex, of each hair.

The skin is modified in many ways in the different vertebrates. The tough, protective scales of fish, although appearing to be surface structures, are actually bony plates produced in the dermis (Fig. 7-2). The scales may protrude through the overlying epidermis, but usually they are completely covered by a thin, often transparent, layer of epidermis, even though the scales overlap one another. The scales of reptiles, however, are horny, keratinized plates of epidermal origin. In snakes the ventral scales form transverse bands that can be raised and lowered by muscles in a coordinated manner to assist in locomotion. Snakes periodically shed their skins as they grow. The keratinized epidermis of the entire body surface separates from the living cells beneath. By first loosening the skin around

FIG. 7-3. Chief differences between horns and antlers. Horns are tough sheaths of keratin surrounding a bony core and are not shed. Antlers are entirely bone and develop beneath skin (called velvet) that is torn off before the breeding season; after breeding the antlers are shed.

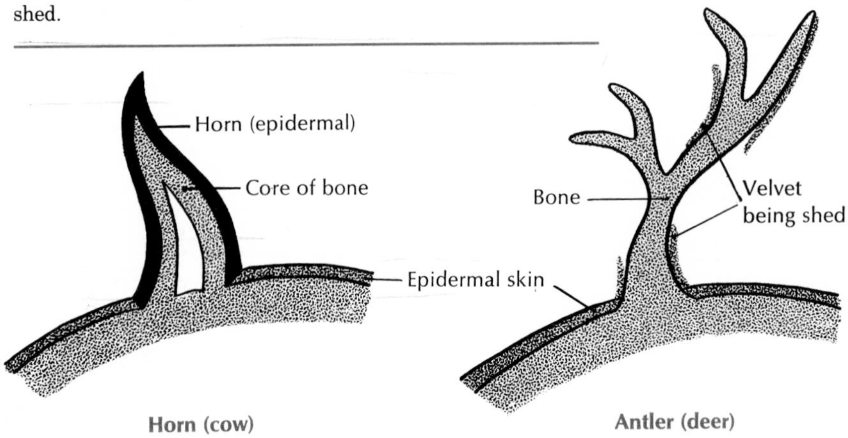

the lips, the snake works the skin back over his head; then as he crawls along the ground, the old skin is peeled off inside out, revealing the glistening new scales beneath.

Three kinds of horns or hornlike structures are found in mammals (Fig. 7-3). **True horns** found in ruminants, for example, sheep and cattle, are hollow sheaths of keratinized epidermis that embrace a core of bone arising from the skull. Horns are not shed, are not branched (although they may be greatly curved), and are found in both sexes. **Antlers** of the deer family are entirely bone when mature. During their annual growth, antlers develop beneath a covering of highly vascular soft skin called "velvet." When growth of the antlers is complete just prior to the breeding season, the blood vessels constrict and the stag tears off the velvet by rubbing the antlers against trees. The antlers are dropped after the breeding season. New buds appear a few months later to herald the next set of antlers. For several years each new pair of antlers is larger and more elaborate than was the previous set. A third kind of horn is that of the rhinoceros. Hairlike horny fibers arise from dermal papillae and are cemented together to form a single horn.

Sunburning and tanning. In general, man lacks the special body coverings that protect other land vertebrates from the damaging action of ultraviolet rays of the sun; he must depend on thickening of the outer layer of the epidermis (**stratum corneum**) and on epidermal pigmentation for protection from the sun's spectrum. Sunburn and suntanning are caused largely by exposure to the ultraviolet area of the sun's spectrum (wavelength about 300 nm.). This spectral band acts almost entirely on epidermis; very little penetrates to the dermis beneath. The ultraviolet rays photochemically decompose nucleoproteins within the nuclei of cells of the deeper layer of the epidermis. Blood vessels then enlarge and other tissue changes occur, producing the red coloration of sunburn. Light skins suntan through the formation in the deeper epidermis of the pigment **melanin,** and by "pigment darkening," that is, the photooxidative blackening of bleached pigment already present in the epidermis.

SKELETAL SYSTEMS

Skeletons are supportive systems that provide rigidity to the body, surfaces for muscle attachment, and protection for vulnerable body organs.

The familiar bone of the vertebrate skeleton is only one of several kinds of supportive and connective tissues, serving various binding and supportive functions, that we will discuss in this section.

Exoskeleton and endoskeleton. Although animal supportive and protective structures take many forms, there are two principal types of skeletons: the **exoskeleton,** typical of mollusks and arthropods, and the **endoskeleton,** characteristic of vertebrates. The invertebrate exoskeleton is mainly protective in function and may take the form of shells, spicules, and calcareous or chitinous plates. It may be rigid, as in mollusks, or jointed and movable as in arthropods. Unlike the endoskeleton, which grows with the animal, the exoskeleton is often a limiting coat of armor, which must be periodically shed (molted) to make way for an enlarged replacement. Some invertebrate exoskeletons, such as the shells of snails and bivalves, grow with the animal. Vertebrates, too, have traces of exoskeleton that serve to remind us of our invertebrate heritage. These are, for example, scales and plates of fishes, fingernails and claws, hair, feathers, and other cornified integumentary structures.

The vertebrate endoskeleton is formed inside the body and is composed of bone and cartilage surrounded by soft tissues. It not only supports and protects, but it is also the major body reservoir for calcium and phosphorus. In the higher vertebrates the red blood cells and certain white blood cells are formed in the bone marrow.

CONNECTIVE TISSUES

The connective tissues are a heterogenous complex of cells and cell products that serve to bind cells, tissues, and organs together. They are both a kind of intercellular skeleton and body "stuffing." All connective tissues are derived from the embryonic **mesenchyme,** which are wandering, ameboid, irregularly shaped cells that proliferate from the embryonic mesoderm. These cells move freely between the cracks and spaces of other differentiating tissue, then settle and differentiate into one of several kinds of connective tissues.

The simplest kind of connective tissue is **loose** connective tissue that invests many body organs and structures. It is mostly a loose network of **col-** lagenous fiber containing protein, water, and a variety of specialized cells (Fig. 6-4, *A,* p. 71).

Dense, or **compact,** connective tissue is a second kind, which contains collagenous fibers arranged in dense interlacing layers. Cellular elements are less abundant in dense than in loose connective tissue. The great tensile strength and toughness of dense connective tissue makes it ideally suited for **tendons** and **ligaments.** Tendons form muscle attachments; one end of the tendon is attached to bone or cartilage, and the other to the connective tissue elements of the muscle. Ligaments are similar to tendons but serve as attachments between skeletal elements only. Ligaments contain enough elastic fibers woven in among the collagenous fibers to provide some stretch and elasticity for movement.

CARTILAGE

Cartilage and bone are the characteristic vertebrate supportive tissues. The **notochord,** the semi-rigid axial rod of protochordates and vertebrate larvae and embryos, is also a primitive vertebrate supportive tissue. Except in the most primitive vertebrates, for example, *Amphioxus* and the cyclostomes, the notochord is surrounded or replaced by the backbone during embryonic development. The notochord is composed of large, vacuolated cells and is surrounded by layers of elastic and fibrous sheaths. It is a stiffening device, preserving body shape during locomotion.

Vertebrate cartilage is the major skeletal element of primitive vertebrates. Cyclostomes and elasmobranchs have purely cartilaginous skeletons. In contrast, higher vertebrates have principally bony skeletons as adults, with some cartilage interspersed. Cartilage is a soft, pliable, characteristically deep-lying tissue. Unlike connective tissue, which is quite variable in form, cartilage is basically the same wherever it is found. The basic form, **hyaline cartilage** (Fig. 6-4, *B,* p. 71), has a clear, glassy appearance. It is composed of cartilage cells **(chondrocytes)** surrounded by firm complex protein gel interlaced with a meshwork of collagenous fibers. Blood vessels are virtually absent. In addition to forming the cartilagenous skeleton of the primitive vertebrates and that of all vertebrate embryos, hyaline cartilage makes up the articulating surfaces of many bone joints of higher adult vertebrates and the supporting tracheal, laryngeal, and bronchial rings. The basic cartilage has several variants. Among these is **calcified cartilage,** where calcium salt deposits produce a bonelike structure. **Fibrocartilage,** resembling connective tissue, and **elastic** cartilage, containing many elastic fibers, are other variations of basic hyaline cartilage found among the vertebrates.

BONE

Bone differs from other connective and supportive tissues by having significant deposits of inorganic calcium salts laid down in an extracellular matrix. Its structural organization is such that bone has nearly the tensile strength of cast iron, yet is only one third as heavy. Most bones develop from cartilage (**endochondral bone**) by a complex replacement of embryonic cartilage with bone tissue. A second type of bone is **membrane bone** that develops directly from sheets of embryonic cells. In higher vertebrates membrane bone is restricted to bones of the face and cranium; the remainder of the skeleton is endochondral bone. In the fishes the dermal scales and plates that may cover most of the body are formed from membrane bone.

Two kinds of bone structure are distinguishable— **spongy** (or **cancellous**) and **compact** (Fig. 7-4). Spongy bone consists of an open, interlacing framework of bony tissue, oriented to give maximum strength under the normal stresses and strains that the bone receives. Compact bone is dense, appearing absolutely solid to the naked eye. Both structural

FIG. 7-4. Section of proximal end of human humerus, showing appearance of spongy and compact bone. (From Bloom, W., and Fawcett, D. W.: A textbook of histology, Philadelphia, 1968, W. B. Saunders Co.)

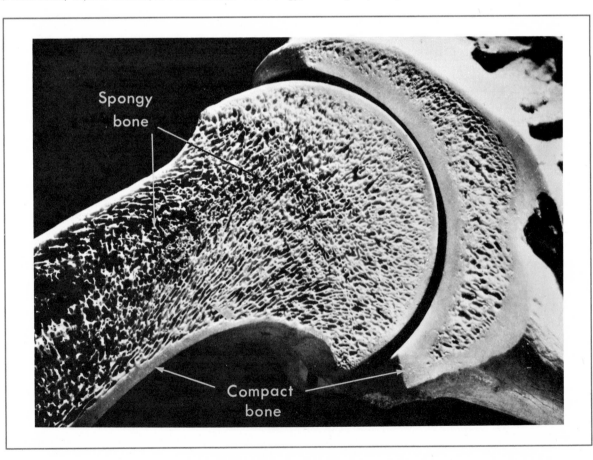

kinds of bone are found in the typical long bones of the body such as the humerus (upper arm bone) (Fig. 7-4).

Microscopic structure of bone

Compact bone is composed of a calcified bone matrix arranged in concentric rings. The rings contain cavities **(lacunae)** filled with bone cells (osteocytes) that are interconnected by many minute passages **(canaliculi).** These serve to distribute nutrients throughout the bone. This entire organization of lacunae and canaliculi is arranged into an elongated cylinder called an **haversian system** (Fig. 7-5). Bone consists of bundles of haversian systems cemented together and interconnected with blood vessels. Bone growth is a complex restructur-

ing process, involving both its destruction internally by bone-resorbing cells **(osteoclasts)** and its deposition externally by bone-building cells **(osteoblasts).** Both processes occur simultaneously so that the marrow cavity inside grows larger by bone resorption while new bone is laid down outside by bone deposition. Bone growth responds to several hormones, in particular **parathyroid hormone,** which stimulates bone resorption, and **calcitonin,** which

FIG. 7-5. Structure of bone, showing the dense calcified matrix and bone cells arranged into haversian systems. Bone cells are entrapped within the cell-like lacunae, but receive nutrients from the circulatory system via tiny canalculi that interlace the calcified matrix. Bone cells were known as osteoblasts when they were building bone, but in mature bone shown here, they become resting osteocytes. Bone is covered with a compact connective tissue called periosteum.

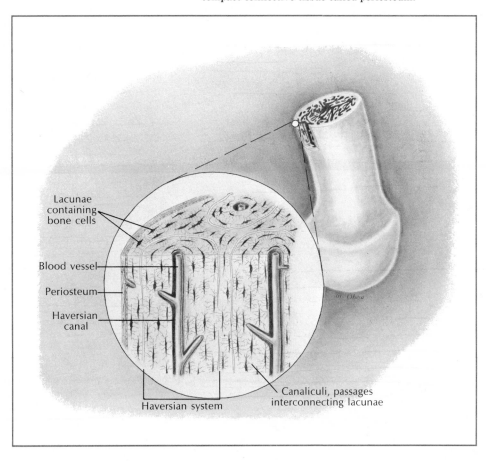

Lacunae containing bone cells
Blood vessel
Periosteum
Haversian canal
Haversian system
Canaliculi, passages interconnecting lacunae

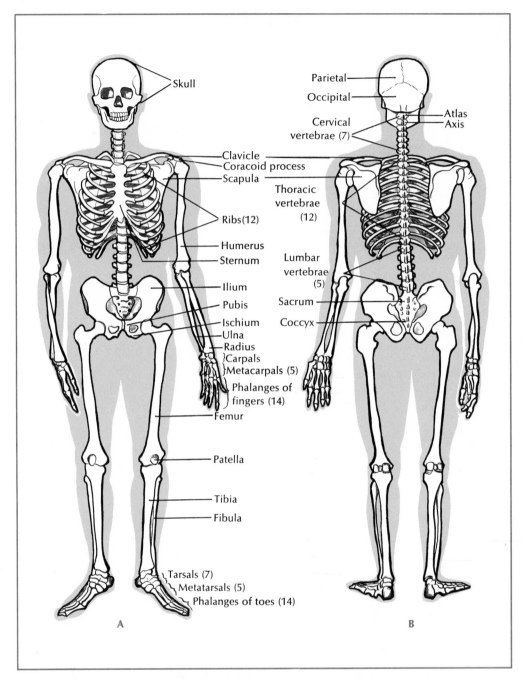

FIG. 7-6. Human skeleton. **A,** Ventral view. **B,** Dorsal view. Numbers in parentheses indicate number of bones in that unit. In comparison with other mammals, man's skeleton is a patchwork of primitive and specialized parts. Erect posture brought about by specialized changes in legs and pelvis enabled primitive arrangement of arms and hands (arboreal adaptation of man's ancestors) to be used for manipulation of tools. Development of skull and brain followed as consequence of premium natural selection put upon dexterity, better senses, and ability to appraise environment.

inhibits bone resorption. These two hormones are responsible for maintaining a constant level of calcium in the blood (p. 183).

PLAN OF THE VERTEBRATE SKELETON

The vertebrate skeleton has undergone a great transformation in the course of evolution. This is hardly surprising. The move from water to land forced dramatic changes in respiration and body form. The most pronounced differences are found in the bones of the gill apparatus, called the **visceral skeleton,** and bones of the skull. The early vertebrates tend to have a larger number of skull bones than more recently evolved forms. Some fish may have 180 skull bones; amphibia and reptiles, 50 to 95; and mammals, 35 or fewer. Man has 29.

The vertebral column varies greatly with different animals and with different regions of the vertebral column in the same animal. In fish it is differentiated only into trunk and caudal vertebrae; the column in many of the other vertebrates is differentiated into **cervical** (neck), **thoracic** (chest), **lumbar** (back), **sacral** (pelvic), and **caudal** (tail) vertebrae. In birds and also in man the caudal vertebrae are reduced in number and size, and the sacral vertebrae are fused. The number of vertebrae varies among the different animals. The python seems to lead the list with 435. In man (Fig. 7-6) there are 33 in the child, but in the adult 5 are fused to form the **sacrum** and 4 to form the **coccyx.** Besides the sacrum and coccyx, man has 7 cervical, 12 thoracic, and 5 lumbar vertebrae. The first cervical vertebra is modified for articulation with the skull and is called the **atlas.** The number of cervical vertebrae (7) is constant in nearly all mammals.

Ribs show many variations among the vertebrates. Primitive forms have a pair of ribs for each vertebra from head to tail, but higher forms tend to have fewer ribs. Certain fish have two ventral ribs for each vertebra, and in some fish there are dorsal and ventral (pleural) ribs on the same vertebra. In tetrapods the single type of rib is supposed to correspond to the dorsal one of fish. The ribs of many land vertebrates are joined to the sternum. The sternum is lacking in snakes. The ribs of vertebrate animals are not all homologous, for they do not all arise in the same way. Ribs, in fact, are not even universal among vertebrates; many, including the leopard frog, do not have them at all. Others, such as the elasmobranchs and some amphibians, have very short ribs. Man has twelve pairs of ribs, although evidence indicates that his ancestors had more. The ribs together form the thoracic basket that supports the chest wall and keeps it from collapsing.

Most vertebrate animals have paired appendages. None are found in cyclostomes, but both the cartilaginous and bony fishes have pectoral and pelvic fins that are supported by the pectoral and pelvic girdles, respectively. Forms above the fish (except snakes) have two pairs of appendages, also supported by girdles. The basic plan of the land vertebrate limb (tetrapod) is called **pentadactyl,** terminating in five digits. Among the various vertebrates there are many modifications in the girdles, limbs, and digits that enable animals to meet specific modes of life. For instance, some amphibians have only three or four toes on each foot, and the horse has only one. Also, the bones of the limbs may be separate or they may be fused in various ways. Whatever the modification, the girdles and appendages in forms above fish are all built on the same plan and their component bones can be homologized. In man the pectoral girdle is made up of 2 scapulae and 2 clavicles; the arm is made up of humerus, ulna, radius, 8 carpals, 5 metacarpals, and 14 phalanges. The pelvic girdle consists of 2 innominate bones, each of which is composed of 3 fused bones — ilium, ischium, and pubis; the leg is made up of femur, patella, tibia, fibula, 7 tarsals, 5 metatarsals, and 14 phalanges. It will be noted that each bone of the leg has its counterpart in the arm, with the exception of the patella. This kind of correspondence between anterior and posterior parts is called **serial homology.**

8

Animal movement

Movement is a unique characteristic of animals. Plants may show movement, but this usually results from changes in turgor pressure or growth rather than from specialized contractile proteins as in animals. Recently it has become evident that virtually all animal movement, whether of muscle, cilia, or a single-celled protozoan, depends on a single fundamental mechanism: contraction of special protein systems that are powered by the high-energy phosphate compound **adenosine triphosphate** (ATP).

AMEBOID MOVEMENT

Ameboid movement is a form of movement especially characteristic of the freshwater *Amoeba* and other sarcodine protozoa; it is also found in many wandering cells of higher animals, such as white blood cells, embryonic mesenchyme, and numerous other mobile cells that move among the tissue spaces. Ameboid cells constantly change their shape by sending out and withdrawing **pseudopodia** (false feet) from any point on the cell surface. Such cells are surrounded by a delicate, highly flexible, membrane called **plasmalemma.** Beneath this lies a nongranular layer, the **hyaline ectoplasm,** which encloses the **granular ectoplasm.** Optical studies of *Amoeba* in movement suggest that the outer cytoplasm (also called **ectoplasm**) contracts in some way to pull the inner cytoplasm (called **endoplasm**) forward where it is squeezed into the advancing pseudopod (Fig. 8-1). The ectoplasm now slips posteriorly under the plasmalemma and joins the endoplasm at the rear to begin another cycle. There are other theories of ameboid movement; at present no completely satisfactory analysis exists.

CILIARY MOVEMENT

Cilia are minute hairlike motile processes that occur on the surfaces of the cells of many animals; they are a distinctive feature of ciliate protozoans, but are also commonly found on the epithelial surfaces of tubular organs in higher animals, where they function to propel fluids and materials. There may be hundreds on the surface of a single cell. A **flagellum** is a whiplike structure, larger than a cilium, and usually present singly at one end of a cell. They are found in members of flagellate protozoans, in animal spermatozoa, and in sponges. Cilia are of remarkably uniform diameter (0.1 to 0.5 μ) wherever they are found. The electron microscope has shown that each cilium contains a peripheral circle of nine double filaments and an additional two filaments in the center (Fig. 8-2). (Exceptions to the 9 + 2 arrangement have been noted; certain

FIG. 8-1. Structure of *Amoeba* in active locomotion. Animal is moving in the direction of the advancing pseudopod. Although there is no entirely satisfactory explanation of ameboid movement, scheme shown is based on a recent analysis of cytoplasmic flow.

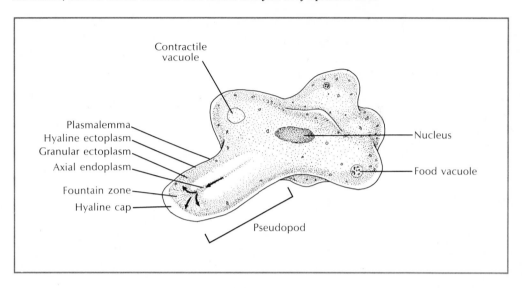

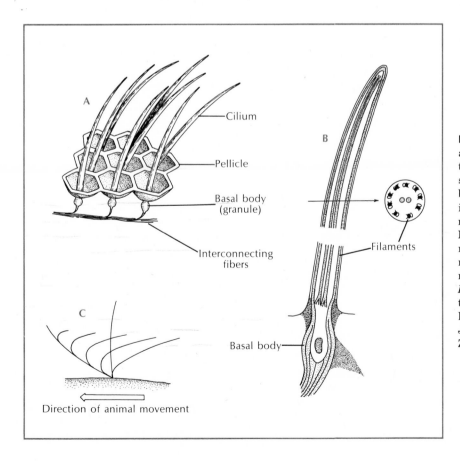

FIG. 8-2. Ciliary structure and movement. **A,** Section of the pellicle of *Paramecium*, showing arrangement of cilia, basal granules, and interconnection fibers (so-called neuromotor system). **B,** Structure of a cilium as revealed by the electron microscope. **C,** Sequence of movements of a cilium of *Paramecium*. Power stroke is to the right and recovery to the left. (**B** adapted from Rhodin, J., and Dalhamn, T.: Z. Zellforsch. **44:**345, 1956.)

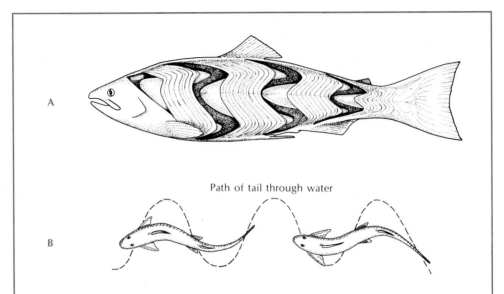

Path of tail through water

FIG. 8-3. A, Trunk musculature of a salmon. Segmental myotomes are W shaped when viewed from the surface. Musculature has been dissected away in four places to show internal anterior and posterior deflections of myotomes that improve muscular efficiency for swimming. **B,** Motion of swimming fish. Noncompressible water must be pushed aside by the forward motion of the head, driven by the snakelike stroke of the body. (**A** after Greene from Romer, A.: The vertebrate body, ed. 4, Philadelphia, 1970, W. B. Saunders Co.; **B** modified from Marshall, P. T., and Hughes, G. M.: The physiology of mammals and other vertebrates, New York, 1967, Cambridge University Press.)

sperm tails have but one central fibril.) It is believed that the driving stroke of the cilium is due to the simultaneous contraction of five of the peripheral fibrils. During recovery, the other four fibrils, which are idle during the drive stroke, slowly contract to bring the cilium back to its starting position. Cilia contract in a highly coordinated way, the rhythmic waves of contraction moving across a ciliated epithelium like windwaves across a field of grain. The columns of cilia are coordinated by an interconnected fiber system through which the excitation wave passes with each stroke.

MUSCULAR MOVEMENT

Contractile tissue reaches its highest development in muscle cells called **fibers.** Although muscle fibers themselves can only shorten, they can be arranged in so many different configurations and combinations that almost any movement is possible.

Types of vertebrate muscle. Vertebrate muscle is broadly classified on the basis of the appearance of muscle cells (fibers) when viewed with a light microscope (Fig. 6-5, p. 72). **Striated muscle** appears transversely striped (striated), with alternating dark and light bands. We can recognize two types of striated muscle: **skeletal** and **cardiac muscle.** A third kind of vertebrate muscle is **smooth** (or visceral) **muscle,** which lacks the characteristic alternating bands of the striated type.

Skeletal muscle is typically organized into sturdy, compact bundles or bands. It is called skeletal muscle because it is attached to skeletal elements and is responsible for movements of the trunk, appendages, the respiratory organs, eyes, mouthparts, and so on. Skeletal muscle fibers are extremely long, cylindrical, multinucleate cells, which may reach from one end of the muscle to the other. They are packed into bundles called **fascicles,** which are enclosed by tough connective tissue. The fascicles are in turn grouped into a discrete **muscle** surrounded by a thin connective tissue layer. Most skeletal muscles taper at their ends, where they connect by tendons to bones. Other muscles, such as the ventral abdominal muscles, are flattened sheets.

In most fish, amphibians, and, to some extent, reptiles, there is a segmented organization of muscles alternating with the vertebrae. The bulk of the trunk of a fish is axial, locomotory musculature, which develops into zigzag bands that take the shape of a W toward the surface (Fig. 8-3). Internally the muscle bands (called myotomes) are deflected forward and backward in a complex fashion that apparently promotes efficiency of movement. In fish these muscles do not attach directly to bones, but are bound to broad sheets of tough connective tissue, which in turn tie to the vertebral column. This musculature arrangement, together with streamlined body form, beautifully adapts fish for rapid movement through water, their noncompressible medium.

The skeletal muscles of higher vertebrates, by splitting, by fusion, and by shifting, have developed into specialized muscles best suited for manipulating the jointed appendages that have evolved for locomotion on land. Skeletal muscle contracts powerfully and quickly but fatigues more rapidly than does smooth muscle. Skeletal muscle is sometimes called **voluntary muscle** because it is innervated by motor fibers and is under conscious cerebral control.

Smooth muscle lacks the striations typical of skeletal muscle (Fig. 6-5, p. 72). The cells are long, tapering strands, each containing a single nucleus. Smooth muscle cells are organized into sheets of muscle circling the walls of the alimentary canal, blood vessels, respiratory passages, and urinary and genital ducts. Smooth muscle is typically slow acting. It is under the control of the autonomic nervous system; thus, unlike skeletal muscle, its contractions are involuntary and unconscious. The principal functions of smooth muscles are to push the contents of a tube, such as the intestine, along its way by active contractions or to regulate the diameter of a tube, such as a blood vessel, by sustained contraction.

Cardiac muscle, the seemingly tireless muscle of the vertebrate heart, combines certain characteristics of both skeletal and smooth muscle. It is fast acting and striated like skeletal muscle, but contraction is under involuntary autonomic control like smooth muscle. Actually, the autonomic nerves serving the heart can only speed up or slow down the rate of contraction; the heart beat originates within specialized cardiac muscle and the heart will continue to beat even after all autonomic nerves are severed. Until very recently, cardiac muscle was thought to be one large unseparated mass (**syncytium**) of branching, anastomosing fibers. Many histologists, their understanding vastly increased by the electron microscope, now consider cardiac muscle to be comprised of closely opposed, but separate, uninucleate cell fibers.

Types of invertebrate muscle. Smooth and striated muscles are also characteristic of invertebrate animals, but there are many variations of both types and even instances where the structural and functional features of vertebrate smooth and striated muscle are combined in the invertebrates. Striated muscle appears in invertebrate groups as diverse as the primitive coelenterates and the advanced arthropods. The thickest muscle fibers known, about 2 mm. in diameter and easily seen with the unaided eye, are those of giant barnacles living along the Pacific coast of North America. Since each fiber is a single cell that can be separated and cultured in isolation, they are much sought after for physiologic studies of muscle.

Invertebrate smooth muscle may closely resemble vertebrate smooth muscle, but commonly the fibers are arranged into a regular spiral course like a single helix. Bivalve mollusk muscles contain fibers of two types. One kind can contract rapidly, enabling the bivalve to snap shut its valves when disturbed. Scallops use these "fast" muscle fibers to swim in their awkward manner. The second muscle type is capable of slow, long-lasting contractions. Using these fibers, a bivalve can keep its valves tightly shut for days or even months. Obviously these are no ordinary muscle fibers! It has been discovered that such retractor muscles use very little metabolic energy and receive remarkably few nerve impulses to maintain the activated state. The contracted state has been likened to a "catch mecha-

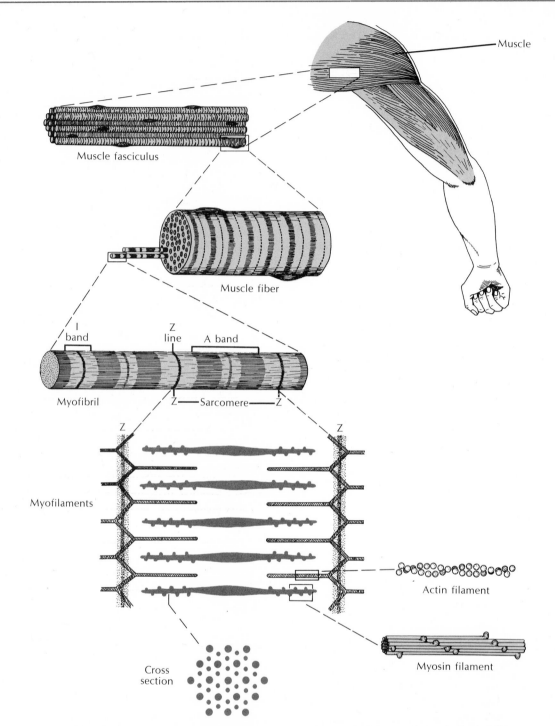

FIG. 8-4. Organization of vertebrate skeletal muscle from gross to molecular level. Actin (thin) and myosin (thick) filaments are enlarged to show supposed shapes of individual molecules, and probable positioning of the cross-bridges (shown as knobs) on myosin molecules which serve to link thick and thin filaments during contraction. Cross section shows that each thick filament is surrounded by six thin filaments and that each thin filament is surrounded by three thick filaments. (Redrawn with slight modifications from Bloom, W., and Fawcett, D. W.: A textbook of histology, Philadelphia, 1968, W. B. Saunders Co.).

nism" involving some kind of stable cross-linkage between the contractile proteins within the fiber. However, despite considerable research, no completely satisfactory explanation for this retractor mechanism exists.

Structure of striated muscle. In recent years the electron microscope and advanced biochemical methods have been focused on the fine structure and function of the striated muscle fiber. These efforts have been so successful that more has been learned of muscle physiology in the last decade than in the previous century. The discussion that follows will be limited to the striated muscle, since its physiology is presently much better understood than is that of smooth muscle.

As we earlier pointed out, striated muscle is so named because of the periodic bands, plainly visible under the light microscope, which pass across the widths of the muscle cells. Each cell, or **fiber,** contains numerous **myofibrils** packed together and invested by the cell membrane, the **sarcolemma** (Figs. 8-4 and 8-7). Also present in each fiber are several hundred nuclei usually located along the edge of the fiber, numerous mitochondria (sometimes called **sarcosomes**), a network of tubules called the **sarcoplasmic reticulum** (to be discussed later), and other cell inclusions typical of any living cell. Most of the fiber, however, is packed with the unique **myofibrils,** each 1 to 2 μ in diameter.

The characteristic banding of the muscle fiber represents the fine structure of the myofibrils that make up the fiber. In the resting fiber are alternating light- and dark-staining bands called the **I bands** and **A bands,** respectively (Fig. 8-4). The functional unit of the myofibril, the **sarcomere,** extends between successive Z lines. The myofibril is actually an aggregate of much smaller parallel units called **myofilaments.** These are of two kinds —thick filaments, 80 to 100 Å in diameter composed of the protein **myosin,** and thin filaments, 50 Å in diameter composed of the protein **actin** (Fig. 8-4). These are the actual contractile proteins of muscle. The thick myosin filaments are confined to the A band region. The thin actin filaments are located mainly in the light I bands but extend some distance into the A band as well. They do not quite meet in the center of the A band, however. The Z line is a dense protein, different

from either actin or myosin, which serves as the attachment plane for the thin filaments and keeps them in register. These relationships are diagrammed in Fig. 8-4.

Contraction of striated muscle. The thick and thin filaments are spatially arranged in a highly symmetric pattern, so that each thick filament is surrounded by six thin filaments; conversely, each thin filament lies among three thick filaments (see cross section at bottom of Fig. 8-4). The two kinds of filaments are linked together by molecular bridges which, it is believed, extend outward from the thick filaments to hook onto active sites on the thin filaments. During contraction the cross-bridges swing rapidly back and forth, alternately attaching and releasing the active sites in succession in a kind of ratchet action (Fig. 8-5).

The thick and thin filaments thus slide past one another; both kinds of filament maintain their original length, but the thin actin filaments now extend farther into the A band as shown in Fig. 8-6. As contraction continues, the Z lines are drawn closer together. During very strong contraction the thin filaments touch and crumple in the center of the A band. Striated muscle contracts so rapidly that each cross-bridge may attach and release fifty to one hundred times per second.

The contractile machinery has been most thoroughly studied in mammals (most of the electron microscopists used a specific thigh muscle, the psoas muscle, of the rabbit), but recent comparative studies indicate a remarkable uniformity of the sliding-filament mechanism throughout the animal kingdom. Even the contractile proteins myosin and actin are biochemically similar in all animals. The actomyosin contractile system evidently appeared very early in animal evolution and proved so flawless that no significant changes occurred thereafter.

Energy for contraction. Muscles perform work when they contract and, of course, require energy to do so. Resting muscles use little energy but consume large amounts during vigorous exercise. Muscles use only 20% of the energy value of food molecules when contracting; the remainder is released as heat. This is a rapid source of body heat as everyone knows; exercising is the quickest way to warm up when one is cold.

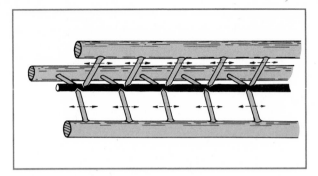

FIG. 8-5. Ratchetlike action of cross-bridges between thick and thin filaments of skeletal muscle fibers. Cross-bridges swing from site to site, pulling thin filaments past the thick. Each thick filament is actually surrounded by six thin filaments and is linked by six sets of cross-bridges. For simplicity, this diagram shows only one set of cross-bridges on each thick filament.

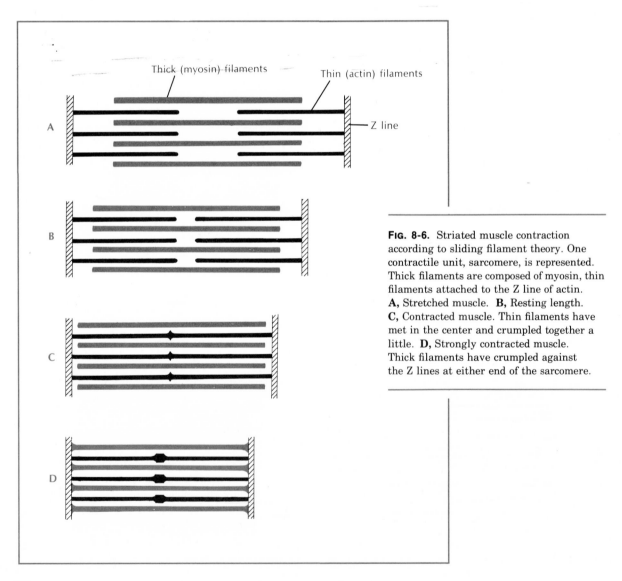

Thick (myosin) filaments

Thin (actin) filaments

Z line

FIG. 8-6. Striated muscle contraction according to sliding filament theory. One contractile unit, sarcomere, is represented. Thick filaments are composed of myosin, thin filaments attached to the Z line of actin. **A,** Stretched muscle. **B,** Resting length. **C,** Contracted muscle. Thin filaments have met in the center and crumpled together a little. **D,** Strongly contracted muscle. Thick filaments have crumpled against the Z lines at either end of the sarcomere.

The immediate source of energy for muscular contraction is ATP (adenosine triphosphate), the energy-rich organic phosphate compound used universally by animals for storing and mobilizing chemical energy (p. 59). When muscle is stimulated to contract, the energy released by ATP powers the ratchetlike mechanism between actin and myosin, causing the filaments to telescope.

Although the ATP stored in muscle supplies the immediate energy for contraction, the supply is limited and quickly exhausted. However, muscle contains a much larger energy storage form, **creatine phosphate,** which can rapidly transfer energy for the resynthesis of ATP. Eventually even this reserve is used up and must be restored by the breakdown of carbohydrate. Carbohydrate is available from two sources—from **glycogen** stored in the muscle and from **glucose** entering the muscle from the bloodstream. If muscular contraction is not too vigorous or too prolonged, glucose can be completely oxidized to carbon dioxide and water by **aerobic glycolysis.** But during prolonged or heavy exercise, the blood flow to the muscles, although greatly increased above the resting level, is not sufficient to supply oxygen as rapidly as required for the complete oxidation of glucose. When this happens, the contractile machinery receives its energy largely by **anaerobic glycolysis,** a process that does not require oxygen (see p. 66). The presence of this anaerobic pathway, although not nearly as efficient as the aerobic one, is of great importance; without it, all forms of heavy muscular exertion such as running would be impossible.

During anaerobic glycolysis, glucose is degraded to lactic acid with the release of energy. This is used to resynthesize creatine phosphate that, in turn, passes the energy to ADP for the resynthesis of ATP. Lactic acid accumulates in the muscle and diffuses rapidly into the general circulation. If the muscular exertion continues, the build up of lactic acid causes enzyme inhibition and fatigue. Thus the anaerobic pathway is a self-limiting one, since continued heavy exertion leads to exhaustion. The muscles incur an **oxygen debt** because the accumulated lactic acid must be oxidized by extra oxygen. Following the period of exertion, oxygen consumption remains elevated until all of the lactic acid has been oxidized, or resynthesized, to glucose.

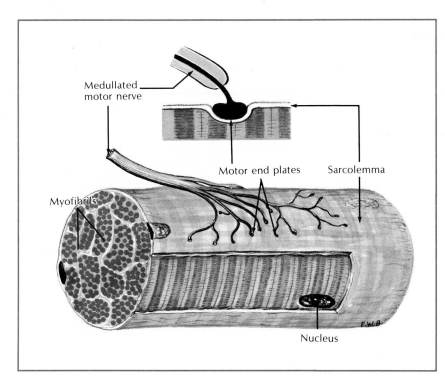

Medullated motor nerve

Motor end plates

Sarcolemma

Myofibrils

Nucleus

FIG. 8-7. Motor end plates (myoneural junctions) of a motor nerve on a single muscle fiber. (From Tuttle, W. W., and Schottelius, B. A.: A textbook of physiology, ed. 16, St. Louis, 1969, The C. V. Mosby Co.)

To summarize, the sequence of chemical sources of energy can be expressed in abridged form as follows:

$$ATP \rightleftharpoons ADP + H_3PO_4 + \text{Energy for contraction}$$

$$\text{Creatine phosphate} \rightleftharpoons \text{Creatine} + H_3PO_4 + $$
$$\text{Energy for resynthesis of ATP (anaerobic)}$$

$$\text{Glucose} \xrightleftharpoons{\text{(anaerobic)}} \text{Lactic acid} + $$
$$\text{Energy for resynthesis of creatine phosphate}$$

$$\text{Glucose} + O_2 \xrightarrow{\text{(aerobic)}} CO_2 + H_2O + $$
$$\text{Energy for resynthesis of creatine phosphate}$$

Stimulation of contraction. Skeletal muscle must, of course, be stimulated to contract. If the nerve supply to a muscle is severed, the muscle will **atrophy,** or waste away. Skeletal muscle fibers are arranged in groups of approximately 100, each group under the control of a single motor nerve fiber. Such a group is called a **motor unit.** As the nerve fiber approaches the muscle fibers, it splays out into many terminal branches. Each branch attaches to a muscle fiber by a special structure, called a **synapse,** or **myoneural junction** (Fig. 8-7). At the synapse is a tiny gap, or cleft, that thinly separates nerve fiber and muscle fiber. In the synapse is stored a chemical, **acetylcholine,** which is released when a nerve impulse reaches the synapse. This substance is a chemical mediator that diffuses across the narrow junction and acts on the muscle fiber membrane to generate an electrical potential. The potential spreads through the muscle fiber, causing it to contract. Thus the synapse is a special chemical bridge that couples together the electrical potentials of nerve and muscle fibers.

9

Nutrition

Nutrition refers to all the processes involved in providing nourishment to living cells. These include the mechanism of feeding, the processes of nutrient storage, digestion, and absorption, and the specific food requirements of animals. Sometimes the term nutrition is used more restrictively, especially in human health, to mean the kinds and amounts of foods needed in the diet to promote health and well-being.

All organisms require energy to maintain their highly ordered and complex structure. This energy is chemical energy that is released by transforming complex compounds acquired from the organism's environment into simpler ones. Obviously, if living organisms must depend on the breakdown of complex foodstuffs to build and maintain their own complexity, these foodstuffs must somehow be synthesized in the first place. Most of the energy for this synthesis is provided by the powerful radiations of the sun, the one great source of energy reaching our planet that is otherwise a virtually isolated system. Organisms capable of capturing the sun's energy by the process of photosynthesis are, of course, the green plants. Green plants are **autotrophic organisms** capable of synthesizing all the essential organic compounds needed for life. Autotrophic organisms need only inorganic com-

pounds absorbed from their surroundings to provide the raw materials for synthesis and growth. Most autotrophic organisms are the chlorophyll-bearing **phototrophs,** although some, the chemosynthetic bacteria, are **chemotrophs,** gaining energy from inorganic chemical reactions.

Almost all animals are **heterotrophic organisms** that depend on already synthesized organic compounds for their nutritional needs. Animals, with their limited capacities to perform organic synthesis, must feed on plants and other animals to obtain the materials they will use for growth, maintenance, and the reproduction of their kind. The foods of animals, usually the complex tissues of other organisms, can seldom be utilized directly. Food is usually too large to be absorbed by the body cells and may contain material of no nutritional value as well. Consequently, food must be broken down, or digested, into soluble molecules sufficiently small to be utilized. One important difference, then, between autotrophs and heterotrophs is that the latter must have digestive systems.

Animals may be divided into a number of categories on the basis of dietary habits. **Herbivorous** animals feed mainly on plant life. **Carnivorous** animals feed mainly on herbivores and other carnivores. **Omnivorous** forms eat both plants and animals. A fourth category is sometimes distinguished, the **insectivorous** animals, which are those birds and mammals that subsist chiefly on insects.

The ingestion of foods and their simplification by digestion are only initial steps in nutrition. Foods reduced by digestion to soluble, molecular form are **absorbed** into the circulatory system and **transported** to the tissues of the body. There they are **assimilated** into the protoplasm of the cells. Oxygen is also transported by the blood to the tissues, where food products are **oxidized,** or burned, to yield energy and heat. Then the wastes produced by oxidation must be **excreted.** Food products unsuitable for digestion are rejected by the digestive system and are **egested** in the form of feces. Much food is not immediately utilized but is stored for future use.

The sum total of all these nutritional processes is called **metabolism.** Metabolism includes both constructive and tearing-down processes. When substances are built into new tissues, or stored in some form for later use, the process is called **anabolism.** The breaking down of complex materials to simpler ones for the release of energy is called **catabolism.** Both processes occur simultaneously in all living cells.

FEEDING MECHANISMS

Only a few animals can absorb nutrients directly from their external environment. Blood and intestinal parasites may derive all their nourishment as primary organic molecules by surface absorption; some aquatic invertebrates may soak up part of their nutritional needs directly from the water. Most animals, however, must work for their meals, and the specializations that have evolved for food procurement are almost as numerous as species of animals. In this brief discussion we will consider some of the major food-gathering devices.

Feeding on particulate matter. Drifting microscopic particles fill the upper few hundred feet of the ocean. Most of this uncountable multitude is **plankton,** plant and animal microorganisms too small to do anything but drift with the ocean's currents. The rest is organic debris, the disintegrating remains of dead plants and animals. Altogether this oceanic swarm forms the richest life domain on earth. It is preyed on by numerous larger animals, invertebrates and vertebrates, using a variety of feeding mechanisms. Some protozoans, such as the ameboid sarcodines, ingest particulate food by a process called **phagocytosis.** The animal, stimulated by the proximity of food, pushes out armlike extensions of the plasmalemma (cell membrane) and engulfs the particle into a food vacuole, in which it is digested. Other protozoans have specialized openings, called **cytostomes,** through which the food passes to be enclosed in a food vacuole.

By far the most important method to have evolved for particle feeding is **filter feeding** (Fig. 9-1). It is a primitive, but immensely successful and widely employed mechanism. The great majority of filter feeders employ ciliated surfaces to produce currents that draw drifting food particles into their mouths. Most filter-feeding invertebrates, such as the tube-dwelling worms and bivalve mollusks, entrap the particulate food in mucus sheets that convey the food into the digestive tract. Filter feeding is characteristic of a sessile way of life, the ciliary currents serving to bring the food to the immobile or slow-moving animal. However, active feeders such as tiny copepod crustaceans and herring are also filter feeders, as are immense baleen whales. The vital importance of one component of the plankton, the diatoms, in supporting a great pyramid of filter feeding animals is stressed by N. J. Berrill*: "A hump-back whale . . . needs a ton of herring in its stomach to feel comfortably full—as many as five thousand individual fish. Each herring, in turn, may well have 6,000 or 7,000 small crustaceans in its own stomach, each of which contains as many as 130,000 diatoms. In other words, some 400 billion yellow-green diatoms sustain a single medium-sized whale for a few hours at most." Filter feeding utilizes the abundance and extravagance of life in the sea.

Filter feeders are as a rule nonselective and omnivorous. Sessile filter feeders take what they can get, having only the options of continuing or ceasing to filter. Active filter feeders, however, such as fish and baleen whales, are much more selective in their feeding.

*Berrill, N. J.: You and the universe, New York, 1958, Dodd, Mead & Co.

FIG. 9-1. Some filter feeders and their feeding mechanisms. **A,** Marine fan-worm (class Polychaeta, phylum Annelida). **B,** Marine clam (class Pelecypoda, phylum Mollusca). **C,** Right whale (class Mammalia, phylum Chordata). **D,** Herring (class Pisces, phylum Chordata). Fan-worms, **A,** have a crown of tentacles. Numerous cilia on the edges of the tentacles draw water (solid arrows) between pinnules where food particles are entrapped in mucus; particles are then carried down a "gutter" in the center of the tentacle to the mouth (broken arrows). Bivalve mollusks, **B,** use their gills as feeding devices as well as for respiration. Water currents created by cilia on the gills carry food particles into the inhalant siphon, and between slits in the gills where they are entangled in a mucus sheet covering the gill surface. Particles are then transported by ciliated food grooves to the mouth (not shown). Arrows indicate direction of water movement. Whalebone whales, **C,** filter out plankton, principally large copepods called "krill," with whalebone, or baleen. Water enters the swimming whale's open mouth by the force of the animal's forward motion and is strained out through the more than 300 horny baleen plates that hang down like a curtain from the roof of the mouth. Krill and other plankters caught in the baleen are periodically wiped off with the huge tongue and swallowed. Herring, **D,** and other filter-feeding fishes use gill rakers that project forward from the gill bars into the pharyngeal cavity to strain off plankters. Herring swim almost constantly, forcing water and suspended food into the mouth; food is strained out by the gill rakers and the water passes on out the gill openings.

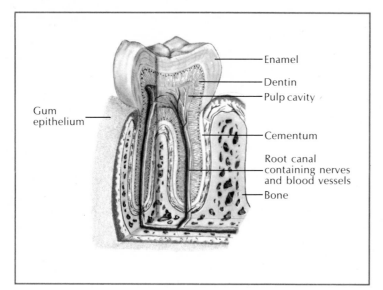

FIG. 9-2. Structure of human molar tooth. Tooth is built of three layers of calcified tissue covering: enamel, which is 98% mineral, and hardest material in the body; dentin, which composes the mass of the tooth, and is about 75% mineral; cementum, which forms a thin covering over the dentin in the root of the tooth, and is very similar to dense bone in composition. Pulp cavity contains loose connective tissue, blood vessels, nerves, and tooth-building cells. Roots of the tooth are anchored to the wall of the socket by a fibrous connective tissue layer called the periodental membrane. (After Netter, F. H.: The Ciba collection of medical illustrations, vol. 3, Summit, N. J., 1959, Ciba Pharmaceutical Products, Inc.)

Feeding on food masses. Some of the most interesting animal adaptations are those that have evolved for procuring and manipulating solid food. Such adaptations, and the animals bearing them, are partly shaped by what the animal eats.

Predators must be able to locate prey, capture it, hold it, and swallow it. Most animals use teeth for this purpose (Fig. 9-2). Although teeth are variable in size, shape, and arrangement, vertebrates, as different as fish and mammals sometimes have remarkably similar tooth arrangements for seizing the prey and cutting it into pieces small enough to swallow (Fig. 9-3). Mammals characteristically have four different types of teeth, each adapted for specific functions. **Incisors** are for biting and cutting; **canines** are designed for seizing, piercing, and tearing; **premolars** and **molars,** at the back of the jaws, are for grinding and crushing (Fig. 9-4). This basic pattern is greatly modified, as the examples in Fig. 9-3 show. Herbivores have suppressed canines but well-developed molars with enamel ridges for grinding. Such teeth are usually high crowned, in contrast to the low-crowned teeth of carnivores. The well-developed, self-sharpening incisors of rodents grow throughout life and must be worn away by gnawing to keep pace with growth. Some teeth have become so highly modified that they are no longer useful for biting or chewing food. An elephant's tusk is a modified upper incisor used for defense, attack, and rooting, whereas the male wild boar has modified canines used as weapons.

Many carnivores among the fishes, amphibians, and reptiles swallow their prey whole. Snakes and some fishes can swallow enormous meals. This, together with the absence of limbs, is associated with some striking feeding adaptations in these groups— recurved teeth for seizing and holding the prey, and distensible jaws and stomachs to accommodate their large and infrequent meals.

Teeth are not vertebrate innovations; biting, scraping, and gnawing devices are common in the invertebrates. Insects, for example, have three pairs of appendages on their heads that serve variously as jaws, teeth, chisels, tongues, or sucking tubes. Usually the first pair is crushing teeth; the second, grasping jaws; and the third, a probing and tasting tongue.

Herbivorous, or plant-eating animals, whether vertebrate or invertebrate, have evolved special devices for crushing and cutting plant material. Despite its abundance on earth, the woody cellulose that encloses plant cells is to many animals an indigestible and useless material; herbivores, however, make use of intestinal microorganisms to digest cellulose, once it is ground up. Certain invertebrates such as snails have rasplike, scraping mouthparts. Insects such as locusts have grinding and cutting mandibles; herbivorous mammals such

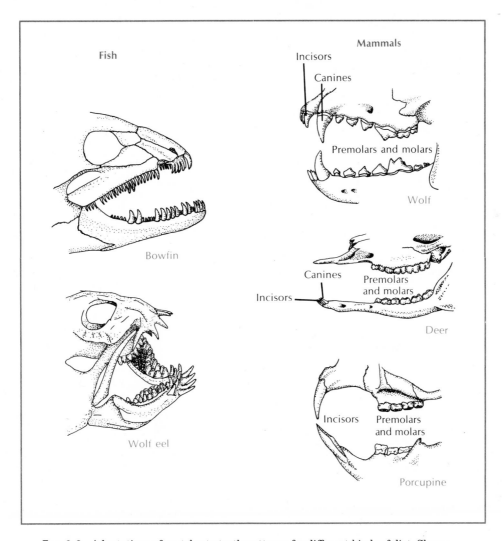

FIG. 9-3. Adaptations of vertebrate tooth patterns for different kinds of diet. Sharp canines of wolf are designed for stabbing, and premolars and molars for cutting rather than grinding. Browsing deer has predominantly grinding teeth; lower incisors and canines bite against a horny pad in the upper jaw. Porcupine has no canines; self-sharpening incisors are used for gnawing. Bowfin uses its many sharp teeth for grasping and holding the prey until it can swallow it whole. Wolf eel uses its incisor-like front teeth for grasping and piercing; prey is then crushed between the powerful jaws with its molarlike teeth. (Wolf, deer, and porcupine drawings modified from Carrington, R.: The mammals, New York, 1968, Life Nature Library.)

as horses and cattle use wide, corrugated molars for grinding. All these mechanisms serve to disrupt the tough cellulose cell wall, to accelerate its digestion by intestinal microorganisms, as well as to release the cell contents for direct enzymatic breakdown.

Feeding on fluids. Fluid feeding is especially characteristic of parasites, but is certainly practiced among free-living forms as well. Most internal parasites (endoparasites) simply absorb the nutrient surrounding them, unwittingly provided by the host. External parasites (ectoparasites) such as leeches, lampreys, parasite crustaceans, and insects use a variety of efficient piercing and sucking mouthparts to feed on blood or other body fluid. Unfortunately for man and other warm-blooded animals, the ubiquitous mosquito excels in its blood-sucking habit. Alighting gently, the mosquito sets about puncturing its prey with an array of six needlelike mouthparts. One of these is used to inject an anticoagulant saliva (responsible for the irritating itch that follows the "bite," and serving as vector for microorganisms causing malaria, yellow fever, encephalitis, and other diseases); another mouthpart is a channel through which the blood is sucked. It is of little comfort that only the female of the species dines on blood. Far less troublesome to man are the free-living butterflies, moths, and aphids that suck up plant fluids with long, tubelike mouthparts.

DIGESTION

In the process of digestion, which means literally "to carry asunder," organic foods are mechanically and chemically broken down into small units for absorption. Animal foods vary enormously, having in common only their organic composition. Even though food solids consist principally of carbohydrates, proteins, and fats, the very components that make up the body of the consumer, these components must nevertheless be reduced to their simplest molecular units before they can be utilized. Each animal reassembles some of these digested units into organic

FIG. 9-4. Human deciduous and permanent teeth. Partly dissected skull of a 5-year-old child, showing milk (deciduous) teeth and permanent teeth. Milk teeth begin to erupt at 6 months and are gradually replaced by the permanent teeth beginning at about 6 years of age. There are twenty deciduous teeth, five on each side of each jaw, and thirty-two permanent teeth, eight on each side of each jaw. These eight are arranged as follows: two incisors, one canine (also called cuspid), two premolars (bicuspids), three molars. The last molar, known as the wisdom tooth, erupts between ages of 17 and 25 or not at all. Upper permanent molars are not seen in this frontal view. (Adapted from Arey, L. B.: Developmental anatomy, Philadelphia, 1965, W. B. Saunders Co.).

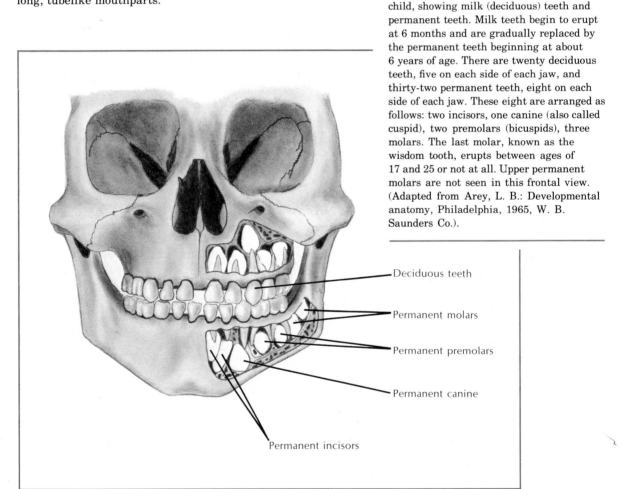

Deciduous teeth

Permanent molars

Permanent premolars

Permanent canine

Permanent incisors

compounds of the animal's own unique pattern. Cannibals enjoy no special metabolic benefit from eating their own kind; they digest their victims just as thoroughly as they do food of another species!

The digestive tract is actually an extension of the outside environment into, or through, the animal. Since most animals eat all manner of organic and inorganic materials, the gut's lining must be something like the protective skin; yet it must be permeable so that foodstuffs can be absorbed. Once in the gut the foods are digested and absorbed as they are slowly moved through it. Movement is either by **cilia** or by **musculature.** In general, the filter feeders that use cilia to feed, such as bivalve mollusks, also use cilia to propel the food through the gut. Animals feeding on bulky foods rely upon well-developed gut musculature. As a rule the gut is lined with two opposing layers of muscle—a longitudinal layer, in which the smooth muscle fibers run parallel with the length of the gut, and a circular layer, in which the muscle fibers embrace the circumference of the gut. This arrangement is ideal for mixing and propelling foods. The most characteristic gut movement is **peristalsis** (Fig. 9-5). In this movement a wave of circular muscle contraction sweeps down the gut for some distance, pushing the food along before it. The peristaltic waves may start at any point and move for variable distances. Also characteristic of the gut are **segmentation** movements that divide and mix the food.

INTRACELLULAR VS. EXTRACELLULAR DIGESTION

Man and other vertebrates and the higher invertebrates digest their food **extracellularly** by secreting digestive juices into the intestinal lumen. There, foodstuffs are enzymatically split into molecular units small enough to be selectively absorbed by the intestinal epithelium, transported by the circulation, and utilized by all body cells. Digestion, then, occurs outside the body's tissues. **Intracellular** digestion is a primitive process typical of the lower invertebrates. This type of digestion is best illustrated by the single-celled Protozoa, which capture food particles by phagocytosis, enclose these particles within food vacuoles, and then digest them. Obviously the big limitation to intracellular digestion is that only small particles of food can be handled. Nevertheless, many multicellular invertebrates have intracellular digestion. Intracellular digestion is typical of filter-feeding marine animals such as brachiopods, rotifers, bivalves, and cephalochordates, as well as the coelenterates and flatworms. In all of these forms the food particle, phagocytized by the cell, is enclosed within a membrane as a food vacuole (Fig. 9-6). Digestive enzymes are then added. The products of digestion, the simple sugars, amino acids, and other molecules, are absorbed into the cell cytoplasm where they may be utilized directly or may be transferred to other cells. The inevitable food wastes are extruded from the cell.

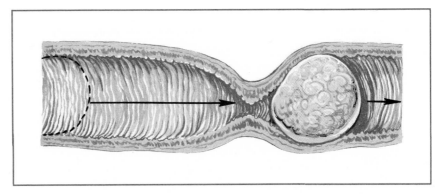

FIG. 9-5. Peristalsis. Food is pushed along before a wave of circular muscle contraction. (From Tuttle, W. W., and Schottelius, B. A.: A textbook of physiology, ed. 16, St. Louis, 1969, The C. V. Mosby Co.)

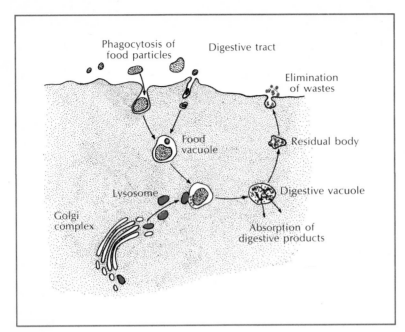

FIG. 9-6. Intracellular digestion. Lysosomes containing digestive enzymes (lysozymes) are produced within the cell, possibly by the Golgi complex. Lysosomes fuse with food vacuoles and release enzymes that digest the enclosed food. Usable products of digestion are absorbed into the cytoplasm and indigestible wastes are expelled to the outside.

It is believed that cellular enzymes are packaged into membrane-bound vacuoles called **lysosomes** (Fig. 9-6). These somehow join with, and discharge their enzymes into, the food vacuoles. All cells seem to contain lysosomes, even those of higher animals that do not practice intracellular digestion. Lysosomes have been called "suicide-bags" because they rupture spontaneously in dying or useless cells, digesting the cell contents. Lysosomes also play a role in the lives of healthy cells in cleaning up residues left by growth processes.

It is probable that the obvious limitations of intracellular digestion were responsible for shaping the evolution of extracellular digestion. Extracellular digestion offers several advantages: bulky foods may be ingested; the digestive tract can be smaller, more specialized, and more efficient; and food wastes are more easily discarded. Only with extracellular digestion could the enormous variation in feeding methods of the higher animals have evolved.

DIGESTION IN VERTEBRATES

The vertebrate digestive plan is similar to that of the higher invertebrates. Both have a highly differentiated alimentary canal with devices for increasing the surface area, such as increased length, inside folds, and diverticula. The more primitive fishes (lampreys and sharks) have longitudinal or spiral folds in their intestine. Higher vertebrates have developed elaborate folds and small fingerlike projections (**villi**). Also, the electron microscope reveals that each cell lining the intestinal cavity is bordered by hundreds of short, delicate processes called **microvilli** (Fig. 9-7). These processes, together with larger villi and intestinal folds, may increase the internal surface of the intestine more than a million times compared to a smooth cylinder of the same diameter. The absorption of food molecules is enormously facilitated as a result.

The herbivorous mammals have a number of interesting adaptations for dealing with their massive diet of plant food. Cellulose, the structural carbohydrate of plants, is a potentially nutritious foodstuff, being comprised of long chains of glucose. However, the glucose molecules in cellulose are linked by a type of chemical bond that few enzymes can attack. No vertebrates synthesize cellulose-splitting enzymes. Instead the herbivorous vertebrates harbor a microflora of anaerobic bacteria in huge fermen-

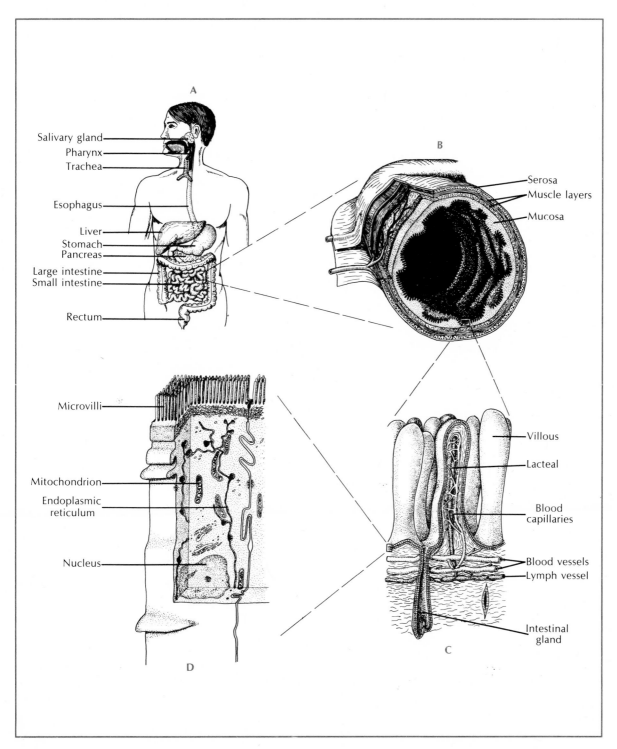

FIG. 9-7. A, Digestive system of man. **B,** Portion of small intestine. **C,** Portion of mucosa lining of intestine, showing fingerlike villi. **D,** Optical section of single lining cell, as shown by electron microscope.

tation chambers of the gut. These bacteria break down the cellulose, releasing a variety of fatty acids, sugars, and starches that the host animal can absorb and utilize. In some herbivores, such as horse and rabbit, the gut has a capacious sidepocket, or diverticulum, called a **cecum,** which serves as a fermentation chamber and absorptive area. Hares and rabbits often eat their fecal pellets, giving the food a second pass through the fermenting action of the intestinal bacteria. The ruminants (cattle, sheep, antelope, deer, giraffe, and other hooved mammals) have a huge four-chambered stomach. When a ruminant feeds, grass passes down the esophagus to the first two chambers, where it is broken down by the rich microflora and then formed into small balls of cud. At its leisure the ruminant returns the cud to its mouth where it is deliberately chewed at length to crush the fiber. Swallowed again, the food returns to the first chamber where it is digested by the cellulolytic bacteria. Finally, the pulp passes to the third chamber and then to the "true" stomach (fourth chamber) where proteolytic enzymes are secreted and normal digestion takes place. Herbivores in general have large and long digestive tracts and must eat a large amount of plant food to survive. A large African elephant weighing 6 tons must consume between 300 and 400 pounds of rough fodder each day to obtain sufficient nourishment for life. In contrast, carnivores feed intermittently on a concentrated, rapidly digestible food and consequently do not require so large a digestive tract.

DIGESTION IN MAN, AN OMNIVORE

We will focus now on man's digestive system because it reveals principles representative of many other vertebrates and invertebrates and because of our natural interest in our own body.

Anatomy of man's digestive system. The structural plan of the digestive system of man is shown in Fig. 9-7. The **mouth** is provided with teeth and a tongue for grasping, masticating, manipulating, and swallowing the food. Three pairs of salivary glands lubricate the food and, in man at least, perform limited digestion. In man and other mammals two sets of teeth are formed during life—the temporary, or "milk," teeth (also

called deciduous teeth) and the permanent teeth (Fig. 9-4).

The **pharynx** is the throat cavity that serves for the passage of food. It is actually a complex reception chamber, receiving openings from (1) the nasal cavity, (2) the mouth, (3) the middle ear by way of two eustachian tubes, (4) the esophagus, and (5) the trachea via the glottis.

The **esophagus** is a muscular tube connecting pharynx and stomach. It opens into the stomach by the **cardiac** opening.

The **stomach** is an enlargement of the gut between esophagus and intestine. In man it is divided into the **cardiac** region (adjacent to esophagus), **fundus** (central region), and **pyloric** region (adjacent to intestine). The stomach is principally a storage organ but aids in digestion.

The **small intestine** is the principal digestive and absorptive area of the gut. It is divided grossly into three regions—**duodenum, jejunum,** and **ilium.** Two large digestive glands, the **liver** and **pancreas,** empty into the duodenum by the **common bile duct.**

The **large intestine (colon)** in man is divided into **ascending, transverse,** and **descending** portions, with the posterior end terminating in the **rectum** and **anus.** At the junction of the large and small intestines is the **colic cecum** and its vestigial **vermiform appendix.** The large intestine lacks villi but contains glands for lubrication.

The stomach, small intestine, and large intestine are all suspended by **mesenteries,** thin sheets of tissue that are modified from the **peritoneum,** or lining, of the coelom and the abdominal organs. Organs such as liver, spleen, and pancreas are also held in place by mesenteries, which carry blood and lymph vessels as well as nerves to the various abdominal organs.

Action of digestive enzymes. We have already pointed out that digestion involves both mechanical and chemical alterations of food. Mechanical processes of cutting and grinding by teeth and muscular mixing by the intestinal tract are important in digestion. However, the reduction of foods to small absorbable units relys principally on chemical breakdown by **enzymes.** Enzymes, the highly specific organic catalysts essential to the orderly progression of virtually all life processes, have been discussed earlier (p. 55).

It is well to state that although digestive enzymes are probably the best known and most studied of all enzymes, they represent but a small fraction of the numerous, perhaps thousands, of enzymes that ultimately regulate all processes in the body. The digestive enzymes are **hydrolytic** enzymes (**hydrolases),** so called because food molecules are split by the process of **hydrolysis,** that is, the breaking of a chemical bond by adding the components of water across it:

$$R - R + H_2O \xrightarrow[\text{enzyme}]{\text{digestive}} R - OH + H - R$$

In this general enzymatic reaction, R — R represents a food molecule that is split into two products, R — OH and R — H. Usually these reaction products must in turn be split repeatedly before the original molecule has been reduced to its numerous subunits. Proteins, for example, are composed of hundreds, or even thousands, of interlinked amino acids, which must be completely separated before the individual amino acids can be absorbed. Similarly, carbohydrates must be reduced to simple sugars. Fats (lipids) are reduced to molecules of glycerol and fatty acids, although some fats, unlike proteins and carbohydrates, may be absorbed without being completely hydrolyzed first. There are specific enzymes for each class of organic compounds. These enzymes are located in various regions of the alimentary canal in a sort of "enzyme chain," in which one enzyme may complete what another has started, the product moving along posteriorly for still further hydrolysis.

Digestion in the mouth. In the mouth, food is broken down mechanically by the teeth and is moistened with saliva from the salivary glands. In addition to **mucin,** which helps to lubricate the food for swallowing, saliva contains the enzyme **amylase.** Salivary amylase is a carbohydrate-splitting enzyme that begins the hydrolysis of plant and animal starches (Fig. 9-8). Starches, as we have seen previously (p. 17), are long polymers of glucose. Salivary amylase does not completely hydrolyse starch, but breaks it down mostly into 2-glucose fragments called **maltose.** Some free glucose as well as longer fragments of starch are also produced. When the food mass (bolus) is swallowed, salivary amylase

continues to act for some time, digesting perhaps half of the starch before the enzyme is inactivated by the acid environment of the stomach. Further starch digestion resumes beyond the stomach in the intestine.

Swallowing. Swallowing is a reflex process involving both voluntary and involuntary components. Swallowing begins with the tongue pushing the moistened food bolus toward the pharynx. The nasal cavity is reflexly closed by raising the soft palate. As the food slides into the pharynx, the epiglottis is tipped down over the windpipe, nearly closing it. Some particles of food may enter the opening of the windpipe, but are prevented from going further by contraction of laryngeal muscles. Once in the esophagus, the bolus is forced smoothly toward the stomach by peristaltic contraction of the esophageal muscles.

Digestion in the stomach. When food reaches the stomach, the **cardiac sphincter** opens reflexly to allow entry of the food, then closes to prevent regurgitation back into the esophagus. The stomach is a combination storage, mixing, digestion, and release center. Large peristaltic waves pass over the stomach at the rate of about three each minute; churning is most vigorous at the intestinal end where food is steadily released into the duodenum. About 3 liters of **gastric juice** are secreted each day by deep, tubular glands in the stomach wall. Two types of cells line these glands: (1) **chief cells** secrete an enzyme precursor called **pepsinogen** and (2) **parietal cells** secrete **hydrochloric acid.** Pepsinogen is an inactive form of enzyme that is converted into the active enzyme **pepsin** by hydrochloric acid and by other pepsin already present in the stomach. Pepsin is a **protease** (protein-splitting enzyme) that acts only in an acid medium— pH 1.6 to 2.4. It is a highly specific enzyme that splits large proteins by preferentially breaking down certain peptide bonds scattered along the peptide chain of the protein molecule. Although pepsin, because of its specificity, cannot completely degrade proteins, it effectively breaks them up into a number of small polypeptides. Protein digestion is completed in the intestine by other proteases that can together split all peptide bonds.

Rennin is another enzyme found in the stomachs of the suckling newborn of many mammals, al-

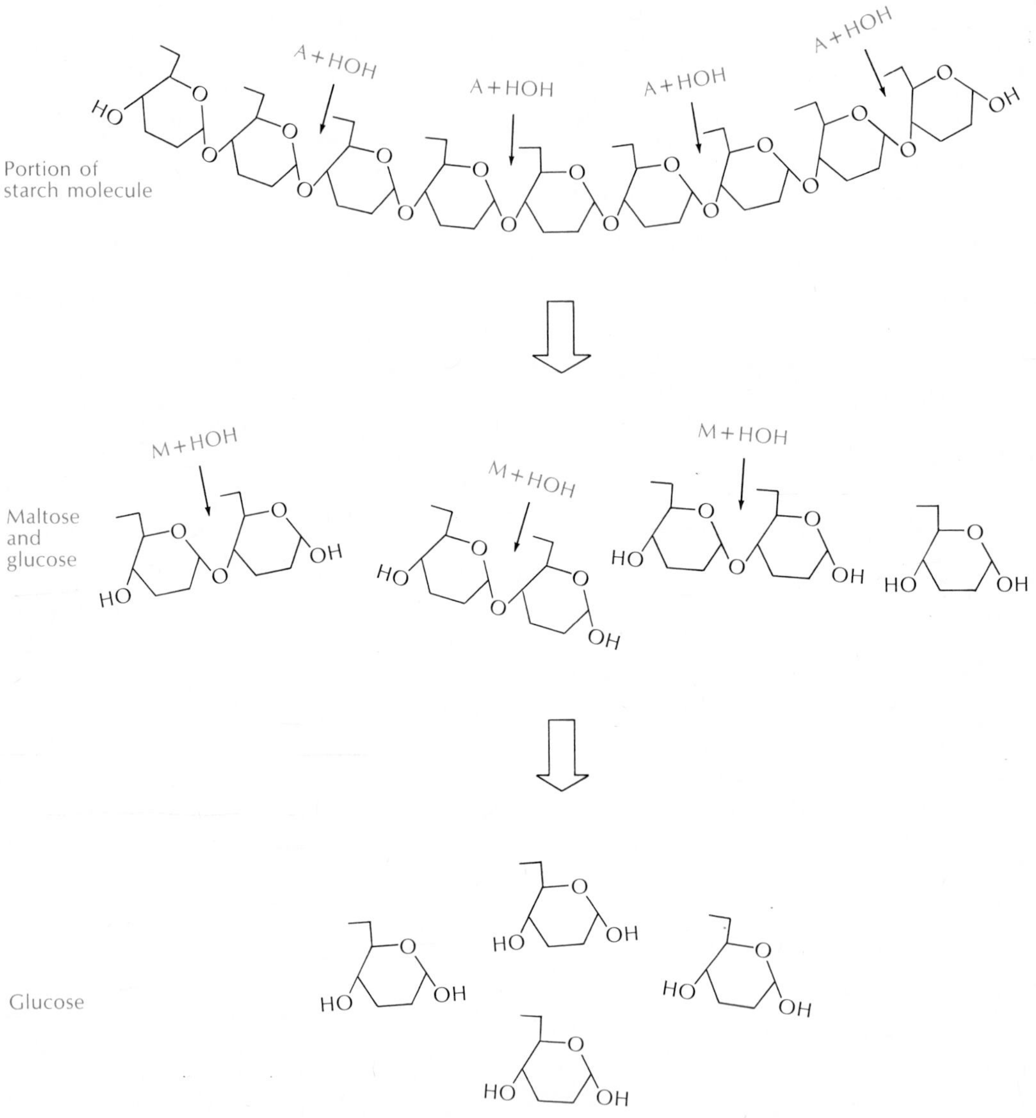

FIG. 9-8. Digestion (hydrolysis) of starch. Long chains of glucose molecules, linked together through oxygen, are first cleaved into disaccharide residues (maltose) by the salivary enzyme amylase. Some glucose may also be split off at the ends of starch chains. The intestinal enzyme maltase then completes the hydrolysis by cleaving the maltose molecules into glucose. A molecule of water is inserted into each enzymatically split bond.

A + HOH = Action of amylase

M + HOH = Action of maltase

though not of man. Rennin is a milk-curdling enzyme that transforms the proteins of milk into a finely flocculent form that is more readily attacked by pepsin. Rennin extracted from the stomachs of calves is used in cheese-making. Human infants, lacking rennin, digest milk proteins with acidic pepsin, the same as adults do.

The unique ability of the stomach to secrete a strong acid is still an unsolved problem in biology, in part because it has not been possible to collect pure parietal cell secretion or to isolate these cells in culture for study. It is well known that acid solutions can readily destroy organic matter. Since the stomach contains not only a strong acid but also a powerful proteolytic enzyme, it seems remarkable that the stomach mucosa is not digested by its own secretions. That it is not is due to another protective gastric secretion mucin, a highly viscous organic compound that coats and protects the mucosa from both chemical and mechanical injury. Sometimes, however, the protective mucus coating fails, allowing the gastric juices to begin digesting the stomach. The result is a peptic ulcer. We should note that despite the popular misconception that "acid stomach" is unhealthy, a notion that is carefully nourished by the makers of patent medicine, stomach acidity is normal and essential.

The secretion of the gastric juices is intermittent. Although a small volume of gastric juice is secreted continuously, even during prolonged periods of starvation, secretion is normally increased by the sight and smell of food, by the presence of food in the stomach, and by emotional states such as anger and hostility. The most unique and classical investigation in the field of digestion was made by the U. S. Army surgeon, William Beaumont, during the years 1825 to 1833. His subject was a young, hard-living French Canadian voyageur, named Alexis St. Martin, who in 1822 had accidentally shot himself in the abdomen with a musket, the blast "blowing off integuments and muscles of the size of a man's hand, fracturing and carrying away the anterior half of the sixth rib, fracturing the fifth, lacerating the lower portion of the left lobe of the lung and the diaphragm, and perforating the stomach." Miraculously, the wound healed, but a permanent opening, or fistula, was formed which permitted Beaumont to see directly into the stomach. St.

Martin became a permanent, although temperamental, patient in Beaumont's care, which included food and housing. Over a period of eight years, Beaumont was able to observe and record how the lining of the stomach changed under different psychic and physiologic conditions, how foods changed during digestion, the effect of emotional states on stomach motility, and many other facts about the digestive processes of his famous patient.

Digestion in the small intestine. The major part of digestion occurs in the small intestine. Three secretions are poured into this region—**pancreatic juice, intestinal juice,** and **bile.** All of these secretions have a high bicarbonate content, especially the pancreatic juice, which effectively neutralizes the gastric acid, raising the pH of the liquefied food mass, now called **chyme,** from 1.5 to 7 as it enters the duodenum. This change in pH is essential because all the intestinal enzymes are effective only in a neutral or slightly alkaline medium.

About 2 liters of **pancreatic juice** are secreted each day. The pancreatic juice contains several enzymes of major importance in digestion. Two powerful proteases, **trypsin** and **chymotrypsin,** are secreted in inactive form as **trypsinogen** and **chymotrypsinogen.** Trypsinogen is activated in the duodenum by **enterokinase,** an enzyme present in the intestinal juice. Chymotrypsinogen is activated by trypsin. These two proteases continue the enzymatic digestion of proteins begun by pepsin, but now inactivated by the alkalinity of the intestine. Trypsin and chymotrypsin, like pepsin, are highly specific proteases that split apart peptide bonds deep inside the protein molecule. Pancreatic juice also contains **carboxypeptidase** that splits amino acids off the ends of polypeptides, **pancreatic lipase** that hydrolyzes fats into fatty acids and glycerol, and **pancreatic amylase** that is a starch-splitting enzyme identical to salivary amylase in its action.

Intestinal juice from the glands of the mucosal lining furnishes several enzymes. **Aminopeptidase** splits off terminal amino acids from polypeptides; its action is similar to the pancreatic enzyme carboxypeptidase. Three other enzymes are present that complete the hydrolysis of carbohydrates: **maltase** converts maltose to glucose (Fig. 9-8); **sucrase** splits sucrose to glucose and fructose; and **lactase**

breaks down lactose (milk sugar) into glucose and galactose.

Bile is secreted by the cells of the **liver** into the **bile duct,** which drains into the upper intestine (duodenum). Between meals the bile is collected into the **gallbladder,** an expansible storage sac that releases the bile when stimulated by the presence of fatty food in the duodenum. Bile contains no enzymes. It is made up of water, bile salts, and pigments. The bile salts (sodium taurocholate and sodium glycocholate) are essential for the complete absorption of fats, which, because of their tendency to remain in large, water-resistant globules, are especially resistant to enzymatic digestion. **Bile salts** reduce the surface tension of fats, so that they are broken up into small droplets by the churning movements of the intestine. This greatly increases the total surface exposure of fat particles, giving the fat-splitting lipases a chance to reduce them. The characteristic golden yellow of bile is produced by the **bile pigments** that are breakdown products of hemoglobin from worn-out red blood cells. The bile pigments also color the feces.

It is well to emphasize the great versatility of the liver. Bile production is only one of the liver's many functions, which include the following: storehouse for glycogen, production center for the plasma proteins, site of protein synthesis and detoxification of protein wastes, destruction of worn-out red blood cells, center for metabolism of fat and carbohydrate, and many others.

Digestion in the large intestine. The liquefied material, now called **chyle,** reaching the large intestine, or **colon,** is low in nutrients, since most important food materials have already been absorbed into the bloodstream from the small intestine. The main function of the colon is the absorption of water and some minerals from the intestinal chyle that enters. In removing more than half the water from the chyle, the colon forms a semisolid feces consisting of undigested food residue, bile pigments, secreted heavy metals, and bacteria. The feces are eliminated from the rectum by the process of **defecation,** a coordinated muscular action that is part voluntary and part involuntary.

The colon contains enormous numbers of bacteria entering the sterile colon of the newborn infant early in life. In the adult about one third of the dry weight of feces is bacteria; these include both harmless bacilli as well as cocci that can cause serious illness if they should escape into the abdomen or bloodstream. Normally the body's defenses prevent invasion of such bacteria. The bacteria break down organic wastes in the feces and provide some nutritional benefit by synthesizing certain vitamins (biotin and vitamin K), which are absorbed by the body.

Absorption. Most digested foodstuffs are absorbed from the small intestine, where the numerous finger-shaped **villi** provide an enormous surface area through which materials can pass from intestinal lumen into the circulation. Little food is absorbed in the stomach because digestion is still incomplete and because of the limited surface exposure. Some materials, however, such as drugs and alcohol are absorbed in part there, which explains their rapid action.

The villi (Fig. 9-7) contain a network of blood and lymph capillaries. The absorbable food products (amino acids, simple sugars, fatty acids, glycerol, and triglycerides as well as minerals, vitamins, and water) pass first across the epithelial cells of the intestinal mucosa and then into either the blood capillaries or the lymph vessel. Small molecules can enter either system, but since blood flow is several hundred times greater than lymph flow, it is unlikely that the lymph system carries much absorbed material out of the intestine. However, any materials that do enter the lymph system will eventually get into the blood by way of the thoracic duct.

Carbohydrates are absorbed almost exclusively as simple sugars (for example, glucose, fructose, and galactose) because the intestine is virtually impermeable to polysaccharides. Proteins, too, are absorbed principally as their subunits, amino acids, although it is believed that very small amounts of small proteins or protein fragments may sometimes be absorbed. Simple sugars and amino acids are transferred across the intestinal epithelium by both passive and active processes.

Immediately after a meal these materials are in such high concentration in the gut that they readily diffuse into the blood, where their concentration is initially lower. However, if absorption were passive only, we would expect transfer to cease as soon as the concentrations of a substance became equal on both sides of the intestinal epithelium. This would leave much valuable foodstuff to be lost in the feces. In fact, very little is lost because passive transfer is supplemented by an **active transport** mechanism located in the epithelial cells, which pick up the food molecules and transfer them into the blood. Materials are thus moved **against** their concentrated gradient, a process that requires the expenditure of energy. Although not all food products are actively transported, those which are, such as glucose, galactose, and most of the amino acids, are handled by transport mechanisms that are specific for that molecule.

NUTRITIONAL REQUIREMENTS

The food of animals must include **carbohydrates, proteins, fats, water, mineral salts,** and **vitamins.** Carbohydrates and fats are required as fuels for energy demands of the body and for the synthesis of various substances and structures. Proteins, or actually the amino acids of which they are composed, are needed for the synthesis of the body's specific proteins and other nitrogen-containing compounds. Water is required as the solvent for the body's chemistry and as the major component of all the body fluids. The inorganic salts are required as the anions and cations of body fluids and tissues and form important structural and physiologic components throughout the body. The vitamins are accessory food factors that are frequently built into the structure of many of the enzymes of the body.

All animals require these broad classes of nutrients, although there are differences in the amounts and kinds of food required. The student should note that of the basic food classes listed above, some nutrients are used principally as fuels (carbohydrates and lipids), whereas others are required principally as structural and functional components (proteins, minerals, and vitamins). Any of the basic foods (proteins, carbohydrates, fats) can serve as fuel to supply energy requirements, but, conversely, no animal can thrive on fuels alone. A **balanced diet** must satisfy all metabolic requirements of the body—requirements for energy, growth, maintenance, reproduction, and physiologic regulation.

The recognition many years ago that many diseases of man and his domesticated animals were caused by, or associated with, dietary deficiencies led biologists to search for specific nutrients that would prevent such diseases. These studies eventually yielded a list of **"essential" nutrients** for man and other animal species studied. The essential nutrients are those that are needed for normal growth and maintenance and which **must** be supplied in the diet. In other words, it is "essential" that these nutrients be in the diet because the animal cannot synthesize them from other dietary constituents. For man more than twenty organic compounds (amino acids and vitamins) and more than ten elements have been established as essential (Table 9-1). Considering that the body contains thousands of different organic compounds, the list in Table 9-1 is remarkably short. Animal cells have marvelous powers of synthesis enabling them to build compounds of enormous variety and complexity from a small, select group of raw materials.

In the average diet of Americans and Canadians, about 50% of the total calories (energy content)

TABLE 9-1. Nutrients required by man*

Amino acids	Elements	Vitamins
Established as essential		
Isoleucine	Calcium	Ascorbic acid (C)
Leucine	Chlorine	Choline
Lysine	Copper	Folic acid
Methionine	Iodine	Niacin
Phenylalanine	Iron	Pyridoxine (B_6)
Threonine	Magnesium	Riboflavin (B_2)
Tryptophan	Manganese	Thiamine (B_1)
Valine	Phosphorus	Vitamin B_{12}
	Potassium	Vitamins A, D, E,
	Sodium	and K
Probably essential		
Arginine	Fluorine	Biotin
Histidine	Molybdenum	Pantothenic acid
	Selenium	Polyunsaturated
	Zinc	fatty acids

*From White, A., Handler, P., and Smith, E. L.: Principles of biochemistry, New York, 1968, McGraw-Hill Book Co., Inc.

comes from carbohydrates and 40% comes from lipids. Proteins, essential as they are for structural needs, supply only a little more than 10% of the total calories of the average North American's diet. Carbohydrates are widely consumed because they are more abundant and cheaper than proteins or lipids. Actually, man and many other animals can subsist on diets devoid of carbohydrates, provided sufficient total calories and the essential nutrients are present. Eskimos, for example, live on a diet that is high in fat and protein and very low in carbohydrate.

Lipids are needed principally to provide energy. In recent years much interest and research has been devoted to lipids in our diets because of the association between fatty diets and the disease **arteriosclerosis** (hardening and narrowing of the arteries). The matter is complex, but evidence suggests that arteriosclerosis may occur when the diet is high in saturated lipids but low in polyunsaturated lipids. For unknown reasons such diets, which are typical of middle-class and affluent North Americans, promote a high blood level of cholesterol, which may deposit in platelike formations in the lining of the major arteries. For this reason the polyunsaturated fatty acids are often considered essential nutrients for man. Generally speaking, animal fat is more saturated, whereas fat from plants is more unsaturated.

Proteins are expensive foods and restricted in the diet. Proteins, of course, are themselves not the essential nutrients, but rather they contain essential amino acids. Of the twenty amino acids commonly found in proteins, eight and possibly ten are essential to man (Table 9-1). The rest can be synthesized. It must be kept in mind that the terms "essential" and "nonessential" relate only to dietary requirements and to which amino acids can and cannot be synthesized by the body. All twenty amino acids are essential for the various cellular functions of the body. In fact, some of the so-called nonessential amino acids participate in more crucial metabolic activities than the essential amino acids. In general, animal proteins have more of the essential amino acids than do proteins of plant origin. An adult man would require about 67 grams of whole wheat bread each day to meet his amino acid requirement, but only 19 grams of beefsteak to meet these same requirements. This is because proteins are less concentrated in plants and contain relatively less of the essential amino acids. We should note, however, that beefsteak, because it is also high in saturated lipids, is a much less desirable food for man than fish or chicken.

Because animal proteins are so nutritious, they are in great demand by all countries. North Americans eat far more animal proteins than do Asians and Africans; on the average a North American eats 66 grams of animal protein a day, supplemented by milk, eggs, and cereals. In the Middle East, the individual consumption of protein is 14 grams, in Africa 11 grams, and in Asia 8 grams. Protein deficiency is a vital factor in the 10,000 deaths estimated by the United Nations as occurring daily from malnutrition. A protein and calorie deficiency disease of children, **kwashiorkor,** is the world's major health problem today and is growing more serious daily. The name kwashiorkor literally means "displaced child" because it occurs especially in nursing infants displaced from the breast by a newborn sibling. The disease is characterized by retarded growth, anemia, liver and pancreas degeneration, renal lesions, acute diarrhea, and a mortality of 30% to 90%. Because animal proteins are relatively scarce and expensive, there is presently a great effort by scientists to find cheap, plentiful sources of plant proteins. With the world's population expected to double to over 7 billion in just thirty years, at the present unchecked rate of growth, the search for protein takes on a desperate urgency. It is altogether fitting that the 1970 Nobel Peace Prize should go to the man, Dr. Norman Borlaug, who developed a dwarf wheat variety that has vastly increased the wheat yield in India, West Pakistan, and Mexico, producing what is called the "green revolution." However, Dr. Borlaug himself emphasizes that the green revolution only defers for a few years the mass famine that seems inevitable unless the human population is stabilized.

Vitamins. Vitamins are relatively simple organic compounds that are required in small amounts in the diet for specific cellular functions. They are not sources of energy, but are often associated with the activity of important enzymes that have vital metabolic roles. Plants and many microorganisms synthesize all the organic compounds they need; animals, however, have lost certain synthetic abilities during their long evolution and depend ultimately on plants to supply these compounds. Vitamins, therefore, represent synthetic gaps in the metabolic machinery of animals. We have seen that several amino acids are also dietary essentials. These are not considered vitamins, however, because they usually enter into the actual **structure** of tissues or proteinaceous tissue secretions. Vitamins, on the other hand, are essential **functional** components of enzyme catalytic systems. Nevertheless, the distinction between certain vitamins (A, D, E, and K) and other dietary essentials not classified as vitamins is not as clear as it was once thought to be.

Vitamins are usually classified as fat-soluble (soluble in fat solvents such as ether) or water-soluble. The water-soluble ones include the B complex and vitamin C. The family of B vitamins, so grouped because the original B vitamin was subsequently found to consist of several distinct molecules, tends to be found together in nature. Almost all animals, vertebrate and invertebrate, require the B vitamins; they are "universal" vitamins. The dietary need for vitamin C and the fat-soluble vitamins A, D, E, and K tends to be restricted to the vertebrates, although some are required by certain invertebrates.

10

Respiration

The energy bound up in food must be released by oxidative processes. As oxygen is used by the body cells, carbon dioxide is produced; this process is called **respiration.** Most animals are **aerobic,** meaning that they require and receive the necessary oxygen directly from their environment. A few animals, called **anaerobic,** are able to live in the absence of oxygen. Forms, such as worms and arthropods dwelling in the oxygen-depleted mud of lakes and parasites living anaerobically in the intestine, derive the necessary oxygen from the metabolism of carbohydrates and fats. As we noted earlier, however, anaerobic metabolism often occurs in the muscles of basically aerobic animals during vigorous muscle contraction.

EVOLUTION OF RESPIRATORY ORGANS

Small aquatic animals such as the one-celled protozoans obtain what oxygen they need by direct diffusion from the environment. Carbon dioxide, the gaseous waste of metabolism, is also lost by diffusion to the environment. Such a simple solution to the problem of gas exchange is really only possible for animals of very small size (less than 1 mm. in diameter) or those having very low rates of metabolism. As animals became larger and evolved a waterproof covering, specialized devices such as lungs and gills were developed that greatly increased the effective surface for gas exchange. But because gases diffuse so slowly through protoplasm, a circulatory system was necessary to distribute the gases to and from the deep tissues of the body. Even these adaptations were inadequate for advanced animals with their high rates of cellular respiration. The solubility of oxygen in the blood plasma is so low that plasma alone could not carry enough to satisfy metabolic demands. So special oxygen-transporting blood proteins such as hemoglobin evolved, greatly increasing the oxygen-carrying capacity of the blood. Thus, what began as a simple and easily satisfied requirement led to the evolution of several complex and essential respiratory and circulatory adaptations.

PROBLEMS OF AQUATIC AND AERIAL BREATHING

How an animal respires is largely determined by the kind of environment the animal is living in. The two great arenas of animal evolution—water and land—are vastly different in their physical characteristics. The most obvious difference is that air contains far more oxygen—at least twenty times more—than does water. Atmospheric air contains oxygen (about

21%), nitrogen (about 79%), carbon dioxide (0.03%), a variable amount of water vapor, and very small amounts of inert gases (helium, argon, neon, etc.). These gases are variably soluble in water. The amount of oxygen dissolved depends on the concentration of oxygen in the air and on the water temperature. Water at 5° C. fully saturated with air contains about 9 ml. of oxygen per liter. (Note that by comparison, air contains about 210 ml. of oxygen per liter.) The solubility of oxygen in water decreases as the temperature rises. For example, water at 15° C. contains about 7 ml. of oxygen per liter, and at 35° C., only 5 ml. of oxygen per liter. The relatively low concentration of oxygen dissolved in water is the greatest respiratory problem facing aquatic animals. Unfortunately it is not the only one. Oxygen diffuses much more slowly in water than in air, and water is much denser and more viscous than air. All of this means that successful aquatic animals must have evolved very efficient ways of removing oxygen from water. Yet even the most advanced fishes with highly efficient gills and pumping mechanisms may use as much as 20% of their energy just extracting oxygen from water. By comparison, a mammal uses only 1% to 2% of its resting metabolism to breathe.

Despite these difficulties life in water does offer certain advantages. It is essential that respiratory surfaces be kept thin and always wet to allow diffusion of gases between the environment and the underlying circulation. This is hardly a problem for aquatic animals, immersed as they are in water, but it is a very real problem for air breathers. To keep the respiratory membranes moist and protected from injury, air breathers have in general developed invaginations of the body surface and then added pumping mechanisms to move air in and out. The lung is the best example of a successful solution to breathing on land. In general, **evaginations** of the body surface, such as gills, are most suitable for aquatic respiration and **invaginations,** such as lungs, are best for air breathing. We will now consider the specific kinds of respiratory organs employed by animals.

Cutaneous respiration. Protozoa, sponges, coelenterates, and many worms respire by direct diffusion of gases between the organism and the environment. We have noted that this kind of in-

tegumentary respiration is not adequate when the mass of living protoplasm exceeds about 1 mm. in diameter. But by greatly increasing the surface of the body relative to the mass, many multicellular animals respire in this way. Integumentary respiration frequently supplements gill or lung breathing in larger animals such as amphibians and fish. For example, an eel can exchange 60% of its oxygen and carbon dioxide through its highly vascular skin. During their winter hibernation, frogs exchange all their respiratory gases through the skin while submerged in ponds or springs.

Gills. Gills are unquestionably the most effective respiratory device for life in water. Gills may be simple **external** extensions of the body surface, such as the **dermal branchiae** of starfish or the **branchial tufts** of marine worms and aquatic amphibians. Most efficient are the **internal** gills of fishes and arthropods. Fish gills are composed of thin filaments covered with a thin epidermal membrane that is folded repeatedly into platelike **lamellae** (Fig. 10-1). These are richly supplied with blood vessels. The gills are located inside the pharyngeal cavity and covered with a movable flap, the **operculum.** This arrangement provides excellent protection to the delicate gill filaments, streamlines the body, and makes possible a pumping system for moving water through the mouth, across the gills, and out the operculum. Instead of opercular flaps as in bony fishes, the elasmobranches have a series of **gill slits** out of which the water flows. In both elasmobranchs and bony fishes the branchial mechanism is arranged to pump water continuously and smoothly over the gills, even though to an observer it appears that fish breathing is pulsatile. The flow of water is opposite to the direction of blood flow (countercurrent flow), the best arrangement for extracting the greatest possible amount of oxygen from the water. Some bony fish can remove as much as 85% of the oxygen from the water passing over their gills. Very active fish, such as herring and mackerel, can obtain sufficient water for their high oxygen demands only by continually swimming forward to force water into the open mouth and across the gills. Such fish will be asphyxiated if placed in an aquarium that restricts free swimming movements, even though the water is saturated with oxygen.

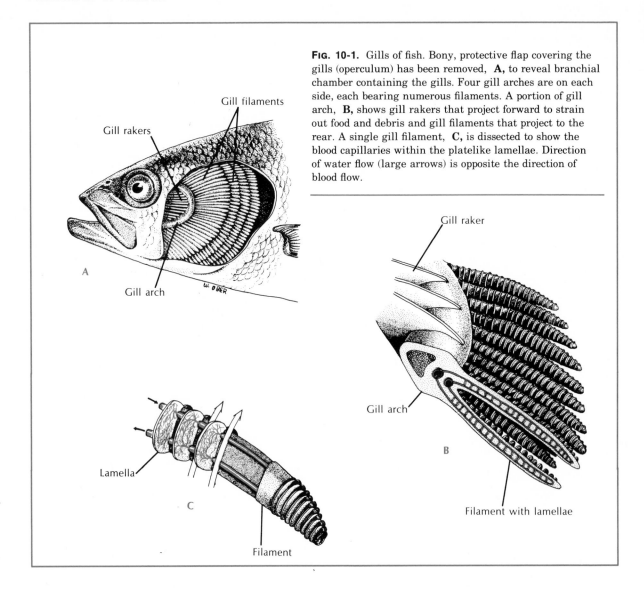

FIG. 10-1. Gills of fish. Bony, protective flap covering the gills (operculum) has been removed, **A,** to reveal branchial chamber containing the gills. Four gill arches are on each side, each bearing numerous filaments. A portion of gill arch, **B,** shows gill rakers that project forward to strain out food and debris and gill filaments that project to the rear. A single gill filament, **C,** is dissected to show the blood capillaries within the platelike lamellae. Direction of water flow (large arrows) is opposite the direction of blood flow.

Lungs. Gills are unsuitable for life in air because when removed from the buoying water medium, the gill filaments collapse and stick together; a fish out of water rapidly asphyxiates despite the abundance of oxygen around it. Consequently air-breathing vertebrates possess lungs, highly vascularized internal cavities. Lungs of a sort are found in certain invertebrates (pulmonate snails, scorpions, some spiders, some small crustaceans) but these structures cannot be ventilated and consequently are not very efficient. Lungs that can be ventilated effi-

ciently are characteristic of the terrestrial vertebrates. The most primitive vertebrate lungs are those of lungfishes (**Dipnoi**), which use them to supplement, or even replace, gill respiration during periods of drought. Although of simple construction, the lungfish lung is supplied with a capillary network in its largely unfurrowed walls, a tubelike connection to the pharynx, and a primitive ventilating system for moving air in and out of the lung. Amphibians also have simple baglike lungs, but in higher forms the inner surface area is vastly in-

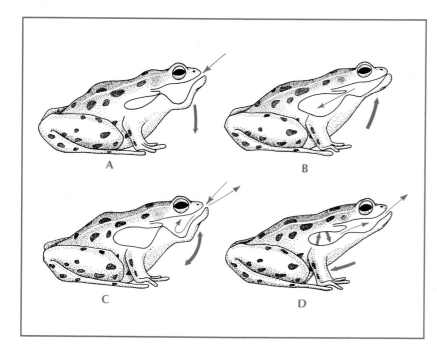

FIG. 10-2. Breathing in frog. The frog, a positive-pressure breather, must fill his lungs by forcing air into them. After filling the lungs **A** and **B**, the frog rhythmically ventilates the mouth cavity for a short period, **C**. Finally the lung is emptied, **D**, and the cycle repeated. (Adapted from Gordon, M. S., et al.: Animal function: principles and adaptations, New York, 1968, The Macmillan Co.)

creased by numerous lobulations and folds. This increase is greatest in the mammalian lung, which is complexly divided into many millions of small sacs **(alveoli)**, each veiled by a rich vascular network. It has been estimated that the human lungs have a total surface area of from 50 to 90 m.^{2}—fifty times the area of the skin surface—and contain 1,000 miles of capillaries.

Getting air into and out of lungs is accomplished in several ways by the different vertebrate groups. Frogs force air into the lungs by first lowering the floor of the mouth to draw air into the mouth through the external nares (nostrils); then by closing the nares and raising the floor of the mouth, air is driven into the lungs (Fig. 10-2, *A* and *B*). Much of the time, however, frogs rhythmically ventilate only the mouth cavity, which serves as a kind of auxiliary "lung" (Fig. 10-2, *C*). Unlike amphibians, which employ a **positive pressure** action to fill their lungs, most reptiles, birds, and mammals breathe by sucking air into the lungs **(negative pressure** action).

Tracheae. Insects and certain other terrestrial arthropods (centipedes, millipedes, and some spiders) have a highly specialized type of respiratory system; in many respects it is the simplest, most direct and most efficient respiratory system to be found in active animals. It consists of a system of tubes **(trachea)** that branch repeatedly and extend to all parts of the body. The smallest end channels **(air capillaries)**, less than 1 μ in diameter, sink into the plasma membranes of the body cells. Oxygen enters the tracheal system through valvelike openings **(spiracles)** on each side of the body and diffuses directly to all cells of the body. Carbon dioxide diffuses out in the opposite direction. Some insects can ventilate the tracheal system with body movements; the familiar telescoping movement of the bee abdomen is an example. The tracheal system is simple because blood is not needed to transport the respiratory gases; the cells have a direct pipeline to the outside.

RESPIRATORY SYSTEM IN MAN AND OTHER MAMMALS

In mammals the respiratory system is made up of the following: the nostrils (external nares); the **nasal chamber,** lined with mucus-secreting epithelium; the **posterior nares,** which connect to the

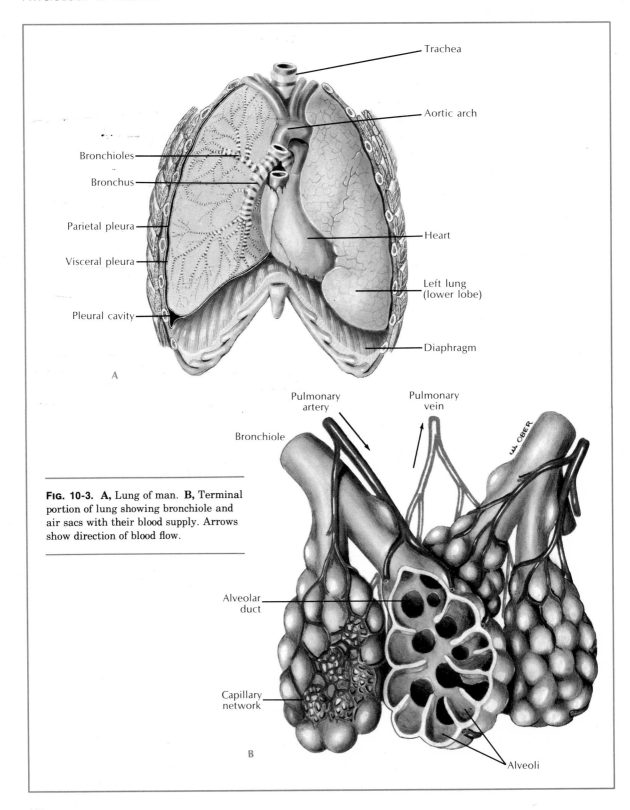

Trachea

Aortic arch

Bronchioles

Bronchus

Parietal pleura

Visceral pleura

Heart

Left lung
(lower lobe)

Pleural cavity

Diaphragm

A

Pulmonary
artery

Pulmonary
vein

Bronchiole

FIG. 10-3. A, Lung of man. **B,** Terminal portion of lung showing bronchiole and air sacs with their blood supply. Arrows show direction of blood flow.

Alveolar
duct

Capillary
network

Alveoli

B

pharynx, where the pathways of digestion and respiration cross; the **epiglottis,** a flap that folds over the **glottis** (the opening to the larynx) to prevent food from going the wrong way in swallowing; the **larynx,** or voice box; the **trachea,** or windpipe; and the two **bronchi,** one to each lung (Fig. 10-3). Within the lungs each bronchus divides and subdivides into smaller tubes (**bronchioles**) that lead to the air sacs (**alveoli**). The walls of the alveoli are thin and moist to facilitate the exchange of gases between the air sacs and the adjacent blood capillaries. Air passageways are lined with mucus-secreting ciliated epithelium and play an important role in conditioning the air before it reaches the alveoli. There are partial cartilage rings in the walls of the trachea, bronchi, and even some of the bronchioles to prevent those structures from collapsing.

In its passage to the air sacs the air undergoes three important changes: (1) it is filtered free from most dust and other foreign substances, (2) it is warmed to body temperature, and (3) it is saturated with moisture.

The lungs consist of a great deal of elastic connective tissue and some muscle. They are covered by a thin layer of tough epithelium known as the **visceral pleura.** A similar layer, the **parietal pleura,** lines the inner surface of the walls of the chest. The two layers of the pleura are in contact and slide over one another as the lungs expand and contract. The "space" between the pleura, called the **pleural cavity,** contains a partial vacuum. Actually, no real pleural space exists; the two pleura rub together, lubricated by lymph. The chest cavity is bounded by the spine, ribs, and breastbone, and floored by the **diaphragm,** a dome-shaped, muscular partition between chest cavity and abdomen.

Mechanism of breathing. The chest cavity is an air-tight chamber. In **inspiration** the ribs are elevated, the diaphragm contracted and flattened, and the chest cavity is enlarged. The resultant increase in volume of chest cavity and lungs causes the air pressure in the lungs to fall below atmospheric pressure; air rushes in through the air passageways to equalize the pressure. **Expiration** is a less active process than inspiration. When their muscles relax, the ribs and diaphragm return to their original position and the chest cavity size decreases. The elastic lungs then contract and force the air out.

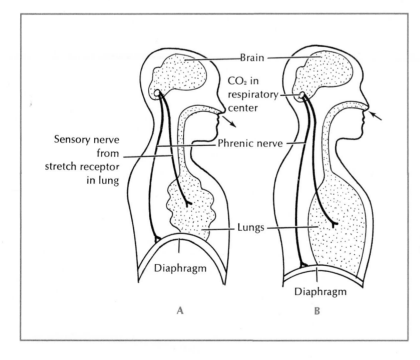

FIG. 10-4. Mechanism of breathing. In inspiration, **B,** CO_2 in blood stimulates respiratory center to send impulses by way of phrenic nerves to diaphragm, which, with elevation of ribs, produces inhalation of air. In **A,** impulses from stretch receptors in lungs inhibit respiratory center and exhalation occurs. (See text for explanation.)

Control of breathing. Respiration must adjust itself to the varying needs of the body for oxygen. Respiration is normally involuntary and automatic but may come under voluntary control. The rhythmic inspiratory and expiratory movements are controlled by a nervous mechanism centered in **medulla oblongata** of the brain (Fig. 10-4). By placing tiny electrodes in various parts of the medulla of experimental animals, neurophysiologists located separate **inspiratory** and **expiratory neurons** that act reciprocally to stimulate the inspiratory and expiratory muscles of the diaphragm and rib cage (intercostal) muscles. The rate of breathing is determined by the amount of carbon dioxide in the blood: a slight rise in the blood CO_2 stimulates respiration; a fall will decrease breathing.

Composition of inspired, expired, and alveolar airs. The composition of expired and alveolar airs is not identical. Air in the alveoli contains less oxygen and more carbon dioxide than does the air that leaves the lungs. Inspired air has the composition of atmospheric air. Expired air is really a mixture of alveolar and inspired airs. The variations in the three kinds of air are shown in Table 10-1. The water given off in expired air depends on the relative humidity of the external air and the activity of the person. At ordinary room temperature and with a relative humidity of about 50%, an individual in performing light work will lose about 350 ml. of water from the lungs each day.

Gaseous exchange in lungs. The diffusion of gases, both in internal as well as external respiration, takes place in accordance with the laws of physical diffusion; that is, the gases pass from regions of high pressure to those of low pressure. The pressure of a gas refers to the partial pressure that that gas exerts in a mixture of gases. If the atmospheric pressure at sea level is equivalent to 760 mm. of mercury (Hg), the partial pressure of O_2 will be 21% (percentage of O_2 in air) of 760, or 159 mm. Hg. The partial pressure of oxygen in the lung alveoli is greater (100 mm. Hg pressure) than it is in venous blood of lung capillaries (40 mm. Hg pressure) (Fig. 10-5). Oxygen then naturally diffuses into the capillaries. In a similar manner the carbon dioxide in the blood of the lung capillaries has a higher concentration (46 mm. Hg) than has this

TABLE 10-1. Variation in respired air

	Inspired air (vol. %)	Expired air (vol. %)	Alveolar air (vol. %)
Oxygen	20.96	16	14.0
Carbon dioxide	0.04	4	5.5
Nitrogen	79.00	80	80.5

FIG. 10-5. Exchange of respiratory gases in lungs and tissue cells. Numbers present partial pressures in millimeters of mercury.

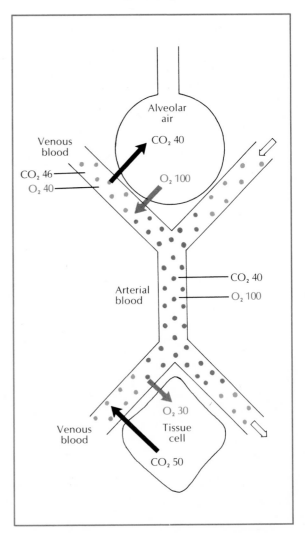

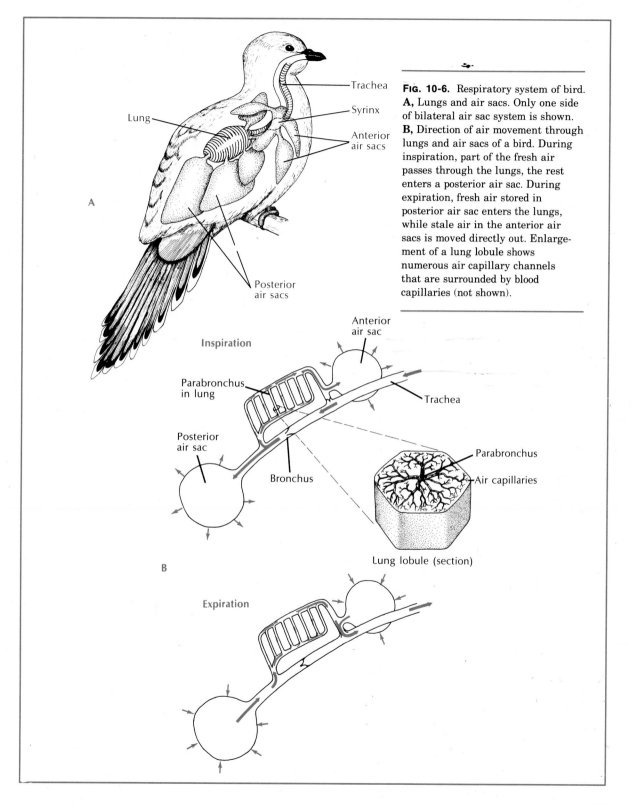

FIG. 10-6. Respiratory system of bird. **A,** Lungs and air sacs. Only one side of bilateral air sac system is shown. **B,** Direction of air movement through lungs and air sacs of a bird. During inspiration, part of the fresh air passes through the lungs, the rest enters a posterior air sac. During expiration, fresh air stored in posterior air sac enters the lungs, while stale air in the anterior air sacs is moved directly out. Enlargement of a lung lobule shows numerous air capillary channels that are surrounded by blood capillaries (not shown).

Trachea

Syrinx

Anterior air sacs

Lung

Posterior air sacs

A

Inspiration

Anterior air sac

Parabronchus in lung

Trachea

Posterior air sac

Bronchus

Parabronchus

Air capillaries

Lung lobule (section)

B

Expiration

same gas in the lung alveoli (40 mm. Hg), so that carbon dioxide diffuses from the blood into the alveoli.

In the tissues respiratory gases also move according to their concentration gradients (Fig. 10-5). Here the concentration of oxygen in the blood (100 mm. Hg pressure) is greater than in the tissues (0 to 30 mm. Hg pressure), and the carbon dioxide concentration in the tissues (45 to 68 mm. Hg pressure) is greater than that in blood (40 mm. Hg pressure). The gases in each case will go from a high to a low concentration.

The bird lung. The respiratory system of birds differs radically from the lungs of reptiles and mammals and is marvelously adapted for meeting the high metabolic demands of flight. The lungs, which are relatively inexpansible because of their direct attachment to the body wall, are filled with numerous tiny **air capillaries** instead of alveoli of the mammalian type. Most unique, however, is the extensive system of interconnecting **air sacs** that are located in pairs in the thorax and abdomen and even extend by tiny tubes into the centers of the long bones (Fig. 10-6, *A*). The air sacs are connected to the lungs in such a way that perhaps 75% of the inspired air bypasses the lungs and flows directly into the air sacs, which serve as reservoirs for fresh air. On expiration, some of this fully oxygenated air is shunted through the lung, while the rest passes directly out. The advantage of such a system is obvious—the lungs receive fresh air during both inspiration and expiration. Rather than locating the respiratory exchange surface deep within blind sacs which are difficult to ventilate as in mammals, birds have arranged to pass a continuous stream of fully oxygenated air through a system of richly vascularized air capillaries (Fig. 10-6, *B*). Although many details of the bird's respiratory system are not yet understood, it is clearly the most efficient of any vertebrate.

Transport of oxygen in the blood. In some invertebrates the respiratory gases are simply carried dissolved in the body fluids. However, the solubility of oxygen is so low in water that this means is adequate only for animals having low rates of metabolism. For example, only about 1% of man's oxygen requirement can be transported in this way. Consequently in all the advanced invertebrates and all

vertebrates, nearly all the oxygen and a significant amount of the carbon dioxide are transported by special colored proteins, or **respiratory pigments,** in the blood. In most animals (all vertebrates) these respiratory pigments are packaged into blood corpuscles. This is necessary because if this amount of respiratory pigment were free in blood, the blood would have the viscosity of syrup and would barely flow through the blood vessels, if at all.

The two most widespread respiratory pigments are **hemoglobin,** a red, iron-containing protein present in all vertebrates and many invertebrates, and **hemocyanin,** a blue, copper-containing protein present in the crustaceans and cephalopod mollusks. Hemoglobin is a complex protein. Each molecule is made up of 5% **heme,** an iron-containing compound giving the red color to blood, and 95% **globin,** a colorless protein. The heme portion of the hemoglobin has a great affinity for oxygen; each gram of hemoglobin (there are about 15 grams of hemoglobin in each 100 ml. of human blood) can carry a maximum of approximately 1.3 ml. of oxygen; each 100 ml. of fully oxygenated blood contains about 20 ml. of oxygen. Of course, for hemoglobin to be of value to the body it must hold oxygen in a loose, reversible chemical combination so that it can be released to the tissues. The actual amount of oxygen bound to hemoglobin depends on the oxygen partial pressure surrounding the blood corpuscles, a relationship expressed in the oxygen dissociation curve in Fig. 10-7. When the oxygen tension is high, as it is in the capillaries of the lung alveoli, hemoglobin becomes almost fully saturated to form oxyhemoglobin. Then, when the oxygenated blood leaves the lung and is distributed to the systemic capillaries in the body tissues, it enters regions of low oxygen partial pressure because oxygen is continuously consumed by cellular oxidative processes. The oxyhemoglobin now releases its bound oxygen, which diffuses into the cells. As the oxygen dissociation curve shows (Fig. 10-7), the lower the surrounding oxygen tension, the greater the quantity of oxygen released. This is an important characteristic because it allows more oxygen to be released to those tissues that need it most (have the lowest oxygen pressure). Another characteristic facilitating the release of oxygen to the tissues is the sensitivity of oxyhemoglobin to carbon dioxide. Carbon

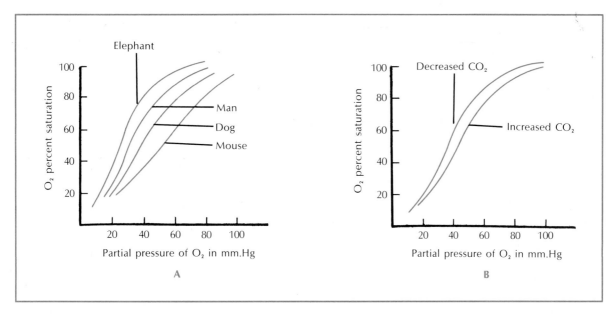

FIG. 10-7. Oxygen dissociation curves. Curves show how the amount of oxygen bound to hemoglobin (oxyhemoglobin) is related to oxygen pressure. **A,** Small animals have blood that gives up oxygen more readily than does the blood of large animals. **B,** Oxyhemoglobin is sensitive to carbon dioxide pressure; as carbon dioxide enters blood from the tissues, it shifts the curve to the right, decreasing affinity of hemoglobin for oxygen.

dioxide shifts the oxygen dissociation curve to the right (Fig. 10-7, *B*). Therefore as carbon dioxide enters the blood from the respiring tissues, it encourages the release of additional oxygen from the hemoglobin. The opposite event occurs in the lungs; as carbon dioxide diffuses from the venous blood into the alveolar space, the oxygen dissociation curve shifts back to the left, allowing more oxygen to be loaded onto the hemoglobin.

Unfortunately for man and other higher animals, hemoglobin has even a greater affinity for carbon monoxide (CO) than it has for oxygen—in fact, the affinity is about two hundred times greater for CO than for O_2. Carbon monoxide is becoming an atmospheric contaminant of ever-increasing proportions as the world's population and industrialization continues rapidly upward. This odorless and invisible gas displaces oxygen from hemoglobin to form a stable compound called **carboxyhemoglobin.** Air containing only 0.2% CO may be fatal. Children and small animals are poisoned more rapidly than adults because of their higher respiratory rate.

Transport of carbon dioxide by the blood. The same blood that transports oxygen to the tissues from the lungs must carry carbon dioxide back to the lungs on its return trip. However, unlike oxygen that is transported almost exclusively in combination with hemoglobin, carbon dioxide is transported in three major forms.

1. Most of the carbon dioxide, about 67%, is converted in the red blood cells into bicarbonate and hydrogen ions, by undergoing the following series of reactions:

$$CO_2 + H_2O \rightleftharpoons H_2CO_3$$
$$\text{CARBONIC ACID}$$

This reaction would normally proceed very slowly, but an enzyme in the red blood cells, **carbonic anhydrase,** catalyzes the reaction to proceed almost instantly. As soon as carbonic acid forms, it instantly and almost completely ionizes as follows:

$$H_2CO_3 \rightleftharpoons HCO_3 + H^+$$

| CARBONIC ACID | BICARBONATE ION | HYDROGEN ION |

127

The hydrogen ion is buffered by several buffer systems in the blood, thus preventing a severe drop in blood pH. The bicarbonate ion remains in solution in the plasma and red blood cell water, since unlike carbon dioxide bicarbonate is extremely soluble.

2. Another fraction of the carbon dioxide, about 25%, combines reversibly with hemoglobin. It is carried to the lungs, where the hemoglobin releases it in exchange for oxygen.

3. A third small fraction of the carbon dioxide, about 8%, is carried as the physically dissolved gas in the plasma and red blood cells.

Internal fluids
and
their circulation

Single-celled organisms live a contact existence with their environment. Nutrients and oxygen are obtained, and wastes are released, directly across the cell surface. These animals are so small that no special internal transport system, beyond the normal streaming movements of the cytoplasm, is required. Even some primitive multicellular forms, such as sponges, coelenterates and flatworms, have such a simple internal organization and low rate of metabolism that no circulatory system is needed. Most of the more advanced multicellular organisms, because of their size, activity, and complexity, require a specialized circulatory, or vascular, system to transport nutrients and respiratory gases to and from all tissues of the body. In addition to serving these primary transport needs, circulatory systems have acquired additional functions; hormones are moved about, finding their way to traget organs where they assist the nervous system to integrate body function. Water, electrolytes, and the many other constituents of the body fluids are distributed and exchanged between different organs and tissues. An effective response to disease and injury is vastly accelerated by an efficient circulatory system, and the most advanced animals, the warm-blooded birds and mammals, depend heavily on the blood circula-

tion to conserve or dissipate heat as required for the maintenance of constant body temperature.

INTERNAL FLUID ENVIRONMENT

The body fluid of a single-celled animal is the cellular cytoplasm, a fluid substance in which the various membrane systems and organelles of the cell are suspended. In multicellular animals the body fluids are divided into two main phases, the **intracellular** and the **extracellular.** The intracellular phase (also called intracellular fluid) is the fluid inside all the body's cells. The extracellular phase (or fluid) is the fluid outside and surrounding the cells (Fig. 11-1, *A*). The significance of this extracellular fluid as a protective environment of the cells was recognized over a century ago by the great French physiologist, Claude Bernard. Bernard called this extracellular fluid the **milieu interieur,** meaning the body's internal environment, as opposed to the external environment, or outside world. The environment outside the animal he called the **milieu exterieur.** Thus the cells, the sites of the body's crucial metabolic activities, are bathed by their own aqueous environment, the milieu interieur, which buffers

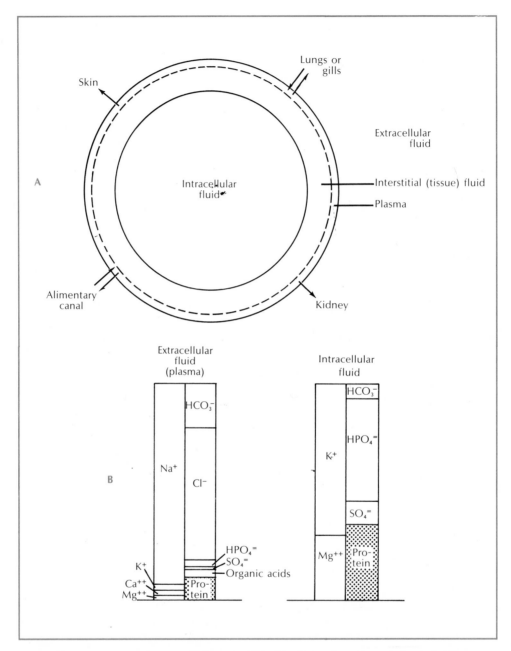

FIG. 11-1. Fluid compartments of body. **A,** All body cells can be represented as belonging to a single large fluid compartment that is completely surrounded and protected by extracellular fluid. The latter is further subdivided into plasma and interstitial fluid. All exchanges with the environment occur across the plasma compartment. **B,** Electrolyte composition of extracellular and intracellular fluids. Total equivalent concentration of each major constituent is shown. Equal amounts of anions (negatively charged ions) and cations (positively charged ions) are in each fluid compartment. Note that sodium and chloride, major plasma electrolytes, are virtually absent from intracellular fluid (actually they are present in low concentration). Note the much higher concentration of protein inside the cells.

them from the often harsh physical and chemical changes occurring outside the body. Even today, English-speaking biologists frequently use the French phrase in referring to the extracellular fluid.

In animals having closed circulatory systems (vertebrates, annelids, and a few other invertebrate groups) the extracellular fluid is further subdivided into blood **plasma** and **interstitial** fluid (Fig. 11-1, *A*). The blood plasma is contained within the blood vessels, while the interstitial fluid, or tissue fluid as it is sometimes called, occupies the space immediately around the cells. Nutrients and gases passing between the vascular plasma and the cells must traverse this narrow fluid separation. The interstitial fluid is constantly formed from the plasma by filtration through the capillary walls.

Composition of the body fluids. All these fluid spaces—plasma, interstitial, and intracellular—differ from each other in solute composition, but all have one feature in common—they are mostly water. Despite their firm appearance, animals are 70% to 90% water. Man, for example, is about 70% water; cell water makes up about 50% of the body weight, interstitial fluid water makes up 15%, and the plasma, 5%. Although an average-sized man contains a hundred pounds of water, it is fortunately so well compartmentalized that he need not worry excessively about springing a disastrous leak. As Fig. 11-1, *A*, diagrammatically shows, it is the plasma space that serves as the pathway of exchange between the cells of the body and the outside world. This exchange of respiratory gases, nutrients, and wastes is accomplished by specialized organs (kidney, lungs, gill, alimentary canal), as well as by the integument.

The body fluids contain many inorganic and organic substances in solution. Principal among these are the inorganic electrolytes and proteins. Fig. 11-1, *B*, shows that **sodium, chloride,** and **bicarbonate** are the chief extracellular electrolytes, whereas **potassium, magnesium, phosphate, sulfate,** and **proteins** are the major intracellular electrolytes. These differences are dramatic; they are always maintained despite the continuous flow of materials into and out of the cells of the body. The two subdivisions of the extracellular fluid—plasma and interstitial fluid—have similar compositions except that the plasma has more proteins

which are too large to filter through the capillary wall into the interstitial fluid.

Composition of blood. Blood is a liquid tissue composed of plasma and formed elements, mostly corpuscles, suspended in the plasma. When the red blood corpuscles and other formed elements are spun down in a centrifuge, the blood is found to be about 55% plasma and 45% formed elements.

The composition of blood is as follows:

Plasma
1. Water 90%
2. Dissolved solids, consisting of the plasma proteins (albumin, globulins, fibrinogen), glucose, amino acids, electrolytes, various enzymes, antibodies, hormones, metabolic wastes, and traces of many other organic and inorganic materials
3. Dissolved gases, especially oxygen, carbon dioxide, and nitrogen

Formed elements (Fig. 11-2)
1. Red blood corpuscles (erythrocytes), for the transport of oxygen and carbon dioxide
2. White blood corpuscles (leukocytes), serving as scavengers and as immunizing agents
3. Platelets (thrombocytes), functioning in blood coagulation

Red blood cells, or **erythrocytes,** are present in enormous numbers in the blood, about 5.4 million per cubic millimeter in man. They are formed continuously from large nucleated **erythroblasts** in the red bone marrow. Here, hemoglobin is synthesized and the cells divide several times. In mammals the nucleus is finally lost and the biconcave, disk-shaped erythrocyte enters the circulation for an average life-span of about four months. During this time it may journey 700 miles, squeezing repeatedly through the capillaries which are sometimes so narrow that the erythrocyte must bend to get through. At last it fragments and is quickly engulfed by large scavenger cells called **macrophages** located in the liver, bone marrow, and spleen. The iron from the hemoglobin is salvaged to be used again; the rest of the heme is converted to **bilirubin,** a bile pigment. It is estimated that 10 million erythrocytes are born, and another 10 million destroyed every second. Only the mammals have nonnucleated erythrocytes. All other vertebrates have nucleated erythrocytes that are usually ellipsoidal, rather than round, disks.

The white blood cells, or **leukocytes,** form a wan-

131

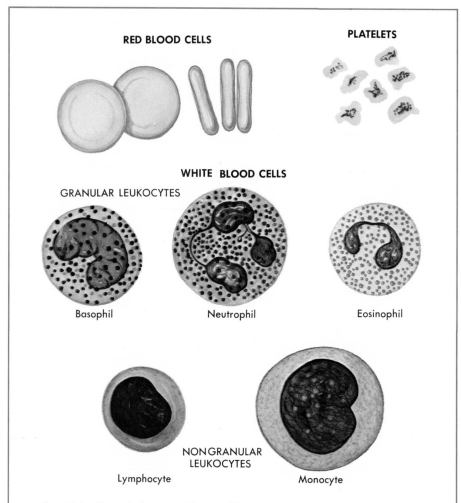

RED BLOOD CELLS

PLATELETS

WHITE BLOOD CELLS

GRANULAR LEUKOCYTES

Basophil

Neutrophil

Eosinophil

NONGRANULAR LEUKOCYTES

Lymphocyte

Monocyte

FIG. 11-2. Formed elements of human blood. Hemoglobin-containing red blood cells of man and other mammals lack nuclei, but those of all lower vertebrates have nuclei. Various leukocytes provide a wandering system of protection for the body. Platelets participate in the body's clotting mechanism. (From Anthony, C. P., and Kolthoff, N. J.: Textbook of anatomy and physiology, ed. 8, St. Louis, 1971, The C. V. Mosby Co.)

dering system of protection for the body. They number only about 7,500 per cubic millimeter, a ratio of 1 white cell to 700 red cells. There are several kinds of white blood cells: **granulocytes** (subdivided into neutrophils, basophils, and eosinophils), **lymphocytes,** and **monocytes** (Fig. 11-2). All have the capacity to pass through the wall of capillaries and wander by ameboid movement through the tissue spaces. Monocytes and the granulocytes have great power to engulf and digest bacteria and other foreign particulate matter, a process called **phagocytosis.** They also clean up and digest the debris of the body's own tissues, such as fragments of worn-out red blood cells, blood clots, or the remains of wounds and disease repair. Lymphocytes are especially important in producing gamma globulins, which act as immune bodies (**antibodies**) that destroy or neutralize toxic molecules (**antigens**).

TABLE 11-1. Major blood groups

Blood group	Antigens in red corpuscles	Antibodies in serum	Can give blood to	Can receive blood from	Frequency in United States (%)		
					Whites	Blacks	Chinese
O	None	a, b	All	O	45	38	46
A	A	b	A, AB	O, A	41	27	28
B	B	a	B, AB	O, B	10	21	23
AB	AB	None	AB	All	4	4	13

The platelets, or thrombocytes, are minute, colorless bodies about one third the diameter of red blood cells. They initiate the coagulation of blood. When blood spills from a vessel, as in a wound, the platelets rapidly disintegrate to release factors that start the formation of a clot. Platelets also readily clump together and plug torn vessels by entangling white blood cells.

Coagulation of blood. It is essential that animals have ways of preventing the rapid loss of body fluids after an injury. Since blood is flowing and is under considerable hydrostatic pressure, it is especially vulnerable to hemorrhagic loss. The most primitive means of preventing hemorrhage, and one used by many soft-bodied invertebrates, is spasmic contraction of body musculature and blood vessels. Vessels in the wound are thus narrowed off or even shut tight, stopping the flow. But most firm-bodied animals (vertebrates and higher invertebrates) and especially those having high blood pressures, have special cellular elements and proteins in the blood capable of forming plugs, or clots, at the site of injury. When tissue cells are damaged, a substance called **thromboplastin** is released from the injured tissue as well as from the blood platelets (see above). This substance, in the presence of calcium, converts a normally inactive protein, **prothrombin,** into an active enzyme, **thrombin.** Thrombin then converts **fibrinogen,** a very large plasma protein, into an insoluble threadlike protein, **fibrin.** These fibrin threads form a stringy, tangled network that entangles white and red blood cells, forming a gellike clot. In time the fibrin threads shrink and squeeze out a faintly yellow fluid called **serum.** The difference between **plasma** and **serum** is that the latter lacks fibrinogen and is incapable of clotting.

The conversion of prothrombin into thrombin is a critical step in the clotting reaction. In addition to thromboplastin, many other coagulation factors are required. A recent estimate listed no less than thirty-five compounds that participate in some way in coagulation. A deficiency of a single factor can delay or prevent the clotting process. Why has such a complex clotting mechanism evolved? Probably it is necessary to provide a "failsafe" system capable of responding to any kind of internal or external hemorrhage that might occur, yet not be activated into forming dangerous intravascular clots when no injury has occurred.

Several kinds of clotting abnormalities in man are known. Of these, **hemophilia** is perhaps best known. Hemophilia is a condition characterized by the failure of the blood to clot so that even insignificant wounds can cause continuous severe bleeding. Called the "disease of kings," it once ran through the royal families of Europe, notably those of Queen Victoria of England and Alfonso XIII, the last king of Spain. Hemophilia is caused by an inherited lack of antihemophilic factor. The disorder is transmitted through females, but almost invariably appears only in males.

Blood groups. In blood transfusions the donor's blood is checked against the blood of the recipient. Blood differs chemically from person to person, and when two different (incompatible) bloods are mixed, **agglutination** (clumping together) results. The basis of these chemical differences is the presence in the red blood corpuscle of **agglutinogens (antigens) A** and **B,** and, in the serum of plasma, **agglutinins (antibodies) a** and **b.** According to the way these antigens and antibodies are distributed, there are four main blood groups: A, B, AB, and O (Table 11-1). Group **A** blood cells have **A** antigens on them, but the serum of a group **A** person has no **a** antibody because if it did, it would destroy its own blood cells. It does contain group **b** antibodies, however. It is therefore incompatible with either group **B** or group **AB** blood because these contain **B** antigens. Similarly group **B** blood contains **B** antigens and lacks **b** antibodies. But since it contains **a** antibodies it is incompatible with any blood (group **A** or **AB**) that contains **A** antigens. Group **O** blood contains both antibodies **a** and **b** but no antigens. Group **AB** contains both **A** and **B** antigens but no antibodies. We see then that the blood group names identify their antigen content. Persons with type O blood are called universal donors because, lacking antigens, their blood can be infused into a person with any blood type. Even though it contains **a** and **b** antibodies, these are so diluted during transfusion that they do not react with **A** or **B** antigens in a recipient's blood. In practice, however, clinicians insist on matching blood types to prevent any possibility of incompatibility.

Rh factor. In 1940 there was discovered in the red blood corpuscles a new factor called the Rh factor, named after the Rhesus monkeys in which it was first found. About 85% of individuals have the factor (positive) and the other 15% do not (negative). It was also found that Rh-positive and Rh-negative bloods are incompatible; shock and even death may follow their mixing when Rh-positive blood is introduced into an Rh-negative person who has been sensitized by an earlier transfusion of Rh-positive blood. The Rh factor is inherited as a dominant; this accounts for a peculiar and often fatal form of anemia of newborn infants called **erythroblastosis fetalis.** Although the fetal and maternal bloods are separated by the placenta, this separation is not perfect. Some admixture of fetal and maternal bloods usually occurs, especially right after birth when the placenta ("afterbirth") separates from the uterine wall. This admixture of blood, normally of no consequence, can be serious **if** the father is Rh positive, the mother Rh negative, and the fetus Rh positive (by inheriting the factor from the father). The fetal blood, containing the Rh antigen, can stimulate the formation of Rh-positive antibodies in the blood of the mother. The mother is permanently immunized against the Rh factor. During the following pregnancy these antibodies

may diffuse back into the fetal circulation and produce agglutination and destruction of the fetal red blood cells. Because the mother is usually sensitized at the end of the first pregnancy, subsequent babies are more severely threatened than is the first.

Erythroblastosis fetalis can now be prevented by giving an Rh-negative mother anti-Rh antibodies just after the birth of her first child. These antibodies remain long enough to neutralize any RH-positive fetal blood cells that may enter her circulation, thus preventing her own antibody machinery from being stimulated to produce the Rh-positive antibodies. Active, permanent immunity is blocked.

CIRCULATION

The circulatory system of vertebrates is made up of a system of tubes, the **blood vessels,** and a propulsive organ, the **heart.** This is a **closed circulation** because the circulating medium, the **blood,** is confined to vessels throughout its journey from the heart to the tissues and back again. Many invertebrates have an **open circulation;** the blood is pumped from the heart into blood vessels that open into tissue spaces. The blood then circulates freely in direct contact with the cells, then reenters open blood vessels to be propelled forward again. In invertebrates having open circulatory systems, there is no clear separation of the extracellular fluid into plasma and interstitial fluids, as there is in closed systems (p. 129). Closed systems are more suitable for large and active animals because the blood can be moved rapidly to the tissues needing it. In addition, flow to various organs can be readjusted to meet changing needs by varying the diameters of the blood vessel.

Closed circulatory systems work in parallel with a cooperative system, the **lymphatic system.** This is a fluid "pick-up" system. It recollects tissue fluid (lymph) that has been squeezed out through the walls of the capillaries and returns it to the blood circulation. In a sense "closed" circulatory systems are not absolutely closed because fluid is constantly leaking out into the tissue spaces. However, this leakage is but a small fraction of the total blood flow.

FIG. 11-3. Plan of circulatory system of fish (above) and mammals (below).

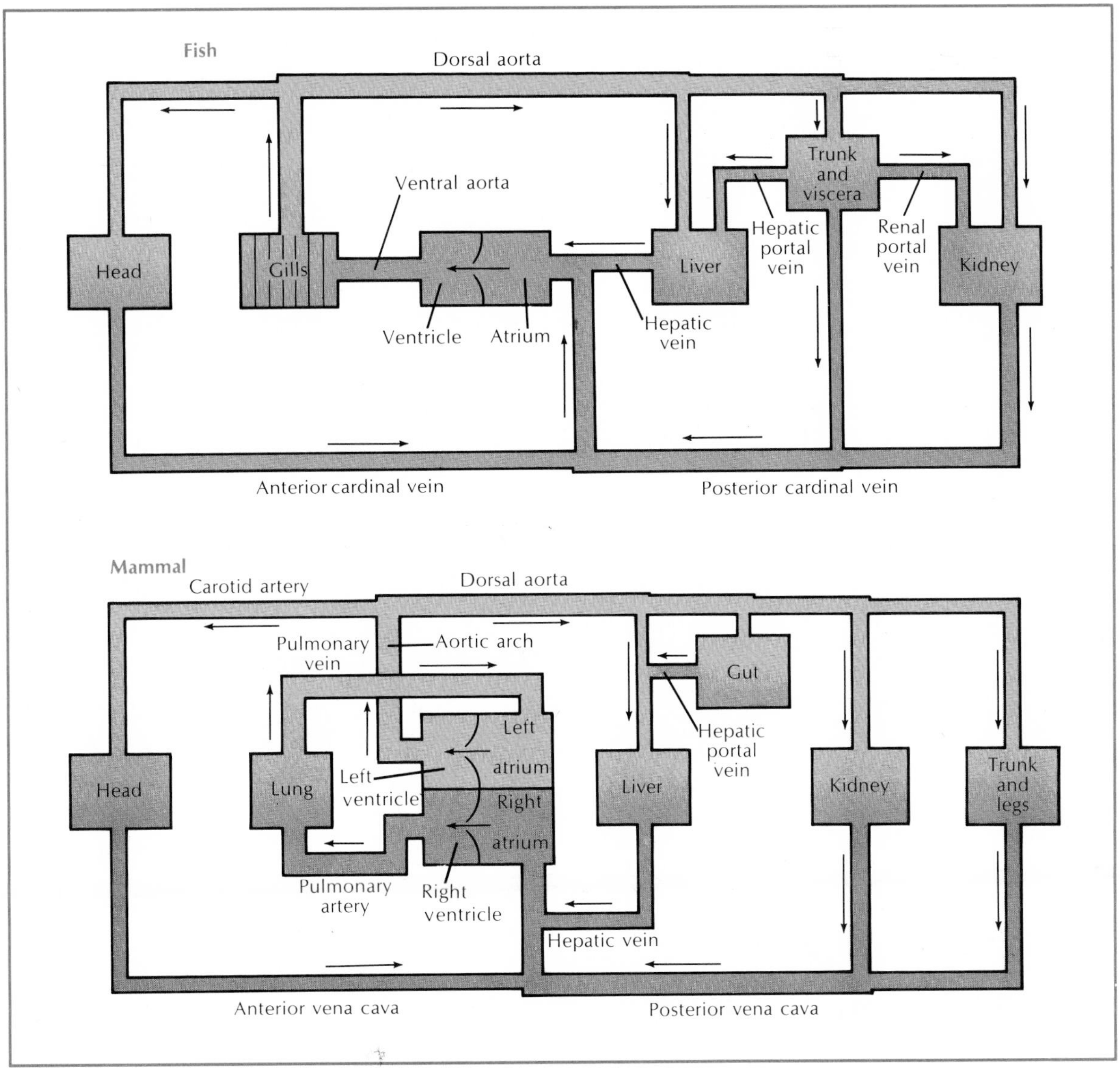

Plan of the circulatory system. All vertebrate vascular systems have certain features in common. A **heart** pumps the blood into **arteries** that branch and narrow into **arterioles** and then into a vast system of **capillaries.** Blood leaving the capillaries enters **venules** and then **veins** that return the blood to the heart. Fig. 11-3 compares the circulatory systems of gill-breathing (fish) and lung-breathing (mammal) vertebrates. The principal differences in circulation involve the heart in the transformation from gill to lung breathing. The fish heart contains two main chambers, the **atrium** (or **auricle**) and the **ventricle.** Although there are also two subsidiary chambers, the **sinus venosus** and **conus arteriosus** (not shown in Fig. 11-3), we still refer to the fish heart as a "two chambered" heart. Blood makes a single circuit through the fish's vascular system; it is pumped from the heart to the gills, where it is oxygenated, and then flows into the dorsal aorta to be distributed to the body organs. After passing through the capillaries of the body organs and musculature, it returns by veins to the heart. In this circuit the heart must provide sufficient pressure to push the blood through two sequential capillary systems, one in the gills and the other in the organ tissues. The principal disadvantage of the single-circuit system is that the gill capillaries offer so much resistance to blood flow that the pressure drops considerably before entering the dorsal aorta. This system can never provide high and continuous blood pressure to the body organs.

Evolving land forms with lungs and their need for highly efficient blood delivery had to solve this problem by introducing a **double** circulation. One **systemic** circuit with its own pump provides oxygenated blood to the capillary beds of the body organs; another **pulmonary** circuit with its own pump sends deoxygenated blood to the lungs. Rather than actually developing two separate hearts, the existing two-chambered heart was divided down the center into four chambers—really two two-chambered hearts lying side-by-side. Needless to say such a great change in the vertebrate circulatory plan, involving not only the heart but

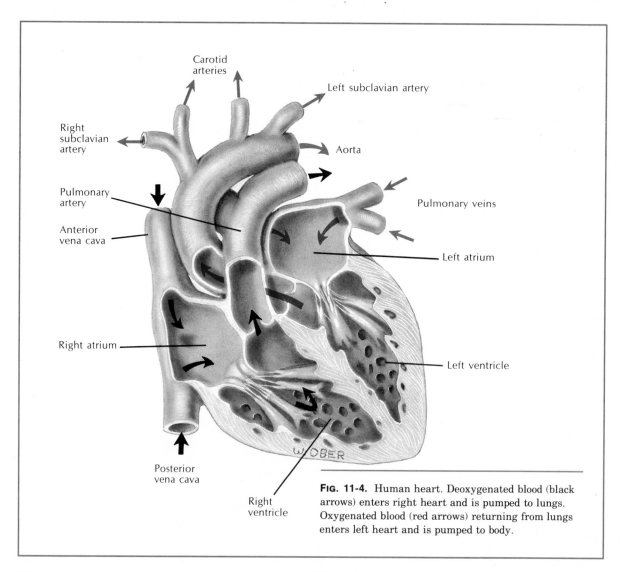

FIG. 11-4. Human heart. Deoxygenated blood (black arrows) enters right heart and is pumped to lungs. Oxygenated blood (red arrows) returning from lungs enters left heart and is pumped to body.

the attendant plumbing as well, took many millions of years to evolve. The partial division of the atrium and ventricle began with the ancestors of present-day lungfish. Amphibians accomplished the complete separation of the atrium, but the ventricle is still undivided in this group. In some reptiles the ventricle is completely divided, and the four-chambered heart appears for the first time. All birds and mammals have the four-chambered heart and two separate circuits—one through the lungs (pulmonary) and the other through the body (systemic). The course of the blood through this double circuit is shown in Fig. 11-3.

The heart. The vertebrate heart is a muscular organ located in the thorax and covered by a tough, fibrous sac, the **pericardium** (Fig. 11-4). As we have seen, the higher vertebrates have a four-chambered heart. Each half consists of a thin-walled atrium and a thick-walled ventricle. Heart (cardiac) muscle is a unique type of muscle found nowhere else in the body. It resembles striated muscle, but

the cells are branched, and dense end-to-end attachments between the cells are called intercalated disks (Fig. 6-5, p. 72). There are four sets of valves. **Atrioventricular valves** (A-V valves) separate the cavities of the atrium and ventricle in each half of the heart. These permit blood to flow from atrium to ventricle but prevent backflow. Where the great arteries, the **pulmonary** from the right ventricle and the **aorta** from the left ventricle, leave the heart, **semilunar** valves prevent backflow.

The contraction of the heart is called **systole,** and the relaxation, **diastole.** The rate of the heartbeat depends on age, sex, and especially, exercise.

FIG. 11-5. Neuromuscular mechanisms controlling beat of the heart. Arrows indicate spread of excitation from the sinoatrial node, across the right atrium, to the A-V node. Wave of excitation is then conducted very rapidly to ventricular muscle over the specialized bundle of His and Purkinje fiber system. (From Tuttle, W. W., and Schottelius, B. A.: A textbook of physiology, ed. 16, St. Louis, 1969, The C. V. Mosby Co.)

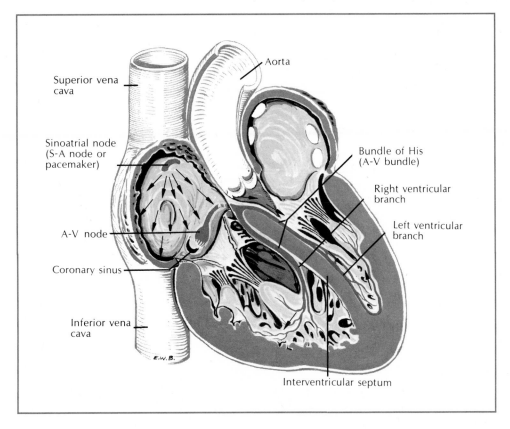

Superior vena cava

Aorta

Sinoatrial node (S-A node or pacemaker)

Bundle of His (A-V bundle)

Right ventricular branch

Left ventricular branch

A-V node

Coronary sinus

Inferior vena cava

E.W.B.

Interventricular septum

Exercise may increase the **cardiac output** (volume of blood forced from either ventricle each minute) more than five-fold. Both the heart **rate** and the **stroke volume** increase. Heart rates among vertebrates vary with the general level of metabolism and the body size. The cold-blooded codfish has a heart rate of about 30 beats per minute; a warm-blooded rabbit of about the same weight has a rate of 200 beats per minute. Small animals have higher heart rates than large animals. The heart rate in an elephant is 25 beats per minute, in a man 70 per minute, in a cat 125 per minute, in a mouse 400 per minute, and in the tiny 4-gram shrew, the smallest mammal, the heart rate approaches a prodigious 800 beats per minute. We must marvel that the shrew's heart can sustain this frantic pace throughout this animal's life, brief as it is. The only rest a heart enjoys is the short interval between contractions. The mammalian heart does an amazing amount of work during a lifetime. Someone has calculated that the heart of a man approaching the end of his life has beat some 2.5 billion times and pumped 300,000 tons of blood!

The heart beat originates in a specialized muscle tissue, called the **sinoatrial node,** located in the right atrium near the entrance of the caval veins (Fig. 11-5). This tissue serves as the "pacemaker" of the heart. The contraction originates in the pacemaker and spreads across the two atria to the **atrioventricular (A-V) node.** At this point the electrical activity is conducted very rapidly to the apex of the ventricle through specialized fibers (bundle of His and Purkinje fiber system), to then spread more slowly up the walls of the ventricles. This arrangement allows the contraction to begin at the apex or "tip" of the ventricles and spread upward to squeeze out the blood in the most efficient way; it ensures that both ventricles will contract simultaneously. Although the vertebrate heart can beat spontaneously—the excised fish or amphibian heart will beat for hours in a balanced salt solution—the heart rate is normally under nervous control. The control (cardiac) center is located in the medulla and sends out two sets of motor nerves. Impulses sent along one set, the **vagus** nerves, apply a brake-action to the heart rate, and impulses sent along the other set, the **accelerator** nerves, speed it up. Both sets of nerves terminate in the sinoatrial node,

thus guiding the activity of the pacemaker. The cardiac center in turn receives sensory information about a variety of stimuli. Pressure receptors (sensitive to blood pressure) and chemical receptors (sensitive to carbon dioxide and pH) are located at strategic points in the vascular system. This information is used by the cardiac center to increase or reduce the heart rate and cardiac output in response to activity or changes in body position. The heart is thus controlled by a series of feedback mechanisms that keep its activity constantly attuned to body needs.

It is no surprise that an organ as active as the heart needs a very good blood supply of its own. The heart muscle of the frog and other amphibians is so thoroughly channeled with spaces between the muscle fibers that sufficient oxygenated blood is squeezed through by the heart's own pumping action. In birds and mammals, however, the heart muscle is very thick and has such a high rate of metabolism that it must have its own vascular **(coronary)** circulation. The coronary arteries break up into an extensive capillary network surrounding the muscle fibers and provide them with oxygen and nutrients. Heart muscle has an extremely high oxygen demand, removing 80% of the oxygen from the blood, in contrast to most other body tissues, which remove only about 30%.

Arteries. All vessels leaving the heart are called arteries whether they carry oxygenated blood (aorta) or deoxygenated blood (pulmonary artery). To withstand high, pounding pressures, arteries are invested with layers of both elastic and tough, inelastic connective tissue fibers. The elasticity of the arteries allows them to yield to the surge of blood leaving the heart during systole, then to squeeze down on the fluid column during diastole. This smooths out the blood pressure. Thus the arterial pressure in man varies only between a high of 120 mm. Hg (systole) and a low of 80 mm. Hg (diastole), rather than dropping to zero during diastole as we might expect in a fluid system with an intermittent pump. As the arteries branch and narrow into **arterioles,** the walls become mostly smooth muscle (Fig. 11-6). Contraction of this muscle narrows the arterioles and reduces the flow of blood. The arterioles thus control the blood flow to body organs, diverting it to where it is needed most.

FIG. 11-6. Cross section of vein and corresponding artery.

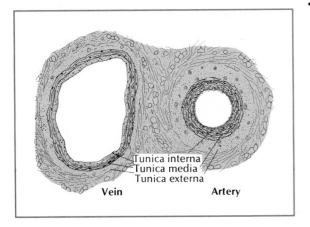

Tunica interna
Tunica media
Tunica externa

Vein   Artery

The blood must be given a hydrostatic pressure sufficient to overcome the resistance of the narrow passages through which the blood must flow. Consequently large animals tend to have higher blood pressures than small animals. The blood pressure was first measured in 1733 by Stephen Hales, an English clergyman with unusual inventiveness and curiosity. He tied his mare "to have been killed as unfit for service" on her back and exposed the femoral artery. This he cannulated with a brass tube, connecting it to a tall glass tube with the windpipe of a goose. The use of the windpipe was both imaginative and practical; it gave the apparatus flexibility "to avoid inconveniences that might arise if the mare struggled." The blood rose 8 feet in the glass tube and bobbed up and down in accordance with the systolic and diastolic beats of the heart. The weight of the 8-foot column of blood was equal to the blood pressure. We now express this as the height of a column of mercury, which is 13.6 times as heavy as water. Hale's figures, expressed in millimeters of mercury, indicate he measured a blood pressure of 180 to 200 mm. Hg, about normal for a horse. Today, blood pressure can be measured with great accuracy with a sensitive pressure transducer; the electronic signal from this instrument is displayed on a graphic recorder.

Capillaries. The capillaries are present in enormous numbers, forming extensive networks in nearly all tissues. In muscle there are more than 2,000 per square millimeter (1,250,000 per square inch), but not all are open at once. Indeed perhaps less than 1% are open in resting skeletal muscle. But when the muscle is active, all the capillaries may open to bring oxygen and nutrients to the working muscle fibers and to carry away metabolic wastes.

Capillaries are extremely narrow, averaging less than 10 μ in diameter, which is hardly any wider than the red blood cells that must pass through them. Their walls are formed of a single layer of thin **endothelial** cells, held together by a delicate basement membrane and connective tissue fibers. Capillaries have a built-in leakiness that allows water and most dissolved substances in the blood plasma to filter through into the interstitial space. The capillary wall is **selectively permeable,** however, which means that it filters some dissolved materials and retains others. In this case the plasma proteins, which are the largest dissolved molecules in the plasma, are held back. These proteins contribute an **osmotic pressure** estimated to be about 25 mm. Hg (Fig. 11-7). Although small, this protein osmotic pressure is of great importance to fluid balance in the tissues. At the arteriole end of the capillaries the blood pressure is about 40 mm. Hg (in man). This **filtration pressure** forces water and dissolved materials through the capillary endothelium into the tissue space where they circulate freely around the cells. As the blood proceeds through the narrow capillary, the blood pressure drops steadily to perhaps 15 mm. Hg. At this point the hydrostatic pressure is less than the osmotic pressure of the plasma proteins, still about 25 mm. Hg. Water now is drawn back into the capillaries. Thus it is the balance between hydrostatic pressure and protein osmotic pressure that determines the direction of capillary fluid shift. Normally water is forced out of the capillary at the arteriole end, where hydrostatic pressure exceeds osmotic pressure, and into the capillary at the venule end where osmotic pressure exceeds hydrostatic pressure. Any fluid left behind is picked up and removed by the **lymph capillaries.**

Veins. The venules and veins into which the capillary blood drains for its return journey to the heart are thinner walled, less elastic, and of considerably larger diameter than their corresponding arteries and arterioles. Blood pressure in the venous system is low, from about 10 mm. Hg where capillaries drain into venules to about zero in the right atrium. Because pressure is so low, the venous return gets assists from valves in the veins, from muscles surrounding the veins, and from the rhythmic pumping action of the lungs. If it were not for these mechanisms, the blood might pool in the lower extremities of a standing animal—a very real problem for people who must stand for long periods. The

veins that lift blood from the extremities to the heart contain valves that serve to divide the long column of blood into segments. When the muscles around the veins contract, as in even slight activity, the blood column is squeezed upward and cannot slip back because of the valves. The well-known risk of fainting while standing at stiff attention in hot weather can usually be prevented by deliberately pumping the leg muscles. The negative pressure

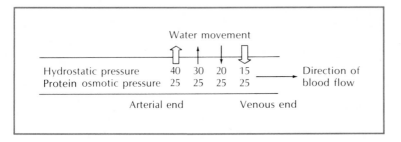

FIG. 11-7. Fluid movement across the wall of a capillary. At arterial end of the capillary, hydrostatic (blood) pressure exceeds protein osmotic pressure contributed by the plasma proteins, and a plasma filtrate (shown as "water movement") is forced out. At venous end, protein osmotic pressure exceeds the hydrostatic pressure, and fluid is drawn back in. In this way plasma nutrients are carried out into the interstitial space where they can enter cells, and metabolic end products from the cells are drawn back into the plasma and carried away.

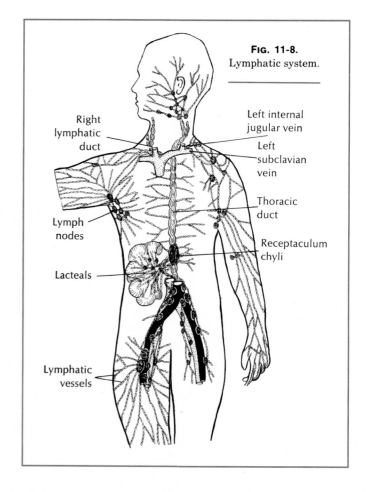

FIG. 11-8. Lymphatic system.

created in the thorax by the inspiratory movement of the lungs also speeds the venous return by sucking the blood up the large vena cava into the heart.

Lymphatic system. The lymphatic system (Fig. 11-8) is an accessory drainage system for the body. As we have seen, the blood pressure in the arteriole end of the capillaries forces a plasma filtrate through the capillary walls and into the interstitial space. This tissue fluid bathing the cells is **lymph**, a clear, nearly colorless liquid. Lymph and plasma are nearly indentical except that lymph contains very little protein, which was screened out as the plasma was squeezed through the capillary walls. Most of the lymph returns to the vascular system at the venous end of the capillaries by the capillary fluid-shift mechanism described earlier. Usually, however, outflow from the capillaries slightly exceeds backflow. This difference is gathered up and returned to the circulatory system by lymphatic vessels. The system begins with tiny, highly permeable lymph capillaries. These lead into larger lymph vessels, which in turn drain into the large **thoracic duct.** This enters the left subclavian vein in the neck region. Lymph flow is very low, a minute fraction of the blood flow.

Located at strategic intervals along the lymph vessels are **lymph nodes** (Fig. 11-8) that have several defense-related functions. They are effective filters that remove foreign particles, especially bacteria, that might otherwise enter the general circulation. They are also germinal centers for **lymphocytes** (p. 132) and they produce gamma globulin **antibodies**—both essential components of the body's defense mechanisms.

12

Body fluid regulation

As we have seen in the previous chapter, the intracellular fluid of animals is the medium in which all metabolic activities occur. Surrounding the cells and protecting them are the extracellular fluids, the plasma and interstitial fluid. The chemical composition of these fluids must be kept constant, despite the continuous movement and turnover of materials within the body. Water, which makes up about two thirds of the body weight of animals, is always entering and leaving the body. Water is also formed within the cells as a by-product of oxidative processes. Ionized inorganic and organic salts are continually moving between the cells and body fluids and also between the animal and its environment. Protein is constantly being formed, transported and broken down again within the tissues, yielding nitrogenous wastes that must be excreted. Obviously, body composition is a dynamic rather than a static thing. It is often described as operating as a **dynamic steady state.** This means that constancy of composition is maintained despite the continuous shifting of components of the system. This kind of internal regulation is **homeostasis.**

It is apparent that body fluid composition can be altered either by metabolic events occurring within the cells and tissues or by events occurring across the surface of the body. In other words, a living system is "open at both ends." On the inside, metabolic activities within the cell require a steady supply of materials, and these activities turn out a continuous flow of products and wastes. On the outside, materials are constantly being exchanged between the plasma and the external environment. What keeps the body working like the finely tuned machine it is, when so many activities threaten to throw the system out of order? This question cannot be answered simply because so many body systems help to regulate body fluid composition. Especially important, however, are those organs that serve as sites of exchange with the external environment: the kidney, the lungs or gills, the alimentary canal, and the skin. Through these organs enter oxygen, foodstuffs, minerals, and other constituents of the body fluids; here water is exchanged and metabolic wastes are eliminated. The kidney is the chief regulator of the body fluids. We tend to regard the kidney strictly as an organ of excretion that serves to rid the body of metabolic wastes. But in fact it is really more a regulatory than an excretory organ. It is responsible for individually monitoring and regulating the concentrations of body water, of the salt ions in the blood, and of other major and minor body fluid constituents. In its task of fine-tuning the composition of the internal environment, the kidney is assisted by the other organs of exchange such as the lungs, skin, and digestive tract, as well as by many internal mechanisms. Several other specialized structures have evolved among the vertebrates that assist in body fluid regulation in various environments; for example, the salt-secreting cells of fish gills and the salt glands of birds and reptiles.

HOW AQUATIC ANIMALS MEET PROBLEMS OF SALT AND WATER BALANCE

Marine invertebrates. Most marine invertebrates are in osmotic equilibrium with their seawater environment. They have body surfaces that are permeable to salts and water so that their body fluid concentration rises or falls in conformity with changes in concentrations of seawater. Because such animals are incapable of regulating their body fluid osmotic pressure, they are referred to as **osmotic conformers.** Invertebrates living in the open sea are seldom exposed to osmotic fluctuations because the ocean is a highly stable environment. Oceanic invertebrates have, in fact, very limited abilities to withstand osmotic change. If they should be exposed to dilute seawater, they are quickly killed because their body cells cannot tolerate dilution and are helpless to prevent it. These animals are restricted to living in a narrow salinity range, and are said to be **stenohaline.** An example is the marine spider crab, represented in Fig. 12-1.

Conditions along the coasts and in estuaries and river mouths are much less constant than those of the open ocean. Here animals must be able to withstand large, and often abrupt, salinity changes as the tides move in and out and mix with fresh water draining from rivers. These animals are referred to as **euryhaline,** which means that they can survive a wide range of salinity change. Most coastal invertebrates also show varying powers of **osmotic regulation.** For example, the brackish water shore crab can resist body fluid dilution by dilute (brackish) seawater (Fig. 12-1). Although the body fluid concentration falls, it does so less rapidly than the fall in seawater concentration. This crab is said to be a **hyperosmotic regulator** because in a dilute environment it can maintain the concentration of its blood above that of the surrounding water.

What is the advantage of hyperosmotic regulation

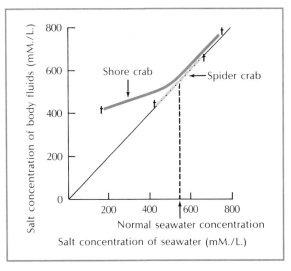

FIG. 12-1. Salt concentration of body fluids of two crabs as affected by variations in the seawater concentration. The 45-degree line represents equal concentration between body fluids and seawater. Since the spider crab cannot regulate its body-fluid salt concentration, it conforms to whatever changes happen in the external seawater environment. The shore crab, however, can regulate osmotic concentration of its body fluids to some degree because in dilute seawater the shore crab can hold its body fluid concentration above the seawater concentration. For example, when the seawater is 200 mM. per liter, the shore crab's body fluids are about 430 mM. per liter. Crosses at ends of lines indicate tolerance limits of each species.

over osmotic conformity, and how is this regulation accomplished? The advantage is that by regulating against excessive dilution, thus protecting the body cells from extreme changes, these crabs can then successfully live in the physically unstable but **biologically rich** coastal environment. Their powers of regulation are limited, however, since if the water is highly diluted, their regulation fails and they die. To understand how the brackish-water shore crab and other coastal invertebrates achieve hyperosmotic regulation, let us examine the problems they face. First, the salt concentration of the internal fluids is greater than in the dilute seawater outside. This causes a steady osmotic influx of water. As with the membrane osmometer placed in a sugar solution (p. 54), water diffuses inward because it is more concentrated outside than inside. The shore

143

crab is not nearly as permeable as a membrane os-mometer—most of its shelled body surface is in fact almost impermeable to water—but the thin respiratory surfaces of the gills are highly permeable. Obviously the crab cannot insulate its gills with an impermeable hide and still breathe. The problem is solved by removing the excess water through the action of the kidney (the antennal gland located in the crab's thorax). The second problem is salt loss. Again, because the animal is saltier than its environment, it cannot avoid loss of ions by outward diffusion across the gills. Salt is also lost in the urine. This problem is solved by special salt-secreting cells in the gills that can actively remove ions from the dilute seawater and move them into the blood, thus maintaining the internal osmotic concentration. This is an **active transport** process that requires energy because ions must be transported against a concentration gradient, that is, from a lower salt concentration (in the dilute seawater) to an already higher one (in the blood).

Freshwater animals. Some 400 million years ago our primitive marine vertebrate ancestors began to penetrate the estuaries, then gradually the rivers and creeks. Before them lay a new unexploited habitat already stocked with food, in the form of insects and other invertebrates, that had preceded them into fresh water. However, the advantages of this new habitat were traded off for a tough physiologic challenge: the necessity of developing effective osmotic regulation. Freshwater animals must keep the salt concentration of their body fluids higher than that of the water. Water therefore enters their bodies osmotically and salt is lost by diffusion outward (Fig. 12-2). Their problems are similar to the brackish-water shore crab, but more severe and unremitting. Fresh water is much more dilute than are coastal estuaries, and there is no retreat, no salty sanctuary into which the freshwater animal can retire for osmotic relief. He must, and has, become a permanent and highly efficient hyperosmotic regulator. The scaled and mucus-covered body surface of a fish is about as waterproof as any flexible surface can be. The water that inevitably enters across the gills is pumped out by the kidney (Fig. 12-2). Even though the kidney is able to make a very dilute urine some salt is lost; this is replaced by salt in food and by active absorption of salt (primarily

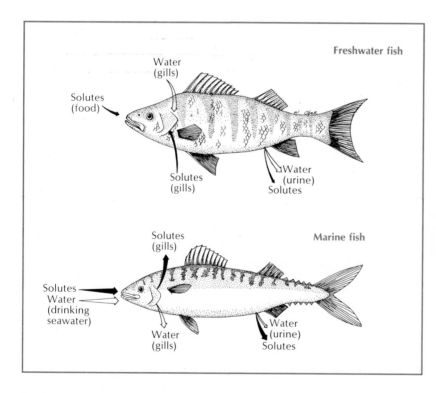

FIG. 12-2. Osmotic regulation in marine and freshwater fish. Freshwater fish (yellow perch) maintains osmotic and ionic balance in its dilute environment by actively absorbing salt across the gills (some salt enters with food) and by pumping out excess water via the kidney. Marine fish (represented by a mackeral) must drink seawater to replace osmotic water loss to its concentrated environment. Excess salt is actively secreted outward by the gills (sodium, chloride, and potassium) and kidney (magnesium, sulfate, and calcium).

sodium and chloride) across the gills. The teleost fishes that inhabit our lakes and streams today are so well adapted to their dilute surroundings that they need expend very little energy to regulate themselves osmotically.

Crayfish, aquatic insect larvae, mussels, and other freshwater animals are also hyperosmotic regulators and face the same hazards as freshwater fish; they tend to gain too much water and lose too much salt. Not surprisingly, all of these forms solved these problems in the same direct way that fish did. They excrete the excess water as urine, and they actively absorb salt from the water by some salt-transporting mechanism on the body surface.

Amphibians, when they are living in water, also must compensate for salt loss by absorbing salt from the water. They use their skin for this purpose. Physiologists learned some years ago that pieces of frog skin will continue to actively transport sodium and chloride for hours when removed and placed in a specially balanced salt solution. Fortunately for biologists, but unfortunately for frogs, these animals are so easily collected and maintained in the laboratory that frog skin has become a favorite membrane system for studies of ion-transport phenomena.

Marine fishes. The great families of bony fishes that inhabit the seas today maintain the salt concentration of their body fluids at about one third that of seawater. Obviously, they must be osmotic regulators. It may seem odd that the marine bony fishes do not take the easy way out by osmotically conforming with their environment, as do most marine invertebrates. The reason is apparently that the marine bony fish are descended directly from early freshwater fishes rather than from marine invertebrate forms. During the many millions of years that the freshwater fishes were adapting themselves so well to their environment, they established a body fluid concentration equivalent to about one-third seawater, thus setting the pattern for all the vertebrates that were to evolve later, whether aquatic, terrestrial, or aerial. The ionic composition of vertebrate body fluid is remarkably like dilute seawater too, a fact that is doubtless related to their marine heritage.

Thus it was that when some of the freshwater bony fishes ventured back to the sea, they encoun-

tered a new set of problems. Having a much lower internal osmotic concentration than the seawater around them, they lost water and gained salt. Indeed, the marine bony fish quite literally risks drying out, much like a desert mammal deprived of water. To compensate for water loss, the marine teleost drinks seawater (Fig. 12-2). This is absorbed from the intestine, and the major sea salt, sodium chloride, is carried by the blood to the gills, where it is secreted outward, back into the surrounding sea. Other ions, especially magnesium, sulfate, and calcium, are excreted by the kidney. In this roundabout way, marine fishes rid themselves of the excess sea salts they have drunk, resulting in a net gain of water, which replaces the water lost by osmosis. Samuel Taylor Coleridge's ancient mariner, surrounded by "water, water, everywhere" would doubtless have been tormented even more had he known of the marine fishes' simple solution for thirst. A marine fish carefully regulates the amount of seawater it drinks, consuming only enough to replace water loss and no more.

The cartilagenous sharks and rays (elasmobranchs) solved their water balance problems in a completely different way. This primitive group is almost totally marine. The salt composition of shark's blood is similar to the bony fishes, but also contains a large amount of organic compounds, especially urea and trimethylamine oxide. Urea is, of course, a metabolic waste that most animals quickly excrete in the urine. The shark kidney, however, conserves urea, causing it to accumulate in the blood. The blood urea, added to the usual blood electrolytes, raises the blood osmotic pressure to slightly exceed that of seawater. In this way the sharks and their kin turn an otherwise useless waste material into an asset, eliminating the osmotic problem encountered by the marine bony fishes.

HOW TERRESTRIAL ANIMALS MAINTAIN SALT AND WATER BALANCE

The problems of living in an aquatic environment seem small indeed compared to the problems of life on land. Remembering that our bodies are mostly water, that all metabolic activities proceed in water, and that the origins of life itself were conceived in water, it seems obvious that animals were meant to stay in water. Nevertheless, many animals (like the plants before them) inevitably moved onto land, carrying their watery composition with them. On land the greatest threat to life is desiccation. Water is lost by evaporation in expired air, by evaporation from the body surface, by excretion in the urine, and by elimination in the feces. These losses must be replaced by water in the food or by drinking water. Water is also formed in the cells by the metabolic oxidation of foods. This gain from **oxidation water,** as it is called, can be very significant, since water is not always available for drinking. In fact some desert mammals, such as the kangaroo rats and ground squirrels of American deserts, sand rats of the Sahara Desert, and gerbils of Old World deserts can, if necessary, derive all the water they need from their food, drinking no water at all. Such animals can produce a highly concentrated urine and form nearly solid feces. They are also nocturnal, avoiding the drying heat of the day.

The excretion of wastes presents a special problem in water conservation. The primary end-product of protein catabolism is ammonia, a highly toxic material. Fish can easily excrete ammonia across their gills, since there is an abundance of water to wash it away. The terrestrial insects, reptiles, and birds, having no convenient way to rid themselves of toxic ammonia, instead convert it into uric acid, a nontoxic, almost insoluble compound. This enables them to excrete a semisolid urine with little water loss. The use of uric acid has another important benefit. All of these animals lay eggs that are impermeable bags enclosing the embryos, their stores of food and water, and whatever wastes that accumulate during development. By converting ammonia to uric acid, the developing embryo's waste can be precipitated into solid crystals, which are stored harmlessly within the egg until hatching.

Marine birds and turtles have evolved a unique solution for excreting the large loads of salt eaten with their food. Located above each eye is a special **salt gland** (Fig. 12-3). In birds these glands are capable of excreting a highly concentrated solution of sodium chloride—up to twice the concentration of seawater. The salt solution runs out the nares, giving gulls, petrels, and other sea birds a perpetual runny nose. Marine turtles and lizards shed their salt gland secretion as salty tears. Salt glands are

FIG. 12-3. Salt glands of a marine bird (gull). One salt gland is located above each eye. Each gland consists of several lobes arranged in parallel. One lobe is shown in cross section, much enlarged. Salt is secreted into many radially arranged tubules, then flows into a central canal that leads into the nose.

important accessory organs of salt excretion to these animals, since their kidney cannot produce a concentrated urine, as can the mammalian kidney.

INVERTEBRATE EXCRETORY STRUCTURES

In such a large and varied group as the invertebrates it is hardly surprising that there is a great variety of morphologic structures serving as excretory organs. Many protozoa and some freshwater sponges have special excretory organelles called contractile vacuoles. The more advanced invertebrates have excretory organs that are basically tubular structures which form urine by first producing an ultrafiltrate or fluid secretion of the blood. This enters the proximal end of the tubule and is modified continuously as it flows down the tubule. The final product, urine, frequently has a very different composition than the blood from which it originated.

Contractile vacuole. This tiny spherical intracellular vacuole of protozoans and freshwater sponges is not a true excretory organ, since ammonia and other nitrogenous wastes of metabolism readily leave the cell by direct diffusion across the cell membrane into the surrounding water. The contractile vacuole is really an organ of water balance. Because the cytoplasm of freshwater Protozoa is considerably saltier than their freshwater environment, they tend to draw water into themselves by osmosis. In *Paramecium* (Fig. 12-4) this excess water is collected by minute canals within the cytoplasm and conveyed to the contractile vacuole. This grows larger as water accumulates within it. Finally the vacuole is emptied through a pore on the surface, and the cycle is rhythmically repeated. Although the contractile vacuole has been carefully studied, it is not yet known how this system is able to pump out pure water while retaining valuable salts within the animal. Contractile vacuoles are common in freshwater Protozoa but rare or absent from marine Protozoa, which are isosmotic with seawater and

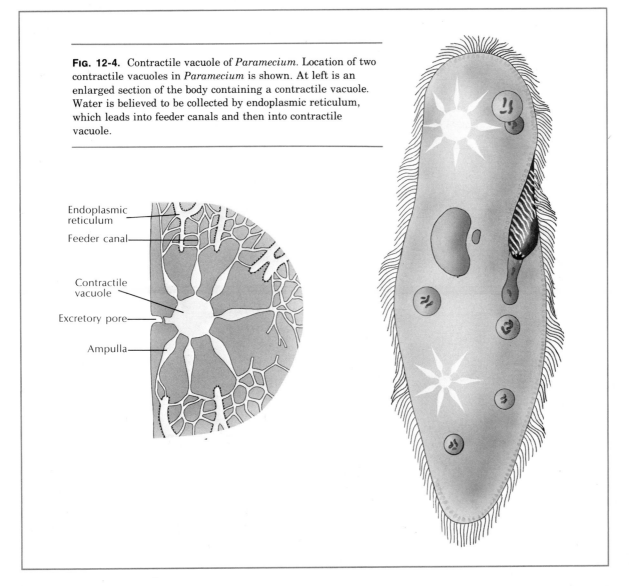

FIG. 12-4. Contractile vacuole of *Paramecium*. Location of two contractile vacuoles in *Paramecium* is shown. At left is an enlarged section of the body containing a contractile vacuole. Water is believed to be collected by endoplasmic reticulum, which leads into feeder canals and then into contractile vacuole.

Endoplasmic reticulum

Feeder canal

Contractile vacuole

Excretory pore

Ampulla

consequently neither lose nor gain too much water.

Nephridia. The nephridium is the most common type of invertebrate excretory organ. All nephridia are tubular structures, but there are large differences in degree of complexity. One of the simplest arrangements is the flame cell system (or **protonephridia**) of the flatworm. In *Planaria* and other flatworms this takes the form of two highly branched systems of tubules distributed throughout the body (Fig. 12-5). Fluid enters the system through special-ized "flame" cells, moves slowly into and down the tubules, and is excreted through pores that open at intervals on the body surface. The flame cell is a relatively large cell that is hollowed out like a cup. Each contains a bundle of cilia projecting from the base of the cup. When the living, semitransparent, animal is viewed through a microscope, the constant beating of the ciliary tufts resembles the flickering of numerous tiny flames within the animal.

Careful observations of the flame cell system have

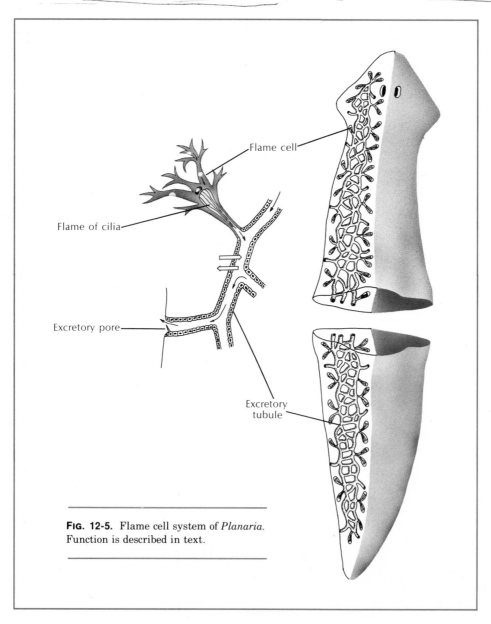

Flame cell

Flame of cilia

Excretory pore

Excretory tubule

FIG. 12-5. Flame cell system of *Planaria*. Function is described in text.

suggested that fluid containing wastes enters the flame bulb from the surrounding tissues by **pinocytosis** (cell drinking). In this process fluid-filled vesicles are formed just under the outer cell surface. The vesicles are transported across the cell and set free into the ciliated lumen of the flame cell (Fig. 12-5). Here, the rhythmic flame beat creates a negative fluid pressure that drives the fluid into the tubular portion of the system. It is believed that as the fluid passes down the tubules, the tubular epithelial cells add certain waste materials to the tubular fluid (secretion) and withdraw valuable materials from it (reabsorption) to complete the formation of urine. However, the flame cell system is so minute that no one has yet successfully collected a sample of urine for chemical analysis. It is probable that the flame cell system, like the contractile vacuole of protozoans, is primarily a water balance system, since it is best developed in

free-living freshwater forms. Branched flame cell systems are typical of primitive invertebrates that lack circulatory systems. Since there is no circulation to carry wastes to a compact excretory organ such as the kidney of higher invertebrates and vertebrates, the flame cell system must be distributed to reach the cells directly.

The protonephridium just described is a **"closed"** system, that is, the urine is formed from a fluid that must first enter the tubule by being transported across the flame cells. Another type of nephridium, typical of many annelids (segmented worms), is the **open,** or "true," nephridium. In the earthworm *Lumbricus* there are paired nephridia in every segment of the body, except the first three and the last one (Fig. 12-6). Each nephridium occupies parts of two successive segments. In the earthworm each nephridium is a tiny, self-contained "kidney" that independently drains to the outside through pores

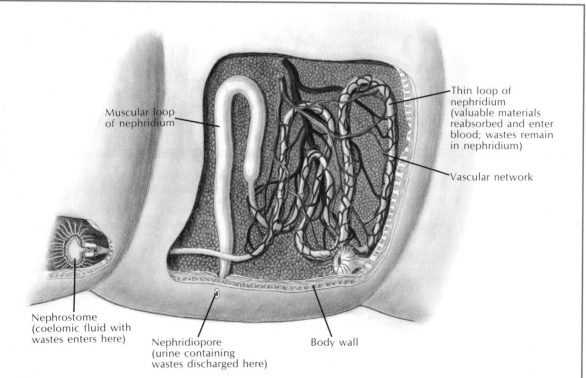

FIG. 12-6. Nephridium of earthworm. Wastes are drawn into ciliated nephrostome in one segment, then passed through loop of nephridium and expelled through nephridiopore of next segment.

(**nephridiopores**) in the body wall; in other worm species the nephridia may drain into a common duct that empties posteriorly in the gut.

Coelomic fluid containing wastes to be excreted is swept into a ciliated, funnellike opening (**nephrostome**) of the nephridium and carried through a long, twisted tubule of increasing diameter. It then enters a bladder and is finally expelled through a nephridiopore to the outside. The nephridial tubule is surrounded by an extensive network of blood vessels. Solutes, especially sodium and chloride, are reabsorbed from the formative urine during its travel through the tubule.

By puncturing the nephridial tubule with minute glass pipettes and withdrawing and analyzing the urine using ultramicro procedures, it was shown that by the time the urine reached the bladder, it was diluted to about 20% of its original concentration. This means that about 80% of the original urine solutes, mostly salts, were reabsorbed. Since the earthworm lives in an unusually moist environment, the ability to form a dilute (hypoosmotic)

FIG. 12-7. Malpighian tubules of insect. Malpighian tubules are located at juncture of mid- and hindgut (rectum) as shown in the cutaway of a wasp, **A.** Function of malpighian tubules is shown in **B.** Solutes, especially potassium, are actively secreted into the tubules. Water and wastes follow. This fluid moves into the rectum where solutes and water are actively reabsorbed, leaving wastes to be excreted.

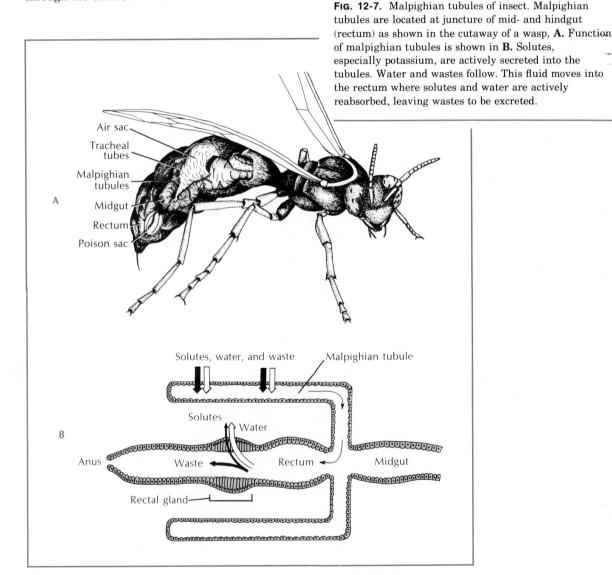

urine is advantageous because excess water can be excreted without too great a loss of salt. The vascular network that surrounds the nephridium functions to carry away solutes reabsorbed by the tubule. The tubules may also pick up wastes directly from the blood and secrete them into the tubule. The addition of the blood vascular network to the annelid nephridium makes it a much more versatile and effective system than the flame cell system. However, the basic process of urine formation is the same: fluid flows continuously through a tubule while materials are added here and taken away there, until urine is formed.

Crustacean kidneys. The excretory organs of crustaceans are usually a single pair of tubular structures located in the ventral part of the head anterior to the esophagus. They are frequently called **antennal glands** because they open at the base of the antennae. The antennal glands resemble nephridia in that they are long tubular structures surrounded by blood capillaries, but unlike nephridia they lack open nephrostomes. Each antennal gland is composed of a closed end sac, an excretory tubule, and an enlarged bladder. A protein-free filtrate of the blood (ultrafiltrate) is formed in the end sac by the hydrostatic pressure of the blood. The filtrate passes from the end sac into the coiled excretory tubule, where it is modified by the selective reabsorption of certain salts and the active secretion of others. As with most aquatic invertebrates, the crustacean excretory organ functions principally to regulate the ionic and osmotic composition of the body fluids. Freshwater crustaceans such as the crayfish are constantly threatened with overdilution of water, which enters osmotically across the gills and other water-permeable surfaces of the body. The kidney, by forming a dilute, low-salt urine, acts as an effective flood control device. In marine crustaceans such as lobster and crabs the kidney functions to adjust the salt composition of the blood by selective modification of the salt content of the tubular urine. In these forms the urine remains isosmotic to the blood.

Insect malpighian tubules. Insects and spiders have a unique excretory system consisting of **malpighian tubules** that operate in conjunction with specialized glands in the wall of the rectum. The malpighian tubules, variable in number, are thin, elastic, blind tubules attached to the juncture be-

tween the mid- and hindguts (Fig. 12-7, *A*). The free ends of the tubules lie free in the hemocoele and are bathed in blood (hemolymph).

Since the malpighian tubules are closed and lack an arterial supply, urine formation cannot be initiated by blood ultrafiltration as in the crustaceans and vertebrates. Instead potassium is actively secreted into the tubules (Fig. 12-7, *B*). This primary secretion of ions pulls water along with it by osmosis to produce a potassium-rich fluid. Other solutes and waste materials also are secreted or diffuse into the tubule. The fluid, or "urine," then drains from the tubules into the intestine. In the rectum specialized rectal glands actively reabsorb most of the potassium and water, leaving behind wastes such as uric acid. By cycling potassium and water in this way, insects living in dry environments may reabsorb nearly all water from the rectum, producing a nearly dry mixture of urine and feces. We cannot judge to what extent this unique excretory system has contributed to the great success of this most abundant and widespread group of land animals. However, the great diversity of habitats occupied by insects is certainly mute testimony to the adaptability of the malpighian tubule system.

VERTEBRATE KIDNEY EVOLUTION

How did the vertebrate kidney evolve? Unfortunately, the historic record is virtually nonexistent because the kidney, like other soft internal organs, does not fossilize. However, a living embryologic record does exist. During its development, the embryo of a species retraces a remarkably accurate record of its evolutionary history. An embryo is an archive, so to speak, of that species' past. To be sure, the record is a rather slurred and modified one. Nevertheless, embryonic development is often the best evidence we have of the evolution of an organ system that leaves no fossil record for our benefit. We will have more to say about the convenient conservatism of embryology in Chapter 19.

From studies of vertebrate embryos, as well as adults, biologists have pieced together the following story. The ancestral vertebrate kidney extended the length of the coelomic cavity and was made up of segmentally arranged uriniferous tubules. Each tubule opened at one end into the coelom by a

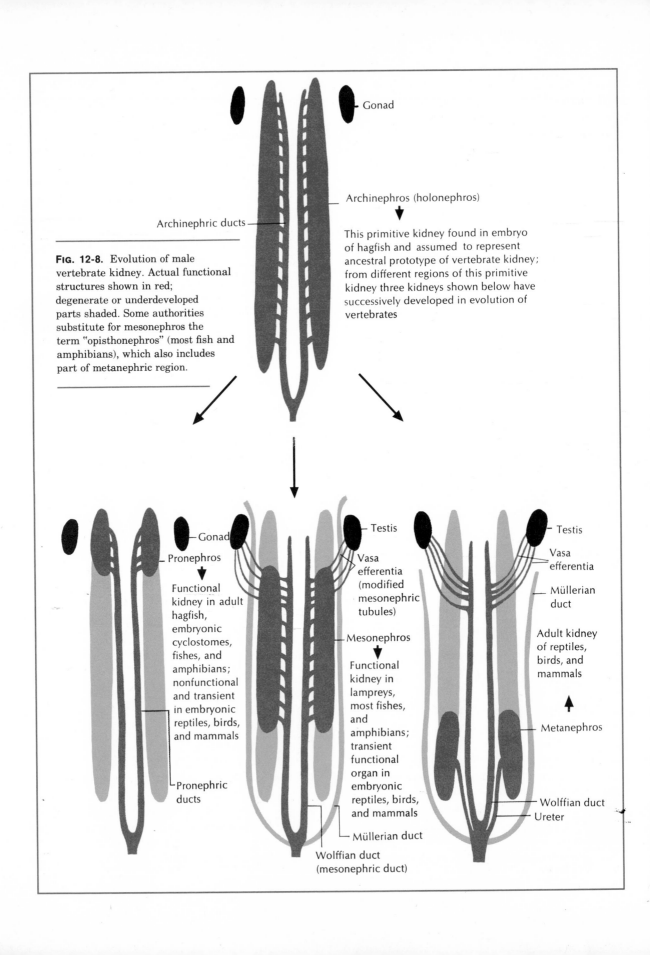

Gonad

Archinephros (holonephros)

Archinephric ducts

FIG. 12-8. Evolution of male vertebrate kidney. Actual functional structures shown in red; degenerate or underdeveloped parts shaded. Some authorities substitute for mesonephros the term "opisthonephros" (most fish and amphibians), which also includes part of metanephric region.

This primitive kidney found in embryo of hagfish and assumed to represent ancestral prototype of vertebrate kidney; from different regions of this primitive kidney three kidneys shown below have successively developed in evolution of vertebrates

Gonad

Pronephros

Functional kidney in adult hagfish, embryonic cyclostomes, fishes, and amphibians; nonfunctional and transient in embryonic reptiles, birds, and mammals

Pronephric ducts

Testis

Vasa efferentia (modified mesonephric tubules)

Mesonephros

Functional kidney in lampreys, most fishes, and amphibians; transient functional organ in embryonic reptiles, birds, and mammals

Müllerian duct

Wolffian duct (mesonephric duct)

Testis

Vasa efferentia

Müllerian duct

Adult kidney of reptiles, birds, and mammals

Metanephros

Wolffian duct

Ureter

nephrostome and at the other end into the common archinephric duct. This ancestral kidney has been called an **archinephros,** and it can still be seen in embryos of hagfish, the most primitive living vertebrate (Fig. 12-8). Kidneys of higher vertebrates developed from this primitive plan. Embryologic evidence indicates that there are three generations of kidneys: pronephros, mesonephros, and metanephros. In all vertebrate embryos, the pronephros is the first and most primitive kidney to appear. As its name implies, it is located anteriorly in the body. It becomes the persistent kidney of adult hagfishes. In all other vertebrates it degenerates during development and is replaced by a more centrally located and more structurally advanced kidney, the mesonephros. The mesonephros becomes the persistent kidney of adult fishes and amphibians. But in the developing embryos of amniotes (reptiles, birds, and mammals) the mesonephros is replaced in turn by the metanephros. The metanephros develops behind the mesonephros and is structurally and functionally the most advanced of the three kidney types. Thus three kidneys are formed in succession, each more advanced and each located more caudally than its predecessor. Kidney evolution defies the general rule that best things should move to the front! Of course, there was little choice. The pronephros occupied the most anterior area of the archinephric region, making it necessary for subsequent kidneys to develop in less exclusive sites to the rear.

In summary, we may state that the evolutionary sequence of adult vertebrate kidneys has been archinephros, pronephros, mesonephros, and metanephros.

VERTEBRATE KIDNEY FUNCTION

The kidneys of man and other vertebrates play a critical role in the body's economy. As vital organs their failure means death; in this respect they are neither more nor less important than are the heart, lungs, or liver, for example. The kidney is part of many interlocking mechanisms that maintain **homeostasis** — constancy of the internal environment. However, the kidney's share in this regulatory council is an especially large one. It must, and does, individually monitor and regulate most of the major

constituents of the blood and several minor constituents as well. In addition it silently labors to remove a variety of potentially harmful substances that animals deliberately or unconsciously eat, drink, or inhale.

Perhaps even more remarkable than the job the kidney does is the way in which it does it. These small organs, which in man weigh less than 0.5% of the body's weight, receive nearly 25% of the total blood flow (cardiac output), amounting to about 2,000 quarts of blood per day. This vast blood flow is channeled to approximately 2 million nephrons, which comprise the bulk of the two human kidneys. Each nephron is a tiny excretory unit consisting of a pressure filter **(glomerulus)** and a long **nephric tubule.** Urine formation begins in the glomerulus where an ultrafiltrate of the blood is squeezed into the nephric tubule by the hydrostatic blood pressure. The ultrafiltrate then flows steadily down the twisted tubule. During its travel some substances are added to, and others are subtracted from, the ultrafiltrate. The final product of this process is urine.

All mammalian kidneys are paired structures that lie embedded in fat, anchored against the dorsal abdominal wall. Each kidney contains a medial indentation, the **hilus,** which is the site of entry and exit of blood vessels and which leads to the **ureter.** The two ureters, 10 to 12 inches long in man, extend to the dorsal surface of the urinary bladder. Urine is discharged from the bladder by way of the **urethra** (Fig. 12-9). In the male the urethra is the terminal portion of the reproductive system as well as of the excretory system. In the female the urethra is solely excretory in function, opening to the outside just anterior to the vagina.

Urine formed by the kidney nephrons drains into the renal pelvis and then into the muscular ureter, which conveys it to the urinary bladder by rhythmic muscular contractions. Once urine has entered the bladder, a small flap of epithelial tissue acts as a valve to prevent backflow into the urethra. The urinary bladder is a highly expansible muscular bag. As the bladder becomes distended, stretch receptors located in the muscular walls are stimulated, producing nerve impulses that travel to the brain to signal the sensation of fullness. The release of urine from the bladder is regulated by a muscular sphinc-

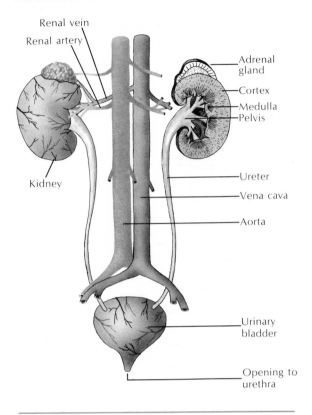

Renal vein
Renal artery
Adrenal gland
Cortex
Medulla
Pelvis
Kidney
Ureter
Vena cava
Aorta
Urinary bladder
Opening to urethra

FIG. 12-9.
Anatomy of
human urinary system.

ter located at the juncture of bladder and urethra. Voiding of urine is involuntary in infants but comes under voluntary control during early childhood.

Since each of the thousands of nephrons in the kidney forms urine independently, each is, in a way, a tiny, self-contained kidney that produces a miniscule amount of urine—perhaps only a few nanoliters per hour. This amount, multiplied by the number of nephrons in the kidney, produces the total urine flow. The kidney is an "in parallel" system of independent units. However, as we will see later, these "independent" nephrons actually work together to create large osmotic gradients in the kidney medulla. This makes it possible for the mammalian kidney to concentrate urine well above the salt concentration of the blood.

As indicated above, the nephron, with its pressure filter and tubule, is intimately associated with the

blood circulation (Fig. 12-10). Blood from the aorta is delivered to the kidney by way of the large **renal artery,** which breaks up into a branching system of smaller arteries. The arterial blood flows to each nephron through an **afferent arteriole** to the **glomerulus,** which is a tuft of blood capillaries enclosed within a thin, cuplike **Bowman's capsule.** Blood leaves the glomerulus via the **efferent arteriole.** This vessel immediately breaks up again into an extensive system of capillaries, the **peritubular capillaries,** which completely surround the nephric tubules. Finally, the blood from these many capillaries is collected by veins that unite to form the **renal vein.** This vein returns the blood to the vena cava.

Let us return now to the glomerulus, where the process of urine formation begins. The glomerulus acts as a specialized mechanical filter in which a protein-free filtrate resembling plasma is driven by the blood pressure across the capillary walls and into the fluid-filled space of Bowman's capsule. This plasma filtrate (called an **ultrafiltrate** because the plasma proteins have been filtered out) begins to flow down the nephric tubule. The nephric tubule consists of several segments. The first segment, the **proximal convoluted tubule,** leads into a long, thin-walled, hairpin loop called the **loop of Henle** (Fig. 12-10). This loop drops deep into the medulla of the kidney, then returns to the cortex to join the third segment, the **distal convoluted tubule.** The collecting duct empties into the kidney **pelvis,** a cavity that collects the urine before it passes into the **ureter,** on its way to the **urinary bladder** (Fig. 12-9).

The ultrafiltrate that enters this complex tubular system must undergo extensive modification before it becomes urine. About 200 L. of filtrate are formed each day by the average person's kidneys. Obviously the loss of this volume of body water, not to mention the many other valuable materials present in the filtrate, cannot be tolerated. How does tubular action convert the plasma filtrate into urine? To answer this question researchers developed tubular micropuncture techniques. After exposing the kidneys of anesthetized amphibians or mammals, tiny glass pipettes with sharpened tips too small to be seen with the naked eye were used to withdraw samples of formative urine from different segments

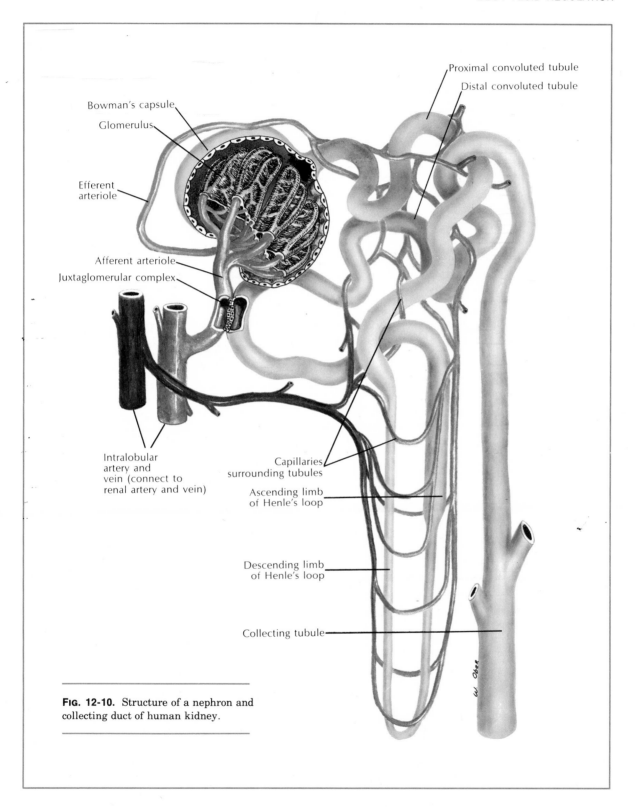

Proximal convoluted tubule
Distal convoluted tubule
Bowman's capsule
Glomerulus
Efferent arteriole
Afferent arteriole
Juxtaglomerular complex
Intralobular artery and vein (connect to renal artery and vein)
Capillaries surrounding tubules
Ascending limb of Henle's loop
Descending limb of Henle's loop
Collecting tubule

FIG. 12-10. Structure of a nephron and collecting duct of human kidney.

PHYSIOLOGY OF ANIMALS

of the tubule. The samples were chemically analyzed with specially developed ultramicro analytical techniques. Such experiments showed that most of the water and solutes were reabsorbed by the proximal tubule. Some vital materials such as glucose and amino acids were completely reabsorbed. Others such as sodium, chloride, and most other minerals underwent variable reabsorption. That is, some were strongly reabsorbed, others weakly reabsorbed, depending on the body's need to conserve each min-

eral. Much of this reabsorption is by **active transport,** in which cellular energy is used to transport materials from the tubular fluid, across the cell, and into the peritubular blood that will return them to the general circulation. For most substances there is an upper limit to the amount of substance that can be reabsorbed. This upper limit is termed the **transport maximum** for that substance. For example, glucose is normally completely reabsorbed by the kidney because the transport maximum for the

FIG. 12-11. Mechanism of countercurrent multiplier system of mammalian kidney. Relative concentration of sodium and other solutes is represented by density of red dots inside tubule, and intensity of color outside tubule. Concentration increases from top (cortex) to bottom (medulla). Final water absorption occurs in collecting duct.

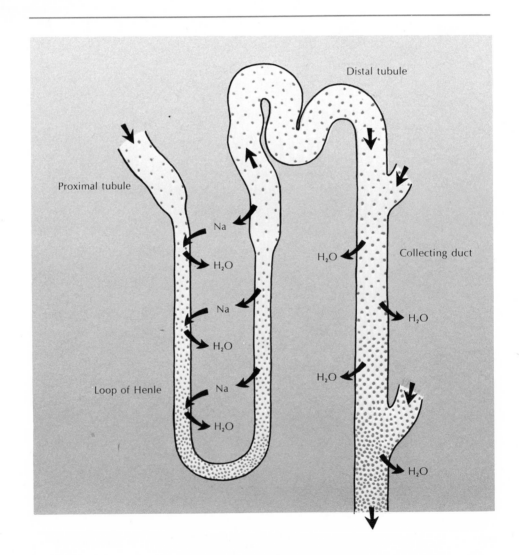

glucose reabsorptive mechanism is well above the amount of glucose normally present in the plasma filtrate. However, in the untreated disease diabetes mellitus the blood-sugar level may rise so high that the sugar reabsorptive mechanism becomes saturated and unable to reabsorb all the glucose in the filtrate. Glucose then appears in the urine.

In addition to reabsorbing large amounts of materials from the plasma filtrate, the kidney tubules are able to secrete certain substances into the tubular fluid. This process, which is the reverse of tubular reabsorption, enables the kidney to build up the urine concentrations of materials to be excreted, such as hydrogen and potassium ions, drugs, and various foreign organic materials. The distal tubule is the site of most tubular secretion.

The total osmotic pressure of the blood is carefully regulated by the kidney. When fluid intake is high, the kidney excretes a dilute urine, saving salts and excreting water. When fluid intake is low, the kidney conserves water by forming a concentrated urine. A dehydrated man can concentrate his urine to four times his blood concentration. Yet certain desert rodents, such as the gerbil, can produce a urine fourteen times more concentrated than the blood!

The ability to form a concentrated urine is closely correlated with the length of the loops of Henle that reach down into the kidney medulla. This loop system functions as a **countercurrent multiplier sys-** tem (Fig. 12-11). Sodium is actively reabsorbed from the water-impermeable ascending limb of Henle's loop, enters the surrounding tissue fluid, and diffuses into the descending loop. By cycling sodium between the two opposing limbs, the concentration of urine becomes multiplied in the bottom of the loop. A tissue fluid osmotic concentration is established that is greatest at the bottom of the loop deep in the medulla and lowest at the top of the loop in the cortex. The actual concentrating of the urine, however, does not occur in the loops of Henle, but in the collecting ducts that lie parallel to the loops of Henle. As the urine flows down the collecting duct into regions of increasing sodium concentration, water is osmotically withdrawn from the urine. The amount of water saved, and the final concentration of the urine, depends on the permeability of the walls of the collecting duct. This is controlled by the antidiuretic hormone (ADH) that is released by the posterior pituitary gland (neurohypophysis). The release of this hormone is governed in turn by special receptors in the brain that constantly sense the osmotic pressure of the blood. When the blood osmotic pressure drops, as during dehydration, more antidiuretic hormone is released, collecting duct permeability is increased, more water is withdrawn from the urine, and the urine becomes concentrated. An opposing sequence of events occurs during overhydration.

13

Nervous coordination and integration

NERVOUS SYSTEM

The origin of the nervous system is based on one of the fundamental principles of protoplasm—irritability. Each cell responds to stimulation in a manner characteristic of that type of cell. But certain cells have become highly specialized for receiving stimuli and for conducting impulses to various parts of the body. Through evolutionary changes, these cells have become the most complex of all body systems—the nervous system. The endocrine system is also used for coordination, but the nervous system has a wider and more direct control of body functions than does the endocrine system.

The evolution of the nervous system has been correlated with the development of bilateral symmetry and cephalization. Along with this development, animals acquired exteroceptors and associated ganglia. The basic plan of the nervous system is to code the sensory information, internally or externally, and transmit it to regions of the central nervous system where it is processed into appropriate action. This action may be of several types, such as simple reflexes, automatic behavior patterns, conscious perception, or learning processes.

NERVOUS SYSTEM OF VERTEBRATES

Vertebrates have, as a rule, a brain much larger than the spinal cord. In lower vertebrates this difference is not marked, but higher in the vertebrate kingdom the brain increases in size, reaching its maximum in mammals, especially man, in which it may be larger than all the rest of the nervous system. Along with this enlargement has come an increase in complexity, bringing better patterns of coordination, integration, and intelligence. The nervous system is commonly divided into central and peripheral parts; the central division is chiefly concerned with integrative activity and the peripheral part with the reception of stimuli from the internal and external environment. The peripheral division also transmits impulses to and from the central nervous system.

THE NEURON: STRUCTURAL AND FUNCTIONAL UNIT OF THE NERVOUS SYSTEM

The neuron is a nerve cell body and all its processes. The structure of a neuron was shown in Fig. 6-7, p. 72. From the nucleated nerve cell body extend several **dendrites,** rootlets that carry impulses toward the cell body, and a single, branchlike **axon,** that carries impulses away. The axons are usually covered with an insulation—a soft, white myelin sheath. This in turn is enclosed by an outer membrane, the neuralemma.

Neurons are commonly divided into three types— **motor, sensory,** and **association** or **connector.** The dendrites of sensory neurons are connected to a **receptor,** and their axons are connected to other neurons; associators are connected only to other neurons; and motor neurons are connected by their axons to an **effector.** Nerves are actually made up of many nerve processes—axons or dendrites or both—bound together with connective tissue. The cell bodies of these bundles of nerves are located either in ganglia or somewhere in the central nervous system (brain or spinal cord).

NATURE OF THE NERVE IMPULSE

The nerve impulse is the chemical-electrical message of nerves, the common functional denominator of all nervous system activity. Despite the incredible complexity of the nervous system of advanced animals, nerve impulses are basically alike in all nerves and in all animals. It is an **all-or-none** phenomenon; either the axon is conducting an impulse, or it is not. The only way an axon can vary its effect on the tissue it innervates is by changing the **frequency** of impulse conduction. Frequency change is the language of a nerve fiber. A fiber may conduct no impulses at all, or a very few per second up to a maximum approaching 1,000 per second. The higher the frequency (or rate) of conduction, the greater is the level of excitation. The same general rule applies to sense organs as well; that is, the more a sense organ is excited by a stimulus, the greater is the frequency of impulses sent out over the axons of the sensory nerves.

Resting potential. To understand what happens when an impulse is transmitted down a fiber, we need to know something about the resting, undisturbed fiber. Nerve cell membranes, like all cell membranes, have special permeability properties that create ionic imbalances. Sodium and chloride predominate on the outside, whereas potassium ions are more common inside (Fig. 13-1). These differences are quite dramatic; there is about ten times more sodium outside than in and twenty-five to thirty times more potassium inside than out. However, the nerve cell membrane is much more permeable to potassium than to sodium. The result is that potassium ions tend to leak outward much more rapidly than sodium ions tend to leak inward. Since both potassium and sodium ions are positively charged, they repel each other. Thus the outward movement of potassium is checked by the sodium, causing the outside of the membrane to become positively charged (Fig. 13-1). This, then, is the origin of the resting transmembrane potential, which is positive outside, negative inside. This potential difference called the **resting potential** is usually about -70 mV., inside negative.

Action potential. The nerve impulse is a rapidly moving change in electrical potential called the **action potential** (Fig. 13-2). This is a very rapid and brief depolarization of the axon membrane; in fact, not only is the resting potential abolished, but in most nerves the potential actually reverses for an instant so that the outside becomes negative as compared to the inside. Then, as the action potential moves ahead, the membrane returns to its normal resting potential, ready to conduct another impulse. The entire event occupies only a fraction of a millisecond. Perhaps the most significant property of the nerve impulse is that it is **self-propagating;** that is, once started, the impulse moves ahead automatically, much like the burning of a fuse.

What causes the reversal of polarity in the cell membrane during passage of an action potential? Careful studies have shown that when the action potential arrives at a given point, the cell membrane suddenly becomes much more permeable to sodium ions than before. Sodium rushes in. Actually only an extremely small amount of sodium traverses the membrane in that instant—less than one one-millionth of the sodium outside—but this brief shift of positive ions inward causes the membrane potential to disappear, even reverse. An electrical "hole" is created. Potassium, now finding its electrical barrier gone, begins to move out. Then, as the action potential passes on, the membrane quickly regains its resting properties. It becomes once again practically impermeable to sodium, and the outward movement of potassium is checked. The rising phase of

159

the action potential is associated with the rapid influx (inward movement) of sodium (Fig. 13-2). When the action potential reaches its peak, the somewhat slower outflux (outward movement) of potassium causes the potential to reverse again, and the action potential falls toward the resting level.

Sodium pump. The resting cell membrane has a very low permeability to sodium. Nevertheless, some sodium ions leak across, even in the resting condition. When the axon is active, sodium flows inward with each passing impulse, and although the amount is very small, it is obvious that the ionic gradient would eventually disappear if the sodium ions were not moved back out again. This is done

by a **"sodium pump"** located in the axon plasma membrane. Although no one has ever actually seen the "pump," we do know quite a bit about it because it has been the object of intense biochemical and biophysical studies. The sodium pump is an active transport device capable of combining with sodium on the inside surface of the membrane, then moving to the outside surface where the sodium is released. It is probably composed of phosphate-containing carrier molecules. The sodium pump requires energy, since it is moving sodium "uphill" against the sodium electrical and concentration gradient. This energy is supplied by ATP through cellular metabolic processes. There is evidence that in some cells sodium transport outward is linked to potassium transport inward; the same carrier molecule may act as a two-way shuttle, carrying ions on both trips across the membrane. This kind of pump is called a sodium-potassium pump.

Synapses: junction points between nerves. A synapse is found at the end of a nerve axon, where it connects to the dendrite or cell body of the next neuron. At this point the membranes are separated by a narrow gap having a very uniform width of about 20 millimicrons. The synapse is of great functional importance because it acts as a one-way valve that allows nerve impulses to move in one direction only. It is also through the many synapses that information is modulated from one nerve to the next.

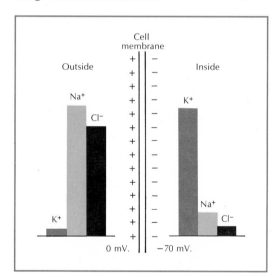

FIG. 13-1. Ionic composition inside and outside a resting nerve cell. An active sodium pump located in the cell membrane drives sodium to the outside, keeping its concentration low inside. Potassium concentration is high inside, and although the membrane is "leaky" to potassium, this ion is held inside by repelling positive charge outside the membrane.

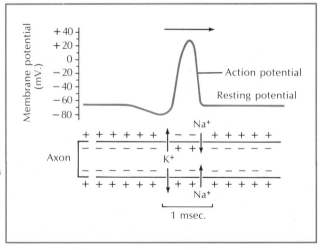

FIG. 13-2. Action potential of nerve impulse. The electrical event, moving from left to right, is associated with rapid changes in membrane permeability to sodium and potassium ions. When the impulse arrives at a point, sodium ions suddenly rush in, making the axon positive inside and negative outside. Then the sodium holes close and potassium holes open up. This restores the normal negative resting potential.

The axon of most nerves divides at its end into many branches, each of which bears a synaptic knob that sits on the cell body of the next nerve (Fig. 13-3). The axon terminations and knobs of several nerves may almost cover a nerve cell body and its dendrites. An impulse coming down a nerve axon sprays out into the many branches and synaptic endings on the next nerve cell. Many impulses therefore converge at the cell body at one moment. But these may not excite the cell body enough to fire off an impulse. A neuron requires much prompting to fire. Usually many impulses must arrive at the cell body simultaneously, or within a very brief interval, to raise the cell body to its firing threshold

level. This is called **summation;** it is the cumulative excitatory effect of many incoming impulses that pushes a nerve cell up to firing threshold.

The synapse is a kind of chemical bridge. The electron microscope shows that each synaptic knob contains numerous synaptic vesicles. These are filled with molecules of **chemical transmitter** (Fig. 13-3, *B*). When an impulse arrives at the knob, it induces some of the vesicles to move to the base of the knob and release their contents of transmitter molecules into the synaptic cleft (Fig. 13-3, *C*). These move rapidly across the narrow cleft to act on the nerve-cell membrane below. A small potential is produced in the postsynaptic membrane. If reinforced by the

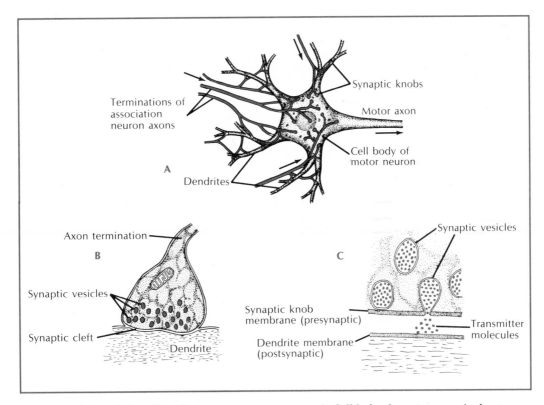

FIG. 13-3. Transmission of impulses across nerve synapses. **A,** Cell body of a motor nerve is shown covered with the terminations of association neurons. Each termination ends in a synaptic knob; hundreds of synaptic knobs may be on a single nerve cell body and its dendrites. **B,** Synaptic knob enlarged sixty times over **A.** An impulse traveling down the axon will cause some synaptic vesicles to move down to the synaptic cleft and rupture, releasing transmitter molecules into the cleft. **C,** Synaptic cleft as it might appear under a high-resolution electron microscope. Transmitter molecules from a ruptured synaptic vesicle move quickly across the gap to produce an electrical potential change in the postsynaptic membrane.

arrival of more impulses and by the release of more packets of transmitter molecules, either at the same or adjacent synapses, the small potential may be built up into a large one, sufficient to fire off an impulse in a nerve cell.

Synapses, then, are critical determinants in nervous system function. Although a nerve impulse is an all-or-none event, the synapses act like variable gates that may or may not allow impulses to proceed from one neuron to the next.

REFLEX ARC AS FUNCTIONAL UNIT

Neurons work in groups called **reflex arcs** (Fig. 13-4). There must be at least two neurons in a reflex arc, but usually there are more. The parts of a typical reflex arc consist of (1) a **receptor,** a sense organ in the skin, muscle, or other organ; (2) an **afferent** or sensory neuron, which carries the impulse toward the central nervous system; (3) a **nerve center,** where synaptic junctions are made between the sensory neurons and the association neurons; (4) the **efferent** or motor neuron, which makes synaptic junction with the association neuron and carries impulses out from the central nervous system; and (5) the **effector,** by which the animal responds to its environmental changes. Examples of effectors are muscles, glands, cilia, nematocysts of coelenterates, electric organs of fish, and chromatophores.

A reflex arc at its simplest consists of only two neurons—a sensory (afferent) neuron and a motor

(efferent) neuron. Usually, however, association neurons are interposed (Fig. 13-4). Association neurons may connect afferent and efferent neurons on the same side of the spinal cord, connect them on opposite sides of the cord, or connect them on different levels of the spinal cord, either on the same or opposite sides. In almost any reflex act a number of reflex arcs are involved. For instance, a single afferent neuron may make synaptic junctions with many efferent neurons. In a similar way an efferent neuron may receive impulses from many afferent neurons. In this latter case the efferent neuron is referred to as the **final common path.**

A **reflex act** is the response to a stimulus carried over a reflex arc. It is **involuntary** and may involve the cerebrospinal or the autonomic nervous divisions of the nervous system. Many of the vital processes of the body such as breathing, heartbeat, diameter of blood vessels, sweat glands, and others are reflex actions. Some reflex acts are inherited and innate; others are acquired through learning processes (conditioned).

ORGANIZATION OF THE NERVOUS SYSTEM

The basic plan of the vertebrate nervous system is a dorsal longitudinal hollow nerve cord that runs from head to tail. During early embryonic development the anterior end of this cord expands to form a series of vesicles, at first three and later five in number (Table 13-1). The three-part **brain** is made

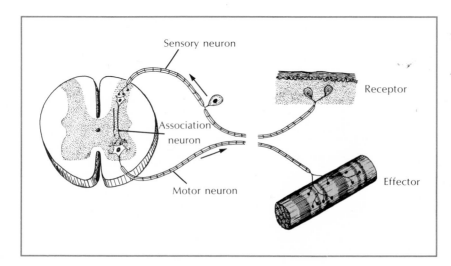

FIG. 13-4. Reflex arc. Impulse generated in the receptor is conducted over a sensory nerve to the spinal cord, relayed by an association neuron to a motor nerve cell body, and by the motor axon to an effector.

TABLE 13-1. Division of the vertebrate brain

Embryonic vesicles		Main components in adults
Forebrain	Telencephalon	Olfactory bulbs
		Cerebrum
	Diencephalon	Thalamus
		Hypothalamus
Midbrain	Mesencephalon	Optic lobes
Hindbrain	Metencephalon	Cerebellum
		Pons
	Myelencephalon	Medulla

up of prosencephalon, mesencephalon, and rhombencephalon (forebrain, midbrain, and hindbrain). The prosenecephalon and rhombencephalon each divide again to form the five-part brain characteristic of the adults of all vertebrates. The five-part brain includes the telencephalon, diencephalon, mesencephalon, metencephalon, and myelencephalon. From these divisions the different functional brain structures arise.

Central nervous system. The central nervous system is composed of the brain and spinal cord.

Spinal cord. The spinal cord, averaging in man about 18 inches long, is protected by three layers of **meninges** (Fig. 13-5)—the **dura mater, arachnoid,** and **pia mater.** Spaces between these protective layers contain cerebrospinal fluid that forms a protective cushion.

In cross section the cord shows two zones—an inner H-shaped zone of gray matter, made up of nerve cell bodies, and an outer zone of white matter, made up of nerve bundles of axons and dendrites (Fig. 13-5). The gray matter contains association neurons and the cell bodies of motor neurons. Just outside the cord on each side of the dorsal region are the **dorsal root ganglia,** which contain the cell bodies of the sensory neurons.

The white matter of the cord is made up of bundles of nerves of similar functions—the **ascending tracts,** carrying impulses to the brain, and the **descending tracts,** carrying impulses away from the brain. The sensory (ascending) tracts are located

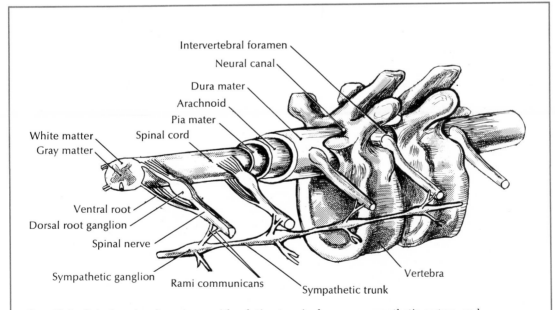

FIG. 13-5. Spinal cord and meninges with relation to spinal nerves, sympathetic system, and vertebrae. Three coats of meninges have been partly cut away to expose spinal cord. Only two vertebrae are shown in position.

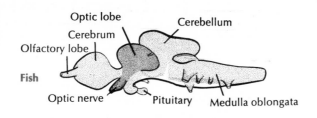

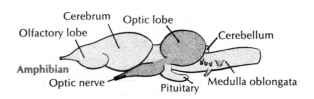

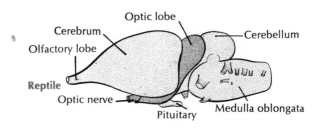

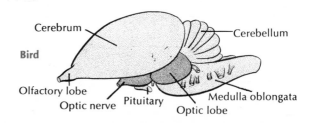

FIG. 13-6. Comparative structure of principal brain divisions in different vertebrate groups—fish, amphibian, reptile, bird, and mammal. Note progressive enlargement of cerebrum in higher groups. Stubs of certain cranial nerves are shown.

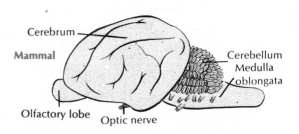

mainly in the dorsal part of the cord; the motor (descending) tracts are found ventrally and laterally in the cord. Fibers also cross over from one side of the cord to the other, the sensory fibers crossing at a higher level than the motor fibers.

Brain. The brain in vertebrates shows an evolution from the linear arrangement in lower forms (fish and amphibians) to the much-folded and enlarged brain found in higher vertebrates (birds and mammals) (Fig. 13-6). The brain is really the enlarged anterior end of the spinal cord. The ratio between the weight of the brain and spinal cord affords a fair criterion of an animal's intelligence. In fish and amphibians this ratio is about 1:1, in man the ratio is 55:1 — in other words, the brain is fifty-five times heavier than the spinal cord.

The evolutionary trend toward a dominant brain has been correlated with better integration and better mastery over the environment. The brain of vertebrates has steadily advanced, with improved sensory reception and a better nervous adjustment for living in a complex environment. The cerebral cortex, also called the pallium, has shown the greatest evolutionary advancement. In fish and most amphibians the cerebrum (roof of the telencephalon) is almost nonexistent, consisting only of a thin layer (the archipallium) chiefly concerned with smell. This region represented mainly by olfactory lobes in higher vertebrates was instrumental in food finding. A rudimentary cerebrum is found in reptiles, but in mammals an additional formation, the neocerebrum, is the largest and most important part of the central nervous system.

The brain is made up of both white and gray matter, with the gray matter on the outside, mostly in the convoluted **cortex.** In the deeper parts of the brain the white matter of nerve fibers connects the cortex with lower centers of the brain and spinal cord or connects one part of the cortex with another. Also in deeper portions of the brain are collections of nerve cell bodies (gray matter) that provide synaptic junctions between the neurons of the higher centers and those of lower centers.

There are also nonnervous elements in the nervous system such as connective, supporting, and capsule cells and the **neuroglia.** Neuroglia cells greatly outnumber the neurons and play various vital roles in the functioning of the neurons. One of their functions is to bind together the nervous tissue proper. Another is their activity in pathologic processes of regeneration. Neuroglia cells are unfortunately the chief source of tumors of the central nervous system.

The human brain is the most complex structure known to man. It has no parallel in the living or nonliving world. This "great ravelled knot," as the British physiologist Sir Charles Sherrington called man's brain, may in fact be so complex that it will never be able to understand its own function.

The main divisions of the brain are given in Table 13-1. The **medulla,** the most posterior division of the brain, is really a conical continuation of the spinal cord. The medulla contains nerve centers that control many vital processes such as heartbeat, respiration, vasomotor control, swallowing, and others.

The **pons,** between the medulla and the midbrain, is made up of a thick bundle of fibers that carries impulses from one side of the cerebellum to the other.

Between the pons and the thalamus and in front of the cerebellum lies the **midbrain.** The white matter of the midbrain consists of ascending and descending tracts that go to the thalamus and cerebrum. On the upper side of the midbrain are the rounded **optic lobes,** serving as centers for visual and auditory reflexes. The midbrain has undergone little evolutionary change in size among vertebrates but has changed in function. It is responsible for the most complex behavior of fish and amphibians; the midbrain serves the higher integrative functions in these lower vertebrates that the cerebrum serves in higher vertebrates.

The **thalamus,** above the midbrain, contains masses of gray matter surrounded by the cerebral hemispheres on each side. This is the relay center for the sensory tracts from the spinal cord. Centers for the sensations of pain, temperature, and touch are supposedly located in the thalamus. In the **hypothalamus** are centers that regulate body temperature, water balance, sleep, and a few other body functions. The hypothalamus also has the neurosecretory cells that produce neurohormones. These pass down fiber tracts to the posterior pituitary where the hormones are released into the circulation.

The largest division of the brain is the **cerebrum,**

which is concerned with learned behavior. It forms the most anterior part of the brain and, in man and most mammals, overlies most of the other parts of the brain. It is incompletely divided into two hemispheres by a longitudinal fissure. The outer **cerebral cortex** is much folded and is made up of gray matter. The uplifted folds are called **gyri**, the depressions, **sulci.** Deeper fissures divide the brain into regions. Within the hemispheres are the first and second ventricles. Both gray and white matter are found in the cerebrum. The cortex contains many billions of neurons and their synaptic junctions with other neurons. In the deeper parts of the cerebral hemispheres are other masses of gray matter that function as centers or relay stations for neurons running to or from the cortex. The white matter, which consists of nerve fiber tracts, lies deeper in the cerebrum.

To a certain extent there is localization of function in the cerebrum (Fig. 13-7). This knowledge has been obtained by direct experimentation, by the removal of parts of the brain, by checking of the locations of brain lesions, and by the sensations experienced by patients during operations, etc. It has been possible to locate the visual center (back of cerebrum [Fig. 13-7]), the center for hearing (side of brain or temporal lobe), the motor area that controls the skeletal muscle (anterior to central sulcus), and the area for skin sensations of heat, cold, and touch (posterior to the central sulcus). Large regions of the frontal lobe of the brain are the "silent areas," or **association areas.** These regions are not directly connected to sense organs or muscles and are for the higher faculties of memory, reasoning, and learning. Only forms that have well-developed cerebral cortices are able to learn and modify their behavior by experience.

The **cerebellum,** lying beneath the posterior part of the cerebrum, consists of two hemispheres and a central portion called the vermis. In man it is deeply folded. The chief function of the cerebellum is regulation of muscular coordination.

Peripheral nervous system. This system is made up of the paired cranial nerves that run to and from the brain, and the paired spinal nerves that run to and from the spinal cord. They consist of bundles of axons and dendrites and connect with the receptors and effectors in the body.

Cranial nerves. In the higher vertebrates, including man, there are 12 pairs of cranial nerves. They are primarily concerned with the sense organs, glands, and muscles of the head and are more specialized than spinal nerves. Some are purely sensory (olfactory, optic, auditory); some are mainly, if not entirely, motor (oculomotor, trochlear, abducens, spinal accessory, hypoglossal); and the others are mixed with both sensory and motor neurons (trigeminal, facial, glossopharyngeal, vagus). The majority of the cranial nerves arise from or near the medulla. Some bear autonomic nerve fibers, especially the facial and vagus. For convenience the various cranial nerves are also designated by the Roman numerals I to XII, as well as by specific names. The numbering begins at the anterior end of the brain and proceeds to the posterior end of the brain.

Spinal nerves. The spinal nerves contain both sensory and motor components in approximately equal numbers. In higher vertebrates and man there are 31 pairs: cervical, 8 pairs; thoracic, 12 pairs; lumbar, 5 pairs; sacral, 5 pairs; and caudal, 1 pair. Each nerve has two roots by which it is connected to the spinal cord (Fig. 13-5). All the sensory fibers enter the cord by the dorsal root, and all the motor fibers leave the cord by the ventral root. The nerve cell bodies of motor neurons are located in the ventral horns of the gray zone of the spinal cord; the sensory nerve cell bodies are in the dorsal spinal ganglia just outside the cord. Near the junction of the two roots, the spinal nerve divides into a small dorsal branch (ramus) that supplies structures in the back, a larger ventral branch that supplies structures in the sides and front of the trunk and in the appendages, and an autonomic branch that supplies structures in the viscera. To supply a large area of the body, the ventral rami of several spinal nerves may join to form a network (plexus). These are the **cervical, brachial,** and **lumbosacral plexuses.**

Autonomic nervous system. The autonomic nerves govern the involuntary functions of the body that do not ordinarily affect consciousness. The cerebrum has no direct control over these nerves, thus one cannot by volition stimulate or inhibit their action. Autonomic nerves control the movements of the alimentary canal and heart, the contraction of

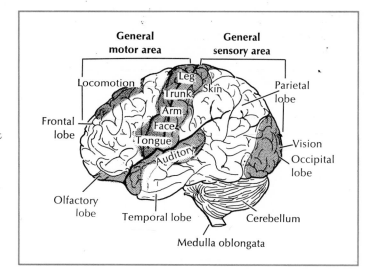

FIG. 13-7. Localization of brain function in left cerebrum.

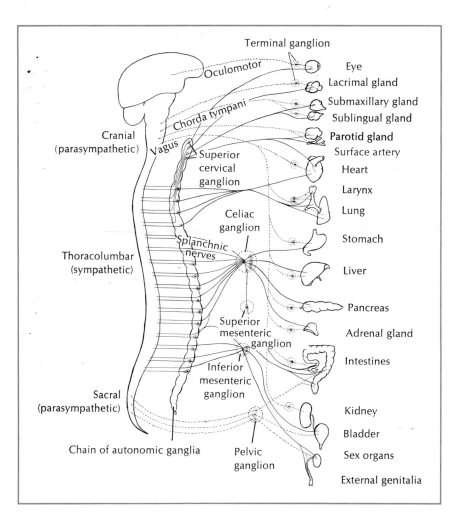

FIG. 13-8. Autonomic nervous system in mammals. Parasympathetic fibers are indicated by dotted lines; sympathetic fibers, by continuous lines. Note that most organs are innervated by both parasympathetic and sympathetic fibers.

the smooth muscle of the blood vessels, urinary bladder, iris of eye, etc., and the secretions of various glands.

Subdivisions of the autonomic system are the **parasympathetic** and the **sympathetic.** Most organs in the body are innervated by both sympathetic and parasympathetic fibers, and their actions are antagonistic (Fig. 13-8). If one speeds up an activity, the other will slow it down. However, neither kind of nerve is exclusively excitatory or inhibitory. For example, parasympathetic fibers inhibit heartbeat but will excite peristaltic movements of the intestine; sympathetic fibers will increase heartbeat but slow down peristaltic movement.

The **parasympathetic** system consists of motor nerves, some of which emerge from the brain by certain cranial nerves and others from the pelvic region of the spinal cord by certain spinal nerves. Parasympathetic fibers **excite** the stomach and intestine, urinary bladder, bronchi, constrictor of iris, salivary glands, and coronary arteries. They **inhibit** the heart, intestinal sphincters, and sphincter of the urinary bladder.

In the **sympathetic** division the nerve cell bodies are located in the thoracic and upper lumbar areas of the spinal cord. Their fibers pass out through the ventral roots of the spinal nerves, separate from these, and go to the sympathetic ganglia, which are paired and form a chain on each side of the spinal column. From these ganglia some of the fibers run through spinal nerves to the limbs and body wall, where they innervate the blood vessels of the skin, the smooth muscles of the hair, the sweat glands, etc.; and some run to the abdominal organs as the splanchnic nerves. Sympathetic fibers **excite** the heart, blood vessels, sphincters of the intestines, urinary bladder, dilator muscles of the iris, and others. They **inhibit** the stomach, intestine, bronchial muscles, and coronary arterioles.

All preganglionic fibers, whether sympathetic or parasympathetic, release **acetylcholine** at the synapse for stimulating the ganglion cells. The terminations of the parasympathetic and sympathetic nervous systems release different types of chemical transmitter substances. The parasympathetic fibers release **acetylcholine** at their endings while the sympathetic fibers release **norepinephrine.** These chemical substances produce characteristic physiologic reactions. Since there is some physiologic overlapping of sympathetic and parasympathetic fibers, it is now customary to describe nerve fibers as either adrenergic (norepinephrine effect) or cholinergic (acetylcholine effect).

SENSE ORGANS

Animals require a constant inflow of information from the environment to regulate their lives. Sense organs are specialized receptors designed for detecting environmental status and change. An animal's sense organs are its first level of environmental perception; they are data input channels for the brain.

A **stimulus** is some form of energy—electrical, mechanical, chemical, or radiant. The task of the sense organ is to transform the energy form of the stimulus it receives into nerve impulses, the common language of the nervous system. In a very real sense, then, sense organs are **biologic transducers.** A microphone, for example, is a man-made transducer that converts mechanical (sound) energy into electrical energy. And like the microphone that is sensitive only to sound, sense organs are, as a rule, quite specific for one kind, or **modality,** of stimulus energy. Thus eyes respond only to light, ears to sound, pressure receptors to pressure, and chemoreceptors to chemical molecules. But, again, all of these different forms of energy are converted into nerve impulses. Since all nerve impulses are qualitatively alike, how do animals perceive and distinguish the different **sensations** of varying stimuli? The answer is that the real perception of sensation is done in localized regions of the brain, where each sense organ has its own hookup. Impulses arriving at a particular sensory area of the brain can be interpreted in only one way. This is why a blow on the eye causes us to "see stars"; the sudden mechanical distortion of the eye initiates impulses in the optic nerve fibers that are perceived as light sensations. Although the operation hopefully has never been done, the deliberate surgical switching of optic and auditory nerves would cause the recipient to quite literally see thunder and hear lightning!

CLASSIFICATION OF RECEPTORS

Receptors are classified on the basis of their location. Those near the external surface are called **exteroceptors** and are stimulated by changes in the external environment. Internal parts of the body are provided with **interoceptors,** which pick up stimuli from the internal organs. Muscles, tendons, and joints have **proprioceptors,** which are sensitive to changes in the tension of muscles and provide the organism with a sense of position.

Another way of classifying receptors is on the basis of the energy form used to stimulate them, such as **chemical, mechanical, photo,** or **thermal.**

Chemoreception

Chemoreception is the most primitive and most universal sense in the animal kingdom. The most primitive animals, protozoans, use chemical receptors to locate food and adequately oxygenated water and to avoid harmful substances. This kind of sensory response is called **chemotaxis.** The chemical sense may be developed to a truly amazing degree of sensitivity in many animals. It probably guides the behavior of animals more than any other sense.

In all vertebrates and in insects as well, the senses of **taste** and **smell** are clearly distinguishable. Although there are similarities between taste and smell receptors, in general the sense of taste is more restricted in response and is less sensitive than the sense of smell. Taste and smell centers are also located in different parts of the brain. In higher forms, **taste buds** are found on the tongue and in the mouth cavity (Fig. 13-9). A taste bud consists of a few sensitive cells surrounded by supporting cells and is provided with a small external pore through which the slender tips of the sensory cells project. The basal ends of the sensory cells contact nerve endings from cranial nerves. Taste bud cells in vertebrates have a short life of about ten days and are continually being replaced.

The four basic taste sensations—sour, salt, bitter, and sweet—are each due to a different kind of taste bud. The tastes for salt and sweet are found mainly at the tip of the tongue, bitter at the base of the tongue, and sour along the sides of the tongue. Taste

buds are more numerous in ruminants (mammals that chew the cud) than in man. They tend to degenerate with age; the child has more buds widely distributed over the mouth.

Sense organs of **smell** (olfaction) are found in a specialized mucous membrane called the olfactory epithelium located high in the nasal cavity. The sense of smell is much more complex than taste. Some humans can detect many thousands of different odors, and it is obvious that many other vertebrates can easily outdo man. Gases must be dissolved in a fluid to be smelled; therefore the nasal cavity must be moist. The sensory cells with projecting hairs are scattered singly through the olfactory epithelium. Their basal ends are connected to fibers of the olfactory cranial nerve that runs to one of the olfactory lobes. The sensitivity to certain odors approaches the theoretical maximum for the chemical sense. The human nose can detect 1/25 millionth of 1 mg. of mercaptan, the odoriferous principal of the skunk. This averages out to about 1 molecule per sensory ending. Since taste and smell are stimulated by chemicals in solution, their sensations may be confused. The taste of food is dependent to a great extent on odors that reach the olfactory membrane through the throat. All the various "tastes" other than the four basic ones (sweet, sour, bitter, salt) are really due to the fla-

FIG. 13-9. Taste buds in rabbit tongue. Buds are little oval bodies lined up on each side of slitlike recesses. (Courtesy J. W. Bamberger.)

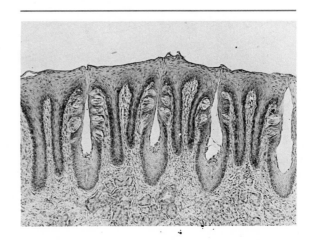

vor's reaching the sense of smell in this manner. A favorite current theory of olfaction is that all odors belong to seven primary odors. Similar odor molecules have similar stereochemical shapes, and the range of detectable odors is due to differences in the way the molecule of the substance smelled fits the receptor site.

Mechanoreception

Mechanoreceptors are sensitive to quantitative forces such as touch, pressure, stretching, sound, and gravity. Many receptors in and on the body constantly monitor information about conditions within the body (muscle position, body equilibrium, blood pressure, pain, etc.) and conditions in the environment (sound and other vibrations such as water currents).

Touch and pain. Although superficial touch receptors are distributed over all the body, they tend to be concentrated in the few areas especially important in exploring and interpreting the environment. In most animals these areas are on the face and limb extremities. Of the more than half a million separate sensitive spots on man's body surface, most are found on his lips, tongue, and fingertips. Many touch receptors are bare nervefiber terminals, but there is an assortment of other kinds of receptors of varying shapes and sizes. Each hair follicle is crowded with receptors that are sensitive to touch.

The sensation of deep touch and pressure is registered by relatively large receptors called pacinian corpuscles. They are common in deep layers of skin (Fig. 7-1, *B*, p. 84), in connective tissue surrounding muscles and tendons, and in the abdominal mesenteries. Each corpuscle, easily visible to the naked eye, is built of numerous layers like an onion. Any kind of mechanical deformation of the pacinian corpuscles is converted into nerve impulses that are sent to sensory areas of the brain.

Pain receptors are relatively unspecialized nerve fiber endings that respond to a variety of stimuli that signal possible or real tissue damage. It is still uncertain whether pain fibers respond directly to

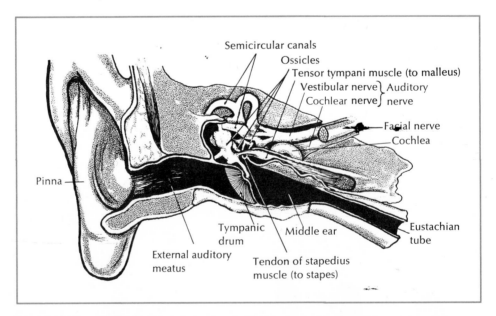

FIG. 13-10. Longitudinal section of ear of man. Note two muscles: tensor tympani and stapedius. Very loud noises cause these muscles to contract by reflex, thus stretching tympanic membrane and oval window and preventing damaging effect of loud, low-pitched sounds.

injury or indirectly to some substance, such as histamine, which is released by damaged cells.

Hearing. The ear is a specialized receptor for detecting sound waves in the surrounding ear (Fig. 13-10). The sense of equilibrium also is associated with the ear in all vertebrate animals. In fact the ear really evolved first as an organ of equilibrium; hearing was added later. In higher vertebrates the ear is typically made up of three parts—an outer ear for collection of sound waves, a middle ear for their transmission, and an inner ear specialized for sensory reception.

Outer ear. The outer, or external, ear of higher vertebrates is made up of two parts: (1) the **pinna,** or skin-covered flap of elastic cartilage and muscles, and (2) the **auditory canal.** In many mammals, such as the rabbit and cat, the pinna is freely movable and is effective in collecting sound waves. The auditory canal condenses the waves and passes them to the eardrum. The walls of the auditory canal are lined with hair and wax-secreting glands as a protection against the entrance of foreign objects.

Middle ear. The middle ear is separated from the external ear by the eardrum, or tympanic membrane, which consists of a tightly stretched connective membrane. Within the air-filled middle ear a chain of three tiny bones, **malleus** (hammer), **incus** (anvil), and **stapes** (stirrup), conduct the sound waves across the middle ear. This bridge of bones is so arranged that the force of sound waves pushing against the eardrum is amplified as many as ninety times where the stapes contacts the oval window of the inner ear. Muscles attached to the middle ear bones contract when the ear receives very loud noises, thus protecting the inner ear from damage. However, these muscles cannot contract quickly enough to protect the inner ear from the damaging effects of a sudden blast. The middle ear communicates with the pharynx by means of the eustachian tube, which acts as a safety device to equalize pressure on both sides of the eardrum.

Inner ear. The inner ear consists essentially of two labyrinths, one within the other. The inner one is called the **membranous labyrinth** and is a closed ectodermal sac filled with the fluid, **endolymph.** The part involved with hearing (**cochlea**) is coiled like a snail's shell, making two and a half turns in man. Surrounding the membranous labyrinth is the **bony labyrinth,** which is a hollowed-out part of the temporal bone and conforms to the shape and contours of the membranous labyrinth. In the space between the two labyrinths, perilymph, a fluid similar to endolymph, is found.

The cochlea is divided into three longitudinal canals that are separated from each other by thin membranes (Figs. 13-11 and 13-12). These canals become progressively smaller from the base of the cochlea to the apex. One of these canals is called the **vestibular canal** (scala vestibuli); its base is closed by the oval window. The **tympanic canal** (scala tympani), which is in communication with the vestibular canal at the tip of the cochlea, has its base closed by the round window. Between these two canals is the **cochlear duct,** which contains the organ of hearing, the **organ of Corti** (Fig. 13-11). The latter organ is made up of fine rows of hair cells that run lengthwise from the base to the tip of the cochlea. There are at least 24,000 of these hair cells in the human ear, each cell with many hairs projecting into the endolymph of the cochlear canal and each connected with neurons of the auditory nerve. The hair cells rest on the **basilar membrane,** which separates the tympanic canal and cochlear duct, and are covered over by the **tectorial membrane** found directly above them.

Sound waves picked up by the external ear are transmitted through the auditory canal to the eardrum, which is caused to vibrate. These vibrations are conducted by the chain of ear bones to the oval window, which transmits the vibrations to the fluid in the vestibular and tympanic canals (Fig. 13-12). The vibrations of the endolymph cause the basilar membrane, with its hair cells, to vibrate so that the latter are bent against the tectorial membrane. This stimulation of the hair cells causes them to initiate nerve impulses in the fibers of the auditory nerve, with which they are connected. In the **place theory** of pitch discrimination it is stated that when sound waves strike the inner ear the entire basilar membrane is set in vibration by a traveling wave of displacement, which increases in amplitude from the oval window toward the apex of the cochlea. This displacement wave reaches a maximum at the region of the basilar membrane, where the natural frequency of the membrane corresponds to the sound

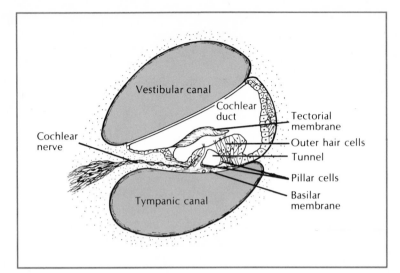

FIG. 13-11. Section through axis of cochlear spiral showing cellular types in organ of Corti.

FIG. 13-12. Cochlea straightened out to show pathway of vibrations through vestibular canal and tympanic canal.

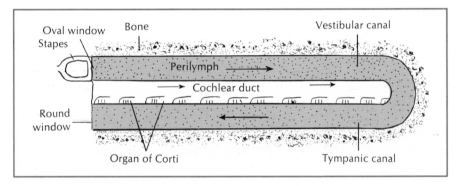

frequency. Here, the membrane vibrates with such ease that the energy of the traveling wave is completely dissipated. Hair cells in that region will be stimulated and the impulses conveyed to the fibers of the auditory nerve. Those impulses that are carried by certain fibers of the auditory nerve are interpreted by the hearing center as particular tones. The **loudness** of a tone depends on the number of hair cells stimulated, whereas the **timbre,** or quality, of a tone is produced by the pattern of the hair cells stimulated by sympathetic vibration. This latter characteristic of tone enables one to distinguish between different human voices and different musical instruments, even though the notes in each case may be of the same pitch and loudness.

Sense of equilibrium. Closely connected to the inner ear and forming a part of it are two small sacs, the **saccule** and **utricle,** and three **semicircular canals.** Like the cochlea, the sacs and canals are filled with endolymph. They are concerned with the sense of balance and rotation. They are well developed in all vertebrates, and in some lower forms they represent about all there is of the internal ear, for the cochlea is absent in fish. They are innervated by the nonacoustic branch of the auditory nerve. The utricle and saccule are hollow sacs lined with sensitive hairs and contain small stones, **otoliths,** of calcium carbonate. Whatever way the head is tipped, the otolith presses against certain hair cells to stimulate them; this is interpreted in a certain way with reference to position.

The three semicircular canals are at right angles

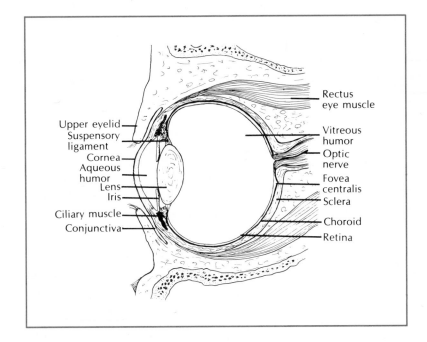

Upper eyelid
Suspensory ligament
Cornea
Aqueous humor
Lens
Iris
Ciliary muscle
Conjunctiva

Rectus eye muscle
Vitreous humor
Optic nerve
Fovea centralis
Sclera
Choroid
Retina

FIG. 13-13. Section through eye of vertebrate.

to each other, one in each plane of space. They are filled with fluid, and at the opening of each canal into the utricle there is a bulblike enlargement, the **ampulla,** which contains hair cells but no otoliths. Whenever the fluid moves, these hair cells are stimulated. Rotating the head will cause a lag, due to inertia, in certain of these ampullae. This lag produces consciousness of movement. Since the three canals of each internal ear are in different planes, any kind of movement will stimulate at least one of the ampullae.

Vision. Light sensitive receptors are called **photoreceptors.** These receptors range all the way from simple light-sensitive cells scattered randomly on the body surface of the lowest invertebrates (dermal light sense) to the exquisitely developed vertebrate eye. Many of the invertebrates can do little more than record differences in light intensities and perhaps determine the direction of the light source. The arthropods, however, have **compound** eyes composed of many independent visual units called **ommatidia.** The eye of a bee contains about 15,000 of these units, each of which views a separate narrow sector of the visual field. Such eyes form a mosaic of images from the separate units. The compound eye probably does not produce a very distinct

image of the visual field, but it is extremely well suited to picking up motion, as anyone knows who has tried to swat a fly.

The vertebrate eye is built like a camera – or rather we should say a camera is modeled somewhat after the vertebrate eye. It contains a light-tight chamber with a lens system in front that focuses an image of the visual field on a light-sensitive surface (the retina) in back. Because eyes and cameras are based on the same laws of optics, we can wear glasses to correct optical defects in our eyes. But here the similarity between eye and camera ends. The human eye is actually replete with optical shortcomings; projected on the retina of the normal eye are more colored fringe halos, apparitions, and distortions than would be produced by even the cheapest camera lens. Yet the human brain corrects for this "poor design" so completely that we perceive a perfect image of the visual field. It is in the retina and the optic center of the brain that the marvel of vertebrate vision can be understood.

The spherical eyeball is built of three layers: (1) a tough outer white **sclerotic** coat (sclera) serving for support and protection, (2) the middle **choroid** coat containing blood vessels for nourishment, and (3) the light-sensitive **retinal** coat (Fig. 13-13). The

173

cornea is a transparent modification of the sclera. A circular curtain, the **iris,** regulates the size of the light opening, the **pupil.** Just behind the iris is the **lens,** a transparent, elastic ball that bends the rays and focuses them on the retina. In land vertebrates the cornea actually does most of the bending of light rays, while the lens adjusts the focus for near and far objects. Between the cornea and the lens is the outer chamber filled with the watery **aqueous humor;** between the lens and the retina is the much larger inner chamber, filled with the viscous **vitreous humor.** Surrounding the margin of the lens and holding it in place is the **suspensory ligament.** This, together with the **ciliary muscle,** a ring of radiating muscle fibers attached to the suspensory ligament, makes possible the stretching and relaxing of the lens for close or distant vision (accommodation).

The **retina** is composed of photoreceptors, the **rods** and **cones.** Approximately 125 million rods and 7 million cones are present in each human eye. Cones are primarily concerned with color vision in ample light; rods, with dim or colorless vision. The retina is actually made up of three sets of neurons in series with each other: (1) photoreceptors (rods and cones), (2) intermediate neurons, and (3) ganglionic neurons whose axons form the optic nerve.

The **fovea centralis,** the region of keenest vision, is located in the center of the retina, in direct line with the center of the lens and cornea. It contains only cones. The acuity of an animal's eyes depends on the density of cones in the fovea. The human fovea and that of a lion contain about 150,000 cones per square millimeter. But many water and field birds have up to 1 million cones per square millimeter. Their eyes are as good as man's eyes would be if aided by eight-power binoculars.

At the peripheral parts of the retina only rods are found. This is why we can see better at night by looking out of the corners of our eyes because the rods, adapted for high sensitivity with dim light, are brought into use.

Chemistry of vision. Each rod contains a light sensitive pigment known as **rhodopsin.** Each rhodopsin molecule consists of a large, colorless protein, **opsin,** and a small carotenoid molecule, called **retinal,** a derivative of vitamin A. When a quantum of light strikes a rod and is absorbed by the rhodopsin molecule, the latter undergoes a chemical bleaching process that causes it to split into separate opsin and retinal molecules. In some way not yet understood this change triggers the discharge of a nerve impulse in the receptor cell. The impulse is relayed to the optic center of the brain. Rhodopsin is then enzymatically resynthesized so that it can respond to a subsequent light signal. The amount of intact rhodopsin in the retina depends on the intensity of light reaching the eye. The dark-adapted eye contains much rhodopsin and is very sensitive to weak light. Conversely most of the rhodopsin is broken down in the light-adapted eye. It takes about half an hour for the light-adapted eye to accommodate to darkness, while the rhodopsin level is gradually built up. The remarkable ability of the eye to dark- and light-adapt vastly increases the versatility of the eye; it enables us to see by starlight as well as by the noonday sun, 10 billion times brighter.

The light-sensitive pigment of cones is called **iodopsin.** It is similar to rhodopsin, containing **retinal** combined with a special protein, **cone opsin.** Cones function to perceive color and require fifty to one hundred times more light for stimulation than rods. Consequently night vision is almost totally rod vision; this is why the landscape illuminated by moonlight appears in shades of black and white only. Unlike man, who has both day and night vision, some vertebrates specialize for one or the other. Strictly nocturnal animals such as bats and owls have pure rod retinas. Purely diurnal forms such as the common gray squirrel and some birds have only cones. They are, of course, virtually blind at night.

Color vision. How does the eye see colors? According to the trichromatic theory of color vision, there are three different types of cones that react most strongly to red, green, and violet light. Colors are perceived by comparing the levels of excitation of the three different kinds of cones. This comparison is made both in nerve circuits in the retina and in the visual cortex of the brain. Color vision is present in all vertebrate groups with the possible exception of the amphibians. Bony fish and birds have particularly good color vision. Surprisingly, most mammals are color blind; exceptions are primates and a very few other species such as squirrels.

14

Chemical coordination and integration

The endocrine system is the second great integrative system controlling the body's activities. Endocrine glands, or specialized tissues, secrete **hormones** (from the Greek root meaning "to excite") that are transported by the blood for variable distances to some part of the body where they produce definite physiologic effects. Hormones are effective in minute quantities; some are active when diluted several billion times in the blood. The endocrine system is a slow-acting integrative system as compared to the nervous system, and, in general, hormonal effects are long lasting. Some hormones are excitatory, others inhibitory. Many physiologic processes are governed by antagonistic hormones (one which stimulates, the other which inhibits the process). Such combinations are very effective in maintaining homeostatic conditions.

Endocrinology, the science of the endocrine system, is a young field. Only within the last thirty years has endocrinology been able to break away from its purely clinical approach concerned with midgets, giants, circus freaks, hermaphrodites, and other unfortunate recipients of endocrine disorders. This unhappy image eventually disappeared when biologists realized that the complex array of hormones in the body is vitally and continually engaged in numerous integrative activities in normal health, just as is the nervous system.

The outstanding discovery in endocrinology in recent years was the recognition of a close functional interrelationship between the endocrine and nervous systems. It was found that certain nerve cells can secrete hormones into the bloodstream, when directly stimulated by other nerves. Such specialized nerve cells are called **neurosecretory cells** and their secreted products are called **neurosecretory hormones**. Subsequent studies demonstrated that these neurosecretory cells are crucial links between the body's two great integrative systems. Such knowledge made it possible to understand how, for example, increasing day length in the spring stimulates the breeding cycle of birds: increasing amounts of light received by the eyes are relayed via nerve tracts in the brain to neurosecretory cells that release hormones which set the reproductive cycle into motion.

Since hormones are transported in the blood, they reach virtually all body tissues. This makes it possible for certain hormones, such as the growth hormone of the pituitary gland, to have a very widespread action, affecting most, if not all cells during growth of an animal. However, most hormones, despite their ubiquitous distribution, are highly specific in their action. Usually only certain cells will respond to the presence of a given hormone. For example, only the pancreatic cells respond to the intestinal hormone **secretin,** even though secretin is carried throughout the body by the circulation. All other cells simply ignore its presence. The cells that respond to a particular hormone are called **target-organ cells.**

MECHANISMS OF HORMONE ACTION

How do hormones exert their effects? This question has been the object of intense research in recent years. The student can readily appreciate that it is much easier to observe the physiologic effect of a hormone than to determine how the specific hormone acts to produce the effect. Although we have known for years that insulin lowers the blood glucose level, we are still uncertain as to **how** insulin does this.

It now seems that there may be no more than two basic mechanisms of hormone action.

1. Stimulation of protein synthesis. Several hormones, such as the thyroid hormones and the sex hormones, stimulate the synthesis of specific enzymes and other proteins, by causing the transcription of particular kinds of messenger RNA. These hormones therefore act directly on specific genes. The response is then mediated by the enzymes that are synthesized.

2. Stimulation of cyclic adenosine monophosphate (AMP) formation. Several other hormones, such as adrenalin and vasopressin, induce an enzyme **adenyl cyclase** to catalyze the formation of cyclic AMP from the body's high-energy compound ATP. Cyclic AMP, found in all tissues, appears to serve as a second messenger that accepts the hormonal message it receives, then mediates the hormonal action by altering the cellular enzymes that control cell processes.

Endocrinologists have long searched for a single, fundamental mechanism through which all hormones act. Although it is true that certain hormones in the same cell can act via both of the mechanisms listed above, most hormones in fact do not. Most produce their physiologic response by modifying the activity of cellular enzymes that in turn regulate cellular processes or change membrane permeability. At present it looks as though hormones achieve this either by stimulating gene activity or by changing the amount of cyclic AMP.

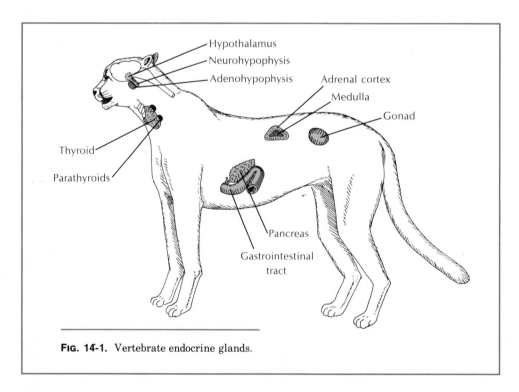

FIG. 14-1. Vertebrate endocrine glands.

VERTEBRATE ENDOCRINE ORGANS

The vertebrate endocrine glands are small, well-vascularized organs located in certain parts of the body (Fig. 14-1). **Endocrine** glands are ductless groups of cells arranged in cords or plates; their hormonal secretions enter the bloodstream and are carried throughout the body. **Exocrine** glands, in contrast, are provided with ducts for discharging their secretions onto a free surface. Examples of exocrine glands are sweat glands and sebaceous glands of skin, salivary glands, and the various enzyme-secreting glands lining the wall of the stomach and intestine. Since the endocrine glands have no ducts, their only connection with the rest of the body is by the bloodstream; they must capture their raw materials from the blood and secrete their finished hormonal products into it. Consequently it is not surprising that the endocrine glands receive enormous blood flows. The thyroid gland is said to have the highest blood flow per unit of tissue weight of any organ in the body.

In the remainder of this section we will describe some of the best understood and most important of the vertebrate hormones. Space does not permit us to deal with all the hormones and hormonelike substances that have been discovered. The mammalian hormonal mechanisms are the best understood, since laboratory mammals and man have always been the objects of the most intensive research. Research with the lower vertebrates has revealed that all vertebrates share similar endocrine organs. All vertebrates have a pituitary gland, for example, and all have thyroid glands, adrenal glands (or the special cells of which they are composed), and gonads. There are some important differences, nevertheless, as we will seek to point out.

HORMONES OF THE PITUITARY GLAND AND HYPOTHALAMUS

The pituitary gland, or **hypophysis,** is a small gland (0.5 gram in man) lying in a well-protected position between the roof of the mouth and the floor of the brain (Fig. 14-2). It is a two-part gland having a double embryologic origin. The **anterior pituitary** is derived embryologically from the roof of the mouth. The **posterior pituitary** arises from a ventral portion of the brain, called the **hypothalamus,** and is connected to it by a stalk, the **infundibulum** (Fig. 14-2, *A*). Although the anterior pituitary lacks any **anatomic** connection to the brain, it is nonetheless **functionally** connected to it by a special portal circulatory system (Fig. 14-2, *B*).

The anterior pituitary consists of an anterior lobe and an intermediate lobe as shown in Fig. 14-2. The anterior lobe, despite its minute dimensions, produces at least six protein hormones. All but one of these six are **tropic hormones,** that is, they regulate other endocrine glands (Fig. 14-3). Because of the strategic importance of the pituitary in influencing most of the hormonal activities in the body, the pituitary has been called the body's "master gland." This name is misleading, however, since the tropic hormones are themselves regulated by neurosecretory hormones from the hypothalamus, as well as by hormones from the target glands they stimulate. The **thyrotropic hormone** (TSH) regulates the production of thyroid hormones by the thyroid gland. The **adrenocorticotropic hormone** (ACTH) stimulates the adrenal cortex. Two of the tropic hormones are commonly called **gonadotropins** because they act on the gonads (ovary of the female, testis of the male). These are the **follicle-stimulating hormone** (FSH) and the **luteinizing hormone** (LH). The fifth tropic hormone is **prolactin,** which stimulates milk production by the female mammary glands and has a variety of other effects in the lower vertebrates. We will discuss the functions of the two gonadotropins and prolactin later in connection with the hormonal control of reproduction.

The sixth hormone of the anterior pituitary is the growth hormone (also called **somatotropic hormone**). This hormone performs a vital role in governing body growth through its stimulatory effect on cellular mitosis and protein synthesis, especially in new tissue of young animals. If produced in excess, the growth hormone causes giantism. A deficiency of hormone in the child or young animal causes dwarfism. Recent research indicates that in man and other primates growth hormone and prolactin are the same hormone.

As pointed out above, the pituitary gland is not the top director of the body's system of endocrine glands, as endocrinologists once believed. The pituitary serves higher masters, the neurosecretory centers of the hypothalamus; and the hypothalamus is itself

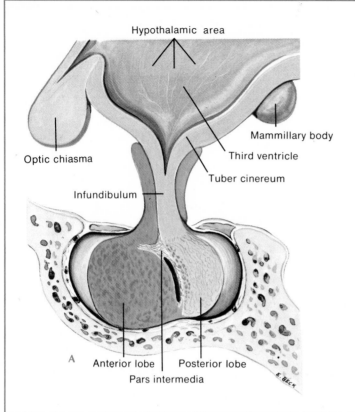

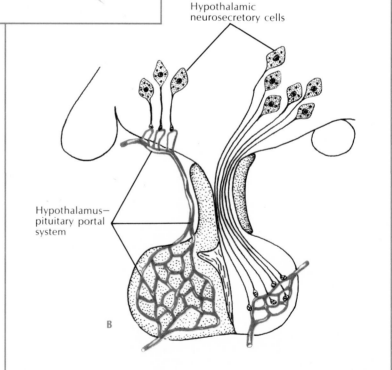

FIG. 14-2. Structure of human pituitary gland and its relationship to hypothalamus. **A,** Posterior lobe is connected directly to hypothalamus by neurosecretory fibers. Anterior lobe is indirectly connected to hypothalamus. **B,** Neurosecretory fibers end in the infundibulum, in contact with a portal circulation that conveys hormones to the anterior pituitary. (**A** from Tuttle, W. W., and Schottelius, B. A.: Textbook of Physiology, ed. 16, St. Louis, 1969, The C. V. Mosby Co.)

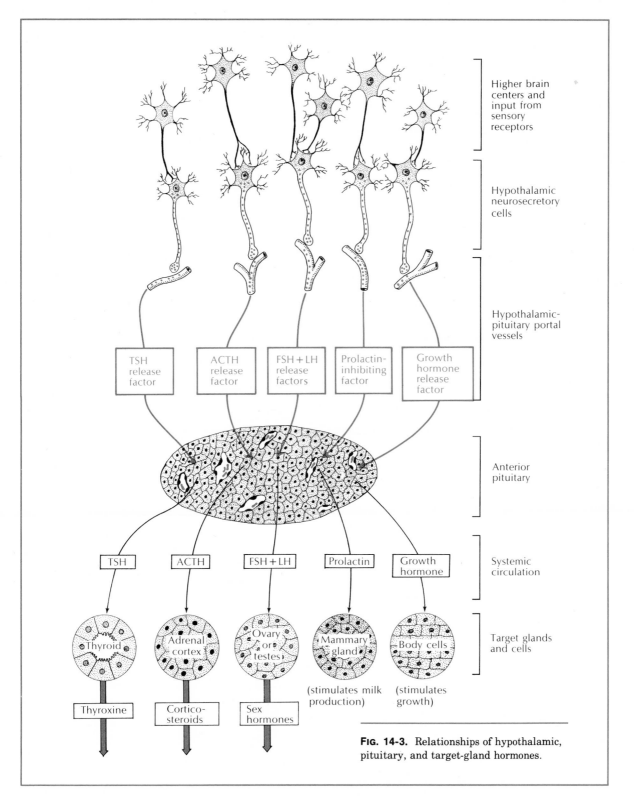

FIG. 14-3. Relationships of hypothalamic, pituitary, and target-gland hormones.

under the ultimate control of the brain. The hypothalamus contains groups of neurosecretory cells, which are specialized giant nerve cells. Polypeptide hormones are manufactured in the cell bodies, then travel down the nerve fibers to their endings where the hormones are stored until released into the blood. The discharge of neurosecretory hormones may occur when a nerve impulse travels down the same neurosecretory fiber (these specialized nerve cells can, in most cases, still perform their original impulse-conducting function) or the release may be activated by ordinary fibers traveling alongside them. Both the anterior and posterior lobes of the pituitary are under hypothalamic control, but in different ways. Neurosecretory fibers serving the posterior lobe travel down the infundibular stalk and into the posterior lobe, ending in close proximity to blood capillaries, into which the hormones enter when released. In a sense the posterior lobe is not a true endocrine gland, but a storage-release center for hormones manufactured entirely in the hypothalamus.

The anterior pituitary's relationship to the hypothalamus is quite different. Neurosecretory fibers do not travel to the anterior lobe, but end some distance above it at the base of the infundibular stalk (Fig. 14-2, *B*). Neurosecretory hormones released here enter a capillary network and complete their journey to the anterior lobe via a short, but crucial, pituitary portal system. These hormones are called **releasing factors** because they govern the release of the anterior pituitary hormones. There appears to be a specific releasing factor for each pituitary tropic hormone (TSH-releasing factor, ACTH-releasing factor, FSH-releasing factor, etc.), although the releasing factors have not yet been isolated in pure form for study.

The hormones of the posterior lobe are chemically

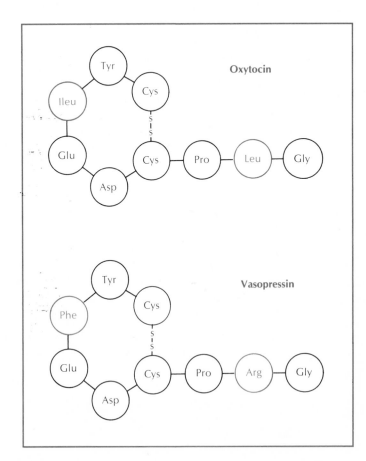

FIG. 14-4. Posterior lobe hormones of man. Both oxytocin and vasopressin consist of eight amino acids (the two sulfur-linked cysteine molecules are considered to be a single amino acid, cystine). Oxytocin and vasopressin are identical except for amino acid substitutions in the red positions.

similar polypeptides consisting of eight amino acids (referred to as octapeptides). All vertebrates, except the most primitive fishes, secrete two posterior lobe octapeptides. However, their chemical structure has changed slightly in the course of evolution. The two posterior lobe hormones secreted, for example, by fish are not identical to those secreted by mammals. Altogether, seven different posterior lobe hormones have been identified from the different vertebrate groups. The two mammalian hormones are **oxytocin** and **vasopressin** (Fig. 14-4). They are formed, as we have seen, in the cell bodies of neurosecretory cells in the hypothalamus, then transported down the nerve cell axons to the posterior lobe. These hormones are among the fastest-acting hormones in the body, since they are capable of producing a response within seconds of their release from the posterior lobe.

Oxytocin has two important specialized reproductive functions in adult female mammals. It causes contraction of uterine smooth muscles during parturition (birth of the young). Doctors sometimes use oxytocin clinically to induce labor and facilitate delivery and to prevent uterine hemorrhage after birth. The second and most important action of oxytocin is that of milk ejection by the mammary glands in response to suckling.

Vasopressin, the second posterior lobe hormone, acts on the kidney to restrict urine flow. It is therefore often referred to as the **antidiuretic hormone** (ADH). The release of ADH is controlled by special receptors in the hypothalamus (osmoreceptors) that are sensitive to minute changes in osmotic pressure of the blood circulating through the hypothalamus. A slight rise in blood osmolality, produced by desiccation, stimulates the release of ADH. Carried to the kidney by the bloodstream, it acts immediately to promote the reabsorption of water from the tubular urine. Thus the urine is concentrated, and body water is conserved. If water intake is high, the pituitary reduces ADH output, and the urine becomes dilute. Vasopressin has a second, weaker effect of increasing the blood pressure through its generalized constrictor effect on the smooth muscles of the arterioles. Although the name vasopressin unfortunately suggests that the vasoconstrictor action is the hormone's major effect, it is probably of little physiologic importance, except perhaps to help sustain the blood pressure during a severe hemorrhage.

HORMONES OF METABOLISM

Many hormones act to adjust the delicate balance of metabolic activities in the body. Metabolism includes the **anabolic** activities of tissue synthesis, building up of energy reserves, and maintenance of tissue organization and the **catabolic** activities of energy release and tissue destruction. Such activities are mediated almost entirely by enzymes. The numerous enzymatic reactions proceeding within cells are complex, but each step in a sequence is in large part self-regulating, as long as the equilibrium between substrate, enzyme, and product remains stable. However, hormones may alter the activity of crucial enzymes in a metabolic process, thus accelerating or inhibiting the entire process. We must emphasize that hormones never initiate enzymatic processes. They simply alter their rate, speeding them up or slowing them down. The most important hormones of metabolism are those of the thyroid, parathyroid, adrenal, and pancreas.

Thyroid hormones. The two thyroid hormones **thyroxine** and **triiodothyronine** are secreted by the thyroid gland. This largest of endocrine glands is located in the neck region of all vertebrates; in many animals, including man, it is a bilobed structure. The thyroid is made up of thousands of tiny spheres, called **follicles;** each follicle is composed of a single layer of epithelial cells enclosing a hollow, fluid-filled center. This fluid contains stored thyroid hormone that is released into the bloodstream as it is needed.

One of the unique characteristics of the thyroid is its high concentration of iodine; in most animals this single gland contains well over half the body store of iodine. The epithelial cells actively trap iodine from the blood and combine it with the amino acid tyrosine, creating the two thyroid hormones. Each molecule of thyroxine contains four atoms of iodine as indicated by the following structural formula:

THYROXINE

Triiodothyronine is identical to thyroxine except that it has three instead of four iodine atoms. Thyroxine is formed in much greater amounts than triiodothyronine, but both hormones have two important similar effects. One is to promote the normal growth and development of growing animals. The other is to stimulate the metabolic rate.

It is not known exactly how the thyroid hormones promote growth, although there is evidence that they stimulate protein synthesis through their effect on messenger RNA. Certainly the undersecretion of thyroid hormone dramatically impairs growth, especially of the nervous system. The human **cretin**, a mentally retarded dwarf, is the tragic product of thyroid malfunction from a very early age. Conversely the oversecretion of thyroid hormones causes precocious development, particularly in lower vertebrates. In one of the earliest demonstrations of hormone action, Gudernatsch in 1912 induced precocious metamorphosis of frog tadpoles by feeding them bits of horse thyroid. The tadpoles quickly resorbed their tails, grew limbs, changed from gill to lung respiration, and became froglets about one-third normal size. The result of a similar experiment is shown in Fig. 14-5.

FIG. 14-5. Precocious metamorphosis of frog tadpoles (*Rana pipiens*) caused by adding thyroid hormone to the water. When frog tadpoles had developed hindlimb buds, small amounts of thyroxine were added to aquarium water. In only 3 weeks tadpoles metamorphosed to normal, but miniature, adults about one third the size of the mother. (From Turner, C. D.: General endocrinology, ed. 4, Philadelphia, 1966, W. B. Saunders Co.)

The control of oxygen consumption and heat production in birds and mammals is the best known action of the thyroid hormones. The thyroid enables warm-blooded animals to adapt to cold by increasing their heat production. Acting as a kind of biologic thermostat, the thyroid senses and controls the body temperature by releasing more or less thyroxine, as required. Thyroxine in some way causes cells to produce more heat and store less chemical energy (ATP); in other words, thyroxine **reduces** the efficiency of the cellular oxidative phosphorylation system (p. 59). This is why cold-adapted animals eat more food than warm-adapted ones, even though they are doing no more work; the food is being converted directly to heat, thus keeping the body warm.

The synthesis and release of thyroxine and triiodothyronine is governed by **thyrotropic hormone** (TSH) from the anterior pituitary gland (Fig. 14-3). Thyrotropic hormone controls the thyroid through a **negative feedback mechanism.** If the thyroxine level in the blood falls, more thyrotropic hormone is released. Should the thyroxine level rise too high, less thyrotropic hormone is released. This sensitive feedback mechanism normally keeps the blood thyroxine level very steady, but certain neural stimuli, as might arise from exposure to cold, can directly increase the release of thyrotropic hormone.

Some years ago, a condition called **goiter** was common among people living in the Great Lakes region of the United States and Canada, as well as in other parts of the earth such as the Swiss Alps. Goiter is an enlargement of the thyroid gland caused by a deficiency of iodine in the food and water. In striving to produce thyroid hormone with not enough iodine available, the gland hypertrophies, sometimes so much that the entire neck region becomes swollen. Goiter is seldom seen today because of the widespread use of iodized salt.

Parathyroid hormone, calcitonin, and calcium metabolism. Closely associated with the thyroid gland, and often buried within it, are the parathyroid glands. These tiny glands occur as two pairs in man but vary in number and position in other vertebrates. They were discovered at the end of the nineteenth century when the fatal effects of "thyroidectomy" were traced to the unknowing removal of the parathyroids as well as the thyroid. Removal of the parathyroids causes the blood calcium to drop rapidly. This leads to a serious increase in nervous system excitability, severe muscular spasms and tetany, and finally death.

The parathyroid glands are vitally concerned with the maintenance of the normal level of calcium in the blood. Actually two hormones are involved: **parathyroid hormone** (parathormone), produced by the parathyroid glands, and **calcitonin,** produced by special cells within the thyroid gland (not the same follicle cells that synthesize thyroxine). These two hormones have opposing but cooperative actions. Between them they stabilize both the calcium and phosphorus levels in the blood through their action on bone. Bone is a densely packed storehouse of these elements, containing about 98% of the body calcium and 80% of the phosphorus. Although bone is second only to teeth as the most durable material in the body, as evidenced by the survival of fossil bones for millions of years, it is in a state of constant turnover in the living body. Bone-building cells **(osteoblasts)** withdraw calcium and phosphorus (as phosphate) from the blood and deposit them in a complex crystalline form around previously formed organic fibers (Fig. 14-6). Bone resorbing cells **(osteoclasts),** present in the same bone, tear down bone by engulfing it and releasing the calcium and phosphate into the blood. These conflicting activities are not as pointless as they may seem. First, they allow bone to constantly remodel itself, especially in the growing animal, for structural improvements to counter new mechanical stresses on the body. Second, they provide a vast and accessible reservoir of minerals that can be withdrawn as the body needs them for its general cellular requirements.

If the blood calcium should drop slightly, the parathyroid gland increases its output of parathormone. This stimulates the osteoclasts to destroy bone adjacent to these cells, thus releasing calcium and phosphate into the bloodstream and returning the blood calcium level to normal. Parathormone also acts on the kidney to decrease the excretion of calcium and this, of course, also helps to increase the blood calcium level. Should the calcium in the blood rise above normal, the parathyroid gland decreases its output of parathormone. In addition, the thyroid is stimulated to release calcitonin. These relationships are shown in Fig. 14-6. Although the action of this sec-

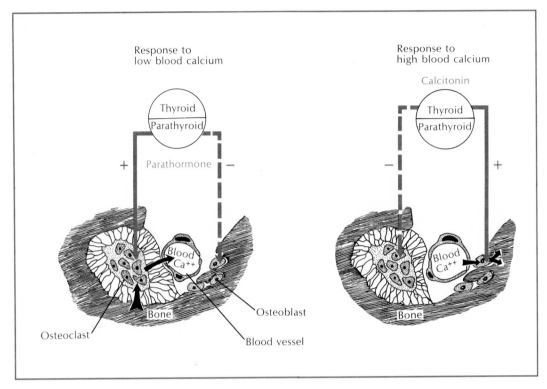

FIG. 14-6. Action of parathormone and calcitonin on calcium resorption and deposition in bone. When blood calcium is low (left), parathyroid gland secretes parathormone that stimulates large bone-destroying osteoclasts to resorb calcium. Bone-building osteoblasts are inhibited. Calcium and phosphate (not shown in diagram) enter the blood and restore the blood calcium level to normal. When blood calcium rises above normal (right), thyroid gland secretes calcitonin that inhibits osteoclasts. Osteoblasts then remove calcium (and phosphate) from the blood and use it to build new bone. Calcitonin may directly stimulate osteoblastic activity. A third hormone, calciferol (often called "vitamin" D), is necessary for calcium deposition in bones.

ond recently discovered (1962) hormone is yet imperfectly understood, evidence suggests that it inhibits bone resorption by the osteoclasts. Calcitonin thus protects the body against a dangerous rise in the blood calcium level, just as parathormone protects it from a dangerous fall in blood calcium. The two act together to smooth out oscillations in blood calcium. A third hormone **calciferol** (commonly called "vitamin" D) is also necessary for calcium deposition in bone. Calciferol is a steroid hormone in skin formed from a steroid precursor by the action of ultraviolet rays of sunlight.

There is increasing evidence that parathormone is one of those hormones that acts through the "second messenger," cyclic AMP. Parathormone ac-

tivates the enzyme adenyl cyclase in the bone-resorbing osteoclasts, causing increased formation of cyclic AMP. Cyclic AMP in some way brings about the resorption of bone, but its precise action is still a mystery.

Adrenal cortical hormones. The vertebrate adrenal gland is a double gland consisting of two very different kinds of tissue: **interrenal** tissue, called **cortex** in mammals, and **chromaffin** tissue, called **medulla** in mammals. The mammalian terminology of cortex (meaning "bark") and medulla (meaning "core") arose because in this group of vertebrates the interrenal tissue completely surrounds the chromaffin tissue like a cover. Although in the lower vertebrates the interrenal and chromaf-

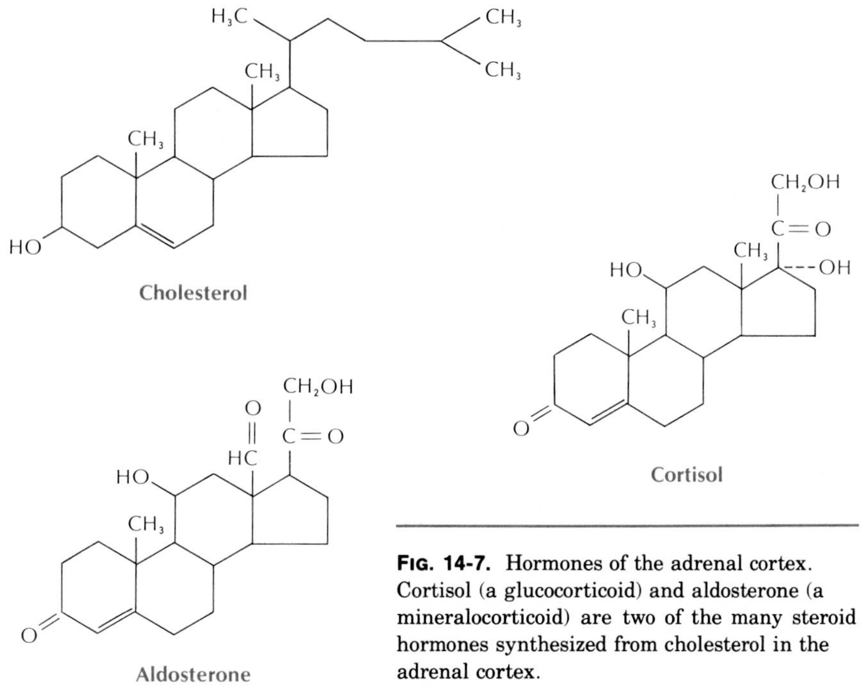

FIG. 14-7. Hormones of the adrenal cortex. Cortisol (a glucocorticoid) and aldosterone (a mineralocorticoid) are two of the many steroid hormones synthesized from cholesterol in the adrenal cortex.

fin tissue are usually separated, the mammalian terms "cortex" and "medulla" are so firmly fixed in our vocabulary that we commonly use them for all vertebrates instead of the more correct terms "interrenal" and "chromaffin."

Biochemists have found that the adrenal cortex contains at least thirty different compounds, all of them closely related lipoid compounds known as steroids. Only a few of these compounds, however, are true steroid **hormones;** most are various intermediates in the synthesis of steroid hormones from **cholesterol** (Fig. 14-7). The cortical steroid hormones are commonly classified into three groups, according to their function:

1. **Glucocorticoids** such as **cortisol** (Fig. 14-7) and **corticosterone** have a number of important effects concerned with food metabolism and inflammation. They cause the conversion of nonglucose compounds, particularly amino acids and fats, into glucose. This process, called **gluconeogenesis,** is extremely important, since most of the body's stored energy reserves are in the form of fats and proteins that must be converted to glucose before they can be burned for energy. Cortisol, cortisone, and corticosterone are also **antiinflammatory.** Because several diseases of man are inflammatory diseases (for example, allergies, hypersensitivity, arthritis), these cortical steroids have important medical applications. They must be used with great care, however, since if administered in excess, they may suppress the body's normal repair processes and lower resistance to infectious agents.

2. **Mineralocorticoids,** the second group of cortical steroids, are those that regulate salt balance. **Aldosterone** (Fig. 14-7) and **deoxycorticosterone** are the most important steroids of this group. They promote the tubular reabsorption of sodium and chloride and the tubular excretion of potassium by the kidney. Since sodium usually is in short supply in the diet, and potassium in excess, it is obvious

185

that the mineralocorticoids play vital roles in preserving the correct balance of blood electrolytes. We may also note that the mineralocorticoids **oppose** the antiinflammatory effect of cortisol and cortisone. In other words, they promote the **inflammatory** defense of the body to various noxious stimuli. Although these opposing actions of the cortical steroids seem self-defeating, they actually are not. They are necessary to maintain readiness of the body's defenses for any stress or disease threat, yet prevent these defenses from becoming so powerful that they turn against the body's own tissues.

3. **Sex hormones** (for example, testosterone, estrogen, progesterone) are produced primarily by the ovaries and testes. The adrenal cortex is also a minor source of certain steroids that mimic the action of testosterone. This sex hormonelike secretion is of little physiologic significance, except in certain disease states of man.

The synthesis and secretion of the cortical steroids are controlled principally by the **adrenocorticotropic hormone** (ACTH) of the anterior pituitary (Fig. 14-3). As with pituitary control of the thyroid, a negative feedback relationship exists between ACTH and the adrenal cortex: a rise in the level of cortical steroids suppresses the output of ACTH; a fall in the blood steroid level increases ACTH output.

Adrenal medulla hormones. The adrenal medulla secretes two structurally similar hormones, **epinephrine** (adrenaline) and **norepinephrine** (noradrenaline). These same compounds are also released at the endings of sympathetic nerve fibers throughout the body, where they serve as "transmitter" substances to carry neural signals across the gap that separates the fiber and the organ it innervates. The adrenal medulla has the same embryologic origin that sympathetic nerves have; in many respects the adrenal medulla is nothing more than a giant sympathetic nerve ending. It is not surprising then that the adrenal medulla hormones have the same general effects on the body that the sympathetic nervous system has. These effects center around emergency functions of the body, such as fear, rage, fight, and flight, although they have important integrative functions in more peaceful times as well. We are all familiar with the increased heart rate, tightening of the stomach, dry mouth, trem-

bling muscles, and general feeling of anxiety, and the increased awareness that attends sudden fright or other strong emotional states. These effects are due both to the rapid release into the blood of epinephrine from the adrenal medulla and to increased activity of the sympathetic nervous system. Epinephrine and norepinephrine have many other effects that we are not so aware of, including constriction of the arterioles (which, together with the increased heart rate, increases the blood pressure), mobilization of liver glycogen to release glucose for energy, increased oxygen consumption and heat production, hastening of blood coagulation, and inhibition of the gastrointestinal tract. All of these changes in one way or another tune up the body for emergencies. Epinephrine and norepinephrine are among those hormones that activate the enzyme **adenyl cyclase,** causing increased production of **cyclic AMP.** Cyclic AMP then becomes the "second messenger" that produces the many observed effects of the adrenal medulla hormones.

Insulin from the islet cells of the pancreas. The pancreas is both an exocrine and an endocrine organ. The **exocrine** portion produces pancreatic juice, a mixture of digestive enzymes that is conveyed by ducts to the digestive tract. Scattered among the extensive exocrine portion of the pancreas are numerous small islets of tissue, called **islets of Langerhans.** This is the **endocrine** portion of the gland. The islets are without ducts and secrete their hormones directly into blood vessels that extend throughout the pancreas. Two polypeptide hormones are secreted by different cell types within the islets: **insulin,** produced by the **beta cells,** and **glucagon,** produced by the **alpha cells.** Insulin and glucagon have antagonistic actions of great importance in the metabolism of carbohydrates and fats. Insulin is essential for the utilization of blood glucose by cells, especially skeletal muscle cells. Insulin somehow allows glucose in the blood to be transported into the cells. Without insulin, the blood glucose levels rise (hyperglycemia) and sugar appears in the urine. Insulin also promotes the uptake of amino acids by skeletal muscle and inhibits the mobilization of fats in adipose tissue. Failure of the pancreas to produce enough insulin causes the disease **diabetes mellitus** that afflicts 2% of the population. It is attended by serious alterations in carbohydrate, lipid, protein, salt, and water metabolism, which, if left untreated, may cause death. The first extraction of insulin in 1921 by two Canadians, Frederick Banting and Charles Best, was one of the most dramatic and important events in the history of medicine. Many years earlier, two German scientists, Von Mering and Minkowski, discovered that surgical removal of the pancreas of dogs invariably caused severe symptoms of diabetes resulting in the animal's death within a few weeks. Many attempts were then made to isolate the diabetes preventive factor, but all failed because powerful protein-splitting digestive enzymes in the exocrine portion of the pancreas destroyed the hormone during extraction procedures. Following a hunch, Banting in collaboration with Best and his physiology professor J. J. R. Macleod tied off the pancreatic ducts of several dogs. This caused the exocrine portion of the gland with its hormone-destroying enzyme to degenerate, but left the islet's tissue healthy, since they were independently served by their own blood supply. Banting and Best then successfully extracted insulin from these glands. Injected into another dog, the insulin immediately lowered the blood sugar level. Their experiment paved the way for the commercial ex-

traction of insulin from slaughterhouse animals. It meant that millions of diabetics, previously doomed to invalidism or death, could now look forward to nearly normal lives.

Glucagon, the second hormone of the pancreas, has several effects on carbohydrates and fat metabolism that are opposite to the effects of insulin. For example, glucagon raises the blood glucose level, whereas insulin lowers it. Glucagon and insulin do not have the same effects in all vertebrates, and in some, glucagon is lacking altogether. Glucagon is another example of a hormone that appears to operate through the cyclic AMP "second messenger" system.

HORMONES OF DIGESTION

Several hormones assist in coordinating the secretion of digestive enzymes. Of these, we will discuss three of the best understood (Fig. 14-8). **Gastrin** is a small polypeptide hormone produced in the mucosa of the pyloric portion of the stomach. When food enters the stomach, gastrin stimulates the secretion of hydrochloric acid by the stomach wall. Gastrin is an unusual hormone in that it exerts its action on the same organ from which it is secreted.

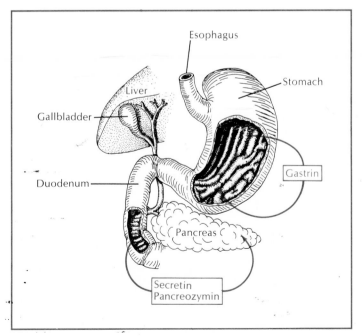

FIG. 14-8. Three hormones of digestion. Arrows show source and target of three gastrointestinal hormones.

Two other hormones of digestion are **secretin** and **pancreozymin.** Both are polypeptide hormones secreted by the intestinal mucosa in response to the entrance of acid and food into the duodenum from the stomach; both stimulate the secretions of pancreatic juice, but their effects differ somewhat. Secretin stimulates a pancreatic secretion rich in bicarbonate that rapidly neutralizes stomach acid. Pancreozymin stimulates the pancreas to release an enzyme-rich secretion. Although secretin was the first hormone discovered (by the British scientists Bayliss and Starling in 1903), the gastrointestinal hormones as a group have received much less attention than the other vertebrate endocrine glands. Only secretin has been obtained in pure form.

HORMONES OF REPRODUCTION

The male and female gonads are endocrine glands as well as gamete-forming glands. Reproduction is a complex process requiring the coordinated action of many hormones, especially in the female. Although the principal features of reproductive endocrinology are understood, the recent search for effective and safe birth control devices has revealed some disturbing gaps in our knowledge. To cite a single but telling example, no one knows precisely how the popular birth control pill works.

The female reproductive activities of almost all animals are cyclic. Usually breeding cycles are seasonal and coordinated so that the young are born at a time of year when conditions for growth are most favorable. Mammals have two major reproductive patterns, the **estrous cycle** (characteristic of most mammals) and the **menstrual cycle** (characteristic of primates only). These two cyclic patterns differ in two ways. First, in the estrous cycle, but not in the menstrual cycle, the female is receptive to the male, that is, she is in "heat," only at restricted periods of the year. Second, the menstrual cycle, but not the estrous cycle, ends with the collapse and sluffing of the uterine lining (endometrium). In the estrous animal each cycle ends with the uterine lining simply reverting to its original state, without the bleeding characteristic of the menstrual cycle. But these differences are really minor variations on a basic theme. The hormonal regulation of the

reproductive cycles is so much alike for all mammals that a great deal of what we know about the human reproductive hormones has come from laboratory studies of the ubiquitous white rat, a rodent having an estrous cycle. Differences become more apparent when we study the lower vertebrates, but even fish share most of the reproductive hormones found in man. Since our discussion of this vast area must be brief, we will restrict our consideration to a favorite vertebrate, man.

The ovaries produce two kinds of steroid **sex hormones**—**estrogens** and **progesterone.** Estrogens are responsible for the development of the female accessory sex structure (uterus, oviducts, and vagina) and the female secondary sex characters, such as breast development, and the characteristic bone growth, fat deposition, and hair distribution of the female. Progesterone is responsible for preparing the uterus to receive the developing embryo. These hormones are controlled by the pituitary **gonadotropins,** FSH (follicle-stimulating hormone), and LH (luteinizing hormone) (Fig. 14-3).

The menstrual cycle begins with the release into the bloodstream of FSH from the anterior pituitary (Fig. 14-9). Reaching the ovaries, it stimulates the growth of one of the several thousand follicles present in each ovary. The follicle swells as the egg matures, until it bursts, releasing the egg onto the surface of the ovary. This event, called **ovulation,** normally occurs on about the fourteenth day of the cycle. Now follows the most critical period of the cycle, for unless the mature egg is fertilized within a few hours it will die. During this period, the egg is swept into an oviduct (fallopian tube) and begins its journey toward the uterus, pushed along by the numerous cilia that line the oviduct walls. If intercourse occurs at this time, the sperm will traverse the uterus and find their way into the oviducts, where one may meet and fertilize the egg. The developing embryo continues down the oviduct, enters the uterine cavity, where it dwells for a day or two, then implants in the prepared uterine endometrium. This is the beginning of pregnancy.

Let us now examine the intricate series of events that occurs both before and after the beginning of pregnancy. We have seen that the pituitary hormone FSH begins the reproductive cycle by stimulating the growth of at least one of the ovarian

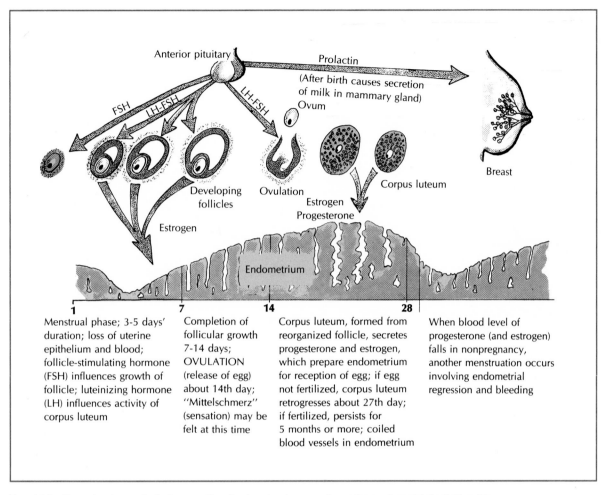

FIG. 14-9. Reproductive cycle in human female showing hormonal-ovarian-endometrial relationships. It will be noted that endometrium undergoes cyclical chain of events (thickness, congestion, etc.) under influence of hormones from ovary—estrogen and progesterone—and from pituitary gland—FSH and LH. Another hormone, prolactin, stimulates secretion of milk.

follicles. As the follicle enlarges, it releases **estrogens** (principally estradiol) that prepare the uterine lining (endometrium) for reception of the embryo.

The rise in blood estrogen is sensed by the pituitary, which responds by stopping the production of FSH. Estrogen also encourages the production of the second pituitary hormone, LH. Ovulation now occurs. LH causes the cells lining the ruptured follicles to proliferate rapidly, filling the cavity with a characteristic spongy, yellowish body called a

corpus luteum. The corpus luteum, responding to the continued stimulation of LH, manufactures **progesterone** in addition to estradiol. Progesterone, as its name suggests, stimulates the uterus to undergo the final maturation changes that prepare it for gestation. The uterus is thus fully ready to house and nourish the embryo by the time the latter settles out onto the uterine surface, usually about 7 days after ovulation. If fertilization has **not** occurred, the corpus luteum disappears, and its hormones are no longer secreted. Since the uterine endometrium depends on progesterone and estrogen for its maintenance, their disappearance causes the endometrial lining to dehydrate and slough off, producing the menstrual discharge. However, if the egg has been fertilized and has implanted, the corpus luteum continues to supply the essential sex hormones needed to maintain the mature uterine endometrium. During the first few weeks of pregnancy the developing placenta itself begins to produce the sex hormones progesterone and estrogen and soon replaces the corpus luteum in this function. As pregnancy advances, progesterone and estrogen prepare the breasts for milk production. The actual secretion and release of milk after birth (lactation) is the result of two other hormones, prolactin and oxytocin. Milk is not secreted during

FIG. 14-10. Endocrine control of molting in a butterfly. Butterflies mate in the spring or summer, and eggs soon hatch into the first of several larval stages (called instars). After the final larval molt, the last and largest larva (caterpillar) spins a cocoon in which it pupates. The pupa, or chrysalis, overwinters and an adult emerges in the spring to start a new generation. Two hormones interact to control molting and pupation. The molting hormone, produced by the prothoracic gland and stimulated by a separate brain hormone, favors molting and the formation of adult structures. These effects are inhibited, however, by the juvenile hormone, produced by the corpora allata. Juvenile hormone output declines with successive molts, and the larva undergoes adult differentiation.

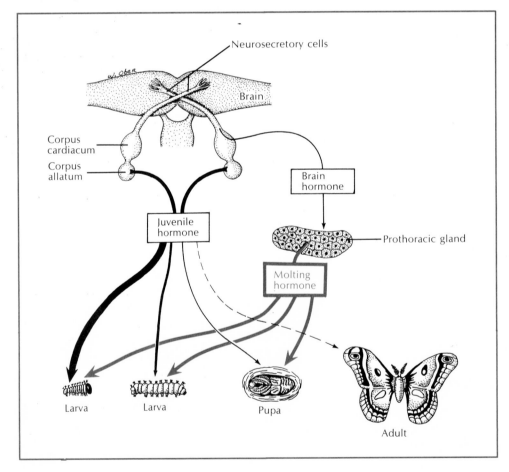

pregnancy because the placental sex hormones inhibit the release of prolactin by the pituitary. The placenta, like the corpus luteum that preceded it, thus becomes a special endocrine gland of pregnancy. After delivery, many mammals eat the placenta (afterbirth), which, because of its hormonal content, encourages the rapid postpartum involution of the uterus. This behavior also serves to remove telltale evidence of a birth from potential predators.

The male sex hormone **testosterone** is manufactured by the **interstitial cells** of the testes. Testosterone is necessary for the growth and development of the male accessory sex structures (penis, sperm ducts, glands), for development of secondary male sex characteristics (hair distribution, voice quality, bone and muscle growth), and for male sexual behavior. The same pituitary hormones that regulate the female reproductive cycle, FSH and LH, are also produced in the male, where they guide the growth of the testes and its testosterone secretion.

INVERTEBRATE HORMONES

Over the last forty years physiologists have shown that the invertebrates have endocrine integrative systems that appear to be as complex and comprehensive as the vertebrate endocrine system. Not surprisingly, however, there are few, if any, homologies between invertebrate and vertebrate hormones. Invertebrates have different functional systems, different growth patterns, and different reproductive processes than do vertebrates. Most studies have been concentrated in the huge phylum Arthropoda, especially the insects and crustaceans. However, recent research has revealed hormonal systems in several other invertebrate phyla, too.

The chromatophores (pigment cells) of shrimp and crabs are definitely known to be controlled by hormones from the **sinus gland** in the eyestalk or in regions close to the brain. Many crustaceans are capable of remarkably beautiful color patterns that change adaptively in relation to their environment;

these changes are governed by an elaborate system of endocrine glands and hormones.

Growth and metamorphosis of arthropods are under endocrine control. As described earlier (p. 83), growth of an arthropod is a series of steps in which the rigid, nonexpansible exoskeleton is periodically discarded and replaced with a new larger one. This process is especially dramatic in insects. In the type of development called **holometabolous** seen in many insect orders (for example, butterflies, moths, ants, bees, wasps, and beetles), there is a series of wormlike larval stages, each requiring the formation of a new exoskeleton; each stage ends with a molt. The last larval stage enters a state of quiescense (pupa) during which the internal tissues are dissolved and rearranged into adult structures (**metamorphosis**). Finally the transformed adult emerges. Insect physiologists have discovered that molting and metamorphosis are controlled by the interaction of two hormones, one favoring growth and the differentiation of adult structures, the other favoring the retention of larval structures. These two hormones are the **molting hormone** (also referred to as ecdysone [ek-di'-son]) produced by the corpora cardiaca and the **juvenile hormone,** produced by the corpora allata (Fig. 14-10). The molting hormone (ecdysone) is under the control of a neurosecretory hormone from the brain, called **brain hormone** (or ecdysiotropin). At intervals during larval growth, brain hormone is released into the blood and stimulates the release of molting hormone. Molting hormone appears to act directly on the chromosomes to set in motion the changes leading to a molt. The molting hormone favors the formation of a pupa and the development of adult structures. It is held in check, however, by the juvenile hormone, which favors the development of larval characteristics. During larval life the juvenile hormone predominates, and each molt yields another larger larva. Finally, the output of juvenile hormone decreases and the final pupal molt occurs.

REFERENCES TO PART III

Barrington, E. J. W. 1967. Invertebrate structure and function. Boston, Houghton Mifflin Co.

Berrill, N. J. 1958. You and the universe. New York, Dodd, Mead & Co.

Chapman, G. 1967. The body fluids and their functions. The Institute of Biology's "Studies in Biology No. 8." New York, St. Martin's Press, Inc.

Clegg, P. C., and H. G. Clegg. 1969. Hormones, cells and organisms. Stanford, Calif., Stanford University Press.

Eccles, J. C. 1957. The physiology of nerve cells. Baltimore, The Johns Hopkins Press.

Eckstein, G. 1969. The body has a head. New York, Harper & Row, Publishers.

Florey, E. 1966. General and comparative physiology. Philadelphia, W. B. Saunders Co.

From Cell to Organism. 1967. Readings from Scientific American, with introductions by D. Kennedy. San Francisco, W. H. Freeman Co., Publishers.

Frye, B. E. 1967. Hormonal control in vertebrates. New York, The Macmillan Co.

Galambos, R. 1962. Nerves and muscles. Science Study Series. Garden City, New York, Doubleday & Co., Inc.

Gordon, M. S. (editor). 1968. Animal function: principles and adaptations. New York, The Macmillan Co.

Griffin, D. R., and A. Novick. 1970. Animal structure and function, ed. 2. Modern Biology Series. New York, Holt, Rinehart & Winston, Inc.

Hoar, W. S. 1966. General and comparative physiology. Englewood Cliffs, N. J., Prentice-Hall, Inc.

Jennings, J. B. 1965. Feeding, digestion and assimilation in animals. Elmsford, N. Y., Pergamon Press, Inc.

Krogh, A. 1941. The comparative physiology of respiratory mechanisms. New York, Dover Publications, Inc.

McCartney, W. 1968. Olfaction and odours. Berlin, Springer Verlag.

Potts, W. T. W., and G. Parry. 1964. Osmotic and ionic regulation in animals. New York, The Macmillan Co.

Ramsay, J. A. 1962. Physiological approach to the lower animals. New York, Cambridge University Press.

Schmidt-Nielsen, K. 1970. Animal physiology, ed. 3. Englewood Cliffs, N. J. Prenctice-Hall, Inc.

Tombes, A. S. 1970. An introduction to invertebrate endocrinology. New York, Academic Press, Inc.

Tuttle, W. W., and B. A. Schottelius. 1969. Textbook of physiology, ed. 16. St. Louis, The C. V. Mosby Co.

Vertebrate Adaptations. Readings from Scientific American, with introductions by N. K. Wessels, 1968. San Francisco, W. H. Freeman & Co. Publishers.

part four

Environmental relations of the animal

Ecology is the study of the relationship of animals to the world around them. It is concerned with how organisms (plant or animal) relate to each other and to the nonliving surroundings in which they exist. It deals with the total economy of organisms and their interdependence in the world. The **biosphere** is the total area of the earth's inhabitable air, water, and land. Within the biosphere is the **ecosystem,** made up of plants and animals together with nonliving materials. The nonliving environment supplies the living with air, water, and minerals. Plants make food for themselves and for animals. Fungi and bacteria, by decomposing dead plants and animals, renew the cycle of life. Keeping the balance permanent and on a workable level produces an optimum condition in the community, and such a state is called a **climax community.** A perfect balance of plants and animals is called a **biome,** which may change by natural or man-made influences.

Ecology is a science that is involved with concepts of energetics, cycling of materials, interdependence, self-regulation, and maintenance of the organization levels of life. It is concerned with food chains and food pyramids and with the dynamics of populations, such as the laws controlling populations, their ranges, their fluctuations, their interactions, and their evolution. Cycles of matter and energy, such as the carbon and nitrogen cycles, conservation of natural resources, and life zones, are all within the realm of ecology. It is concerned with the web of life with all its complex relations. The reactions of the animal, or animal behavior, give an insight into the way animals behave in their surroundings. Ecology looms as one of the most complex disciplines in the whole range of the life sciences.

At the present time the proper use of the environment is one of the most important concerns in man's thinking. Many of the most pressing problems confronting man are ecologic. Population expansion, food demands, problems of pollution of man's environment, wasteful use of natural resources, and man's power to change his environment must be given consideration by legislators if we are to restore and maintain a beneficial biologic and physical environment. Only when all society realizes its responsibility will it be possible to activate solutions for our ecologic problems.

15

The biosphere and animal distribution

The **biosphere** is that part of our planet where animals and plants live. It includes fresh and salt water, surface, depths below the surface, and air. The biosphere may be said to extend down in the ocean to more than 30,000 feet, below the land surface more than 1,000 feet, and vertically in the atmosphere to more than 40,000 feet. With the possible exception of certain arid regions, frozen mountain peaks, restricted toxic sea basins, and a few others, every place on earth is represented by some form of life. Most species are adapted to a particular type of environment that is restricted in size, resources of food, places to live, and general conditions of living.

Weather and climate conditions, temperature, pressure, nature of substratum, physicochemical structure, constant change due to geologic, geochemical, meteorologic, and other factors—all are involved in forming a background to which animals must adjust to survive. The animal itself is part and parcel of the earth's substance, and its evolutionary diversity has been correlated with the changing earth at every level of its existence. As an open system, an animal is forever receiving and giving off materials and energy. Inorganic materials are obtained from the physical environment, either directly by producers, such as green plants, or indirectly by consumers, which return the inorganic substances to the environment by excretion or by the decay and disintegration of their bodies.

The biosphere may be conveniently divided into three major subdivisions—hydrosphere, lithosphere, and atmosphere (Fig. 15-1). The **hydrosphere** refers to the aquatic portions of the biosphere, the streams, rivers, ponds, oceans, and wherever water may be found. The **lithosphere** is made up of the solid portions of the biosphere, such as soil and rock. Surrounding the other two subdivisions is the **atmosphere,** which forms a gaseous envelope. Animals obtain inorganic metabolites from each of these subdivisions. From the hydrosphere, they get water that makes up about 75% of living material. The lithosphere furnishes the essential minerals and chemicals, whereas the atmosphere supplies oxygen, nitrogen, and carbon dioxide. These inorganic substances are needed in all living organisms.

HYDROSPHERE

About 71% of the earth's surface is water, which also forms part of the lithosphere and atmosphere. Not only is water the most abundant constituent of protoplasm, but it is also the source of hydrogen that is so fundamental to the metabolic reactions of all living substance. Water also serves as one of the sources of oxygen in the body of organisms.

195

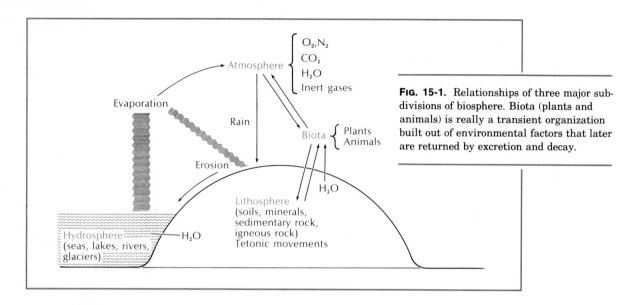

FIG. 15-1. Relationships of three major subdivisions of biosphere. Biota (plants and animals) is really a transient organization built out of environmental factors that later are returned by excretion and decay.

Water is involved in a cycle that consists of evaporation to a gaseous state and a return to a liquid by condensation of the vapor at higher altitudes. Evaporation of water occurs both in the ocean and on the land. About five sixths of this evaporation is in the ocean. It is estimated that about the depth of 1 meter of water evaporates from the ocean's surface annually. Part of the land evaporation of water is the transpiration of plants. An ecologic principle is that the precipitation of water on land exceeds the evaporation. The difference in water between these two factors represents the annual runoff water from the continents, carrying off minerals, producing erosion, and wearing away the surface of the continents. For instance, the average rainfall annually in the United States is about 30 inches of which 21 inches are returned to the atmosphere by evaporation and transpiration. This leveling off process, however, is offset by geologic uplift, thus bringing marine sediments above sea level. Living things also lose water through their bodies in addition to what they need themselves and in doing so speed up the return of water to the atmosphere. The metabolism of organisms thus accelerates the cycle of water and may profoundly influence weather conditions, not only locally, but also over extensive areas such as jungles with their extensive vegetation.

Displacement of water in the ocean is also cyclic in nature. The warm water of tropical seas comes to the surface while the cold polar water sinks. This creates currents between the equator and the pole that are further influenced by the east-west displacements produced by the rotation of the earth. Ocean currents, such as the Gulf Stream, have enormous influences on climatic conditions in all parts of the world.

Water is slow to heat or cool and stores great amounts of thermal energy. Heat radiation from water can produce favorable regions for land organisms in many places that otherwise would be too cold for their survival. Water is also unique in attaining its greatest density at 4° C.; therefore ice floats because it is lighter than the water beneath it.

At least four great glacial periods have occurred in the last million years. These cyclic changes of ice ages involve the advance or retreat of polar ice. Warm interglacial periods have actually freed the poles from ice. Amphibian fossils have been discovered in the Antarctic. Melting of polar ice during the present warm trend in temperature has gradually raised the level of ocean water and made possible a steady advance of biota toward the poles. Deserts are also on the march and localized glaciers are receding. It is estimated that the level of the sea was 136 meters below its present level during the last glacial period (16,000 years ago).

LITHOSPHERE

The lithosphere consists of a number of components. Below the loose soil and subsoil is the solid bedrock of sedimentary rock, such as limestone and sandstone, resting on a thicker base of igneous and metamorphic rocks. Below this layer is a thicker stratum composed mostly of basaltic rock. These two layers form the so-called crust of the earth.

All the mineral metabolites of the animal are received from the lithosphere. The crust of the earth shows striking changes over long periods of time. Uplifting and buckling are always occurring, resulting in mountain building and shifting of land masses. Mountains produce profound changes in climate conditions such as the unequal distribution of thermal energy and moisture. Moisture-laden clouds, unable to pass mountain barriers, may dump their contents on one side alone. This inequality results in favorable rainfall with fertility on one side and in little or no rainfall with desert on the other. These factors influence the distribution of both vegetation and animals that must adapt to these different conditions in order to survive.

Excess water is run off from the land into the rivers and streams, eroding the land surface by carrying with it dissolved soluble mineral matter to the sea. Billions of tons of dissolved inorganic and organic matter as well as much undissolved matter are carried into the oceans each year. Many important chemicals such as phosphorus are lost to animal and plant life and the biosphere is unable to make good the losses, especially to terrestrial forms. Marine forms, on the other hand, may profit from such an economy. Some valuable minerals are replaced by the decay of animals and plants and by the upheaval of the sea floor to form new land. This last is a slow process and life could be greatly affected by the gradual decline of certain key minerals. Many ecologists have been concerned about this decline, especially so because of modern man's misuse of certain resources.

Much of the terrestrial surface of the earth cannot be used to any extent by animal life because of severe climatic conditions. The scarcity of moisture in desert regions requires that activity be restricted to certain periods of the day by those forms adapted for living there.

ATMOSPHERE

The gases present in the atmosphere are (by volume) oxygen about 21%; nitrogen, 78%; carbon dioxide, 0.03%; water vapor, in varying amounts; and small traces of inert gases (neon, helium, krypton, argon, ozone, and xenon). Each of these, except the inert gases, serve as metabolites in living substances. The percentage of gases remains very constant, except in a few places such as volcanoes, underground sources, or industrial plants. Carbon dioxide plays a unique role in the ecology of animals, although it makes up a small percentage (0.03%) of atmospheric air. Among its many functions are its action as a chemical buffer in the maintenance of neutrality in aquatic habitats, its role in photosynthesis, its regulation of, or influence on, respiration, and its general influence on other essential biologic activities.

With the exception of anaerobic animals (which can carry on oxidative processes without free oxygen), oxygen is a necessary prerequisite for respiration in all animals. Its major function in cellular metabolism is its role as the final hydrogen acceptor, resulting in the formation of water. Oxygen is dissolved in water, which makes aquatic life possible. All atmospheric gases dissolve in water in accordance to their individual solubilities. This will depend on the water temperature and the partial pressure of the gas. In a mixture of gases each gas will exert a pressure proportional to its partial pressure in the mixture and each gas will dissolve irrespective of the solution of other gases. If atmospheric air of 760 mm. pressure (sea level) is exposed to 1 L. distilled water at 0° C., at equilibrium the water will contain about 49 cc. of oxygen, 23 cc. of nitrogen, and 1715 cc. of carbon dioxide. Oxygen is lost from water by the respiration of organisms, oxidation of organic matter and decay of dead bodies, consumption by bacteria, bubbling of other gases that carry oxygen with them, and the warming of the surface layer of water. The sparsity of dissolved oxygen in water may be a severe limiting factor in the distribution of aquatic animals.

Nitrogen is a chemically inert gas, but nevertheless is one of the chief constituents of living matter. Atmospheric nitrogen is the ultimate source. It may be fixed as nitrites or nitrates by lightning and later washed to earth by rain or snow. In the

197

nitrogen cycle, nitrogen-fixing bacteria living symbiotically with legumes form nitrites and nitrates by the reactions **ammonification** and **nitrification** and add them to the soil where plants can get their nitrogen supplies. Animals get their nitrogen from eating plants or one another.

The atmosphere has a low degree of buoyancy and cannot be used as a permanent habitat by organisms, although it has been suggested recently (B. C. Parker) that definite atmospheric ecosystems may exist in clouds. Here microorganisms (both animals and plants) undergo metabolic reactions and interrelationships characteristic of regular ecosystems. The presence in rainwater of certain vitamins and other nutrients is evidence for such ecosystems.

The atmosphere is commonly divided into three great strata, the **troposphere, stratosphere,** and **ionosphere.** The troposphere is the layer nearest the earth and extends upwards about six to ten miles above sea level. Above the troposphere is the stratosphere which extends to about fifty miles above the earth, and the ionosphere is the gradually thinning air still above the stratosphere. In the troposphere are the complex wind movements and currents that help produce the great climatic changes. As air masses warmed in the tropical regions rises and cooled polar air sinks, the rotation of the earth shifts the air masses laterally and produces the enormous currents of many types that profoundly influence the lives of both man and beast.

DISTRIBUTION OF ANIMALS (ZOOGEOGRAPHY)

Zoogeography is the study of the geographic distribution of animals and the factors responsible for their distribution. Animals having wide ranges of suitable ecologic conditions are called **eurytopic;** those with narrow boundaries, **stenotopic.** Dispersal of animals usually depends on their means of dispersal and the presence of limiting barriers. As a rule, members of a particular species or closely related species occupy **continuous** ranges. When the same taxonomic unit or group is found in widely separated areas, it is said to have **discontinuous** distribution. For example, marsupials are found

in the Americas and Australia—regions far apart.

Extinction has played a major role in animal distribution, but many of these groups left descendants that migrated to other regions and survived. Camels probably originated in North America, where their fossils are found, but spread to Eurasia by way of Alaska (true camels) and to South America (llamas). Barriers are altered by geologic changes in the earth's surface and by climatic changes. Many places now occupied by land were once covered with seas; regions now plains were formerly mountain ranges; such cold regions as Greenland were once quite warm. Evolution has been responsible for both the historic geographic distribution of animals and their ecologic relationships.

By adaptive radiation, animals tend to spread into regions where they are ecologically fitted. Distinct species found in the same general area are called **sympatric;** those living in different geographic areas are called **allopatric.** Although life is believed to have originated in the sea, terrestrial conditions favor a more varied and more rapid evolution; 80% of the known species are terrestrial. The abundance of oxygen on land has made possible a more active life. There are also many disadvantages of terrestrial life over an aquatic one, such as the necessity of a strong skeletal structure and the threat of desiccation.

Animals tend to spread from their center of origin because of competition for food, shelter, and breeding places. Changes in environment may force them to move elsewhere, although most great groups have spread to gain favorable conditions, not to avoid unfavorable ones. Dispersal movements must be distinguished from seasonal migration, such as that of birds, where there is a regular to-and-fro movement between two regions.

G. G. Simpson has stressed three chief paths of faunal interchange—corridors, filters, and sweepstakes routes. A **corridor** may be a widespread, more or less open stretch of land that usually allows the free movement of animals from one region to another. A **filter route** is defined as one that allows some animals to pass into another area but keeps others from doing so. A mountain may act as a filter; it may prevent mollusks from passing but affords fewer impediments to mammals or birds. Filters may also be narrow land bridges such as that of the Isthmus of Panama. A **sweepstakes route** is one that is highly improbable for animals to pass over, but some manage to do so by some fortuitous event. An example is the populating of oceanic islands by certain species that have floated, or been carried, across from the mainland.

Faunal interchange between continents has occurred many times. This interchange is most likely to occur when regions afford potential ecologic niches to a migrating group. Newcomers to a particular region often had adaptations that did not clash with the groups already there. In this way the fauna of a continent could be greatly enriched and diversified, such as that of South America when North American types passed into that region in the Pleistocene epoch.

Man himself has played a major role in the spread of species, as in the introduction of the English sparrow into America and the rabbit into Australia.

Convergent evolution, or the independent evolution of similar structures, may also account for two similar groups widely separated from each other. This could occur by the repetition of a mutation under similar environmental conditions in the two regions. Often, however, a better explanation in such cases is that the original range has been broken up by environmental changes such as a mountain barrier so that the individuals of the two similar groups became isolated.

KRAKATAO AS EXAMPLE OF ANIMAL DISPERSION

Krakatao is a volcanic island off the Sundra Straits between Java and Sumatra in the East Indies. In 1883 this island was practically destroyed by one of the most terrific volcanic eruptions in modern times. Every living thing was reported to be destroyed, for what remained of the island was covered with many feet of hot volcanic ash. Many naturalists have watched with great interest to see what life would first reappear there. It thus served as an interesting case of animal distribution under direct observation. The position of the island between two great land masses and less than forty miles from the nearest land made it especially favored for receiving a new stock of plants and animals.

Vegetation was the first to appear on the island and in a matter of twenty years completely covered the island. The first animals to appear were flying forms—birds and bats. It took more than twenty years for the first strictly terrestrial animals to gain a foothold. Rats, snakes, and lizards became common there. Insects, centipedes, millipedes, and spiders, especially those that travel by gossamer threads, had become common by 1921. Some of these had reached the island by driftwood, by currents of water, and by wind. There are few species but many individuals on the island, which may be due to the absence of natural enemies. Yet despite the favorable position of the island for receiving new animals, some forms have been slow in reaching the island and many have not reached there at all. This study may indicate how long oceanic islands, far from the mainland, have been in acquiring their flora and fauna.

LAKE BAIKAL AS EXAMPLE OF RESTRICTED DISPERSAL

The large Lake Baikal in eastern Siberia is the deepest lake in the world, being more than a mile deep in some places. It is famed for its characteristic fauna, which is almost unique in the great number of peculiar or endemic species. Among certain groups up to 100% of the species are found nowhere else. Some groups, such as planarian worms, are represented by more species here than in all the rest of the world put together. Only groups such as protozoans and rotifers, which are easily transported from one region to another, are represented by nonendemic species to any extent. The distribution in the lake is also somewhat unique in that animal life is found at nearly all depths, perhaps because the water is well oxygenated throughout. All the evidence indicates that Lake Baikal has never had any connection with the sea but is strictly a freshwater formation. Whence came its unique fauna? Long geographic speciation through isolation, because this lake is very ancient, might account for some species. But the prevailing opinion is that the fauna of this lake mainly represents species that have evolved elsewhere in many ecologic habitats and have been washed into this lake by different river systems over long periods of time. Hence, the present fauna represents the accumulation and survival of ancient freshwater organisms (relicts) that were widely distributed at one time but are now found only here. Other deep freshwater lakes of ancient geologic history, such as Tanganyika and Nyasa in East Africa, also have unique faunas for the same reasons.

INTERSTITIAL FAUNA (PSAMMON)

Within the past few years much investigation has been made on the interstitial fauna living within the sandy sediments or between the sand grains of marine, brackish, and freshwater beaches. The region or biotope of this type of fauna is commonly considered to be between the tide lines and below the low tide line. The biotope is a mixture of sand grains, water, air, and detritus and has often been called psammon (Gr. *psammos*, sand) or psammo-littoral habitat. The microfauna characteristic of the psammon has a wide distribution, not only in ponds, lakes, streams, brackish water, and the sea, but also in subsoil water (phreatic habitat).

The ecology of the fauna involves a microlabyrinth, or passageway, between the sand grains where the water is often deficient in oxygen, and a great deal of organic substance, mostly decayed, is present. In the surf zone considerable disturbance (water displacement, temperature, rain, etc.) also occurs, but the fauna by creeping or gliding in their labyrinth passageways can usually adjust themselves to such a change of conditions.

Representatives of all the animal phyla are probably found in the makeup of the psammon. Many of them are primitive forms, and although populations vary, some habitats have very large populations of bacteria, protozoans, and micrometazoa. It may be that the close relationship between the psammolittoral and the phreatic (Gr. *phrear,* well) habitats makes possible a pathway from marine or brackish water into the phreatic groundwater, as such a pathway would be less rigorous than through an open estuary or salt marsh.

MAJOR FAUNAL REALMS

Numerous regions over the earth have distinctive animal populations. These divisions indicate the influence of land masses and their geologic history, as well as the corresponding evolutionary development of the various animal groups. These realms of distribution have developed and fluctuated during geologic times. The higher vertebrates mainly have been used in working out these broad faunal realms. There are many complications in dividing the earth into such realms in which all groups of animals are involved. Some animals have purely a local origin; others within the same realm show affinities with groups quite remote. To explain many of these discrepancies, it has been necessary to assume various land connections or bridges for which there are no geologic evidences. Such major faunal realms can thus have only a limited significance.

Australian. This realm includes Australia, New Zealand, New Guinea, and certain adjacent islands. Some of the most primitive mammals are found here, such as the monotremes (duckbill) and marsupials (kangaroo and Tasmanian wolf), but few placental mammals. Most of the birds are also different from those of other realms, such as the cassowary, emu, and brush turkey. The primitive lizard *Sphenodon* is found in New Zealand.

Neotropical. This realm includes South and Central America, part of Mexico, and the West Indies. Among its many animals are the llama, sloth, New World monkey, armadillo, anteater, vampire bat, anaconda, toucan, and rhea.

Ethiopian. This realm is made up of Africa south of the Sahara desert, Madagascar, and Arabia. It is the home of the higher apes, elephant, rhinoceros, lion, zebra, antelope, ostrich, secretary bird, and lungfish.

Oriental. This region includes Asia south of the Himalaya Mountains, India, Ceylon, Malay Peninsula, Southern China, Borneo, Sumatra, Java, and the Philippines. Its characteristic animals are the tiger, Indian elephant, certain apes, pheasant, jungle fowl, and king cobra.

Palearctic. This realm consists of Europe, Asia north of the Himalaya Mountains, Afghanistan, Iran, and North Africa. Its animals include the tiger, wild boar, camel, and hedgehog.

Nearctic. This region includes North America as far south as southern Mexico. Its most typical animals are the wolf, bear, caribou, mountain goat, beaver, elk, bison, lynx, bald eagle, and red-tailed hawk.

16

Ecology of populations

BASIC REQUIREMENTS FOR EXISTENCE

For life to exist, certain basic requirements make up what is known as the fitness of the environment. These essentials are **food,** a place to live **(shelter),** and suitable conditions for **reproduction.** To meet these requirements, organisms are provided with adaptive structures of both form and function. Each animal requires a particular combination of factors of food and shelter. Favorable climate, weather, food, and shelter conditions are especially necessary for reproduction, a critical period in the life cycle.

These three requirements—nutrition, shelter, and reproduction—interact with each other as well as with other environmental factors to control the existence of an organism. Meadowlarks nest in meadows and grasslands, not in forests or tilled fields. The food habits of some animals destroy the shelters of others. A flock of goats will clean out the shrubbery of a waste field and thus destroy the nesting sites of birds and meadow mice. This intense competition of many complex relations is called the **web of life.**

The ability of an organism to adjust readily to unfavorable conditions is called **vagility.** Many animals can adapt to a new food when their preference is unavailable. Grass-eating animals will often browse on shrubbery during severe winters.

ENVIRONMENTAL FACTORS OF ECOLOGY

Heredity furnishes in the genes a basic type of germ plasm organization. But how these genes express themselves in the structure and functioning of an animal is conditioned by environmental factors. These include both the nonliving **physical** factors (temperature, moisture, light, etc.) and the living or **biotic** factors (other organisms). Organisms react toward their environment in characteristic ways, either by trying to avoid detrimental situations or by being able to adjust physiologically, within their genetic limits, to adverse factors.

How organisms are influenced by these physical and biotic factors is determined largely by the **law of tolerance.** To describe a species with a narrow range of tolerance for a particular factor, we use the prefix **steno-;** for those with a wide range we use the prefix **eury-.** The terms **stenothermal** and **eurythermal** refer to temperature tolerance. **Stenohaline** and **euryhaline** refer to the salinity tolerance of aquatic forms. A species will be most restricted by the factor for which it has the narrowest range of tolerance; those with the widest range for the most factors are likely to have the widest distribution. In this connection Liebig's "law of the minimum" may apply. According to this principle, an animal's ability to survive depends on those requirements that must be present in at least minimum amounts for the needs of the organism in question, even though all other conditions are fully met.

PHYSICAL FACTORS

Temperature. Temperature is an important factor in the animal's environment. Warm-blooded (homoiothermal or endothermal) animals are more independent of temperature changes than cold-blooded (poikilothermal or ectothermal) animals, although they are restricted by temperature extremes. Usually cold-blooded forms have body temperatures not much higher than that of their surroundings, although some active forms such as insects may have higher temperatures. Some forms can help regulate their body temperature to some extent by fanning their wings to create air currents and increase evaporation, or by living massed together as the bees do to conserve heat.

Many animals have an optimum temperature at which their body processes work best. For some protozoans this is between 24° and 28° C. Others may have a wider range, usually with a lower limit of just above freezing and an upper limit of around 42° C. Stenothermal animals have a very narrow range of temperature tolerance and have restricted distribution, whereas eurythermal animals are more widely distributed. An increase in temperature speeds up body metabolism so that cold-blooded animals, sluggish during cold spells, become more active as the temperature rises.

Many warm-blooded animals **hibernate** during winter months. Cold-blooded forms may retire deep underground or to other snug places to pass the winter. In some of the lower forms the adults die at the approach of winter, and the species is maintained by larval forms or by eggs.

Habitats of animals are greatly influenced by temperature. Herbivorous forms are restricted by the amounts of grass and leaves available. Deer feed high on the mountains during the summer but in winter retreat to the valleys where they find better shelters and more food. Birds that live upon insects are forced to go elsewhere when their food is destroyed by cold weather.

A spot climate, or **microclimate,** near the surface of the ground, in contrast to the **macroclimate** of the higher air levels, may show striking ranges. Air at the 2-inch level, for instance, has a higher daytime and lower nocturnal temperature than the macroclimate. The range may be as great as 10° F. The microclimate also has a greater humidity as well as less wind disturbance.

Light. Many primitive forms express their reaction to light in the form of tropisms or taxes, either tending to move toward the light (positively phototactic) or away from it (negatively phototactic). The effect of different intensities and wavelengths of light varies greatly for different animals. Within the visible spectrum, heat energy is more common at the red end (longer rays), and photochemical influences are greater at the violet end (shorter rays). Positively phototactic animals usually collect near the blue end of the spectrum when they have a choice; negatively phototactic forms collect at the red end.

Plankton, which is composed of small plants and animals in surface waters, is restricted to the upper strata of water. Only about 0.1% of light reaches a region 600 feet below the surface of most marine waters.

Pigment cells (melanophores) in many of the lower vertebrates are affected by the amount of light that enters the eye. Nerve impulses aroused in the eye may cause a contraction of the pigment cells and a lighter color in the animal, whereas less light might cause an expansion of the pigment cells and darker color.

Another way that light affects animals is through seasonal changes in day length. Birds and many mammals become sexually active with lengthening days and can be induced to breed out of season when exposed experimentally to a great amount of light. It is thought that the seasonal northward migration of birds may be induced by the stimulation of their gonads by the greater light associated with longer days in the spring and that the southward migration is caused by the regression of their glands by shorter days. In many other ways animals and plants are influenced by the length of the daylight (photoperiod), such as at the diapause (resting period) in arthropods, various physiologic functions of animals, the seasonal coat changes of birds and mammals, and the growth of trees. These seasonal responses may be controlled by a photoreaction involving the hypothalamus, pituitary, and certain hormones.

Hydrogen ion concentration. The factors that regulate the hydrogen ion concentration of water are numerous and complicated. Some of these are carbon dioxide and carbonates; and these factors, rather than the hydrogen ions, may be responsible for an animal's reactions in a particular medium. Although some animals prefer alkaline surroundings and others acid ones, many forms can endure a wide range of pH concentration. Tapeworms can live in intestinal fluids as acid as pH 4 or as alkaline as pH 11. Some protozoans are limited to a very narrow alkaline medium, whereas others, such as certain species of *Euglena,* live and flourish in water that varies from pH 2 to pH 8. Some mosquito larvae

are normally found in water with a pH of less than 5 and will not live in an alkaline medium. The pH of water has a very limited importance in the distribution of fish, which seem able to adjust to a wide range. Animals with calcium carbonate shells such as clams may be more sensitive to acid media because their shells are corroded by acids.

Substratum and water. The **substratum** is the medium on or in which an animal lives, such as soil, air, water, and bodies of other animals. The wings of bats, insects, and birds are fitted for the air (Fig. 16-1); the streamline form of fish and whales for the water; the digging feet of moles and other mammals for the earth; and the hooks and suckers of parasites for the host. Most organ-

FIG. 16-1. Of four groups of animals that experimented with flying, only reptiles, as group, gave it up entirely.

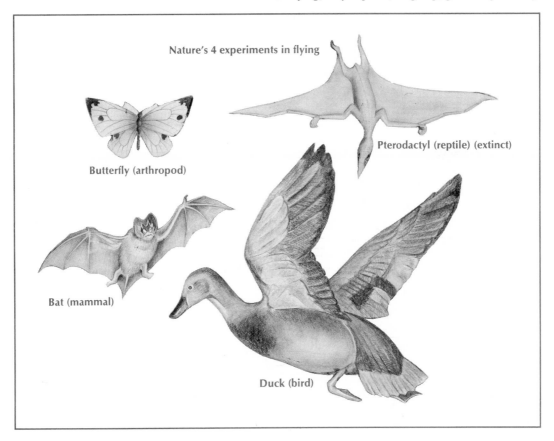

Nature's 4 experiments in flying

Butterfly (arthropod)

Pterodactyl (reptile) (extinct)

Bat (mammal)

Duck (bird)

isms are found on a hard substratum—on the land, at the bottom of a body of water, or in the hole of a tree. Many small forms, such as water striders, whirligig beetles, various larvae, and pulmonate snails, make use of the surface film of water either in locomotion or for clinging.

The soil supports an enormous population of animals, such as nematodes, crustaceans, insects, protozoans, and bacteria. Earthworms prefer soil rich in humus. Whether the soil is acid or alkaline makes little difference if they have abundant food. Land snails are more common on soils rich in calcium because they need this mineral for their shells. Deer also depend on calcium for their antlers and bony skeleton. Many chemicals (cobalt, fluorine, iodine) in trace amounts in the soil are important to animals.

Animals have evolved interesting adaptations for burrowing. Moles and mole crickets have shovellike appendages and broad, sturdy bodies. Earthworms secrete mucus to line their burrows. The trap-door spider lines its burrow with silk and conceals the entrance with a trapdoor that it pops open to seize its prey. Many forms live under stones and other objects on the surface of the soil. Termites and ants build mounds of soil in which they make tunneled passageways.

All animals must preserve a proper water balance. Most animals cannot lose more than one third of their water and live, but some can survive considerable desiccation. Protozoans secrete a cyst around themselves to prevent excessive desiccation; roundworms and rotifers may lose much water and then revive when placed in moist conditions. During prolonged dry spells, some undergo **aestivation,** burying themselves deep in the soil and remaining dormant until the wet season comes again. Breathing systems, such as the tracheal systems of insects and the internal lungs of snails, help cut down the amount of water evaporated. The dry feces of birds and reptiles is another water-saving device.

Water contains many salts and other substances that organisms use for the structure and functioning of their bodies. The salinity of marine water is rather constant in all oceans, being about 3.5%; the salinity of fresh water varies greatly. Some salt lakes may have a salinity of 25% to 30%, which greatly restricts life in them. The freezing point of sea water is about $-1.9°$ C., which is an advantage to animals in colder regions. Water is heaviest at $4°$ C. so that ice at $0°$ C. can float; thus deep lakes and ponds do not have permanent ice on their bottoms. Although shallow lakes in high altitudes or very cold climates do freeze to the bottom in the winter, life may survive there. Water also has a very high heat capacity, which makes for a constancy in temperature during most of the year.

Animals in rain forests can live only where the air is almost saturated with moisture (high humidity); desert forms can survive where the air is extremely dry (low humidity). Amphibians are especially sensitive to humidity changes. Forests are important as shelters, food supplies, and hiding places, for the relative humidity is higher and evaporation much lower than in open fields.

Wind and general weather conditions. Airborne eggs, spores, and adults (insects and snails) are often taken long distances by strong currents of air. Ballooning spiders are carried on their gossamer threads to locations far away. So powerful are wind currents that animal life may be transported from continents to islands and other lands hundreds of miles away.

Insects and other forms on wind-swept regions take advantage of cover to prevent being swept away. Many such insects are wingless; this may be adaptive, for wingless animals might stand less chance of being blown away.

Shelters and breeding places. Many animals, such as fish, squids, deer, and antelopes, that are endowed with speed make limited use of shelters, but others must hide from danger. In rapid streams, where there is danger of being dislodged and washed away, some animals are flat for creeping under stones, others have suckers for attachment, and still others live in firmly attached cases. Aquatic plants serve as cover for small fish, snails, crustaceans, and other forms. Forest areas, grasslands, and shrubbery contain a variety of habitats for terrestrial animals.

The destruction of forests has brought a decline of large birds of prey because the birds have been unable to find suitable nesting sites. The disappearance of sandy beaches due to the growth of aquatic vegetation affects fish that require sandy bottoms in which to spawn.

Other physical factors. Aquatic animals that live at great depths in water are subjected to enormous pressure (more than 14 pounds per square inch for each 33 feet of depth). Many vertebrates and invertebrates do live at such depths because their internal pressure is the same as the external pressure. Fish without swim bladders are less sensitive to deep pressure than those with swim bladders because of buoyancy complications; and most invertebrates are more resistant to pressure than vertebrates. Many marine animals have wide vertical ranges (eurybathic). They make diurnal movements of great amplitude, and can adjust themselves to a wide range of pressures.

Animal life may be lacking from the bottoms of deep bodies of water (Black Sea), where no dissolved oxygen is found. However, because of the currents of sinking cold water from the polar seas, some deeper waters of the sea have more oxygen than regions near the surface. Lakes at high altitudes (Andes) cannot dissolve enough oxygen to support fish life. Atmospheric oxygen can thus be considered a limiting factor, for at altitudes of 18,000 to 20,000 feet, the barometric pressure is less than one half that at sea level, and the absolute amount (but not the percentage) of oxygen is correspondingly reduced.

BIOTIC FACTORS

Nutrition. Many animals such as man have a varied diet and can use the food that happens to be convenient (**omnivores**). Other animals are plant feeders (**herbivores**) or flesh feeders (**carnivores**). Within each of these main types, there are numerous subdivisions. Thus the beaver lives on the bark of willows and aspens, the crossbill lives on pine cones, aphids suck plant juice, leeches suck blood, and the king cobra feeds on other snakes.

The interrelations between animals in their food getting furnish interesting **food chains** (Fig. 16-2). Because plant life is the most abundant food in most localities, herbivorous animals form the base of the animal community. These in turn serve as food for larger predators. In a food chain the successive animals involved usually are larger in size but fewer in numbers. Animals at the end of the food chain are large and few, and usually one or two of them dominate a definite region, jealously keeping out all other members of that species. In a forest, for instance, there are many small insects, a lesser number of spiders and carnivorous insects that prey on the small insects, still fewer small birds that live on the spiders and carnivorous insects, and finally one or two hawks that live on the birds. Such an arrangement of populations in the food chain of a community is often called a **food pyramid** (Fig. 16-3); each successive level of the pyramid shows an increase in size and a decrease in number of animals. Food chains may be more complex than the one cited or may be very short, as, for example, the whale, which lives mainly on plankton that forms the base of that particular pyramid. Because of this pyramid arrangement, one can expect very few large predatory animals within any region, since such a large pyramid of animals is required to support them. Only one grizzly bear can be found on the average of each forty square miles of its territory; and in India, tigers are few in number for the same reason.

Biomass refers to the weight of a species population per unit of area. For instance, Juday (1938) found in a Wisconsin lake that there were 209 pounds of carp per acre (biomass). The biomass of a community would be the sum of the biomasses of the many species that make up the steps of the food pyramid. This might be called a **pyramid of biomass**. However, total biomasses obtained by sampling methods are not always reliable. So far, no complete biomass for a community has been obtained.

A better quantitative pyramid, the **energy pyramid**, expresses both the total amount and the energy loss by rate of production of energy. It indicates the rate of energy loss by the food mass as it moves through the food chain. Life on earth receives energy from the sun, moon, cosmic radiation, meteors, and other sources, but mostly it is solar energy. This amounts to about 13×10^{23} gram-calories annually. Some of this energy is lost by being reflected back to space. Most of the rest is used on earth for purposes other than supporting life; only a small amount actually enters the food chain. To give some idea of how energy is expended in a food chain, this example of L. C. Cole (1958) explains the relationship. Out of each 1,000 calories stored by algae in

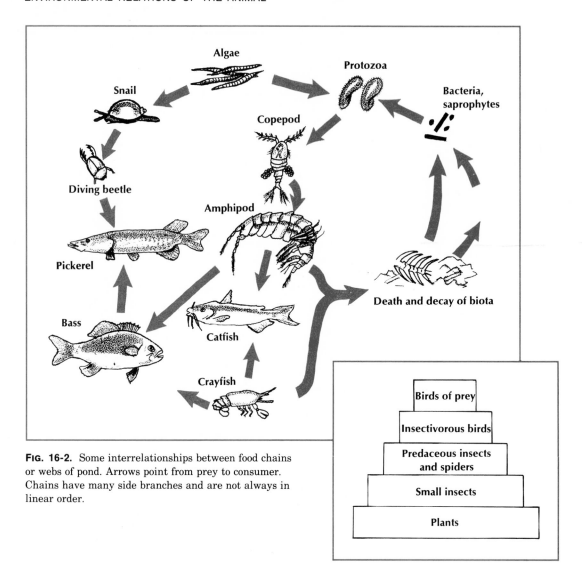

FIG. 16-2. Some interrelationships between food chains or webs of pond. Arrows point from prey to consumer. Chains have many side branches and are not always in linear order.

FIG. 16-3. Simple food pyramid. Size of boxes does not correspond to relative abundance of forms.

Cayuga Lake, 150 calories can be converted to protoplasm in small aquatic animals, 30 calories by smelt in the next step of the food chain, 6 calories by trout that eat the smelt, and 1.2 calories by man who eats the trout. All the rest is lost as heat. It is thus seen that passage through a food chain is very expensive and that there is a definite limit to the amount of life this planet can support.

Carnivorous animals, for instance, are unable to kill animals above a certain size, and many cannot live on forms below a certain size, for they cannot eat enough of them to furnish the necessary amount of energy. A lion, for example, could not catch enough mice under ordinary circumstances to satisfy its

food requirements. Elton showed that the tsetse fly (*Glossina palpalis*) can suck up the blood of only those animals whose blood corpuscles do not exceed 18 μ in diameter. Only man, by his ingenuity, masters any food, regardless of size.

Other biotic factors. These involve social life within a group, cannibalism, mutual assistance, symbiotic relations, mating habits, predatory and parasitic relations, commensalism, mutual dependence of plants and animals, etc.

Various degrees of social life are found among animals, from those that simply band together with no real division of labor (schools of fish and flocks of birds) to those that have worked out complicated patterns of social organization and division of labor (bees, termites, ants). Many predators (grizzly bears, mountain lions, tigers, hawks, owls) tend toward a solitary life and are rarely found together except for mating purposes. Many birds may be more or less solitary most of the year but quite gregarious during their migrations.

There are many examples of **symbiotic relationships,** such as the small fish that obtain shelter in the cloaca of certain sea cucumbers, the tiny crab *Pinnotheres* commensal in the shells of oysters and scallops, or the flagellates that live in the gut of the termite and make its digestion possible. Ants use the secretions of aphids for food and, in return, give protection and shelter to the aphids. Some reef crabs of the tropics fasten an anemone in each claw and use them for protection. Many examples of **cleaning symbiosis** have been found by skin divers among marine organisms. This symbiosis involves the removal of debris or parasites from the teeth or body of one animal by another. Cleaners include many species of small fish, shrimp, and other forms and are usually conspicuous by color or behavior patterns. Cleaners may even have stations where the larger fish go to be cleaned. Parasitism is another common type of symbiosis.

A unique form of symbiosis may exist among certain natives of New Guinea who excrete far more nitrogen than can be explained by their scanty protein diets. The hypothesis has been advanced that nitrogen-fixing bacteria in their intestines may manufacture protein components. Such a symbiotic relationship is analogous to the herbivore stomach where bacteria convert cellulose and nitrogenous materials into proteins.

Cannibalism is not unusual among animals, both low and high. It is common among insects, especially ants and termites.

Many plants depend on insects for transferring their pollen. This is a mutual relation, for insects use nectar or pollen from the blossoms. The classic case of the delicate reciprocal relation between insect and plant in pollination is shown by the yucca moth *(Tegeticula)* and the yucca plant of the southwestern states. This moth collects some pollen from one plant and carries it to another, where it lays its eggs in the ovary of the yucca; after depositing an egg, the moth climbs to the top of the pistil and inserts the pollen into a stigmatic tube. This process is repeated for each egg laid (usually six). Each egg in its development requires a fertilized ovule, but enough ovules are left unmolested by the developing larvae to ensure seed for the plant.

ECOLOGIC ENERGETICS

Ecologic energetics is the study of energy transfer and energy transformations within ecosystems. Since all forms of energy can be converted completely to heat, the calorie, or kilogram-calorie, is considered the basic unit of measurement for comparative purposes. Food chains, as we have seen, can be very complex, and different methods are employed in determining them. Gut analysis is often used to determine what food a member of a food chain uses, but it is impossible to get a complete picture of the food relations unless there are hard parts as key identifying structures. A method of greater value is perhaps the "tagged-atom" method whereby radioactive isotopes (for example, phosphorus 32) are used to label (by spraying or otherwise) the suspected food source. Later animals are checked for the presence of the tagged atoms.

The amount of solar energy per unit area per unit time available to plants varies with the geographic region. In Michigan the amount of solar energy available to plants is considered to be about 4.7×10^8 calories per square meter per year; in Georgia the figure is 6×10^8 calories per square meter per year. Of this energy at least 95% is lost in the form of sensible heat and heat of evaporation, and the remainder (that is, about 5% or less) is used in photosynthesis and transformed into the chemical energy of plant tissue. But not all of this stored energy is available to animals (heterotrophs) that eat plants, for the plant uses up some of its stored organic supply in the process of respiration. The animal that lives directly on plants probably gets only about 80% to 90% of the total energy the plant first stored up.

ANIMAL POPULATIONS

The term "population" is defined by some ecologists as a group of organisms of the same species that live at a given time in a particular area. Others broaden the term to include similar species. Genetically, the members of a population share in a common gene pool. A population has its own characteristics, such as population density, birth rate, death rate, reproductive potential, age distribution, population pressure, population cycles, and growth. Ecologic units, such as communities, are made up of complex population groups and cannot be understood without a study of the interrelations of populations.

Studies of animal populations in most habitats often reveal a large number of different species as well as of individuals in each species. An acre of rich humus soil may contain several hundred thousand earthworms and many million nematode worms. A quart of rich plankton water may have more than a million protozoans and other small forms. The number of insects of all kinds found on an acre of lush meadow in midsummer often reaches millions. On the other hand, there may be only two or three birds per acre and only one or two foxes per several hundred acres. Animals at the top of the food pyramid, having few or no enemies, often regulate their numbers by arbitrarily dividing their territory and keeping out all other members of their species.

The population of any species at a given time and place depends on its birth rate and its mortality or death rate. If more organisms are born than die, the species will increase. Shifting of members of a species from one habitat into an adjacent one (migration, etc.) would affect the local abundance of that species but not the general population of the species involved.

The biotic potential rate is the innate capacity of a population to increase under optimal surroundings and stable age ratios. Ecologists now stress a study of these regulations. The concept of **density-dependent** factors indicates that animal populations are regulated automatically, to a certain extent, by such influences as density, increased mortality or reduced births, food supplies, infectious diseases, and territorial behavior. All these factors operate to produce a general overall effect of feedback, either negative or positive. There are many complicated factors that play a role in population regulation and that are only partially understood. Adult males of predators are known to practice cannibalism on the young as a population control mechanism. Other examples are the dominance hierarchies and the fluctuating fecundity of populations in accordance with the degree of density. As density increases, for instance, fertility declines—a good example of feedback control of population growth.

The success of a population is reflected in its **density,** which is the number of individuals per unit area or volume (Fig. 16-4). For small forms such as plankton, estimates may be made from forms found in a liter of water; for larger animals the acre or square mile may be the unit. The complete count of individuals in an area is called a **census.** Usually counts are made on sample plots, from which estimates are made. Small mammals, such as mice and chipmunks, may be trapped in live traps, tagged by clipping toes or ears, and then released. Suppose 100 animals are caught and tagged, and at a later date another lot of animals is caught in the same way on the same area. In the second sample the number of tagged animals is noted. If the second sample showed 5 tagged animals among 100 caught, then the total population (X) would be $100/X = 5/100$, or $X = 2,000$. This assumes that animals caught in the first sample are just as likely to be caught in the second sample. If sample plots are carefully selected and possible sources of error carefully checked, this random sampling method of estimating the population density of the entire area is considered fairly reliable.

All populations undergo what is called **population dynamics,** which refers to the quantitative variations of growth, reproductive rates, death rates, fluctuations in numbers, age distribution, etc. The characteristic growth of a population is represented by a **population growth curve.** This is the mathematic expression of the growth of a population from its early beginning until it arrives at some stabilizing level of density. Such a curve or graph is produced by plotting the number of animals, or its logarithm, against time. In the beginning, if there is no serious competition with other

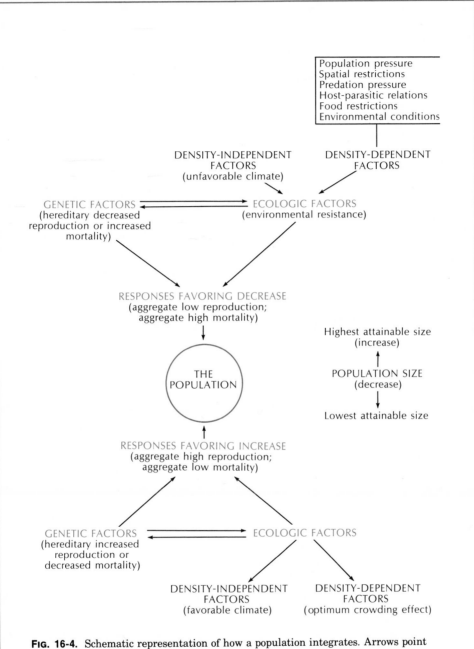

FIG. 16-4. Schematic representation of how a population integrates. Arrows point from pressure factors to those responses that favor decrease or increase of the population size. A population is nearly always in a state of flux—moving upward or downward as a result of the interaction of the two responses. Factors involved in a population study are complex and are often difficult to evaluate. (Modified from Park, T.: Biol. Symposia **8:**123, 1942.)

species and enemies and there is plenty of food, the population grows at about the rate of its potential increase and the curve grows steeply upward. Such curves, however, are rarely realized, except for brief periods, because of the increasing factors of competition, crowding, and higher mortality. One usually starts out with a **lag phase** because it takes time for the few individuals to find each other and start mating. Then it proceeds at a rapid rate, so that a **logarithmic phase** of growth occurs when the population tends to double with each generation, and the curve is fairly straight. But because of more competition for food, losses to enemies, fewer places to live, and greater mortality, the growth rate slows down or levels off into the **stationary phase** (Fig. 16-5).

The **birth** rate (natality) is the average number of offspring produced per unit of time. The theoretic or maximum birth rate is the potential rate of reproduction that could be produced under ideal conditions. This is never realized because not all females are equally fertile, many eggs do not hatch, not all the larvae survive, and for many other reasons. **Mortality** rate is the opposite of birth rate and is measured by the number of organisms that die per unit time. **Minimum mortality** refers to those that die from old age. The actual mortality however, is far different from the minimal one, for as the population increases, the mortality increases.

FIG. 16-5. Population growth curve.

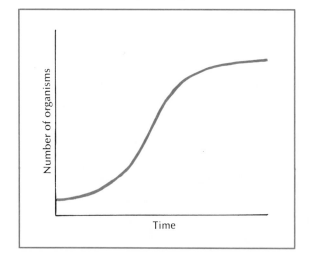

Survival curves (made by plotting the number of survivors against the total life-span) vary among different species. Among many small organisms the mortality is very high early in life; others, such as man, have a higher survival rate at most levels.

Both birth rates and mortality are influenced by the **age distribution** of a population. Asexual forms begin to multiply quite early; many sexual forms (insects, etc.) attain sexual maturity a few days after hatching; others may not become mature for several months or years. Reproductive capacity is usually highest in middle-age groups. Rapidly growing populations have many young members; stationary populations have a more even age distribution. The relative distribution of age groups in a population indicates trends toward stability or otherwise.

Fluctuation in members occurs in all animal populations. There are more birds in early summer than at other times because of the crop of recently hatched members. Later, many of these birds are destroyed by the hazards of the environment, such as **population pressure.** Examples of cycle populations are the lemmings of the northern zones which become so abundant every three or four years that they migrate to the sea and drown; the snowshoe hares of Canada that have approximately a ten-year cycle of abundance (there is a close parallelism in cyclic abundance of the lynx, which feeds on the hare); and meadow mice that usually show a four-year cycle of abundance. No satisfactory theory as yet accounts for cyclic fluctuations. Some irregular fluctuations can be explained by weather and climate changes. The great "dust bowl" of 1933 to 1936 must have affected every aspect of life in the stricken area.

BIOTIC COMMUNITIES

Organisms in a community share the same physical factors and react to them in a similar way. The community and the nonliving environment together form the **ecosystem.**

A **major community** is the smallest ecologic unit that is self-sustaining and self-regulating. It is made up of innumerable smaller **minor communities** that are not altogether self-sustaining. Forests and lakes are major communities; decaying logs and ant hills are minor communities. Members of a major community are relatively independent of other communities, provided they receive radiant energy from the sun.

Stratification is the division of the community into definite horizontal or vertical strata. In a forest community, for instance, there are animals that live on the forest floor, others on shrubbery and low vegetation, and still others in the treetops. Many forms shift from one stratum to another, especially in a diurnal manner. Many of the adjustments and requirements of a particular stratum are very similar in forests widely separated from each other in many parts of the world. The animals that occupy such similar strata, although geographically separated, are called **ecologic equivalents.** The pronghorn antelope of North America and the zebra of South Africa are equivalents.

Between two distinct communities there may be an intermediate transitional zone. This is called an **ecotone,** or tension zone. An example would be the marginal region between a forest and a pasture or open land.

Food relations are a basic aspect of all communities. **Producer organisms,** such as green plants, make their own food by using sunlight as their energy source and inorganic and simple organic com-

pounds as a source of building materials. **Primary consumers** are the various herbivores that feed on plants; **secondary consumers,** such as carnivores, live on the primary consumers, etc. Other organisms such as bacteria are called **decomposer organisms,** for they break down the dead organisms into simpler substances that can then be used by plants.

Generally those organisms with the largest biomasses within their levels of feeding interrelations are the ones that exert a controlling interest. In land communities, plants are usually the **dominants,** and some communities are named from their dominant vegetation, such as beech-maple woods. In the ecologic cycle of a community the removal of a dominant usually causes serious disturbances in the ecologic balance.

ECOLOGIC SUCCESSION

Communities are not static but are continually changing according to well-defined laws. This process, called biotic or **ecologic succession,** may be brought about by physiologic factors, such as the erosion of hills and mountains down to a base level, the filling up of lakes and streams, and the rise and fall of the earth's surface. All organisms die, decay, and become a part of the substratum; vegetation

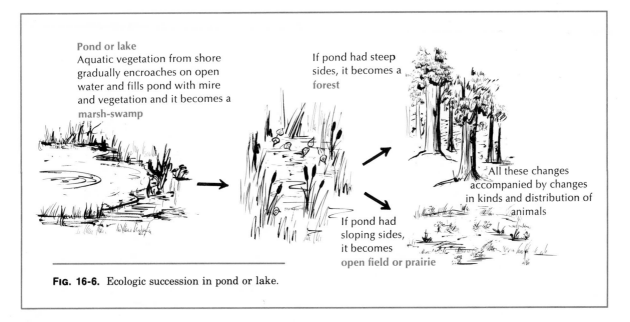

FIG. 16-6. Ecologic succession in pond or lake.

211

invades ponds and lakes; regions of the earth's surface become grasslands, forests, or deserts, according to physical factors of temperature and rainfall. Communities are succeeded by other communities until a fairly stable end product is attained. Such a sequence of communities involves early pioneer communities, transient communities, and finally a **climax community,** which is more or less balanced with its environment.

The sequence of plants follows in a certain order during the evolution of a habitat. For instance, a small lake begins as a clear body of water with sandy bottom and shores more or less free from vegetation. As soil is washed into the lake by the surrounding streams, mud and vegetable muck gradually replace the sandy bottom. Vegetation grows up along the sides of the lake and begins a slow migration into the lake, resulting in a bog or marsh. The first plant life is aquatic or semiaquatic, consisting of filamentous algae on the surface and later of rooted plants, such as *Elodea,* bulrushes, and cattails. As the water recedes and the shore becomes firm, the marshy plants are succeeded by shrubs and trees, such as alders and larches and, later, beeches and maples. Eventually the lake may be replaced by a forest, especially if its sides and slopes are steep; if the sides have gentle slopes, a grassy region may replace the site of the lake. The terminal forest or grassland is a climax community (Fig. 16-6).

In its beginning a lake may contain fish that use the gravelly or sandy bottoms for spawning. When the bottoms are replaced by muck, these fish will be replaced by others that spawn in aquatic vegetation. Eventually, no fish may be able to live in the habitat; but other forms, such as snails, crayfish, many kinds of insects, and birds, are able to live in the swampy, boggy community. As the community becomes a forest or grassland, there will be other successions of animal life.

ECOLOGIC NICHE

The special place an organism has in a community with relation to its food and enemies is called an **ecologic niche.** In every community there are herbivorous animals of several types, some of which feed on one kind of plant and others on other plants. There are also in every community different types

of carnivores that prey on different species of animals. Similar niches in different communities are occupied by forms that have similar food habits or similar enemies, although the species involved may be different in each case. For example, there is the niche in wooded regions occupied by hawks and owls that prey on field mice and shrews, but in regions close to homes this niche is taken over by cats. In this respect the birds of prey and cats occupy the same niche. The arctic fox and the African hyena are other examples. Both are scavengers; the arctic fox eats what the polar bear leaves; the African hyena eats leavings from the lion.

FIG. 16-7. Tree hole habitat. Some arthropods found here, such as pselaphid beetles, are never or rarely found elsewhere.

KINDS OF HABITAT

Habitats, or the places where animals live, tend to be rather sharply defined from each other, each with its own set of physical and biotic factors. An abundance of different habitats may be found in a small region if there is a diversity of physicochemical and other factors. A small lake or pond may have littoral or open water, cove, sandy or pebbly bottom, bulrush or other vegetation, drift, and other kinds of habitats. The small altitudinal life zones of the mountain are similar to the large latitudinal zones of the earth's surface with respect to vegetation and, to some extent, to the distribution of animal life. On the other hand, there may be extensive regions, such as the surface of the open sea or a sandy desert, where there is no such diversity of habitats because ecologic conditions are more or less uniform.

The animals that are distributed among the various habitats may be classified into two groups: (1) **exclusive,** or those that are not found outside a particular habitat (Fig. 16-7), and (2) **characteristic,** or those that are not confined to one habitat but occur in others. Examples of the first are crossbills, which, on account of their peculiar adaptation, are confined to the coniferous forests, and, of the second class, such forms as rabbits, which roam both woods and open fields. Obviously a habitat may contain more than one niche.

FRESHWATER STREAMS

Freshwater habitats are usually divided into those that are found in **running water** and those found in **standing water.** Freshwater streams range from tiny intermittent brooks to rivers. Larger, permanent streams include the swift brooks and the rivers that have reached the level of permanent ground

FIG. 16-8. Brook rapids habitat contains many different forms that have special adaptations for withstanding water currents.

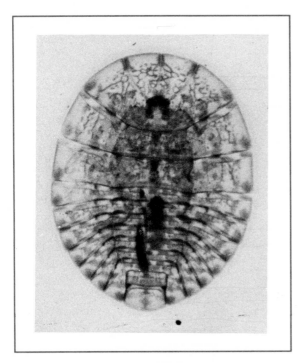

FIG. 16-9. Water penny, larva of riffle beetle *Psephenus*. This flat larva is adapted for clinging to lower surfaces of stones in swift brooks. (Size, 1/3 inch.)

water. Water found in streams differs from ocean water by having smaller volume, greater variations in temperature, lesser mineral content, greater light penetration, greater suspended material content, and greater plant growth. Many animals living in such habitats have organs of attachment, such as suckers and modified appendages, streamlined body shapes for withstanding currents, or shapes adapted for creeping under stones. Various kinds of habitats are found in all streams. Some are found in the swiftly flowing regions where there may be rapids or cataracts; others are found in pools of sluggish waters. Usually the types of animals living in the two regions vary. In rapids (Fig. 16-8) characteristic forms may include the black fly larvae, caddis fly larvae, snails, darters of several species, water penny larvae (Fig. 16-9), miller's thumbs, and stone fly nymphs. All these forms have characteristic behavior patterns, such as positive rheotaxis, high oxygen requirements, and low temperature tolera-

tion. In the pool habitats of streams are found various minnows, mussels, snails, dragonfly nymphs, mayfly larvae, crayfish, flatworms, leeches, water striders, and many others.

The study of fresh waters in all their aspects is called **limnology.**

PONDS

Unlike the streams, ponds have feeble currents or none at all. Most ponds contain a great deal of vegetation that tends to increase with the age of the pond. Many of them have very little open water in the center, for the vegetation, both rooted and floating types, has largely taken over. As ponds fill up, the higher plants become progressively more common. The bottoms of ponds vary all the way from sandy and rocky (young ponds) to deep mucky ones (old ponds). The water varies in depth from a few inches to 8 to 10 feet, although some may be deeper. Ponds are too shallow to be stratified, for the force of the wind is usually sufficient to keep the entire mass of water in circulation. Because of this, the gases (oxygen and carbon dioxide) are uniformly distributed through the water and the temperature is fairly uniform.

Animal communities of ponds are usually similar to those of bays in larger bodies of water (lakes). The large amount of vegetation and plant decomposition products affords an excellent habitat for many forms. Among the common forms found are varieties of snails and mussels, larvae of flies (Fig. 16-10), beetles, caddis flies, dragonflies, many kinds of crustaceans, midge larvae, and many species of frogs. Most of these live on the bottom or among the submerged vegetation and are called the **benthos.** Swimming forms, called **nekton,** include fish, turtles, water bugs, and beetles. Most ponds also have plankton composed of microscopic plants and animals, such as protozoans, crustaceans, worms, rotifers, diatoms, and algae (Fig. 16-11).

Forms that live in ponds ordinarily require less oxygen than those found in streams or rapids.

LAKES

The distinction between lakes and ponds is not sharply defined. Lakes are usually distinguished

Fig. 16-10. Group of larval forms commonly found in ponds. **A,** Crane fly larva found in decaying debris along bank. Beside it is small crustacean, *Asellus.* **B,** Rat-tailed maggot, *Tubifera,* gets its air by means of its caudal respiratory tube ("rat-tail"), which can be extended to four times the length of body. **C,** Larval form of *Dytiscus,* giant diving beetle. These were all found in a tiny midwestern woodland pond in early March.

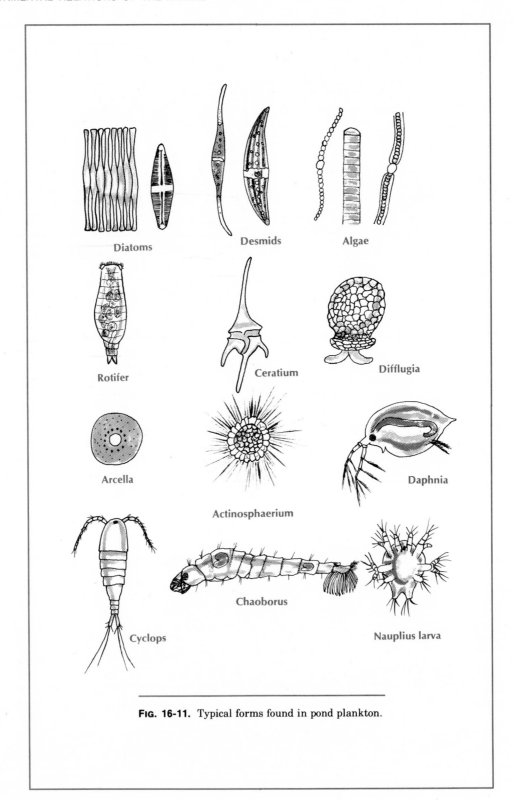

FIG. 16-11. Typical forms found in pond plankton.

from ponds by having permanent water in their centers and by having some sandy shores. When waves are common, the bottoms are usually sandy, but protected areas may contain a great deal of deposited bottom muck and aquatic vegetation. The surface of water in proportion to the total volume of water is less in a lake than in a pond, for lakes are deeper. Oxygen is scarcer in the deeper regions where there is little circulation. Light penetration depends on the sediment in water. Most of the light is absorbed in the first meter of surface water, and little penetrates beyond a few meters.

Water in lakes tends to become stratified because only the surface layers are stirred up by wind action during the summer. Within this surface layer of water, usually about 10 meters deep in medium-sized lakes, the temperature is very uniform (about 20° to 25° C.). Below this the water becomes much colder, reaching 4° C. at the bottom. The level between the surface layer of uniform temperature and that stratum where the temperature falls rapidly is called the **thermocline.** In other words, the waters above the thermocline are agitated; below it they are still. The thermocline is shallow at midsummer and deeper in early autumn. In the autumn when the surface water is colder, the wind agitates the water from surface to bottom and the thermocline disappears so that the temperature is about uniform throughout. In winter the surface water may freeze (0° C.), but the bottom remains at 4° C., the temperature of maximum density. As the surface waters become warmer in the spring, there occurs another complete overturn through wind action. In early summer the thermocline is established again. The spring and fall overturns are of great importance to a lake, since only at these times is the deep water of the lake oxygenated.

TERRESTRIAL

Land habitats are more varied than those of the water because there are more variable conditions on land. Physical differences in the air are expressed in such factors as humidity, temperature, pressure, and winds to which air-dwelling forms must adapt themselves, as well as types of soils and vegetation. Variations from profuse rainfall to none at all; topographic differences of mountains, plains, hills, and valleys; climatic differences from arctic conditions to those of the tropics; temperature variations from those in hot deserts to those of high altitudes and polar zones; and air and sunlight differences from those of daily variations to great storms—all these factors have influenced animal life and have been responsible for directing its evolutionary development.

Land forms have become specially adapted for living in the soil (subterranean), on the open ground, on the forest floor, in vegetation, and in the air. Although more species of animals live on land than in water, there are fewer phyla among terrestrial forms. The chief land organisms are the mammals, birds, reptiles, amphibians, worms, protozoans, and arthropods.

Land habitats are classified on the basis of soil relations, climatic conditions, plant associations, and animal relations.

Subterranean refers to regions within the soil or under the land surface. Subterranean habitats include holes and crevices in or between rocks, burrows in the soil, and caves and caverns. Larval forms of many insects develop there. Ants, nematodes, earthworms, moles, and shrews either have their homes in the soil or spend a part of their time there. Most cave animals originated on the surface and were adapted for existence in caves before they entered them. Caves are unique in having uniform darkness, high humidity, no green plants, and no rain or snow. Meager food supply restricts the number of animals living there. Cave animals are usually small, have little or no pigment, and have degenerate sense organs. Many are totally blind. Most true cave animals illustrate regressive evolution in that selection puts a premium on loss of internal stability, low basal metabolism, and low food requirements.

Another habitat is the ecosystems of snowbanks studied in recent years in the Colorado Rockies. In summer alpine snowbanks are often colored a reddish or greenish hue, caused by the algae that coexist with other organisms in the same interrelationships that are found in other ecosystems. Populations of organisms found here include protozoans, bacteria, and fungi. Such organisms (cryophiles) are adapted to low temperatures because the temperature of the water films (in which the organisms live)

217

of the melting snow crystals is always 32° F. The food chain is rather simple, but the flow of matter through the ecosystem is cyclical.

BIOMES

Biotic communities may be aggregated into biomes, the largest ecologic units. In a biome the climax vegetation is of a uniform type, although many species of plants may be included. Each biome is the product of physical factors, such as the nature of the substratum and the amount of rainfall, light, temperature, etc. They are distributed over the surface of the earth as broad belts from the equator to the poles and each may or may not be continuous. Within each biome are many major communities. Biomes are not always sharply marked off from each other; there are often intermediate zones. Some six important terrestrial biomes are recognized: desert, grassland, equatorial forests, deciduous forest, coniferous forest, and tundra. This succession of biomes from the tropics to the poles may also be found

in condensed form on the vertical zones of a high mountain (Fig. 16-12). The various oceans make up the so-called marine biome, a major community that may be subdivided into a number of minor communities.

MARINE BIOME

The oceans have so many diversified physicochemical conditions that innumerable habitats are found in them.

The conditions of marine existence are many. Salt water makes up about 70% of the earth's surface. The average depth of the sea is about 10,000 feet, but there are much deeper regions. Its average salt content is about 3.5%, which makes its density greater than fresh water but not so great as protoplasm, for the animal body will sink in it. Its pressure in the great depths may reach a thousand atmospheres; yet animals live there and are not crushed, for the pressure is equal inside and out.

FIG. 16-12. Correspondence of life zones or biomes as correlated with latitudinal and altitudinal zones. This succession of biomes over extensive horizontal range from the tropics to polar regions is found in a condensed form within a few miles on a high mountain in the tropics and, to a slightly lesser extent, in temperate zone.

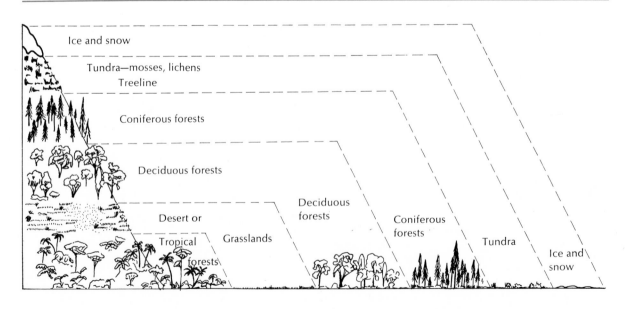

Besides tides and surface wave action, the sea is in continuous circulation. The currents are, in general, due to prevailing trade winds, caused by temperature differences between poles and equator, and due to the rotation of the earth, which can give rise to movements in the deep layers of the ocean as well as near the surface. The ice caps at the poles, particularly the South Pole, form glaciers that eventually melt in the sea and produce a cold but light current of dilute salt water that flows away toward the equator. A similar current of cooled water under the surface layer sinks and flows along the ocean floor, also toward the equator. Other currents of water flow toward the poles between the other two currents to take the place of the displaced water. Off-shore winds push coastal waters away, and their place is taken by the upwelling of water from below, bringing to the surface phosphates and nitrates that encourage the growth of plankton. These restricted areas of upwelling are the only truly productive areas of the ocean.

The temperature of sea water varies with location and with the seasons. In the arctic regions it may go to below 0° C.; in the tropics the surface water may exceed 30° C. These regions may have a relatively uniform temperature the year round, but in the temperate zones there are seasonal variations in the temperature of surface waters that are more pronounced in landlocked bodies of the sea. Deep regions of the sea always contain cold water.

The penetration of light is restricted mainly to the upper 50 to 60 meters, although the blue end of the light spectrum penetrates to greater depths.

The animals that inhabit the sea may be divided into two main groups—pelagic and benthonic.

Pelagic group. The pelagic group, which lives in the open waters, includes (1) the **plankton,** small organisms (protozoans, crustaceans, mollusks, worms, etc.) that float near the surface of the water, and (2) the **nekton,** composed of animals that swim by their own movements (fish, squids, turtles, whales, seals, birds, etc.).

Benthonic group. The benthonic group includes the bottom-dwelling forms, or those that cannot swim about continuously and need some support. This group can be subdivided according to the zones in which they are found.

1. The **littoral,** or lighted zone, is the shore region between the tidelines that is exposed alternately to air and water at each tide cycle. These forms are subjected to high oxygen content and much wave disturbance. In this region originated the ancestors of all aquatic fauna, both fresh water and salt water. It contains a very rich animal life, both in species and numbers of individuals.

2. The **neritic** zone lies below the tide water on the continental shelf and has a depth of 500 to 600 feet. There is some wave action here, and the water is well oxygenated. Many forms, including fish, echinoderms, and protozoans, are found in this habitat.

3. The **bathyal** zone is a stratum of the deeper water from the neritic region down to 5,000 or more feet. It contains small crustaceans, arrowworms, medusae, and fish. Many of the animals in this region have luminescent organs, for this is a dark zone.

4. In the **abyssal** zone, or deeper parts of the oceans, the water is always cold and there is total darkness. Oxygen is scarce or absent, and only a few deep sea forms are found—certain specialized fish and crustaceans. Many of these are provided with light organs.

TERRESTRIAL BIOMES

Tundra. The tundra is characteristic of severe, cold climates, especially that of the treeless arctic regions and high mountain tops. Plant life must adapt itself to a short growing season of about 60 days and to a soil that remains frozen for most of the year. Despite the thin soil and short growing season, the vegetation of dwarf woody plants, grasses, sedges, and lichens may be quite profuse. The plants of the alpine tundra of high mountains, such as the Rockies and Sierra Nevadas, may differ from the arctic tundra in some respects. Characteristic animals of the arctic tundra are the lemming, caribou, musk ox, arctic fox, arctic hare, ptarmigan, and (during the summer) many migratory birds.

Grasslands. This biome includes prairies and open fields and has a wide distribution. It is subjected to all the variations of temperature in the temperate zones, from freezing to extremely hot temperatures. It undergoes all the vicissitudes of seasonal climatic factors of wind, rain, and snow. On the western prairies there will be jack rabbits, antelope, wolves, coyotes, skunks, gophers, prairie chickens, and insects. In the eastern parts of the country, some of these will be replaced by other forms such as meadow mice, meadow larks, and cotton-tailed rabbits.

Desert. Deserts are extremely arid regions where permanent or temporary flowing water is absent. When rain does come, it may do so with a terrific downpour. The temperature becomes very hot during the day but cools off at night. There is some scattered vegetation that quickly revives after a rain. Some of the most characteristic plants are the cacti.

Desert faunas are varied and mostly active at night, so as to avoid the heat of the day. To conserve water they have physiologic devices for passing dry excretions. To the casual visitor the desert fauna may seem somewhat scanty, but actually the desert possesses representatives of many animal groups, especially mammals, birds, reptiles, and arthropods.

Coniferous forests (taiga). The coniferous forests are the evergreens—pines, firs, and spruces—found in various areas of the North American continent. They may occur in mountains or flat country. They bear leaves the year around and afford more cover than deciduous forests. They are often subject to fires that influence the animal habitats. On mountains and in northern regions they undergo severe winters with much snowfall; in southern regions they have milder conditions. A great deal of food—berries, nuts, and cones—is found in evergreen forests, and there is also a great variety of animal life. In the north there are furbearers, moose, bears, many birds, some reptiles, amphibians, and many insects. Southern coniferous forests lack some of these forms but have more snakes, lizards, and amphibians.

Deciduous forests. Trees of the deciduous forests shed their leaves in the fall, leaving them bleak during the winter, especially in northern climates. There may be some low underbrush and vines. Some of these forests have scattered evergreen trees. Among characteristic fauna of these forests are the deer, fox, bear, beaver, squirrel, flying squirrel, raccoon, skunk, wildcat, rattlesnake, copperhead,

and various songbirds, birds of prey, and amphibians. Insects and other invertebrates are common, since decaying logs afford excellent shelters for them.

Tropical rain forests. Vegetation is luxuriant and varied. The trees are mainly broad-leaved evergreens; there are also many vines. These forests have a copious rainfall and a constant high humidity. The forests are divided ecologically into a vertical series of strata, each of which is occupied by characteristic animals. These strata include the forest floor, the shrubs, small trees, lower treetops, and the upper forest canopy. The enormous amount of life found here is represented by monkeys, amphibians, insects, snails, leeches, centipedes, scorpions, termites, ants, reptiles, and birds.

FOOD CYCLE

All energy utilized by animals is derived ultimately from the sun. Plants utilize radiant energy from sunlight and the chlorophyll in their cells to produce carbohydrates from carbon dioxide and water. Plants can also form proteins and fats. Animals, with few exceptions, depend on the plants either directly or indirectly as sources for the basic food substances.

When a plant transforms light into the potential energy of food by the process of photosynthesis, energy is being transformed into another form without being destroyed (first law of thermodynamics). When this plant food is utilized or consumed by other organisms, although there is no loss in total energy, there is a decrease in amount of useful energy, for some energy is degraded or lost as heat in a dispersed form (second law of thermodynamics). Thus in every step in a food chain or pyramid there is a certain loss in useful energy. Energy is used only once by an organism or population, for it is then converted into heat and lost. On the other hand, the nonenergy materials, such as nitrogen, carbon, and water, may be used over and over again.

All plants and animals in their metabolism require certain elements, such as carbon, oxygen, nitrogen, hydrogen, and, to a lesser extent, potassium, magnesium, calcium, sulfur, iron, and a few others (see p. 5). All these are derived from the environment where they are present in the air, soil, rock, or water. When both plants and animals die and their bodies decay, or when organic substances are burned or oxidized, these elements are released and returned to the environment. The elements that are involved in these processes, therefore, pass through cycles that involve relations to the environment, to plants, and to animals. Four of these important cycles will be pointed out here.

Carbon cycle (Fig. 16-13). Both animals and plants respire and give off carbon dioxide to the air. More is released in the bacterial decomposition of organic substances, such as dead plants and animals. Although the percentage of carbon dioxide in air is relatively small (0.03%) as compared with the other gases of air, this small amount is of great importance in nature's economy. Living green plants take carbon dioxide from the air or water and

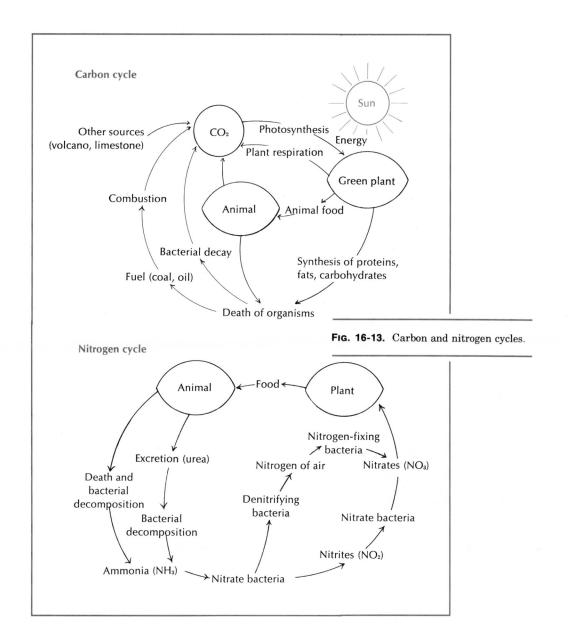

Fig. 16-13. Carbon and nitrogen cycles.

by **photosynthesis,** with the help of sunlight, carbohydrates are formed. This process is complicated but in a highly simplified form may be expressed thus:

$$6H_2O + 6CO_2 \rightarrow C_6H_{12}O_6 + 6O_2$$

WATER CARBON   SUGAR   OXYGEN
DIOXIDE

Carbohydrates thus formed, together with proteins and fats, compose the tissues of plants. Animals eat the plants and the carbon compounds become a part of animal tissue. Carnivorous animals get their carbon by eating herbivorous forms. In either case a certain amount of carbon dioxide is given off to the air in breathing (this also occurs in plants), and when the animal dies, a great deal of the gas is released to the air by bacterial decomposition. This cycle is called the carbon cycle. The importance of carbon in the organic world cannot be overestimated; for of all the chemical elements, it is the one that enters into the greatest number of chemical combinations.

Oxygen cycle. Animals get their oxygen from the air or from oxygen that is dissolved in water and utilize it in their oxidative reactions. They return it to their surroundings in the form of carbon dioxide (CO_2) or water (H_2O). Plants also give off oxygen in photosynthesis, as seen in the formula above. Plants use some oxygen in their own respiration. In the interesting relationship between plants and animals in the plankton life of surface waters, the floral part of these populations gives off the oxygen necessary for the life of the faunal portion and determines the distribution of the latter.

Nitrogen cycle (Fig. 16-13). Atmospheric nitrogen (78% of air) can be utilized directly only by the nitrogen-fixing bacteria that are found in the soil or in the root nodules of leguminous plants. These nitrogen-fixing bacteria combine nitrogen into nitrates (NO_3), and plants form proteins from these nitrates. When animals eat plants, the amino acids of plant proteins are converted into animal proteins. In animal metabolism, nitrogenous waste (urea, etc.) is formed from the breakdown of proteins and is excreted. In the soil or water certain bacteria convert this waste into ammonia and into nitrites; other bacteria (nitrifying) change the nitrites into nitrates. Whenever plants and animals die and

undergo bacterial decomposition, their proteins are converted into ammonium compounds.

Mineral cycle. Many inorganic substances are necessary for plant and animal metabolism. Among these, phosphorus is one of the most important. It is found in the soil and water in the form of phosphoric oxides. It is taken up by plants, is passed on to the animals, and is eventually returned to the soil or water in the form of excreta or upon the decay of their bodies. Many think that phosphorus is the critical resource for the efficient functioning of ecosystems. The supply is normally very much limited and vast amounts are carried into the sea. Man has so disrupted the amount of phosphorus in his disturbance of the soil that erosion can readily carry it away. The runoff of phosphate fertilizers and the wide use of detergents have increased phosphate levels in surface waters 27-fold in recent years. This overfertilization produces large blooms of algae and depletes the water of oxygen.

BEHAVIOR AND ENVIRONMENT

Every kind of organism has its characteristic pattern of response to changes in its environment. A given behavior pattern may start in response to a definite external change or stimulus, or it may originate from internal stimuli. Many animals may initiate a behavior pattern without any apparent reason at all. A response may take place immediately after an animal is disturbed, or it may be delayed for a considerable time after the stimulation. It is obvious that a vertebrate animal has a greater complexity of nervous and other systems involved in behavior than that of many low forms of life. Anything that lives, moves, and has being has some form of behavior, but the same can also be said of nonliving entities that have molecular patterns of action involving movements of atomic constituents. From a mechanistic viewpoint it would be said that there is a unity of behavior plan for both living and nonliving matter, but this has not yet been demonstrated.

Animal behavior can never become static because the animal is always in the process of adjustment to the environment. Just as animals show morphologic evolutionary changes in time, so sequences of behavior patterns also occur. No genetic code can anticipate all adjustments that an animal must face in its life-span. The higher the animal in the levels of life, the more flexible are its adjustments; the lower the animal is, the more it is at the mercy of its environment.

The marked revival of interest in animal behavior in recent years has been due mainly to the researches of two European investigators, K. Lorenz and N. Tinbergen. Their theory of instinctive and innate behavior has attracted attention everywhere because of their fresh outlook. In fact, they call their approach to behaviorism **ethology**, which is the comparative study of the physiologic basis of the organism's reaction to stimuli and its adaptations to its environment. They have tried to explain innate behavior by investigating the stimuli that control it and by studying the animal's internal conditions that are organized for particular patterns. Their studies have greatly stimulated other competent workers, such as D. S. Lehrman, T. C. Schneirla, and W. H. Thorpe, to undertake similar investigations, either in confirmation or in refutal of the theories Lorenz and Tinbergen have proposed.

METHODS OF STUDYING BEHAVIOR

Animal behavior involves the activities of the whole organism with reference to the environment so that animal sociology or group relations must also be understood. Testing must conform to the standard scientific procedure of the control experiment in which conditions are kept as uniform as possible, except in the one environmental factor (stimulus, etc.) that is being studied. The experimentalist must know the nature and normal responses of the animals being studied. A racoon, for instance, is far more adept in manipulating its forelimbs than a cat, and experiments that involve the use of this limb must therefore take this into account. An animal can organize its behavior capacities only within the range of its abilities.

Many investigations have been conducted to determine the neurologic basis of behavior. Several methods may be employed. Certain areas of the brain may be removed surgically or destroyed and the resulting functional deficiencies noted; electric stimulation of brain regions is effective in causing responses in muscles and other effectors; and it is possible to use the electroencephalogram (for detecting electric discharges) to find those parts of the nervous system that are functioning under a given condition. By such methods it has been possible to determine important nerve centers, such as the cocoon-spinning center (corpus pedunculatum) of caterpillars, the satiety center of mammals, motivation centers, and many others.

An exciting new development in understanding animals in their wild state is the method of **biotelemetry**, or radio-tracking. This technique involves the attaching of a transistorized radio transmitter to an animal and then recording the data given off by signals while the animal is undisturbed and freely functioning under its natural conditions. Many physiologic aspects can be obtained this way, such as body temperature, wing beats, and other activity states. Radio-telemetry has been used successfully with ruffed grouse, grizzly bears, rabbits, reptiles, amphibians, etc. Factors of the animal's environment can also be studied by this method.

The greatest pitfall in all behavior studies is interpretation, for there is the tendency to attribute humanlike reasons for an animal's activities. When another animal does something humanlike in nature, it does not mean that the animal is thinking like a man. A bird thinks like a bird, a dog like a dog, and an anthropomorphic interpretation is unjustified in either case. Another pitfall is the ascribing of purpose to an animal's reactions. It is true that many of its behavior patterns are adaptive, but this does not mean that animals perform these acts with an understanding of the end result or with the ability to anticipate what the end is to be.

LEVELS OF NERVOUS ORGANIZATION

Complex behavior is restricted to highly organized nervous systems and superior sensory reception. It is only when the brain has advanced to the role of an organizing center that it can truly be thought of as regulating and controlling behavior organization.

The trend of evolution in the nervous system beyond the protists is centralization. From this standpoint most animals fall into one of three major types of nervous systems—nerve nets of coelenterates, nervous systems with beginning of brains, and centralized nervous systems. The trends that promote the capacity for organized behavior of increasing complexity are (1) the development of centralized control by means of concentrating the nerve cells (neurons) in dominant ganglia (brains) and in ganglia on or near a few nerve cords; (2) the differentiation of various kinds of neurons of more or less polarity (carrying impulses in one direction only), such as afferent, efferent, and association neurons, which are arranged to form a mechanism of coordination (reflex arc); (3) the variety and richness of nerve pathways, connections, and associations that make an organization suitable for precision and variation of specialized behavior; and (4) the development of the sensory capacities that depend on many sense organs of great complexity, sensitivity, and range of response.

HEREDITY AND BEHAVIOR

Behavior patterns are the result of the interaction between hereditary factors and the environment. They must develop out of certain potentialities (or genes) that physiologic influences act on and limit at each stage of development. Behavior patterns involve many factors such as nervous integration, hormone balance, and muscular coordination. Many genes must therefore be responsible for even the simplest activity of an animal. It is often difficult to determine whether heredity or learning experience is more involved.

Some types of activity are rather definitely triggered. Each species of bird builds its typical nest without being taught; the parasitic cowbird raised in a warbler's nest never tries to mate with a warbler but only with its own kind; a spider weaves its web without learned modification; a stickleback fish always performs its courtship ritual the same way. The influence of hereditary factors seems to be much more pronounced in lower than in higher animals. Many of their basic patterns of adaptive behavior seem to be stereotyped.

TAXONOMY AND BEHAVIOR

In recent years emphasis has been placed on the relationship between behavior patterns and taxonomic units at all levels. Evolutionary relationships are clearly expressed by behavior similarities and differences that are correlated with taxonomic subdivision. H. S. Barber was able to separate many species of fireflies on the basis of differences in characteristic flashes emitted by flies of different populations. B. B. Fulton, working on field crickets, found four different populations (supposedly one species) that would not interbreed (a behavior trait) in the laboratory, thus indicating the divergence and formation of four new species. Behavior patterns of *Drosophila* have been shown to conform to the accepted taxonomy of the various species, but in some cases taxonomic revision has been made on a behavior basis.

WHAT IS AN INSTINCT?

The concept of the term "instinct" formerly meant any form of innate behavior that arose independently of the animal's environment, that was distinct from learned behavior, and that followed an inherited pattern of definite responses. At one time there was thought to be a sharp line between instinct and learned experience, and any action of an organism was either instinctive or learned. Psychologists at present believe that most behavior must be interpreted in terms of both innate traits and learning. Few behaviorists are willing to concede that a particular activity is wholly instinctive or is wholly learned.

Some behaviorists argue that behavior cannot be inherited through the genes of the chromosomes but must develop under the influence of environment. Certain types of behavior, it is true, can be modified more than other types. Nest building among birds is unlearned, yet older robins build better nests than younger robins. Some birds reared away from their parents will still sing the song characteristic of their species; other kinds of birds when raised with members of another species will sing the song of that species rather than their own.

Lorenz and Tinbergen have stressed instinctive behavior as a stereotyped action that follows a definite pattern of expression. They believe that there are at least three components involved in an instinct. First, there is an **appetitive behavior,** which may be regarded as a buildup of readiness for the instinctive act. An appetite for the act is generated in the organism so that it gets into a situation in which the instinct can be released. The animal is very restless until the instinct is released. This phase is goal directed, concerned only with the actual performance of the act. Second, an **innate releasing mechanism** is activated. This may be due to something in the environment or to an inner bodily condition. It refers to the removal of any inhibition for the performance of the instinctive behavior. Third, the **final consummatory act,** which might be considered the relief of the animal's tension by the actual discharge of the activity. This pattern of instinctive behavior might be illustrated simply by the reactions of a hungry young bird in a nest when something is waved before it. The bird is in a condition for response (appetitive); the movement of the object activates the release mechanism; and the lunge and gaping that follow is the consummatory act.

SIMPLE BEHAVIOR PATTERNS (TROPISMS AND TAXES)

The simplest form of organized behavior is one in which a specific stimulus gives rise to a specific response. This type belongs to what is called inherited behavior patterns and is best represented perhaps by the **tropisms** of plants and the **taxes** of animals. Plants illustrate this specific stimulus-response behavior, or tropism, for most of their behavior patterns are tropistic responses. A tropism, which literally means a "turning," refers to the bending movements of plants brought about by differences in the stimulation of the two sides of an organ (stem, root, etc.).

The term **taxis** is used by zoologists to describe the movements of free-swimming forms such as protozoans. Phototaxis, or response to light rays, is an example of a taxis.

In the early part of the century J. Loeb and his school tried to interpret all animal behavior, whether low or high, on the basis of tropisms (taxes). He attempted to show that the differences in stimulation intensity on the two sides of an animal toward light, current, etc. caused the animal to orient itself toward or away from the source.

Loeb's theory has met with opposition from many sources. H. S. Jennings, in his now classic book *The*

Behavior of the Lower Organisms, proposed a trial-and-error explanation in place of the tropistic theory. This is really a stimulus-response theory, for Jennings found that most environmental changes will produce a response. The avoiding reaction of paramecia is due not to unequal stimulation of its two sides but to a fixed orientation pattern that enables the animal to find a favorable escape channel. All organisms have the capacity for several different responses to the same external stimulus.

Perhaps in our present state of knowledge the best way to regard an instinct is that it is concerned with activities that depend mainly or wholly on an animal's organic equipment in reaction with the environment, with learning playing a minor role or else being entirely absent in the process.

INSTINCTIVE BEHAVIOR PATTERNS

An instinct is usually considered to differ from a taxis or tropism in being more complicated and involving more separate phases in the performance of the act. However, the two types of behavior may overlap. Both involve reflex action, but a taxis or tropism is less flexible and is based more on a rigidly inherited plan. In studying any instinctive behavior it is necessary to determine the stimuli that control the behavior, to appraise the internal conditions that prepare the organism for the reaction, and to seek out the neural mechanism responsible for the integration of the whole basic pattern of behavior. The organic factors that may be involved in instincts are the sensory equipment, endocrine system, neuromuscular system, etc. Another important influence on instinctive behavior is maturation, or the development of behavior patterns, as correlated with the age and growth of animals.

Most instincts are adaptive in nature and can contribute much to the success of a group, as shown by the arthropods in which innate behavior of great variety predominates over learning. However, one cannot attribute purpose in their performance, for instincts are triggered by definite stimuli and will occur when certain stimuli act on an inherited organic equipment, producing thereby a more or less predictable outcome. In some cases the act may be far from adaptive. The life cycle of the schistosome cercariae, which produce swimmer's itch in man, is terminated by the death of the cercariae when they penetrate man's skin.

Striking examples of automatic behavior are perhaps best shown by the web spinning of spiders, the communication of honeybees, and the cocoon spinning of caterpillars. Each stage of performance seems to serve as a stimulus for the succeeding stage. Environmental influences necessary for a characteristic behavior to appear is demonstrated by the experiment of D. Lack on the European robin. He discovered that a male robin during its breeding season and when holding its territory will attack even a bundle of red feathers (which simulate the red breast of an actual male robin) but will do so only under the conditions mentioned. The same experiment also shows the effect of what is called **"sign stimuli,"** for Lack found that a tuft of red feathers would provoke an attack, whereas a stuffed young robin with a brown breast would not. In this case the red breast is the effective stimulus.

MOTIVATION AND BEHAVIOR

Why do animals perform characteristic patterns of activity? Why do they act at all? A simple answer for many forms of activity is the stimulus-response theory that may involve some change in the environment. But not all behavior can be answered as simply as this, for some actions are related to internal conditions not easily appraised. One of the basic principles of life is the maintenance of stable internal conditions. Hunger, for instance, is accompanied by a low blood glucose level, an imbalance of fluids, etc. Such internal changes stimulate characteristic behavior patterns. A hungry hydra will behave differently from a satiated one.

Animals perform acts that either give them pleasure or else prevent pain or unpleasant conditions. They make their adjustments to these conditions perhaps wholly unconscious of the end results (this certainly would be true of the lower forms). This concept is in line with the maintenance of internal stability as already described. The appetitive phase of instincts that Lorenz and Tinbergen stressed is supposed to furnish the chief source of motivation. According to their theory, animals actively seek out those stimuli that trigger their instinctive acts.

A clear understanding of these instinctive acts is furnished by the rather modern concept of **biologic drives.** A biologic drive may be defined as a motive for stabilizing the organism. Examples of such drives are hunger, which arises from the nervous impulses of an empty, contracting stomach; thirst, which may be due to sensations from a drying pharyngeal mucosa; and sex, which is caused by a release of hormones from sex glands. All of these upset internal stability, and restoration of this stability is a reward or motive. Most drives are also characterized by rhythm patterns in which the drive fluctuates up and down in a periodic manner, such as the estrous cycle of the female mammal. But a certain nervous pattern has to be satisfied, as, for instance, the weaving of a net by a spider.

REFLEX ACTION

A simple reflex act (reflex arc or circuit) is a ready-made behavior response in which a specialized receptor, when stimulated, transmits impulses to a specialized motor cell that arouses an effector (muscle or gland) to act (p. 162). Most reflexes, at least in the higher forms, are more involved and include an association neuron, and the impulse must pass through a number of synapses. A reflex act may or may not involve a conscious sensation. Reflex acts are built into the body mechanism and control the automatic working of the internal organs.

Reflexes are less elaborate than instinctive actions. The latter also have a wider range of adaptability and variability. However, as Lorenz stresses, the releasing mechanism that triggers instinctive acts involves reflex action in its pattern. The idea that an instinct is a chain of reflexes is not rigidly held because of the varying intensity of instinctive behavior patterns.

LEARNED BEHAVIOR

Conditioned or learned behavior differs from inherited instinctive behavior in being acquired or modified from experience. The distinction between unlearned and learned behavior is not easy. Behavior is never exactly predictable. The concept of learning must then have a variety of meanings. Ordinarily it refers to changes in behavior that are brought about by past experience and that involve more or less permanent modifications of the neural basis of behavior.

Learning implies multiplicity of responses so that if an organism is frustrated in one response it can try another. This requires adequate nervous interconnections and dominant nerve centers. In many invertebrates the brain is an efficient transmitter rather than an organizing center. When the principal nerve centers can override local reflex patterns, so characteristic of lower organisms, more diverse types of behavior can occur. The capacity for learning depends, then, on the anatomic and physiologic capacities possessed by each group of animals.

To what extent can learning be determined in the animal kingdom? One of the greatest difficulties encountered in this field is the impossibility of supplementing objective behavior observations with subjective knowledge. The experimenter must depend on the use of stimuli and the way animals react to them for his knowledge of their behavior. By using controlled experiment, observation, physical analysis of the environment, discriminatory learning, etc., the investigator tries to ferret out just what the learning capacity is for a particular animal.

Conditioned reflexes or associative learning. A conditioned reflex is substituting one stimulus for another in bringing about a type of response. It is often considered the simplest form of learning and involves a new stimulus-response connection. We owe this concept mainly to the Russian physiologist I. P. Pavlov, who noted that hungry dogs (and other animals) secrete copious amounts of saliva at the sight or odor of food. By ringing a bell (conditioned stimulus) at the same time the dogs saw food (unconditioned stimulus), it was possible in time for the conditioned stimulus alone to elicit the response of salivation. There was a definite limit to the number of factors to which an animal could be conditioned for a single response. Animals with higher nervous systems can handle more (and more complex) factors of conditioning, which is far more prominent among mammals and some birds.

Selective learning (trial and error). Selective learning is a higher type of learning and is rarely found below the arthropods. With this type the animal does not learn something new but selects a random response on the basis of a reward or punishment. A dog, for instance, finds that a problem box containing food can be opened by pulling on a lever. To get to the food, the animal at first makes many random, useless movements (trial and error), but when it finally succeeds, trials later become fewer until it learns to open the box when confronted with the situation the first time. It has thus mastered a habit of appropriate response. Animals vary greatly in mastering problem boxes. Some do so with few trials; others require many. The widely used Skinner box is based on a reward of food when the animal presses the right lever (Fig. 16-14).

Insight learning. When a process is slowly learned by an animal, it is often called trial and error; when learned rapidly, it is insight, often called "abridged learning." It involves making an initial survey of the elements involved, then achieving a solution to the problem on the first trial. Many cases of selective learning in which animals solve problems after a few trials may involve a test situation not entirely new to the animal; they profit from previous experience. The capacity for short-cut solutions is rare among most invertebrates but is common in higher mammals. A form of insight learning in birds has been illustrated in recent years in England. There, great tits have rather suddenly acquired the habit of opening milk bottles left on the doorsteps of households and feeding on the milk. The widespread nature of the practice poses the question as to how such a habit could be picked up by so many of them in a relatively short time.

Imprinting. This concept refers to a very special type of learning in birds. In the first hours after hatching, a duck or goose is attracted to the first large object it sees and thereafter will follow that object (man, dog, or inanimate object) to the exclusion of all others; such birds show no recognition of their parents. When once accomplished, the behavior pattern is very stable and may be irreversible in some cases. Other species of birds may show imprinting. Some psychologists think the type of bird song young hatched birds acquire when exposed to members of different species may be of this nature. According to Lorenz, when a young bird is imprinted to a member of another species, the imprinted bird will adjust its own functional cycles

FIG. 16-14. Pigeon taught to do work of man. This bird was taught by researchers of a drug company to sort out inadequately coated capsules by pecking the proper key as nonacceptable capsules were brought into view through a tiny window. Birds were trained by rewards of food. Birds varied in time required for training, but most became expert inspectors in 60 to 80 hours and could easily detect minor flaws overlooked by human inspectors. (Courtesy T. Verhave, Eli Lilly & Co., Indianapolis.)

to that of its adopted parent. Other psychologists, however, think that imprinting is merely a strong early habit and that its socialization to another species is very restricted.

TOOL USING AMONG ANIMALS

The ability to use tools is often considered one of the major achievements of man toward the high evolutionary rank he now holds. The need to hunt must have been a great pressure for man's development of tool making. Tool using among animals below man is very restricted. Yerkes found that chimpanzees manipulated certain tools in a manner that indicated they had a clear perception of what

the tools were for and that their use of them was not a chance trial-and-error method. Many animals display amazing feats of craftsmanship, but this usually involves manipulating the materials with bodily parts, beaks, feet, and jaws. The behavior is of an instinctive pattern and no high degree of intelligence need be assigned it.

One of the early observations of true tool using by animals is recorded by the Peckhams in their famous monograph of the solitary wasps. *Ammophila,* a sphegid wasp, seizes a small pebble in her jaws and pounds down (as with a hammer) the earth with which she closes up her burrow. P. and N. Rau reported the same behavior in different species of the same group. In some cases a stone is used only occasionally; more commonly they use only their head and jaws to tamp down the sand.

A widely publicized case of tool using among birds is one of Darwin's finches of the Galapagos Islands. One of the finch species has the habits of a woodpecker in probing into crevices of bark and trees. To overcome the handicap of a short beak the bird holds a stick or thorn in its beak to pry out its prey. The bower bird of Australia uses a crude brush of fibers to paint the walls of its bower with charcoal and saliva.

A recent example of tool using was discovered in the Egyptian vulture (*Neophron percnopterus*). This raven-sized bird breaks open ostrich eggs, which are too big to be seized by its beak, by casting stones at the eggs. Many stones may be thrown before a vulture attains the desired effect. Sea otters hold a stone or large shell upon their chests and then crack open sea urchins or mollusks by banging them against the stone.

SOCIAL BEHAVIOR AMONG ANIMALS

Social behavior refers to groups of animals (usually of the same species) living together and exhibiting activity patterns different from what the members would display when living as separate individuals. In this sense, animals are social when their behavior is modified by living together and when they influence and are influenced by other members of the group.

Animals may be attracted by some common favorable environmental factor (for example, light, shade, moisture). Moths may be attracted to a light or barnacles to a common float. Some aggregations are the result of more positive reactions to others like themselves, such as schools of fish, flocks of birds roosting together, and birds with similar migratory gregarious habits.

Parents and their offsprings may form a closely knit social group while the young are developing; in a wolf pack this family relationship may be more or less permanent.

Social organization seems to be the result of the interaction between an inherited behavior pattern and learned experience. Perhaps among invertebrates, such as bees, ants, and termites, the inherited pattern is the dominant factor. Social drives or

appetites are usually less intense than those of hunger or sex, which may also be considered a form of social behavior.

The roles of the various members of ant and termite societies are determined by structural differences that result in an inflexible division of labor. Such societies have a long evolutionary history and are so integrated that a member has little chance of survival when separated from its society. Survival here depends on the fate of the group in meeting the requirements of the environment.

Social organization is most highly developed among the arthropods and the mammals. The social development in these groups arises from family units rather than from aggregations of individuals. Most species of ants, for instance, start a colony by the queen laying eggs that differentiate into the various caste members of workers, soldiers, sexes, etc.

The social behavior of vertebrates seems to be built around three general principles. These are territorial rights, dominance-subordinance hierarchies, and leadership-followership relations. The concept of **territorial rights** involves the taking over of a restricted area and its vigorous defense against trespass by members of the same species.

The concept of **dominance-subordinance** refers to a type of social behavior in a group in which one animal is dominant over others. It was observed that in poultry flocks one hen was dominant over the others and exerted the right to peck other members without being pecked in return. Further study revealed that flocks were organized on the basis of social hierarchies; that is, there was a series of social levels in which members of the higher levels had peck rights over those of lower levels, etc. Dominance hierarchies have been found in all classes of vertebrates, including fish, lizards, mice, and primates. Some arthropods show dominance behavior patterns to a limited extent, but ordinarily the behavior of arthropods does not fit into this pattern of learned behavior.

Leadership behavior refers to the tendency of the members of a group to follow a certain member. It usually involves the selection of an experienced member that stabilizes the other members of the group. In some cases the leader may be the dominant member of the group; more often it is an old, experienced female that has earned the right because of the larger number of offspring that acquired the habit of following her when they were young. It is apparent that the female is the primary influence in leadership and the center of family life. The aggressive males tend to break up social behavior patterns, and nature, to offset this tendency, either produces males in limited numbers (as in many insects) or the males are kept more or less to themselves except during the breeding season.

In addition to mammals, certain types of leadership are known to exist among fish, lizards, and birds. Some arthropods show leadership behavior. In the termite organization, males play a role almost as prominent as that of the females.

COMMUNICATION AMONG ANIMALS

Communication as applied to animals is simply the influencing of one individual by the behavior of another. This may take the form of bodily contacts (rubbing antennae in bees), scents from glands (mammals at mating season), voice effects (warning cries of birds and mammals), and hosts of others. In some cases, among the higher nonhuman mammals, distinctions between communication and language cannot be rigidly drawn. Yerkes observed in his chimpanzees many sounds and signs that seemed to be understood by other members in specific or symbolic ways. However, it has been impossible to teach any nonhuman animal a true symbolic language with meaningful association. The reproduction of words and phrases by parrots and mynas has no significant meaning to the birds themselves. Bird sounds of a particular species often show a great deal of differentiation. Many of these are thought to be meaningful calls of distress, hunger, warning, etc. It is now known that the bird song is actually a warning cry to others of territorial rights. The varied songs of birds are an integration of both innate and learned song patterns. They sing both for communication and for pleasure.

It has been known from the experience of underwater-sound men during World War II that fish and other marine forms make a variety of noises within the sea. Fish and many other marine forms have no vocal organs but manage to produce a great variety of noises in diverse ways. The chief noisemakers are the toadfish, squirrelfish, sea robin, and triggerfish. Many of them use their air bladders to produce sounds. The toadfish and sea robin cause vibrations in their air bladders by muscle contraction; the triggerfish uses its pectoral fins for beating on a membrane of the air bladder near the body surface; and some, like the squirrelfish, grind together teeth in the back of their mouths and this sound is amplified by the air bladder.

HOW HONEYBEES COMMUNICATE LOCATION OF FOOD

One of the most striking behavior patterns is the ability of honeybees to inform others about the location of a source of food. The experiments conducted over many years by Professor Karl von Frisch have given some clues to this interesting problem. By using glass observation hives and marked bees, he was able to observe with considerable accuracy just what occurs when bees report back to the hive the presence of a source of food. Whenever a foraging bee finds nectar, she returns to the hive and performs a peculiar dancing movement that conveys to others in the hive in what direction and how far they must go to find the nectar. If the food is more than 100 meters away, she performs a characteristic waggle dance (Fig. 16-15). This dance is roughly in the pattern of a figure eight that she makes against the vertical side of the comb. In the performance of this act she waggles her abdomen from side to side in a characteristic manner. She repeats this dance over and over, the number of waggles decreasing per unit of time the farther away the source is. The direction of the food source is also indicated by the direction of the waggle dance in relation to the position of the sun. When the waggle dance is upward on the comb, the source of food is toward the sun. A waggle run downward on the comb indicates that the food is opposite to the position of the sun. In the straight run of its dance the bee emits a low frequency sound by the vibration of its wings. This sound picked up by other bees in contact with the scout bee may indicate the distance to the food source. If the food source is at an angle to the sun, the direction of the waggle dance is at a corresponding angle. During a dance other bees keep in contact with the scout bee with their

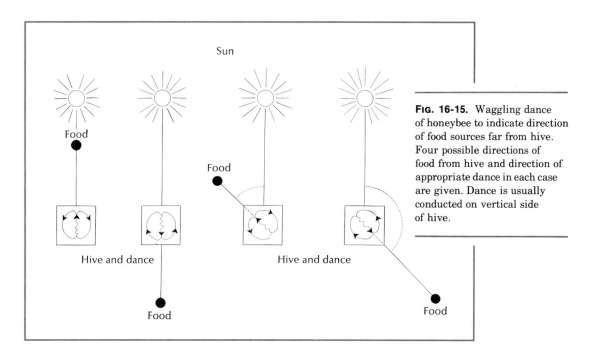

FIG. 16-15. Waggling dance of honeybee to indicate direction of food sources far from hive. Four possible directions of food from hive and direction of appropriate dance in each case are given. Dance is usually conducted on vertical side of hive.

antennae, and each performance results in several bees taking off in search of the food. When they return with the food, they also perform the dance if there is still food there.

The waggle dance is used principally to inform inexperienced bees of the location of food. Experienced collectors do not use the waggle dance but alert older bees by olfactory cues instead. When the source of food is less than 100 meters from the hive, the scout bee simply turns around in a circle first to the right and then to the left, a performance she repeats several times (Fig. 16-16). She is able in this way to convey to the other bees the information to seek around the hive food of the same odor she bears. One other interesting phenomenon is the ability of bees to determine the direction of the sun when only a small area of the sky is visible. They seem to be able to do this by the pattern of polarization in the light from that part of the sky that is still visible.

BIOLOGIC CLOCKS

Nearly all aspects of the behavior and physiology of organisms are rhythmic in nature. Thus there are

FIG. 16-16. Round dance of honeybee. Whenever food supply is within 100 yards or so of hive, scout bee circles first one way and then the other to convey information of food source.

periods when the organism is dormant or sleeping, alternating with periods of wakefulness. There are daily rhythms of higher and lower temperatures, reproductive cycles, cyclic variations in color, and seasonal rhythms for different activities. Some animals are more active at night (nocturnal), others in daylight (diurnal), and still others in dim light (crepuscular). Some marine animals feed in the littoral zone at high tides, and others only feed at low tides. Some marine organisms spawn with reference to specific moon phases. In plants there are flowering periods and seed-bearing periods. Basal metabolic rhythms occur in both plants and animals. These are only a few of the examples that could be cited. Because many rhythms are on a daily or 24-hour periodic basis, they are often called **circadian rhythms** (L. *circum,* about, + *diem,* day).

It is logical that organisms must have their activities geared to the rhythmic cycles of the physical world. The external physical environment of organisms displays such patterns as the solar day of 24 hours, the lunar day of 24 hours and 50 minutes, the annual changes of light and temperature, the cycle of the tides in the oceans, the revolving of the earth about the sun, the cycles of weather, and even the patterns of atomic and molecular structure of matter. In all animals studied, endogenous rhythms of metabolic fluctuations corresponding to the geophysical frequencies are present. Many such rhythmic patterns of organisms are so firmly entrenched that they are not easily altered by prolonged exposure to external agencies, such as temperature differences and drugs, that are known to influence the ordinary metabolic chemical reactions of the body. What is more striking is that many organisms that have rhythms coordinated with environmental changes will also have these same rhythms independent of the stimuli from the physical environment. For instance, plants kept in constant darkness will show the same periodic leaf movements as when they are exposed to the daily light-dark alternation. Fiddler crabs have darkened skin during daylight and are lighter at night. When placed in constant darkness for long periods of time, they will undergo the same cycle of color changes as if they were in a natural day-night environment. However, if the crabs (with normal cycles) are placed on ice for 6 hours, the phases of their cycle will be set back the same length of time so that they now darken and blanch 6 hours earlier.

Biologic clocks enable plants and animals to adapt to the best advantage their own rhythms to the rhythms of the physical environment. In this way their activities can occur at the time of day when their physiologic adjustments are best served. Biologic clocks pose many problems to the many able investigators in this field. Where are these clocks located within the organisms? Are they localized in a particular tissue (for example, nervous system), or do they reside within all cells and involve the total organization of the animal? Their level of organization is known to be as low as the cell because they have been found in protistan organisms. Evidence indicates that biologic clocks are endogenous and innate. However, an alternative theory argues that, despite all precautions to exclude organisms from environmental forces, they do continuously receive from their external environment information about the rhythmic geophysical cycles to which they regulate their activities. Another problem is, what is the mechanism whereby light-dark cycles (photoperiodism) entrain or mediate circadian rhythms within organisms? These and many other problems are far from being resolved.

MIGRATION

Birds. The term "migration," as used with reference to birds, signifies the regular seasonal movement birds make from one region to another. It does not mean the occupation of a new territory by birds in order to extend their ranges.

Homing is the ability of an animal to return to a familiar region when the animal has been removed to some other region. Homing may depend to some extent on topographic memory of landmarks, which could explain certain forms of migration. However, it could not explain the ability of some young birds to migrate without their parents to wholly unknown places.

Most bird migrations are **latitudinal,** that is, the birds move into the northern zones during the summer and return south for the winter. In the southern hemisphere, where the seasons are reversed, there is a similar but more restricted migration of certain forms. Some birds are permanent residents the year round; others shift from regions far in the northern hemisphere to those deep in the southern hemisphere.

What are the advantages of migration? No doubt, one of the principal advantages is that the bird can live in a favorable climate all the time. This not only ensures it an adequate supply of food but also provides the optimum conditions for the rearing of its young. Waterfowl can avoid frozen waters by retiring to the south. The shortened hours of daylight in northern climates also restrict the ability of birds to get enough food. However, many birds leave their breeding grounds while food supplies are still abundant.

How can we explain the evolution of long bird migrations? Much geologic and paleontologic evidence in recent years has been advanced to support a modified view of A. L. Wegener's theory of continental drift (1912), which showed how large lateral displacements of the earth's crust could force the continents to be driven apart to positions where they are today. This theory presupposes two original land masses, a northern one (Laurasia) and a southern one (Gondwanaland). These two great land masses were at times in contact, and birds that had their original home on Gondwanaland drifted into Laurasia for better conditions to rear their young and to avoid overcrowding, returning to Gondwanaland at other times. As these two great land masses drifted apart and broke up into continents as we know them, this habit of going from one land mass to the other persisted. This theory of migration is supposed to account for some of the strange, circuitous routes of certain migratory birds such as the turnstone, Arctic tern, and golden plover (Fig. 16-17).

The direction-finding or orientation of birds has been much investigated in the past decade, for a solution of this problem would get at the very heart

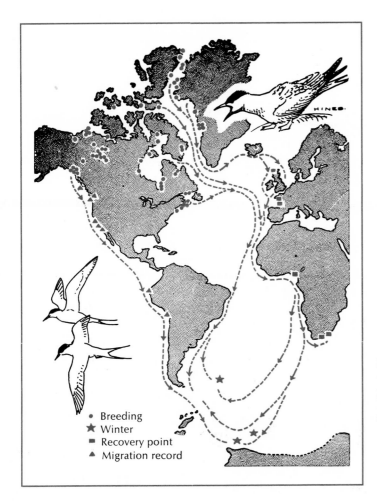

• Breeding
★ Winter
■ Recovery point
▲ Migration record

FIG. 16-17. Map showing migration of arctic tern. Enormous route covered by this bird in one year is probably 25,000 miles. (From drawing by R. W. Hines, U. S. Fish and Wildlife Service.)

of homing and migration. Griffin, who released gannets many miles from their nesting sites and followed their wanderings in an airplane, concluded that the birds wandered aimlessly or in circles at first until they picked up visual clues; then they headed straight for home. However, such an explanation could hardly apply to a shearwater that was released in America after being removed from its home in Wales and was back home a few days later. In recent years Kramer has shown that starlings and pigeons can orient themselves by the sun's position. In specially covered cages provided with six windows, these birds were trained to find food in a definite compass direction at a certain time of day. When tested at another time of day when the sun's position had changed, they compensated for the sun's motion and immediately went to the right window by keeping track of the time of day. But how does one explain nocturnal migration that is very common among birds? Evidence was recently advanced that warblers could orient themselves in a particular geographic position in a planetarium when the stars coincide with the night sky in Germany. By changing the synthetic constellations about, the birds were able, with their amazing time sense, to take that direction which would enable them to reach the point normally taken when they start their migration. These experiments involving the apparent effect of visible celestial bodies on migration may explain the ability of shearwaters to return thousands of miles directly to their home.

Some species are known for their long-distance migrations. The arctic tern (Fig. 16-17), for example, breeds north of the Arctic Circle and in winter is found in the antarctic regions, 11,000 miles away.

Eels. For centuries naturalists had been puzzled about the breeding and development of the common eel. It was known that the adults spent most of their life in freshwater streams in both North America and Europe, but where they spawned or where they underwent their development was not known until the patient work of Dr. Johannes Schmidt brought to light most of the facts in the case.

Both the European species of eel (Anguilla vulgaris) and the American species (Anguilla rostrata) spawn in the Sargasso Sea northeast of Puerto Rico, although in general the breeding grounds of the two do not overlap. The eggs hatch into pelagic larval forms less than 1/4 inch long and are called leptocephali. The adults die immediately after spawning. A year later, the American species, now about 3 inches long, reach the American coasts, where the females distribute themselves through the freshwater streams and rivers; the males usually remain behind in the brackish waters near the coast. It takes from eight to fifteen years for them to grow to maturity. Eventually each female goes down the stream to the sea and joins a male; they go together to their breeding grounds. Their rate of travel through the sea is only about 1/2 mile an hour. It has been estimated that it takes them about 1 to 2 months to reach the place of spawning, although there are no reports of adult eels having ever been taken in the open ocean. The European species with their much greater distance to travel require about 6 months to make the journey. It takes the larval forms of the European species an incredible three years to reach the European coasts.

Salmon. The salmon lives its adult life at sea and returns to the headwaters of streams to spawn. Both the Atlantic (one species) and Pacific (five species) forms have this practice, but there are some differences between the two. After spending three to four years at their feeding grounds at sea, the salmon return and ascend the freshwater streams, both sexes making the journey together. After spawning, the Pacific species invariably die. The Atlantic species frequently survives to spawn a second or third time. From the time they return to fresh water until they spawn, the salmon lives on its reserve food, which it has accumulated in its body in the form of fat. After hatching in the shallow gravel pits, the larval fish live for some time in the streams before going out to sea. Some species stay only a few weeks, but others may remain for as long as two years.

Adult salmon appear to find their way from the sea to the correct river mouth by using the sun's position much as birds do during their migration. Once in the river the salmon use olfactory cues to guide them unerringly to the same spawning ground from which they originated several years before.

Butterflies. A number of the larger and stronger winged insects apparently are able to make long flights, such as the monarch butterfly (Danaüs *plexippus)* of North America. In early autumn immense swarms of these butterflies gather in the northern part of the United States and eastern Canada and make southward flights that may take them 2,000 miles or more to warmer regions, around the Gulf of Mexico and South America. Some of them are known to leave the mainland and journey as far as the Hawaiian Islands.

LIGHT PRODUCTION

The production of light by living organisms (bioluminescence) is widespread, but usually only scattered representatives are found within a particular phylum.

For light production, animals may be divided into two groups—those that produce their own light (self-luminous) and those whose light is produced by symbiotic bacteria. The self-luminous forms will emit light only when they are stimulated. Many of the self-luminous fish have rather complicated organs for the production of light. These may contain, in addition to special photogenic cells, a layer of reflecting cells and a pigmented shield, similar to a vertebrate eye.

The luminescence of this type of luminous animal is due to the interaction of two substances—**luciferin** (Fig. 16-18), which is oxidized in the presence of an enzyme, **luciferase.** Oxygen is therefore necessary for light production. In the firefly the mechanism of light production also involves ATP and magnesium (Fig. 16-19). The stimulation for light production may be merely mechanical, or it may be nervous, as it appears to be in higher organisms. In those whose light is produced by symbiotic bacteria, light is given off continuously, although some of the forms have devices for concealing and showing the light intermittently.

The famed glowworm of literature is usually the wingless female of the lampyrid beetle *(Photinus),* although some are larval forms. Most of them are terrestrial, although one or two live in fresh water. The luminous organs are found on the ventral side of the posterior segments. The light they emit is rather bluish green in appearance and the rays are restricted to a very narrow range of the spectrum, as compared with other luminous animals. The light of the glowworm, like that of all forms that have bioluminescence, is truly "cold light," that is, light without heat.

The best known of all light-producing organisms is the firefly, or "lightning bug." Most of them are

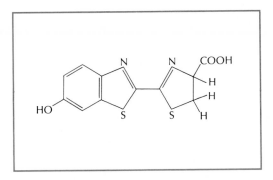

FIG. 16-18. Structure of luciferin in firefly. In bioluminescence, light is produced by a biochemical mechanism that involves oxygen, ATP, luciferin, luciferase, water, and inorganic ions. In presence of oxygen, ions, and enzyme luciferase, luciferin-ATP complex is converted to oxyluciferin with liberation of light. Luciferin and luciferase differ in composition in different species.

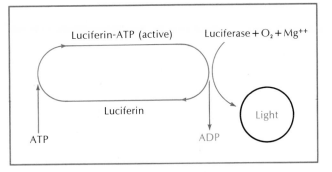

FIG. 16-19. Light production in bioluminescent animals. Luciferin and the enzyme luciferase are the principal light-producing components, but each is nonluminous by itself. ATP is added to luciferin to form an active luciferin-ATP complex, which, in the presence of oxygen, magnesium ions, and the enzyme luciferase, emits light. ATP becomes ADP when it activates luciferin and is the chief source of energy in the process. Light emitted by different animals varies in wavelengths of the visible spectrum.

tropical, but two genera are very common in temperate North America (*Lampyris* and *Photinus*). The light organ of the male is located on the ventral surface of the posterior segments of the abdomen. It is made up of a dorsal mass of small cells (the reflector) and a ventral mass of large cells (the photogenic tissue). Many branches of the tracheal system pass into the organ to ensure an ample supply of oxygen, so necessary for the light. The organ is also supplied with nerves that control the rhythmic flashing. The function of the reflector is to scatter the light produced by the photogenic cells. In the female the light organ is confined to a single abdominal segment.

The light is considered a mating signal in the American fireflies. During the day the fireflies lie hidden under vegetation, but they emerge at dusk. When the males flash, the females respond and they eventually find each other.

ELECTRIC ORGANS

The power to produce strong electric shocks is confined to two groups—teleosts and elasmobranchs.

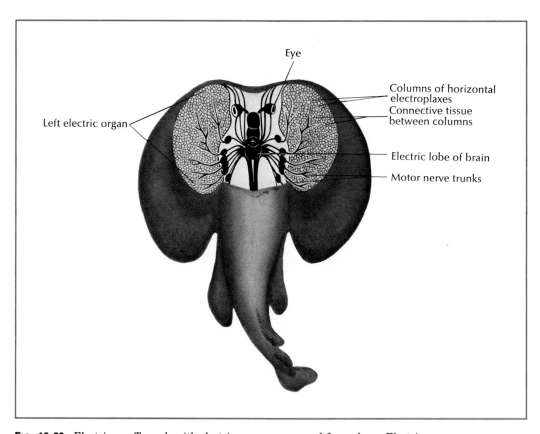

FIG. 16-20. Electric ray *Torpedo* with electric organs uncovered from above. Electric organs are restricted to certain teleost and elasmobranch fish. Electric organs are all modified from striated muscle (with one possible exception). Organs are built up of disklike, multinucleated cells called electroplaxes or electroplates, which are embedded in jellylike substance and enclosed within connective tissue compartments. Each vertical column in electric ray is made up of a stack of electroplaxes piled on top of each other. Nerve fibers run to electroplaxes and blood capillaries course through jelly layer. In addition to defensive and offensive purposes, electric organs may be used for recognition among members of species when other methods of communication are absent. Some fish (especially those with weak electric organs) are able to detect disturbances made by objects in electric dipolar field they form around their bodies, and thus keep informed about their environment.

The actual forms that have this adaptation are the electric eels *Electrophorus* and *Gymnotus,* the electric ray *Torpedo,* the stargazer *Astroscopus,* and the electric catfish, *Malapterurus.* The electric organs in all these forms are modified from skeletal muscles, with the possible exception of *Malapterurus,* in which they have developed from skin muscles. The organs are composed of flattened plates arranged one above another like the alternating layers in a storage battery. Each plate, of which there are thousands, is innervated by a nerve fiber.

In *Torpedo,* the giant electric ray of the Atlantic coast, the plates are stacked in vertical columns like stacks of coins (Fig. 16-20). The many columns are arranged in parallel which permits the ray to generate tremendous amperage. However, in the freshwater electric eel the few horizontal columns, each containing thousands of plates, are arranged in series. This allows the eel to generate great voltages, sometimes more than 500 volts, to overcome the resistance of fresh water. As much as 40% of the electric eel's body is devoted to the electric organ. The electric fishes can deliver either weak or strong shocks at will. The smaller shocks serve as a warning to potential enemies; the strong shocks are used to stun or kill prey or enemies.

ECHOLOCATION IN ANIMALS

Echolocation is a method of orientation whereby animals make use of a sonar system to detect objects at a distance. First worked out precisely in certain bats (by Griffin and Galambos), the method is now known to be employed by a number of other animals. Bats, such as *Myotis* and *Eptesicus,* emit chirps about ten times per second while cruising in the open. These sounds have high frequencies (about 40 to 80 kilocycles) and are inaudible to human ears (ultrasonic), which cannot detect sounds much above 20 kilocycles. Ultrasonic sounds can be picked up by man with a microphone, or sonic detector, and are heard as a series of clicks or buzzes. These sounds are bounced off objects while the bat is chasing its prey or avoiding obstacles and are reflected back to its sensitive ears. The large fruit-eating bats or flying foxes depend on vision or smell for finding their way about and do not have the echolocation method.

Echolocation has been described in whirligig beetles *(Gyrinus),* oil birds *(Steatornis)* that make sounds audible to the human ear, porpoises, etc. and may be widespread in the animal kingdom.

17

Man's relation to the ecosystem

It is agreed among thinking people that one of the greatest problems confronting man today is his relation to his own environment. Man has always been a part of his environment, but only within recent decades has he become aware of the enormity of the environmental problems that have arisen. He has realized that these problems must be solved to ensure his very survival on earth. The first man (probably a few million years ago) lived very much as other animals did. His numbers were relatively few and he could shift from one locality to another. His evolutionary success involved his cultural development from the cumulative efforts of previous generations. He learned the secrets of fire, the protection of clothing for his body, methods of shelter, and the sources of food.

New methods of communication, the speed of modern travel, the development of powerful weapons, the control of hazardous diseases, and above all the terrific impact of technological skills have reached such magnitudes that they pose problems for existence almost beyond our comprehension. Our power needs are increasing ten times as fast as our population. So vast are the crises for which man himself has been responsible that some authorities fear that there is not enough time left to solve them.

MODERN CONDITION OF THE ENVIRONMENT

It cannot be said that our present environment has been totally ruined despite mismanagement and unwise control. At no time in the past has the environment been so productive to man's welfare and needs as the present—a tribute to the growth of sophisticated technology. In the past, problems concerning the environment were solved largely through geographic expansion into new regions. The surface of the planet Earth was large and unimproved. But as man could not indefinitely better himself by moving into new territories, he turned to transforming his biosphere locally. At no time in his long history has man been able to enjoy the fruits of his labors better than at present. It is true that large segments of human populations live in substandard conditions today, but it is doubtful that they were better off in the past.

Why, then, is there so much concern over the environmental conditions at present? Simply because in the midst of all this activity we have been largely insensitive to the damage we have created in our biosphere. Our numbers have increased at a terrific rate, our technology has produced waste beyond our means to dispose of it, and our terrain has been restricted by mismanagement. So far, our capacity to injure our environment has exceeded our power to resolve the problems leading to its destruction.

CRISES THAT CONFRONT US

Our status at present is one of vast expansion, with a socioeconomic system that stresses the development of technology and the consumption of the products of technology. Transformation of raw materials into sophisticated products has produced environmental pollutants of both air and water. Trash and garbage involve problems of disposal or conversion. Our present society not only produces more goods, but it also produces goods of shorter life and usefulness. Products that are useful to an affluent society are quickly supplanted by more sophisticated gadgets. What is new today is old tomorrow.

The development of motorized transportation has contributed enormously to the pollution of the air and the extinction of silence. The growth of cities, where about 80% of our American population now lives, produces more problems of pollution, along with unhealthful living conditions. The crowded conditions of metropolitan communities magnify the problems of garbage and waste disposal as well as air and water pollution.

In 1956 there was published an international symposium entitled *Man's Role in Changing the Face of the Earth.* In this revealing work many able and qualified scholars and specialists pointed out the various ways in which man has modified his environment. One of the problems emphasized in this symposium was the menace of world overpopulation. Uncontrolled birth rates and lower death rates with no widespread positive checks indicate a real change in the human time scale. The average life is longer, and so the lengthening of the time scale by many years has altered the character of human life.

Other areas of concern include soil erosion, soil changes resulting from human use, the problem of food supply, the misuse of mineral resources, the modification of biotic communities, the heavy use of chemical pesticides, rapid raw material consumption, the urbanization of the population, the socioeconomic system of production and consumption, and many others.

POPULATION PROBLEM AND CONTROL

The problems concerned with population growth and control are unique in many ways. Any alterations in the trends in this field are going to require drastic changes in the whole social and cultural pattern of man. Family size is a private family matter and difficult to regulate without the cooperation of the family. It is highly unlikely that great masses of people, very unequal in education and culture, will react alike in these matters. Population growth is exponential, and the increase is geometric. Natural selection at present does not operate as it did in the past, when it tended to keep populations stable.

At present the current rate of increase is about 1.5% to 2% per year. About 60 million persons are added to the world's population each year. The more than 3.5 billion people in the world today will double in the next thirty-five to forty years, so that by the end of this century more than 6 billion individuals will be living on this planet. Only a few hundred years ago it required centuries for the population to double.

This phenomenal population growth has been a result of the striking drop in early death rate, especially a drop in infant mortality. Improved sanitation and medical care, together with a better control over certain devastating diseases such as malaria and yellow fever, have also contributed to the declining death rate. A far greater percentage of persons now live to child-bearing age, which tends to increase the population. Man, in common with other primates, has no relaxation in his sexual urge as do other animals that have seasonal periods for reproduction. Although many persons sincerely desire to have children, most children are conceived because of the compulsion of the sex instinct. With most organisms in nature there is a need to produce many offspring to preserve the species because so many are eliminated by the hazards of existence. Man has eliminated most of the hazards, but he has retained the breeding urge.

The family represents the closely knit attachment of its members to each other, and our whole social structure is built around it. But no longer can we consider large families a necessity as did our pioneering forefathers. Even if greater expansion is feasible, the problem of overpopulation still remains; some authorities believe we are postponing what eventually must come – a compulsory means of some kind for restricting the number of people.

Family planning in the form of birth control has been practiced by enlightened persons for a long time. Although entirely successful devices for preventing conception have not yet been perfected, the current intensive research in this field will no doubt result in completely safe and efficient methods.

The final objective is a more or less stable population that can be supplied with food and the benefits of our modern technology. The practice of human husbandry, or the planned regulation of family size, must be established in the mores of our social life. Recent statistics suggest that this is happening in North America now, and that the United States population is beginning to stabilize. But for the great teeming masses of people who are ignorant or indifferent about birth control methods, for those who have religious or other scruples about such methods of family limitation, the outlook for a practical solution is still dismal. Some writers have even advocated laws for compulsory sterilization if all other methods fail. Sir Charles Darwin, a leading exponent of family restriction and the grandson of Charles Darwin of evolutionary fame, once said that if overpopulation persists, only the superficialities of civilization will survive, and man may again revert to natural selection for controlling his numbers in order to exist.

MODIFICATION OF BIOTIC COMMUNITIES

A balance of nature involves man's relation to his total environment; he is a fellow companion with the other creatures of this planet. Any plan of conservation must imply a creative interest in man, plants, and animals as well as other aspects of nature. We have learned much about the practices of conservation, but so far we have done little to put these practices into action.

In the belief that man was created to have dominion over the earth and its resources, he has done many things that run counter to the concept that he is actually a part of nature himself. His role in changing the face of the earth has been due to his natural desire for expansion and growth. He has developed scientific technology as a tool for solving his problems, but has remained unaware that such technology creates new problems in the form of new desires.

We have destroyed forests to provide agricultural fields and places for human habitation without considering the effects on erosion or on the wildlife whose habitation was obliterated. We have dredged swamps and interfered with natural waterways. Pollution from our industries has made water unfit for fish and other forms to live in. Our economy has been destructive to animal habitats. Even when we planned the preservation of wildlife, we often ignored the basic laws of ecology and gave little consideration to the role of predators in the web of life. Mismanagement has brought about the extinction of many species and endangered the existence of others.

Wild animals and fish are not of esthetic value alone; they also contribute to economic wealth. According to the estimates of the U. S. Fish and Wildlife Service more than 4 billion dollars are spent each year by hunters and fishermen.

Each year the Bureau of Sport Fisheries and Wildlife of the Interior Department of the United States publishes a volume entitled *Rare and Endangered Fish and Wildlife of the United States.* The most recent volume reveals that Americans are about to lose forever approximately 89 species of vertebrate animals. The list of endangered forms includes 21 fishes, 8 reptiles and amphibians, 46 birds, and 14 mammals. Thoughtful naturalists have long pointed out the need for preserving suitable habitats, strict enforcement of protective regulations, and the development of a public attitude of sympathetic understanding of the plight of these species. In 1969 the President of the United States signed the Endangered Species Act, which prevents the importation of endangered species of wildlife and fish into this country and the interstate shipment of reptiles, amphibians, and other wildlife taken contrary to state laws.

The list of endangered species includes such striking forms as the bald eagle, whooping crane, American alligator, panther, manatee, Atlantic salmon, cutthroat trout, condor, peregrine falcon, ivory-billed woodpecker, and many others. The fate of the passenger pigeon and other desirable species that are already extinct should make us realize that whenever a species shows a steady decline it may reach a condition of no return no matter what measures are taken to preserve it.

MISTREATMENT OF OUR LAND HERITAGE

The very abundance of natural resources in a relatively newly settled country such as America has posed some of the exceptional problems of conservation that are of public concern at present. One of these problems is the misuse of our soil. Soil is made up of mineral particles produced by the weathering of parent rock plus humus or the organic detritus of decaying vegetation. The interaction of these two components, which results in organic acids, causes further disintegration of the mineral component, so that the two processes, physical and biologic, are taking place all the time. But it takes a long time for the production of a fertile soil, so that its abuse represents the loss of a priceless natural resource.

Soils regulate the flow of water, serving as a spongy receptacle that prevents excessive flow of water in floods. For instance, each year Arkansas farmers have been clearing 150,000 acres of forest land for agricultural purposes. More than a million acres of forests have been so depleted there in the past eight years. Why not leave forests on rough ground to prevent erosion that sweeps away quickly the very best of soils? This short-sighted policy of removing forests is well illustrated in hilly communities in America. Copper smelters in the Appalachian mountains have completely killed all vegetation in certain communities, leaving the soil to be totally eroded. Growing single crops in rows without attempting to enrich the soil by rotation of grass crops and by live stock is another common practice of soil destruction. Little attention has been given to the surface contour (topography) of the land by many American farmers. The revealing book of J. Richie, *The Influence of Man on Animal Life in Scotland* (1920), shows what can happen when man misuses his soil.

Another example of our misuse of land is our flood-plain development. There are wide flat plains on both sides of most rivers onto which the river overflows occasionally. These plains and marshes serve to spread out flood waters and slow down their force. They also draw out and deposit quantities of precious topsoil that would be carried out to sea. However, man builds roads, dwellings, and factories in these bottomlands, dries up the marshes, and then builds levees to protect these improvements from floods. Now the topsoil is forced on out to sea, and the bottomland is lost to agriculture and to wildlife. Then, having destroyed the natural flood-control function of the plain, man goes upstream and builds giant, expensive flood-control dams that flood out additional farmlands and wildlife habitats. How much simpler and wiser to zone the flood plains for agriculture and recreation, for which they are well suited, and put the dwellings and factories on the bluffs overlooking the plains and out of reach of floods!

Many of the dams built by the Army Corps of Engineers are of doubtful value and often destroy regions of scenic beauty. Dams at best are of temporary value because they fill with silt in a matter of a few years.

The establishment of our national parklands stemmed from a desire to set aside some unspoiled land for recreational purposes and for an appreciation of nature under natural conditions. But these sanctuaries are often threatened too. By diverting the water from Lake Okeechobee (that normally supplies the Everglades National Park) the very existence of the park land is threatened. The Everglades is not really a swamp, but a free-flowing river coursing through saw-grass country from south-central Florida down to the Gulf of Mexico. Recently a decision was made to build a major supersonic jetport in and near the park. The jetport was to be set in the mainstream of this river just north of the park, despite the fact that the whole ecologic balance of southern Florida depended on the purity and quantity of the water in this river. Conservationists pointed out that every takeoff of a jet liner uses 4,000 pounds of fuel, the exhaust from which contains carbon monoxide, unburned hydrocarbons, carbon, and nitrogen oxide. Such pollution would have killed the algae that forms the base of the food chain that includes small animals, fish, birds, and alligators. By January, 1970, an agreement of federal, state, and local authorities was reached, forbidding completion of the international jetport—a real tribute to the thousands of concerned citizens who have worked to prevent further deterioration of their natural ecosystem.

WATER CONSERVATION

The proper management of our water supplies involves an understanding of our water resources, the quantity and quality of water needed by a community, the causes and nature of its pollution, the proper drainage of surface water, the improved treatment of water to make it fit for consumption, the development of new underground water resources, the potential of desalinization, and its general economic and aesthetic values. Increase in human population and increased use of water

FIG. 17-1. Some of the thousands of fish that died in polluted streams of Illinois in 1967 are shown washed up on the bank of a stream. These victims are mostly of the "rough" species, considered most resistant to common water pollutants. (Courtesy Federal Water Programs, Environmental Protection Agency.)

per capita have resulted in greater emphasis on augmenting the world supply of clean water.

Water resources include all forms of water—rain, snow, hail, ice, atmospheric vapors, soil moisture, as well as surface and ground water, and the water of streams, rivers, wells, and the ocean. About 97% of all water is contained in the ocean. Of the remainder about 2% is found in surface water and somewhat less than 1% is ground water. A lesser amount of water occurs as atmospheric water. Water represents a renewable natural resource that is delivered from the atmosphere in the form of rain and condensation and is returned to the at-

mosphere by evaporation and transpiration. Water from streams, rivers, and lakes, seeps into the soil to be taken up by plants. Part of this water becomes ground and surface water that eventually flows through streams and rivers to the ocean. This is the regular cycling of water as propelled by solar radiation (p. 196). It is well known, of course, that this cycling effect varies enormously from region to region, and variations occur even within the same region.

Water quality standards. All states at present have water quality standards that indicate the quality of water necessary for agricultural, industrial,

FIG. 17-2. A vast expanse of dead algae disfiguring Montrose Beach, Chicago, aiding in the pollution of Lake Michigan. Treated and untreated sewage discharged into the lake contains nutrients that encourage algae growth. (Photograph by John Hendry; courtesy Federal Water Programs, Environmental Protection Agency.)

recreational, municipal, and fish and wildlife usage. The United States government has implemented to some degree the Federal Water Quality Act of 1965 and the Clean Waters Act of 1966 as advancements in a program of water conservation. The intent of these programs is to have water that meets certain standards of good water by 1973.

Water quality is determined by many natural conditions, such as man's activities, land use, waste disposal, pesticides, storage facilities (reservoirs), and many other features. Water may be naturally acid or alkaline, hard or soft, low or high in minerals, clear or discolored, and varied in temperature and in other ways. Some water is productive of biologic life, whereas other water may be injurious to most life (Fig. 17-1). Almost any ecologic change can be made in water by development projects. Manure on soil is useful in the production of crops; in water it favors the growth of algae and consequent loss of fish and other useful wildlife. The accumulation of organic and inorganic debris even in some of our larger lakes (for example, Lake Erie) is producing the condition known as **eutrophication** in which an abundant accumulation of nutrients

FIG. 17-3. Power plant, a major source of water and air pollution. (Photograph by Billy Davis; courtesy National Air Pollution Control Administration.)

promotes a dense growth of plants and plankton (Fig. 17-2), the decay of which depletes the water of oxygen. Glaring examples of pollution are too well known to need repeating here. Pure, uncontaminated water at present is found only in restricted mountain regions, usually far from human habitation and industrial development.

Aquatic thermal pollution. Animal life appears to have a temperature tolerance range of somewhere between 0° and 45° C., although the tolerance of some animals may exceed these limits. Most higher aquatic animals are not found in water above 35° C. Fish are rarely found in water above 30° C., and water at a temperature of 30° to 35° C. may be considered a biologic desert for most organisms.

The principal offenders in producing thermal pollution of water are electric-power plants (Fig. 17-3) that daily discharge warm water that has been used to cool their steam generators, into lakes and rivers. In small streams in summer the heated water plus the climatic heat can easily raise the water to a temperature unsuitable for fishes. High temperatures also promote the growth of plants, algae, and plankton, all of which consume oxygen not only while alive but also after they die and decay, so that the oxygen of water may be depleted below the level sufficient for animal life.

Heavy-metal. The threat of mercury pollution in streams and lakes was brought to public attention in North America early in 1969 by Canadians who reported mercury levels in the fish caught in Lake St. Clair to be far in excess of Public Health Service limits. Their investigations blamed the source of the mercury pollution on chlorine-caustic plants that were discharging several hundred pounds of mercury waste each day into Lake St. Clair and Lake Erie. Subsequent investigations revealed the problem to be widespread in North America.

When mercury waste in any chemical form is discharged into water, bacteria in the bottom mud convert it into highly toxic methyl mercury. Methyl mercury may become concentrated in fish and other living things. Mercury concentrations become further biomagnified through the food chain; predators at the top of a food chain (for example, man) may receive the most. Methyl mercury may produce severe and irreversible neurologic damage.

The realization that a mercury problem existed in the United States and Canada, years after the problem was recognized in Japan and Sweden, illustrates only too well how haphazard our discovery of such things can be. The need is great for research on all aspects of the mercury question, as well as on other heavy metals, such as zinc, cadmium, and lead, that pose health problems. For example, virtually nothing is known of the long-term effects of mercury on fish, the group that has been getting the most unremitting exposure to mercury. This is especially so for the vulnerable coastal streams through which man's wastes flow and where most of the world's commercially important fish are concentrated.

Contamination of the ocean. The dumping of waste and raw sewage into the coastal waters of our oceans is worldwide. Nuclear explosions have been responsible for the accumulation of isotopes in ocean waters, and oil pollution is on the increase. The wreck of certain oil tankers has proved one great source of oil spills. One of these was the *Torrey Canyon* off the shore of England that was so devastating to marine life there. The damage to marine life was compounded by the use of detergents (themselves harmful to marine life) in the treatment of the oil spill. Another tanker *Witwater* ruptured on its way to the Atlantic entrance of the Panama Canal. Although these are isolated cases, the potential damage to the intertidal flora and fauna may be long lasting. Spontaneous eruption of oil may also be a source of oil spills. The disposal in August, 1970, of nerve gas (no longer necessary as a chemical warfare agent) in the ocean off the coast of Florida by the U. S. Department of Defense may represent another example of marine water pollution. Although encased in steel and concrete blocks, the gas will be freed eventually into the seawater by corrosion of the containers and may be toxic to sea life until rendered nontoxic by hydrolysis with seawater.

AIR POLLUTION

The air of our cities—and cities throughout the world—is increasingly loaded with particulate matter, smog, colloidal material, noxious gases, and other pollutants. The automobile has been one of

FIG. 17-4. Endive plants. Left, injured by air pollution. Right, healthy plant. (Courtesy National Air Pollution Control Administration.)

the chief culprits in producing air pollution (more than 60%), although there are also other causes. The burning of fuels for home consumption and industrial uses ranks high in pollution from sulfur dioxide. Other causes are the burning of trash and other combustible materials. School children cannot be permitted to play outdoors in Los Angeles on those days when smog causes the ozone of the air to exceed a certain point. In 1969 the school playgrounds were closed several times in midsummer and early fall. In July, 1970, a dangerous pall of smog hung over our eastern seacoast, enveloping several eastern states for a number of days.

Inhalation of contaminated air often causes labored and difficult breathing. Breathing noxious gases (for example, sulfur dioxide) injures delicate tissues of the lung and produces other effects on the body. The exhaust of automobiles discharges a number of harmful substances—noxious gases, and small particles (particulates) that emerge from the combustion chamber of the car. These particles may remain in the air for a long time and when breathed into the lungs can affect the delicate tissue much as coal dust would do. The adverse effects of air pollutants to health develop slowly and cannot be diagnosed easily. Persons also become accustomed to breathing smog and, in time, to feeling little effect. The air pollutants are cumulative and may cause chronic bronchitis and worse disorders. In 1952 more than 4,000 deaths were attributed to a heavy smog in London.

Air pollutants may also injure clothing and fabrics, kill vegetation, and damage buildings and statues (Figs. 17-4 and 17-5). A recent U. S. Forest Service study shows that more than a million trees in southern California are dying from smog effects.

Solar radiation, which influences smog, is most pronounced on cloudless days, and winds often shift the pollutants a long distance from their source. The location of the source is very important. Low river bottoms and elevations (hills, mountains, etc.) may form pockets in which pollutants become concentrated and the general effects are most common. Tall chimneys or stacks are used in industry and often discharge their effluent far above ground level. Unlike water pollution whose dispersion is regulated chiefly by gravity, air pollutants may be carried by wind for long distances (Fig. 17-6).

FIG. 17-5. Alexander Hamilton statue suffers soiling from air pollution at U. S. Treasury Building, Washington, D. C.; taken April, 1966. (Courtesy National Air Pollution Control Administration.)

Automobile manufacturers have developed devices that can be used for cutting down on the nuisance of the exhaust. Some states, such as California, require exacting standards for new model cars and such standards are becoming nationwide. Such progress in new cars, however, does not affect the older models that are still pouring out poisonous exhaust into our air. Federal standards for the future provide for reduction of other harmful substances, carbon monoxide, and nitrogen oxides.

Changes also have been made by industrial plants, especially in the control of sulfur oxides. New York has limited the sulfur content of fuel to 1%. Since soft coal has a sulfur content of about 2% to 3% (and sometimes more), natural gas and clean fuel oil will be the fuels emphasized more in the future. Perhaps more progress has been made toward controlling air pollution than in any other field of our pollution problem.

That rain may be produced by the combination of

247

FIG. 17-6. Effects of fluorine on the bones of this cow produced calcium deposits on legs and ribs. Fluorine can be a serious air pollutant from brick, glass, steel, and aluminum industries. (Photograph by Robertson Studio, Bartow, Fla.; courtesy National Air Pollution Control Administration.)

smog and moisture has been shown by the increase in rainfall in certain Indiana communities, which is caused by the smoke fumes from the steel mills of Gary. The sun's heat may be held unnaturally close to the earth's surface, causing it to be too warm, or the opposite effect may be produced by screening out the sun's rays.

EFFECTS OF PESTICIDES

From time immemorial insects have posed one of the greatest problems of our ecosystem. In many ways they are our greatest competitors for the things we need to survive—our agricultural crops,

our plants, our forests, and our livestock. They also afflict us by acting as vectors of diseases, as sources of irritation, and in other ways. But all living things are interdependent and form a highly integrated web of relationships so that disturbances in a single group may profoundly affect the others in this web of life. The balance of nature is being threatened as never before by the chemical pesticides and herbicides man employs to control insects and weeds.

One of the basic laws of ecologic control is that if the environment is simplified to a few species, the whole web of life and its interwoven relationships are destroyed. Agricultural development with its emphasis on a few select plant crops has greatly encouraged certain insect pests and upset the balance.

The use of pesticides in man's attempt to control insects and plants that he considered harmful probably stemmed back to an early time. But it was not until this century that the development of pesticides mushroomed into fantastic proportions. DDT's effective control of lice and mosquito vectors of diseases (typhus and malaria) during World War II was dramatic, and its use has since been widespread. Millions of lives have been saved by the use of DDT and other pesticides in controlling these and other arthropod-borne diseases.

Prior to the development of DDT most insecticides were obtained from plants and were nonpersistent (rapidly degraded). However, the great impetus given to chemical research during World War II resulted in persistent chlorinated hydrocarbons, such as dieldrin, lindane, endrin, and others (altogether about 300 pesticides are known at present). Millions of pounds of these chemical pesticides have been scattered over the landscape in this and other countries to control insects and weeds. At present it is estimated that more than 1 billion pounds of DDT alone can be found in our biosphere. The result has been that it is concentrated in plants, is washed into the sea from agricultural crops, and has accumulated in sea plankton. It becomes further concentrated in the fish that feed on plankton and finally reaches high concentrations in the birds that feed on insects or fish. Its increase is thus greater in those organisms high in the food chain. Its concentration increases tenfold as it passes from one trophic level to the next, until, after much biomagnification, it is returned to us in animal fat or milk.

In 1962 and 1964 there appeared two books that were instrumental in alerting the public to what is taking place in our ecosystem by the use of pesticides. The first of these books was by Rachel Carson entitled *Silent Spring*. This classic book treats of the widespread results of pesticides and their effects on nontarget organisms in our environment. It has had an immense influence in formulating public opinion about the dangers of pesticides and what they are doing to our ecosystem. It pointed out the other side of the pesticide story. One side — the value of pesticides to our general economy — had already been presented by the pesticide manufacturers. The factual evidence presented by *Silent Spring* and thoroughly documented from the scientific literature showed how fish were poisoned by endrin used by farmers against the boll weevil in the Mississippi valley, how fishes far from our coastlines had traces of the insoluble DDT, how pesticides caused the death and the loss of fertility of birds, and how the

use of pesticides had completely upset the biotic community, resulting in the loss of useful species and the increase of harmful ones, etc. The publication of the book was followed by the report of the President's Scientific Advisory Committee (1965). This report tended to confirm the statements about pesticides made by Miss Carson. Naturally, the book has been the subject of fierce controversy between those who favor the use of pesticides and those who do not.

The other book, *Pesticides and the Living Landscape,* was by Dr. R. L. Rudd. In this work Dr. Rudd has given more attention to scientific details from which he has drawn logical conclusions. It covers a wider range of the ecologic impact of the use of pesticides, but in general confirms the statements made by Carson's *Silent Spring*. Neither author recommends the abolition of pesticides, which have proved extremely useful in the preservation of crops. Both authors have urged greater responsibility in the control of pesticides and ask that prospective users weigh all possible results.

Since the publication of these and other books, an enormous amount of factual information has been steadily accumulating on the general effects of pesticides. In Florida when sand flies (target objects) were sprayed with dieldrin, the nontarget fishes and fiddler crabs were also killed by thousands. Wurster et al. (1956) reported that 70% of the robin population was destroyed when elm trees in the eastern United States were sprayed with DDT for the Dutch elm disease. After Clear Lake, California, was sprayed with pesticides for gnats, an analysis of the plankton showed that it contained the pesticide in a concentration several hundred times as great as the concentration in which it was originally applied. Frogs and fishes showed an even greater concentration. Grebes that ate the frogs and fish were almost completely eliminated from the lake where formerly they had nested by hundreds. Peregrine falcons (duck hawks) have been completely eliminated in many communities where once they were common, and DDT and dieldrin are believed to be the cause. The chief damage of pesticides in these cases is that high concentrations cause fragile egg shells that burst prematurely in the incubation process, killing the embryo. DDT and its metabolite also cause the liver to synthesize enzymes that lower estrogen levels in the blood, thus interfering with normal breeding.

It is not known just what are the effects of pesticides on the human, although some speculations implicate them as a possible cause of cancer. Herbicides or chemical weed killers are not as dangerous in general as are the insecticides. It is possible that heavy dosages in spraying will also affect nontarget or useful plants. The chlorinated herbicide 2,4-D has been widely used, but altogether more than forty different kinds of herbicides are available. The use of many pesticides is being restricted at present. DDT has been outlawed by all states, and dieldrin and other pesticides may follow the same fate.

Biologic control of insect pests whereby certain insects prey upon others may be part of the answer to the pesticide problem. In one study the release of 200,000 aphid lions (parasitic insects of the family Chrysopidae) per acre was as effective in controlling the bollworm as the standard insecticides. Some twenty other parasitic or predator insects have been useful in controlling destructive pests. So far, no damaging effects on the environment (other than their specific action on the pests) have been reported. The technique of "sterile male release" may also be a solution in certain areas. In this method a large number of male insects are reared and made sterile by ionizing radiation and then are released. When many female insects mate with sterile males the fecundity of the population as a whole will be reduced. Such a method has proved successful in some instances (screw-worm fly). Instead of rendering the males 100% sterile (which reduces their mating competitiveness), a method of partially sterilizing the males with reduced gamma radiation has been found to be even more effective. Another method is to treat insects (or their eggs) before they become adults with the juvenile hormone that renders the embryos nonviable.

Efforts are being continued to find insecticides as effective as the wide-spectrum chlorinated hydrocarbons, but having more specific toxicities (killing only insect pests instead of all insects including the beneficial ones). Organophosphates have the advantage of being quickly degraded (nonpersistent), but when first applied are even more poisonous to man and other animals than the chlorinated hydrocarbons.

EFFECT OF POLLUTION ON HEALTH

There are many biologic unknowns in the effects of pesticides and pollutants on our ecosystem. Individuals, for instance, may become adjusted to some extent to smog and polluted air, so that their health and behavior seem unaffected. This feeling, however, is a delusion for the body tissues may be building up contaminants while that person is unaware of what is happening to his body. The aged and the young are perhaps the ones who show the effects of pollution most quickly, especially pollutants affecting the respiratory system. The cumulative effects of irritation resulting in chronic bronchitis and even cancer may demonstrate this delayed response better than some others, although it is often difficult to relate the effect to the cause.

In the industrial communities of America it is firmly established that chronic pulmonary disorders are related to the degree of pollution. The incidence of such diseases increases with the length of exposure and with the concentration of the pollutants so that the worst effects of pollution may not show up until much later. In a quantitative way emphysema has increased from 1.5 persons per 100,000 population in 1945 to 15 persons in 1964. It may be argued that some of this increase is due to the steady rise in numbers of the aged, for emphysema is more common in the higher age groups, but this age correlation does not explain why emphysema is more common in those communities where air pollution is greatest. The death rate (in the United States) from bronchitis and emphysema is now known to be nine times as high as it was in 1949. Contaminated water is responsible for the spread of infectious hepatitis, but little is known about how this operates. Radiation exposure is known to cause tumors of the thyroid gland, as was demonstrated by the Bikini tests of 1954, although this is an example of delayed response to rather high radiation levels. Man can acquire pesticide residues by consuming food chain organisms that have accumulated high levels of such residues. DDT and other chlorinated hydrocarbons have a marked resistance to bacterial action and have been accumulating in the water and soil. However, levels of pesticides evidently harmless to man may adversely affect birds and fish. Many studies show that pesticides are stored in human tissue. Even though many of these pesticides are banned or will be banned, those that have already been used will continue to be accumulated in animal tissue for some time.

Another alarming factor is the noise produced in our technologic society by everything from super-sonic booms of high-powered jets, industrial machinery, and the automobile to sirens, lawn mowers, air conditioners, and household appliances. Decibels are units used to measure the intensity of sound waves. The rustling of a newspaper in a silent room measured 15 decibels and the honk of a horn 90 decibels. We can tolerate up to 80 decibels fairly comfortably, but continuous exposure above 85 makes us uncomfortable and can result in ear damage and loss of hearing. Today's four-engine jets blast off at 155 decibels. The noise in an electric power plant in which men work 8 hours a day was measured at 118 decibels. Music at a discotheque measured 122 decibels. Research indicates that persistent noise, even though supposedly of tolerable levels, can cause ear damage. The heart rate of an unborn child is accelerated by noises to which the mother has become tolerant. The human is very adaptable and may become apparently adjusted to a noise range, but such adjustment may occur at the cost of loss of finer degrees of hearing. Noise not only causes deafness, it also is suspected of contributing to heart disease, high blood pressure, and stomach ulcers. The general effects of noise pollution on mental health are presently unknown.

CHALLENGE OF THE ECOSYSTEM

The foregoing account of our ecosystem presents a gloomy picture. Man has made great strides in his technology, but in his economic zeal he has often been unable to comprehend the overall picture. He may destroy obnoxious insects but in so doing he may destroy beneficial animals as well. He has little understanding of the web of life and its intricate relationships so that when he interferes with one organism in the chain of relations he may unwittingly disrupt the entire web. Much of our precious land is being exploited for technologic development without regard to the consequences.

In a revealing article in *Science* for Nov. 18, 1969, entitled "What we must do," Dr. John Platt suggests ways to check this crisis of transformation. He shows that the human race is undergoing a transition to new technologic power all over the world. Some persons believe that because man does not realize how enormous and disruptive his powers are, he may not be able to harness and control his energies soon enough to insure his survival. Others feel there is reason for optimism. John Platt foresees the possibility of a "world of incredible potentialities for all mankind." He deems it necessary to mobilize competent experts and scientists, as is done on solving great crises in wartime, for solving the problems that we are just now beginning to understand.

Aldo Leopold, a conservationist of the first rank, urged that one of the requisites for an ecologic comprehension of land is an understanding of ecology. Ecologic concepts must be incorporated into our school systems at all levels. The development of what Leopold calls a land ethic is an intellectual as well as an emotional process. Solution of the problems of the ecosystem must reach the ordinary citizen; the role of the specialist in the ecologic discipline can accomplish little without a thoroughly aroused and dedicated mankind as a whole behind the movement. It is everybody's problem. Ecology may be the most important discipline in all sciences for human survival, but at present it is least understood and studied by a too indifferent mankind.

The literature on the problems of the ecosystem is becoming very extensive, but the student interested in the subject could very well start his reading with G. Hardin's "The Tragedy of the Commons," *Science* **162**:1243-1248, Dec. 13, 1968.

REFERENCES TO PART IV

Andrewartha, H. G. 1961. Introduction to the study of animal populations. Chicago, University of Chicago Press.

Benarde, M. A. 1970. Our precarious habitat. New York, W. W. Norton & Co., Inc.

Berland, T. 1970. The fight for quiet. Englewood Cliffs, N. J., Prentice-Hall, Inc.

Blus, L. J. 1970. Measurements of brown pelican eggshells from Florida and South Carolina. Bioscience 20:267-269 (Aug.).

Brown, F. A., Jr., J. W. Hastings, and J. D. Palmer. 1970. The biological clock, two views. New York, Academic Press, Inc.

Bureau of Sport Fisheries and Wildlife. 1969. Rare and endangered fish and wildlife of the United States. Washington D. C., U. S. Department of the Interior.

Burns, W. 1969. Noise and man. Philadelphia, J. B. Lippincott Co.

Calvin, M. 1969. Chemical evolution. New York, Oxford University Press, Inc.

Carson, R. 1962. Silent spring. Boston, Houghton Mifflin Co.

Cheng, T. C. 1970. Symbiosis, organisms living together. New York, Pegasus. (Paperback.)

Ehrlich, P., and A. H. Ehrlich. 1970. Population, resources, environment: issues in human ecology. San Francisco, W. H. Freeman and Co. Publishers.

Eibl-Eibesfeldt, I. 1970. The biology of behavior (translated from the German by Erich Klingshammer). New York, Holt, Rinehart & Winston, Inc.

Elton, C. E. 1958. The ecology of invasions by plants and animals. New York, John Wiley & Sons, Inc.

Ewer, R. F. 1968. Ethology of mammals. New York, Plenum Publishing Corp.

Fossey, D. 1970. Making friends with mountain gorillas. Nat. Geogr. Mag. 137(1):48-67.

Goodall, J. 1963. My life among wild chimpanzees. Nat. Geogr. Mag. 124(2): 272-308.

Graham, F., Jr. 1970. Since silent spring. Boston, Houghton Mifflin Co.

Hardin, G. (editor). 1969. Population, evolution, and birth control, ed. 2. San Francisco, W. H. Freeman and Co. Publishers.

Hardin, G. 1968. The tragedy of the commons. Science 162:1243-1248 (Dec.).

Harker, J. E. 1964. The physiology of diurnal rhythms. Cambridge monographs in experimental biology, No. 13. New York, Cambridge University Press.

Hickey, J. H. (editor). 1969. Peregrine falcon populations. Madison, University of Wisconsin Press.

Howse, P. E. 1970. Termites, a study in social behavior. New York, Hillary House Publishers.

Klopfer, P. H., and J. P. Hallman. 1967. An introduction of animal behavior; ethology's first century. Englewood Cliffs, N. J., Prentice-Hall, Inc.

Lorenz, K. 1966. Evolution and modification of behavior. Chicago, University of Chicago Press.

Marler, P., and W. J. Hamilton. 1966. The mechanism of behavior. New York, John Wiley & Sons, Inc.

Ng, L. K. Y., and S. Mudd (editors). 1965. The population crisis. Bloomington, Indiana University Press.

Odum, H. T. 1970. Environment, power, and society. New York, John Wiley & Sons, Inc.

Oparin, A. I. 1962. Life: its nature, origin and development. New York, Academic Press, Inc.

Peckham, G. W., and E. G. Peckham. 1905. Wasps: social and solitary. Boston, Houghton Mifflin Co.

Ritchie, J. 1920. The influence of man on the animal life in Scotland. Cambridge, Cambridge University Press.

Rosen, W. G. 1970. The environmental crisis: through a glass darkly. Bioscience 20:1209-1216 (Nov.).

Rudd, R. L. 1964. Pesticides and the living landscape. Madison, University of Wisconsin Press.

Schaller, G. B. 1963. The mountain gorilla: ecology and behavior. Chicago, University of Chicago Press.

Singer, S. F. 1971. The population revolution: population growth and environmental change. Bioscience 21:163 (Feb.).

Teal, J., and M. Teal. 1969. Life and death of the salt marsh. Boston, Little, Brown & Co.

Thomas, W. L., Jr. (editor). 1956. Man's role in changing the face of the earth. Chicago, University of Chicago Press.

Tinbergen, N. 1965. Animal behavior. New York, Time-Life Books, a Division of Time, Inc.

Tinbergen, N. 1951. The study of instinct. New York, Oxford University Press, Inc.

White, L., Jr. 1967. The historical roots of our ecological crisis. Science 155:1203-1207.

Wilson, B. R. 1968. Environmental problems. Philadelphia, J. B. Lippincott Co.

Wurster, D. H., C. F. Wurster, and W. N. Strickland. 1965. Bird mortality following DDT spraying for Dutch elm disease. Ecology 46:488-489.

Yerkes, R. M. 1948. Chimpanzees: a laboratory colony. New Haven, Yale University Press.

Zimmerman, A. 1961. A catholic viewpoint on overpopulation. Garden City, Hanover House.

part five

The continuity of life

All living things are mortal, that is, there is a definite limit to their individual existence. This period of existence varies enormously in both plants and animals, but none, so far as we know, enjoy continuous individual perpetuation. Even in protozoans with binary fission individuality is lost when the parent cell divides to form the two daughter cells.

To comprehend the continuity of life it is necessary to understand how hereditary units are transmitted from parent to offspring, how self-replication of hereditary units occurs, how hereditary units communicate structural and functional information, how evolutionary diversity has resulted from the interaction of heredity and environment, and how mechanisms of differentiation give rise to the characteristic adult structure of multicellular animals. Reproduction is universal among all animals. Reproduction differs from other body processes that are necessary for the welfare of the individual organism. Animals can live normally without reproduction. Reproductive organs can be removed from the animal without endangering its general health and life expectancy. Reproduction may or may not be indulged in by the individual. Its own adaptive mechanism, whatever pattern it may be, does not interfere with the general economy of the body. Reproduction is concerned with the survival of the species rather than with survival of the individual. Its primary purpose is to form a link from one generation to another so that the species line may continue.

Heredity is the transmission to the offspring of biochemical information on structure and function. This information is in the genes of the chromosomes, which carry the potentialities for the whole individual. Under the impact of environmental influences, the genes develop the visible traits of the individual animal. Modifications or variations of this information may occur by mutation of the genes (uniparental reproduction), or both by mutation and by the sexual recombination of genes (biparental reproduction).

Inherited variations provided by the hereditary mechanisms and mutations of genes form the basis of the evolutionary concept. Evolution comes about by natural selection (imposed by the environment) acting on the inheritable variations of a population. Evolution may be considered the process whereby existing animals (and plants) have developed by gradual, continuous change from previously existing forms. It may be considered a descent in the course of time with modifications. Organic evolution applies to the origin of life and its historic development.

18

The reproductive process

Reproduction refers to the capacity of all living systems to give rise to new systems similar to themselves. It may also refer to self-duplication in a single cell, in groups of cells or organs, or in a complete organism. Any new living unit resembles the old, and reproduction almost always implies a reasonable facsimile of the unit duplicated. In all cases, reproduction consists of a basic pattern: the conversion of raw materials from the environment into the offspring or into sex cells that develop into offspring of a similar constitution, and the transmission of a hereditary pattern or code from the parents. Reproduction of any pattern requires specific nutrition and controlled metabolic processes. Whenever a duplication of a large unit occurs, there must also be duplication of smaller units. Molecular duplication must happen before cellular duplication, and cellular duplication occurs before that of the whole animal.

Two fundamental types of reproduction may be distinguished—asexual, or vegetative, reproduction and sexual, or gametic, reproduction.

ASEXUAL REPRODUCTION

In asexual reproduction the reproductive unit may be the whole parent body or merely a part of the body, and the unit is not exclusively for reproduc-

tion. Asexual reproduction may be of advantage in allowing beneficial combinations of characteristics in one animal to continue unchanged and in eliminating the early stages of embryonic growth.

Asexual reproduction is found only among the simpler forms of life such as protozoans, coelenterates, bryozoans, and a few others.

The forms of asexual reproduction are fission, budding (both internal and external), fragmentation, and sporulation. **Fission** is common among protozoans and to a limited extent among metazoans. In this method, which is essentially cell division, the body of the parent is divided into two identical daughter individuals. **Budding** is an unequal division of the organism. The new individual arises as an outgrowth (bud) from the parent, develops organs like that of the parent, and then usually detaches itself. The bud is typically external, but in some cases internal buds, or **gemmules,** are produced. Gemmules are collections of many cells surrounded by a dense covering in the body wall. When the body of the parent disintegrates, each gemmule gives rise to a new individual. External budding is common in the hydra and internal budding in the freshwater sponges. **Fragmentation** is a method in which an organism breaks into two or more parts, each capable of becoming a complete animal. This method is found among the Platyhelminthes, Nemertinea, and

255

Echinodermata. **Sporulation** is a method of multiple fission in which many cells are formed and enclosed together in a cystlike structure. Sporulation occurs in a number of protozoan forms.

SEXUAL REPRODUCTION

Sexual reproduction is the general rule in the animal kingdom. It involves, as a rule, two parents, each of which contributes one gamete, or special cell, to a union known as the zygote. There are usually two kinds of gametes—the **ovum** (egg) and the **spermatozoan.** Eggs are produced by the female, are nonmotile, and contain a great amount of yolk. Sperm are formed by the male, are motile, and are relatively small. The union of egg and spermatozoan is called **fertilization,** and the fused cell so formed is known as the **zygote,** which develops into a new individual.

Sexual reproduction is of universal occurrence among all higher forms. One of its advantages is that the characteristics of two organisms can combine and variations can be multiplied. This affords evolution a greater variety of forms to pick from in natural selection. Recombination of characters makes possible wider and more diversified evolution. The chief disadvantages of the sexual method are the hazards involved in the meeting of eggs and sperm and the possibility of unfavorable growing conditions.

A variant form of sexual reproduction is the union of nuclei in the paramecium (**conjugation**). In this method two individuals fuse together temporarily and exchange micronuclear material. Other forms of sexual reproduction occur in protozoans, in which there is a union between two special cells. These cells may be alike (**isogametes**) or they may be different (**anisogametes**). In some cases it is difficult to distinguish sex, for although two parents are involved, they cannot be designated as male and female.

Among metazoans, some individuals can be called male and others female. Organs that produce the germ cells are known as **gonads.** The gonad that produces the sperm is called the **testis** and that which forms the egg, the **ovary** (Fig. 18-1). The gonads represent the **primary sex organs,** the only sex organs found in certain groups of animals.

Most metazoans, however, also have **accessory sex organs.** In the primary sex organs the sex cells undergo many complicated changes during their development, the details of which are described later in this chapter. In our present discussion we shall point out the various types of sexual reproduction—biparental reproduction, parthenogenesis, pedogenesis, and hermaphroditism.

Biparental reproduction. This common method of sexual reproduction involves two separate and distinct individuals—male and female. Each of these has its own reproductive system and produces only one kind of sex cell, spermatozoan or ovum, never both. Nearly all vertebrates and many invertebrates have separate sexes, and such a condition is called **dioecious.**

Parthenogenesis. This is a modification of sexual reproduction in which an unfertilized egg develops into a complete individual. It is found in rotifers, plant lice, certain ants, bees, and crustaceans. Usually parthenogenesis occurs for several generations and is followed by a biparental generation in which the egg is fertilized. In some cases parthenogenesis appears to be the only form of reproduction. The queen bee is fertilized only once by a male (drone) or sometimes by more than one drone. She stores the sperm in her seminal receptacles, and as she lays her eggs, she can either fertilize the eggs or allow them to pass unfertilized. The fertilized eggs become females (queens or workers), and the unfertilized eggs become males (drones).

Pedogenesis. Parthenogenesis among larval forms is called **pedogenesis. Neoteny,** or the retardation of bodily development, has about the same meaning as pedogenesis. The most striking example of this is the tiger salamander (*Ambystoma tigrinum mexicanum)* that in certain parts of its range is found to mate in a larval (axolotl) form. Such larvae can be transformed into adults under the right conditions.

Hermaphroditism. Animals that have both male and female organs in the same individual are called hermaphrodites, and the condition is called **hermaphroditism.** In contrast to the dioecious state of separate sexes, hermaphroditism is called **monoecious.** Many lower animals (flatworms and hydra) are hermaphroditic. Most of them avoid self-

Sperm

Function of testis to produce sperm in its seminiferous tubules and sex hormones in its interstitial tissue; in lower species (fish and salamanders) seminiferous tubules short, lobulated structures and all developing sperm cells in 1 part of tubule tend to be approximately at same stage of development

In higher species (including frog and man) seminiferous tubules are long and their walls contain sperm cells in various stages of development

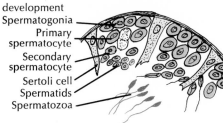

Spermatogonia
Primary spermatocyte
Secondary spermatocyte
Sertoli cell
Spermatids
Spermatozoa

In final maturing of sperm, testes of vertebrates may be divided into 2 types. In 1 type (certain fish, salamanders, and frogs) anterior part of sperm duct does not form convoluted epididymis, and sperm mature and are stored in testis; this type of testis best suited for seasonal activity when sperm discharged at one time. In other type (higher vertebrates) sperm become physiologically functional only in convoluted epididymis where they are also stored; this type best adapted for more or less continuous reproduction

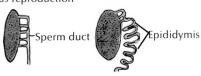

Sperm duct — Epididymis

Ova

Ova formed in ovaries that vary among different vertebrate classes; 2 ovaries of frog saccular because interiors represented by large lymph spaces; each ovarian sac consists of 2 membranes, theca externa and theca interna; between these, young ova in various stages of development; each ovum surrounded by follicle cells (for nourishment), which are left behind when mature egg discharged into body cavity through rupture in theca membranes

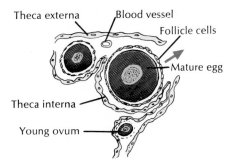

Theca externa — Blood vessel
Follicle cells
Mature egg
Theca interna
Young ovum

In mammalian ovary developing egg with its surrounding cells called graafian follicle. As egg or ovum enlarges, split appears between outer and inner layers of cells and follicular cavity formed; this cavity filled with liquor folliculi, which contains hormone estrogen; when mature follicle ruptures, freed egg surrounded with fuzzy coat of follicle cells

Ruptured follicle   Egg with follicle cells

FIG. 18-1. Comparison of male and female gamete formation in vertebrates.

fertilization by exchanging germ cells with each other. The earthworm ensures that its eggs are fertilized by the copulating mate as well as vice versa. Another way of preventing self-fertilization is by developing the eggs and sperm at different times.

PLAN OF REPRODUCTIVE SYSTEMS

The basic plan of the reproductive systems is similar in all animals, although differences in reproductive habits, methods of fertilization, etc. have produced many variations. In vertebrate animals the reproductive and excretory systems are often referred to as the **urogenital system** because of their close anatomical connection. This association is very striking in their embryonic development. The male urogenital system usually has a more intimate connection than has the female. This is the case with those forms (some fish and amphibians) that have an opisthonephros kidney. In the male of fishes and amphibians the duct that drains the kidney (**wolffian duct**) also serves as the sperm duct. In male reptiles, birds, and mammals in which the kidney develops its own independent duct (**ureter**) to carry away waste, the old wolffian duct becomes exclusively a sperm duct (**vas deferens**). In all these forms, with the exception of mammals, the ducts open into a **cloaca.** In higher mammals there is no cloaca; instead the urogenital system has its own opening separate from the anal opening. The **oviduct** of the female is an independent duct that, however, does open into the cloaca in forms that have a cloaca.

The plan of the reproductive system in vertebrates includes (1) **gonads** that produce the sperm and eggs; (2) **ducts** to transport the gametes; (3) **special organs** for transferring and receiving gametes; (4) **accessory glands** (exocrine and endocrine) to provide secretions necessary for the reproductive process; and (5) **organs** for storage before and after fertilization. This plan is modified among the various vertebrates, and some of the items may be lacking altogether.

Male reproductive system. The male reproductive system in man (Fig. 18-2) includes testes, vasa efferentia, vas deferens, penis, and glands.

The **testes** are paired and are responsible for the production and development of the sperm. Each testis is made up of numerous **seminiferous tubules,** which produce the sperm (Fig. 18-3), and the **interstitial** tissue, lying among the tubules, which produces the male sex hormone (testosterone). The two testes are housed in the scrotal sac, which in many mammals hangs down as an appendage of the body. This arrangement protects the sperm against high temperature, since in at least some forms sperm apparently will not form at body temperatures.

The sperm are conveyed from the seminiferous tubules to the **vasa efferentia,** small tubes passing to a coiled **vas epididymis** (one for each testis). The epididymis is connected by a **vas deferens** to the **urethra.** From this point the urethra serves to carry both sperm and urinary products through the penis, or external intromittent organ.

Three pairs of glands open into the reproductive channels—the **seminal vesicles, prostate glands,** and **Cowper's glands.** Fluid secreted by these glands furnishes food to the sperm, lubricates the passageways of the sperm, and counteracts the acidity of the urine so that the sperm will not be harmed.

Female reproductive system. The female reproductive system (Fig. 18-4) contains ovaries, oviduct, uterus, vagina, and vulva.

The paired ovaries, each about the size of an almond, contain many thousands of developing eggs (ova). Each egg develops within a **graafian follicle** that enlarges and finally ruptures to release the mature egg (Fig. 18-5). During the fertile period of the woman about thirteen eggs mature each year, and usually the ovaries may alternate in releasing an egg. Since the female is fertile for only some thirty years, only about 400 eggs have a chance to reach maturity; the others degenerate and are absorbed.

The **oviducts,** or fallopian tubes, are egg-carrying tubes with funnel-shaped ostia for receiving the eggs when they emerge from the ovary. The oviduct is lined with cilia for propelling the egg in its course. The two ducts open into the upper corners of the **uterus,** or womb, which is specialized for housing the embryo during the nine months of its intrauterine existence. It is provided with thick muscular walls, many blood vessels, and a specialized lining—

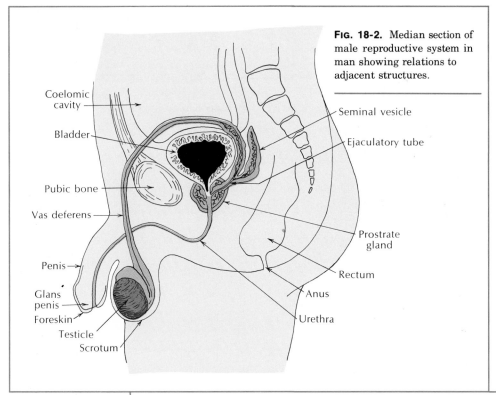

FIG. 18-2. Median section of male reproductive system in man showing relations to adjacent structures.

Coelomic cavity

Bladder

Pubic bone

Vas deferens

Penis

Glans penis

Foreskin

Testicle

Scrotum

Seminal vesicle

Ejaculatory tube

Prostrate gland

Rectum

Anus

Urethra

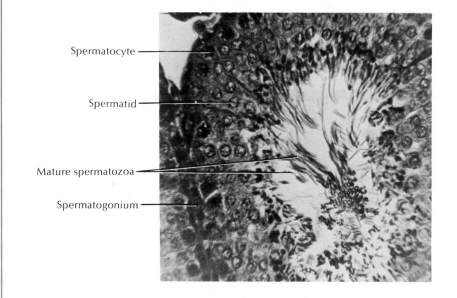

Spermatocyte

Spermatid

Mature spermatozoa

Spermatogonium

FIG. 18-3. Section through testis of rat showing seminiferous tubule with different stages of sperm formation. Mature sperm are nearest fluid-filled center of tubule; between them and outermost part of tubular wall are various stages of sperm formation. In rat each spermatogonium has 42 chromosomes, but by meiosis this number is reduced to 21 in each sperm. Corresponding figures for human being would be 46 and 23. (Courtesy J. W. Bamberger.)

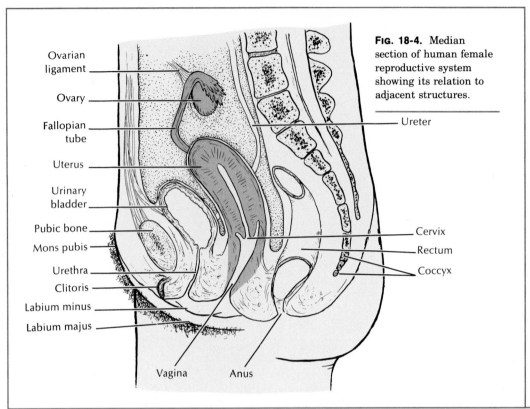

Ovarian ligament

Ovary

Fallopian tube

Uterus

Urinary bladder

Pubic bone

Mons pubis

Urethra

Clitoris

Labium minus

Labium majus

Vagina

Anus

FIG. 18-4. Median section of human female reproductive system showing its relation to adjacent structures.

Ureter

Cervix

Rectum

Coccyx

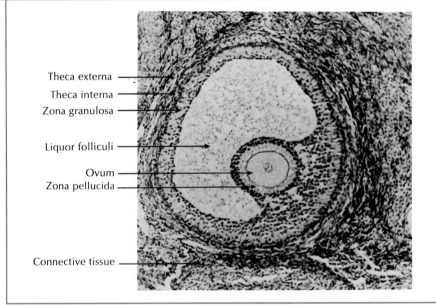

Theca externa

Theca interna

Zona granulosa

Liquor folliculi

Ovum

Zona pellucida

Connective tissue

FIG. 18-5. Enlarged view of graafian follicle showing ovum, or egg, in position. Follicle is nearing maturity and is about ready to rupture and to discharge egg. In human ovary usually only one ovum ruptures each 4 weeks during active life of ovary. (Courtesy J. W. Bamberger.)

the **endometrium.** The uterus varies with different mammals. It was originally paired but tends to fuse in higher forms.

The uterus is connected to the outside of the body by the **vagina.** This muscular tube is adapted for receiving the male's penis and for serving as the birth canal during expulsion of the fetus from the uterus. Where the vagina and the uterus meet, the uterus projects down into the vagina to form the **cervix.**

The external genitalia of the female, or vulva, include folds of skin, the **labia majora** and **labia minora,** and a small erectile organ, the **clitoris.** The opening into the vagina is normally closed in the virgin state by a membrane, the **hymen.**

Homology of sex organs. For every structure in the male system, there is a homologous one in the female. They do not have identical functions in both sexes, however. Many of them are functional in one sex and their homologues in the other may be vestigial and nonfunctional.

Although sex is probably determined at the time of fertilization, it is not until many weeks later that the distinct sex characters associated with one or the other sex are recognized. The animal is a chemical hermaphrodite and can become either sex, depending on the balance of the sex hormones.

ORIGIN OF REPRODUCTIVE CELLS

Protoplasm is commonly divided into two types—**somatoplasm** and **germ plasm.** The body cells are made up of somatoplasm and are called **somatic cells;** reproductive cells are formed of germ plasm and are called **germ cells.** All the somatic cells die when the individual dies. The germ plasm is continuous from generation to generation, whereas the somatoplasm is formed anew at each generation. At the present time this continuity is recognized as residing in the chromosomes, so the chromatin material of the nucleus is considered to be the germ plasm and the cell cytoplasm the somatoplasm. The distinction, however, between somatic and germ cells is not absolutely rigid. Many invertebrates are known to regenerate whole bodies from small parts of themselves.

The actual tissue from which the gonads arise appears in early development as a pair of ridges lying beneath the kidneys. Surprisingly, the primordial ancestors of the cells that are going to form gametes (primordial germ cells) do not arise in the developing gonad but in the yolk-sac endoderm. Later they migrate to the embryonic gonad. The gonad at first is sexually indifferent, but will become a testis or ovary as determined by the hormonal environment.

Meiosis. In ordinary cell division, or mitosis, each of the two daughter cells receives exactly the same number and kind of chromosomes. All body (somatic) cells have two sets of chromosomes, one of paternal and the other of maternal origin. The members of such a pair are called **homologous chromosomes.**

In sexual reproduction, however, the formation of the **gametes,** or **germ cells,** requires a different process than that of somatic cells. The fusion of two gametes (egg and sperm) produces the zygote or fertilized egg from which the new organism arises. If each sperm and egg had the same number of chromosomes as somatic cells, there would be a doubling of chromosomes in each successive generation. To prevent this from happening, germ cells are formed by a special type of cell division called **meiosis,** whereby the chromosome number is reduced by one half. The result is that mature gametes have only **one** member of each homologous pair, or a **haploid** (n) number of chromosomes. In man the zygotes and all body cells normally have the **diploid** number (2n) of 46; the gametes (eggs and sperm) have the haploid number (n) of 23.

Meiosis is similar to mitosis in its morphologic changes and movements of chromosomes, but mitosis has only **one** chromosomal division (Fig. 18-6). Meiosis consists of **two successive** chromosomal (nuclear) divisions, each of which has the same four stages—prophase, metaphase, anaphase, and telophase—found in mitosis. The first meiotic division involves the pairing of homologous chromosomes to form bivalent units, the resolution of each homologous chromosome into two half or sister chromatids, and the separation of homologous chromosomes to opposite poles of the cell. Each resulting daughter cell thus contains half the number of chromosomes characteristic of the diploid chromosomes of the organism. The second meiotic division results in a separation and distribution of the sister chromatids

Somatic cell (interphase) or
Primordial germ cell
(diploid number of 8 chromosomes)

Mitosis ← → Meiosis

Chromosomes begin to appear | Each chromosome consists of 2 sister chromatids | Homologous chromosomes, each of 2 chromatids, pair

Chromosomes arranged on spindle | Paired chromosomes arranged on spindle

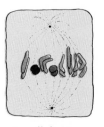 

In cell division, one sister chromatid passes to one daughter cell and its mate to the other

In first meiotic division, homologous chromosomes separate to opposite poles so that each daughter cell has only haploid number of chromosomes

 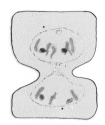 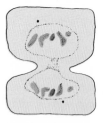

Each cell retains diploid number (8) of chromosomes

Second meiotic division (not shown) involves separation of sister chromatids of each chromosome to their respective daughter cells; each cell has only haploid number of chromosomes

FIG. 18-6. Comparison of mitosis and meiosis. Each process starts with diploid number of chromosomes. In **mitosis** chromosomes replicate, then the chromatids separate, one going to each pole resulting in two identical daughter cells with diploid number of chromosomes. In **meiosis** chromosomes replicate, then arrange themselves in homologous pairs on spindle; each pair separates, one member going to each pole, resulting in daughter cells with haploid number of chromosomes. Each haploid cell now divides (not shown) with one set of chromatids passing to each daughter. Final result of meiosis is four gametes, each with haploid number of chromosomes.

of each chromosome to opposite poles and thus involves no reduction in number of chromosomes. In contrast to mitosis, meiosis is concerned with genetic reassortment. It does this in two ways: (1) by random segregation of homologous chromosomes and (2) by crossing-over, or exchanging segments between each homologous pair. Another source of variation is, of course, the mutation of the gene. This source of inherited variability enables natural selection to bring about evolutionary changes.

Gametogenesis. The series of transformations that results in the formation of mature gametes (germ cells) is called gametogenesis.

Although the same essential processes are involved in the maturation of both sperm and eggs, there are some minor differences. Gametogenesis in the testis is called **spermatogenesis** and in the ovary it is called **oogenesis.**

Spermatogenesis (Fig. 18-7). The walls of the seminiferous tubules contain the differentiating sex

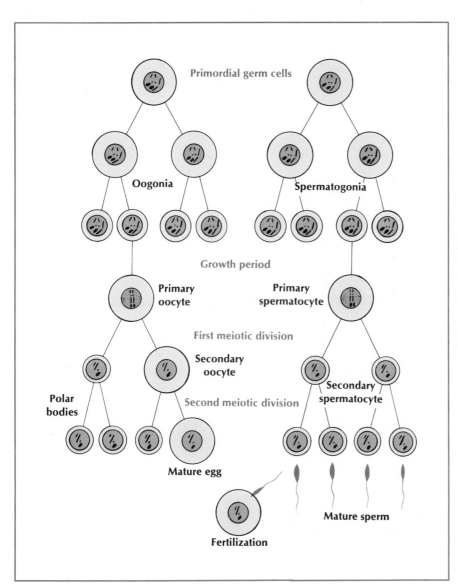

FIG. 18-7. Process of gametogenesis, or formation of germ cells. Oogenesis shown on left, spermatogenesis on right.

cells arranged in a stratified layer five to eight cells deep. The outermost layers contain **spermatogonia** (Fig. 18-3), which have increased in number by ordinary mitosis. Each spermatogonium increases in size and becomes a **primary spermatocyte.** Each primary spermatocyte then undergoes the first meiotic division, as described above, to become two **secondary spermatocytes.**

Each secondary spermatocyte now enters the second meiotic division, without the intervention of a resting period. The resulting cells are called **spermatids,** and each contains the haploid number (23) of chromosomes. A spermatid may have all maternal, all paternal, or both maternal and paternal chromosomes in varying proportions. Without further divisions the spermatids are transformed into mature sperm by losing a great deal of cytoplasm, by condensing the nucleus into a head, and by forming a whiplike tail.

It will be seen from following the divisions of meiosis that each primary spermatocyte gives rise to four functional sperm, each with the haploid number of chromosomes (Fig. 18-7).

Oogenesis (Fig. 18-7). The early germ cells in the ovary are called **oogonia,** which increase in number by ordinary mitosis. Each oogonium contains the diploid number of chromosomes. In the human being, after puberty, typically one of these oogonia develops each menstrual month into a functional egg. After the oogonia cease to increase in number, they grow in size and become **primary oocytes.** Before the first meiotic division, the chromosomes in each pri-

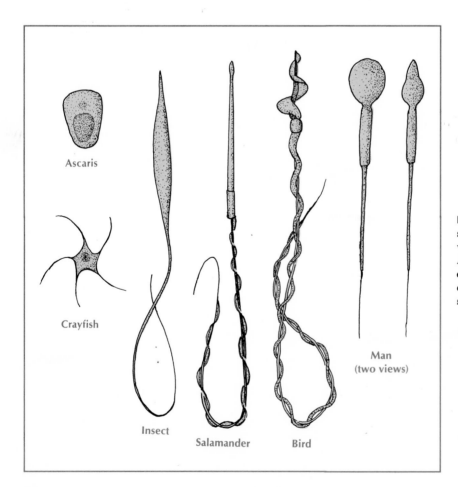

FIG. 18-8. Types of animal sperm, represented by both vertebrates and invertebrates. All belong to flagellate variety except those of *Ascaris* and crayfish that are nonflagellate sperm.

Ascaris

Crayfish

Insect

Salamander

Bird

Man
(two views)

mary oocyte meet in pairs, paternal and maternal homologues, just as in spermatogenesis. When the first maturation (reduction) division occurs, the cytoplasm is divided unequally. One of the two daughter cells, the **secondary oocyte,** is large and receives most of the cytoplasm; the other is very small and is called the **first polar body.** Each of these daughter cells, however, has received half the nuclear material or chromosomes.

In the second meiotic division, the secondary oocyte divides into a large **ootid** and a small polar body. If the first polar body also divides in this division, which sometimes happens, there will be three polar bodies and one ootid. The ootid grows into a functional **ovum;** the polar bodies disintegrate because they are nonfunctional. The formation of the nonfunctional polar bodies is necessary to enable the egg to get rid of excess chromosomes, and the unequal cytoplasmic division makes possible a large cell with sufficient yolk for the development of the young. Thus the mature ovum has the haploid number of chromosomes the same as the sperm. However, each primary oocyte gives rise to only **one** functional gamete instead of four as in spermatogenesis.

Gametes of various animals. Sperm among animals show a greater diversity of form than do ova. A typical spermatozoan is made up of a head, a middle piece, and an elongated tail for locomotion (Fig. 18-8). The head consists of the nucleus containing the chromosomes for heredity and an **acrosome** believed to contain an enzyme that assists in egg penetration. The total length of the human sperm is 50 to 70 μ. Some toads have sperm that exceed 2 mm. (2,000 μ) in length. Most sperm, however, are microscopic in size.

Ova are oval or spherical in shape and are nonmotile. Mammals' eggs are very small (not more than 0.25 mm.) and contain little yolk because the young receive nourishment from the mother. On the other hand, the eggs of some birds, reptiles, and sharks are very large and yolky, to supply nutritive material for the developing young before hatching. Some eggs (reptiles and birds) also contain a great deal of albumin, which also serves for nourishment. Most eggs are provided with some form of protective coating. This may be in the form of a calcified shell (birds), leathery parchment (reptiles), or albuminous coats (amphibians).

19

Principles
of embryology

Embryology is the study of the progressive growth and differentiation that occurs during the early development of an organism. All but the simplest animals pass through a life cycle that typically takes the following form: fertilized egg, embryo, larva, juvenile, adult, and, to complete the circle, another fertilized egg. Although embryology strictly means "study of embryos," the developmental state within the egg or the body of the mother, embryologists are also very much interested in the events that both precede and follow the embryo stage.

The phenomenon of development is a remarkable, and in many ways awesome, process. How is it possible that a tiny, spherical fertilized human egg, scarcely visible to the naked eye, can unfold into a fully formed, unique person, consisting of thousands of billions of cells, each cell performing a predestined functional or structural role? How is this marvelous unfolding controlled? Obviously all the information needed is contained within the egg, principally in the genes of the egg's nucleus. The fabric of genes is deoxyribonucleic acid (DNA). Thus all development originates from the structure of the nuclear DNA molecules and in the egg cytoplasm surrounding the nucleus. But knowing where the blueprint for development resides is very different from understanding how this control system guides the conversion of a fertilized egg into a fully differentiated animal.

This remains a major—many feel **the** major—unsolved problem of biology. We are now beginning to gain some sense of how development is controlled, and this will be discussed on p. 277. First, however, we need to know what the progressive stages of animal development are. In the discussion that follows, we will deal with the metazoan animals that reproduce sexually. Acellular animals that reproduce principally by asexual means (for example, budding and fission) must develop too, of course, but these processes will be discussed separately in Chapter 25.

The stages of embryogenesis are as follows:

1. **Fertilization.** The activation of the egg by a sperm in biparental reproduction. The union of these male and female gametes forms a **zygote;** this is the starting point for development.

2. **Cleavage and blastulation.** The division of the zygote into smaller and smaller cells (cleavage) to form a hollow ball of tiny cells (blastula).

3. **Gastrulation.** The sorting out of cells of the blastula into layers (ectoderm, mesoderm, endoderm) that become **committed** to the formation of future body organs.

4. **Differentiation.** The formation of body organs and tissues, which take on their specialized functions. The basic body plan of the animal becomes established.

5. **Growth.** Increase in size of the animal by cell

division or cell enlargement. Growth depends on the intake of food to supply material for the synthesis of protoplasm.

Although the embryonic development begins with fertilization, this event must obviously be preceded by the preparatory stages of egg and sperm development. This was dealt with in the preceeding chapter.

FERTILIZATION

We have defined fertilization as the formation of a zygote by the union of a spermatozoan and an ovum (egg). This process accomplishes two things: it triggers the process of development and it provides for the recombination of paternal and maternal inheritance units. Thus it restores the original diploid number of chromosomes characteristic of the species.

For a species to survive, it must ensure that fertilization will occur and that enough progeny will result to continue the race. Most marine fish simply set their eggs and sperm adrift in the ocean and rely on the random swimming movements of sperm to make chance encounters with eggs. Even though an egg is a large target for a sperm, the enormous dispersing effect of the ocean, the short life-span of the gametes (usually just a few minutes for fish gametes), and the limited range of the tiny sperm all conspire against an egg and a sperm coming together. Consequently these animals employ a seemingly wasteful but nonetheless effective saturation technique. Each male releases countless millions of sperm at spawning. The odds against fertilization are further reduced by coordinating the time and place of spawning of both parents. Ensuring that some eggs are fertilized, however, is not enough. The ocean is a perilous environment for a developing fish, and most never make it to maturity. Thus, the females produce huge numbers of eggs. Someone with infinite patience once counted 28,361,000 eggs in a single large, female ling cod. The common gray cod of the North American east coast regularly spawns 4 to 6 million eggs, of which only two or three will, on the average, reach maturity.

Fishes and other vertebrates that provide more protection to their young produce fewer eggs than do the oceanic marine fishes. The chances of the eggs and sperm meeting is also increased by courtship and mating procedures and the simultaneous shedding of the gametes in a nest or closely circumscribed area. Internal fertilization, characteristic of the sharks and rays as well as reptiles, birds, and mammals, avoids dispersion of the gametes and protects them. However, even with internal fertilization, vast numbers of sperm must be released by the male into the female tract. Furthermore, the events of ovulation and insemination must be closely synchronized and the gametes must remain viable for several hours to accomplish fertilization. Sperm may have to travel a considerable distance to reach the egg in the female genital tract, many parts of which are rather hostile to sperm. Experiments with rabbits have shown that of the approximately 10 million sperm released into the female vagina, only about 100 reach the site of fertilization.

Once a sperm contacts an egg, it must penetrate the protective egg membranes (Fig. 19-1). Many kinds of sperm produce special enzymes, called **sperm lysins,** which dissolve the egg membranes to clear an entrance path. The egg, which has been quiescent up to this point, undergoes a series of rapid changes which serve to activate it for the period of rapid divisions that lie ahead. The sperm, now immotile, is drawn into the interior of the egg. A special **fertilization membrane** quickly forms around the egg to protect it from polyspermy (entrance of more than one sperm). Once inside the egg, the tail of the sperm is discarded. The highly condensed sperm nucleus, now called a **male pronucleus,** begins to swell. The male and female pronuclei, each carrying half the diploid number of chromosomes, are drawn together by unknown means and fuse. With this union a **zygote** is formed, ready to embark on the remarkably complex embryonic process leading to an adult animal.

We will conclude this section with a word about **parthenogenesis,** the development of young from unfertilized eggs. **Natural parthenogenesis** (virginal reproduction) has long been known to be common in various invertebrate groups (for example, rotifers, aphids, bees, ants, and wasps). The resulting animal may be composed of either haploid or diploid cells, depending on the species. Recently many experiments with artificial parthenogenesis have been done. A great variety of artificial agents (salt solutions, acids, temperature shocks, electric shocks, ultraviolet light, or pricking the egg surface) can

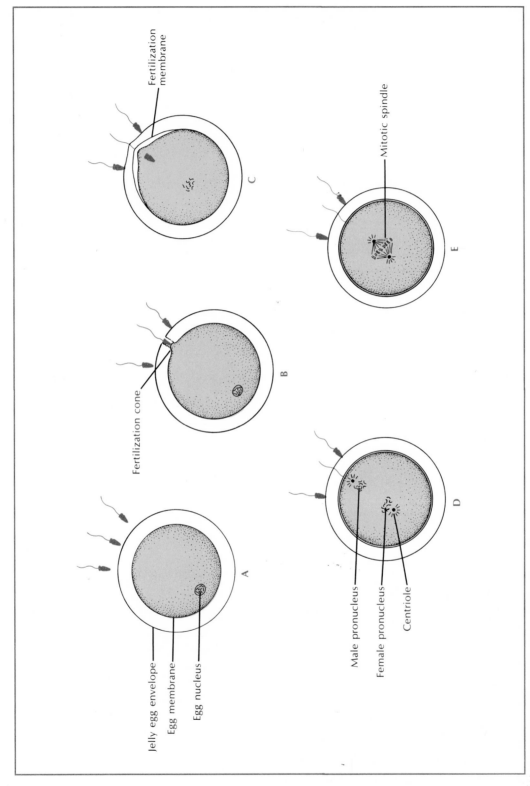

FIG. 19-1. Fertilization of egg. **A,** Many sperm swim to egg. **B,** First sperm to penetrate protective jelly envelope and contact egg membrane causes fertilization cone to rise and engulf sperm head. **C,** Fertilization membrane begins to form at site of penetration and spreads around entire egg, preventing entrance of additional sperm. **D,** Male and female pronuclei approach one another, lose their nuclear membranes, swell, and fuse. **E,** Mitotic spindle forms, signaling creation of a zygote and heralding first cleavage of new embryo.

trigger development of vertebrate embryos. This suggests that the sperm is a triggering agent and that all the equipment needed for embryonic development, the maternal genome, resides in the egg. Of course, in normal bisexual development the sperm also contributes the paternal complement of chromosomes and genome.

CLEAVAGE AND BLASTULATION

The unicellular zygote now begins to divide, first into two cells, those two into four cells, those four into eight. Repeated again and again, these cell divisions soon convert the zygote into a ball of cells. This process, called **cleavage,** occurs by mitosis. But unlike ordinary body-cell mitosis, there is no true growth and no increase in protoplasmic mass. With each subsequent division, the cells are reduced in size by one half. The cleavage process converts a single, very large, unwieldy egg into many small, more maneuverable, ordinary-sized cells.

Cleavage patterns are much affected by the amount of yolk in the egg. In eggs with very little yolk, such as those of mammals, the cytoplasm is uniformly distributed through the egg, and the nucleus is in, or near, the egg center. In such eggs, cleavage is complete (**holoblastic),** and the daughter cells formed at each division are of approximately equal size. The eggs of frogs and other amphibians are richly supplied with yolk that tends to be massed in the so-called vegetal pole of the egg. The opposite, or animal, pole contains the egg cytoplasm and the nucleus. The early cleavage divisions tend to be displaced toward the animal pole because the mass of relatively inactive yolk in the vegetal pole retards the rate of cleavage in that region (Fig. 19-2). Birds and reptiles produce the largest eggs of all animals. Nearly all of this comparatively enormous size is storage food—the part of the egg we commonly call yolk and the investment of albumin, or egg "white." The active cytoplasm, containing the nucleus, is but a tiny disk resting on top of the ball of yolk. Cleavage (called **meroblastic)** is confined to this area of cytoplasm, since the cell divisions cannot possibly cut through the vast bulk of inert yolk.

Cleavage, however modified by the presence of varying amounts of yolk, results in a cluster of cells,

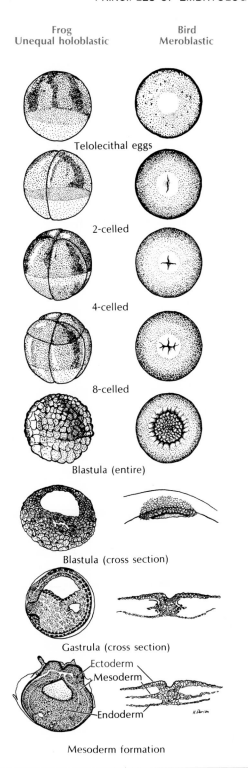

FIG. 19-2. Examples of cleavage in vertebrate eggs.

called a **blastula** (Fig. 19-2). In many animals, such as the amphibian and mammalian embryos, the cells rearrange themselves around a central fluid-filled cavity called the **blastocoel.** We have seen that cleavage has resulted in the proliferation of several thousand maneuverable cells poised for further development. There has been a great increase in total DNA content, since each of the many daughter cell nuclei, by chromosomal replication at mitosis, contains as much DNA as the original zygote nucleus.

GASTRULATION

Gastrulation is a regrouping process in which new and important cell associations are formed. Up to this point the embryo has divided itself up into a multicellular complex; the cytoplasm of these numerous cells is nearly in the same position it was in the original undivided egg. In other words, there has been no significant movement or displacement of the cells from their place of origin. As gastrulation begins the cells become rearranged in an orderly way by morphogenetic movements. In amphibian embryos, as in most forms, cells on the surface begin to sink inward at one point, the **blastopore.** Through the curved groove of the blastopore, surface cells move as a sheet to the interior to form a two-layered embryo. A rodlike **notochord** forms at this time, growing forward to run lengthwise along the dorsal side of the embryo. Continued rearrangements of cells form a third layer; these three layers, called **germ layers,** are the primary structural layers that play crucial roles in the further differentiation of the embryo. The outer layer, or **ectoderm,** will give rise to the nervous system and outer epithelium of the body. The middle layer, or **mesoderm,** will give rise to the circulatory, skeletal, and muscular structures. The inner layer, or **endoderm,** will develop into the digestive tube and its associated structures.

DIFFERENTIATION

With formation of the three primary germ layers, cells continue to regroup and rearrange themselves into primordial cell masses. These masses will continue to differentiate, leading ultimately to the for-

mation of specific organs and tissues. During this process, cells become increasingly committed to specific directions of differentiation. Cells that previously had the potential to develop into a variety of structures, now lose this diverse potential and assume committments to become, for example, kidney cells, intestinal cells, or brain cells.

Derivatives of ectoderm: nervous system and nerve growth. The brain, spinal cord, and nearly all the outer epithelial structures of the body develop from the primitive ectoderm. They are among the earliest organs to appear. Just above the notochord, the **ectoderm** thickens to form a **neural plate** (Fig. 19-3). The edges of this plate rise up, fold, and join together at the top to create an elongated, hollow **neural tube.** The neural tube gives rise to most of the nervous system: anteriorly it enlarges and differentiates into the brain, cranial nerves, and eyes; posteriorly it forms the spinal cord and spinal motor nerves. Sensory nerves arise from special **neural**

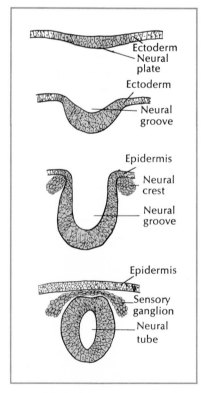

Fig. 19-3. Development of neural tube and neural crest from neural plate ectoderm (cross section).

crest cells pinched off from the neural tube before it closes.

How are the billions of nerve axons in the body formed? What directs their growth? Biologists were intrigued with these questions that seemed to have no easy solutions. Since a single nerve axon may be many feet in length (for example, motor nerves running from the spinal cord to the toes), it seemed impossible that a single cell could spin out so far. It was suggested that nerve fibers grew from a series of preformed protoplasmic bridges along its route. The answer had to await the development of one of the most powerful tools available to biologists, the cell culture technique. In 1907 an embryologist Ross G. Harrison discovered that he could culture living neuroblasts (embryonic nerve cells) for weeks outside the body by placing them in a drop of frog lymph hung from the underside of a cover slip. Watching nerves grow for periods of days, he saw that each nerve fiber was the outgrowth of a single cell. As the fibers extended outward, materials for growth flowed down the hollow axon center to the growing tip, where they are incorporated into new protoplasm. The tissue culture technique is now used extensively by scientists in all fields of active biomedical research, not just by embryologists. The great impact of the technique has been felt only in recent years. Harrison was twice considered for the Nobel Prize (1917 and 1933), but he failed to ever receive the award because, ironically, the tissue culture method was then believed to be "of rather limited value."

The second question—what directs nerve growth—has taken longer to unravel. An idea held well into the 1940's was that nerve growth is a random, diffuse process. It was thought that the nervous system developed as an equipotential network, or blank slate, that later would be shaped by usage into a functional system. The nervous system just seemed too incredibly complex for one to imagine that nerve fibers could find their way selectively to predetermined destinations. Yet it appears that this is exactly what they do! Recent work indicates that each of the billions of nerve cell axons acquires a chemical identification tag that in some way directs it along a correct path. Many years ago Ross Harrison observed that a growing nerve axon terminated in a "growth cone," from which extend numerous

tiny threadlike processes (Fig. 19-4). These are constantly reaching out, testing the environment in all directions, to guide the nerve chemically to its proper destination. This chemical guidepost system, which must, of course, be genetically directed, is just one example of the amazing precision that characterizes the entire process of differentiation.

Derivatives of endoderm: digestive tube and survival of gill arches. In the frog embryo the primitive gut makes its appearance during gastrulation with the formation of an internal cavity, the **archenteron.** From this simple endodermal cavity develops the lining of the digestive tract, lining of the pharynx and lungs, most of the liver and pancreas, the thyroid and parathyroid glands, and the

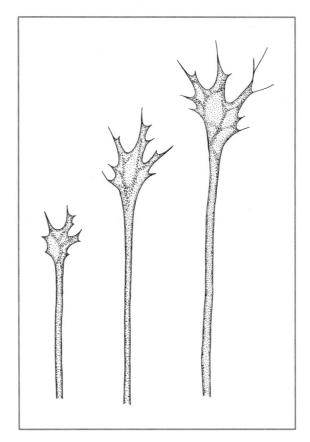

FIG. 19-4. Growth cone at the growing tip of a nerve axon. Materials for growth flow down hollow axon to growth cone from which numerous threadlike pseudopodial processes extend. These appear to serve as a pioneering guidance system for the developing axon.

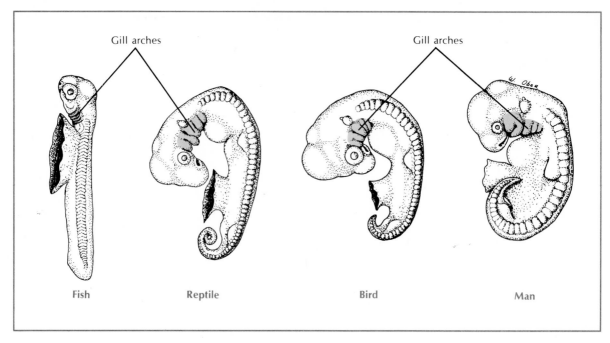

FIG. 19-5. Comparison of gill arches of different vertebrate embryos.

thymus. Among the most intriguing derivatives of the digestive tract are the pharyngeal (gill) arches, which make their appearance in the early embryonic stages of all vertebrates (Fig. 19-5). In fish the embryonic gill arches will later serve as respiratory organs. In the adults of terrestrial vertebrates the gill arches disappear altogether or become modified beyond recognition. Are they useless vestiges that evolution failed to prune off, serving only to remind us of our ancient aquatic heritage? Certainly, the appearance of gill arches in the embryos of terrestrial vertebrates, and other marked similarities between the embryos of fishes and higher vertebrates, can only be explained as an indication of a common vertebrate ancestry. Embryonic development is thus a record, although a considerably modified one, of evolutionary history. Biologists of the last century were so impressed by embryonic similarities between widely separated vertebrate groups that they used embryonic development to reconstruct phylogenies (lines of evolutionary descent) within the animal kingdom. Nevertheless, we can scarcely believe that mammalian embryos retrace

vertebrate evolutionary history for the convenience of biologists. Even though the gill arches serve no respiratory function in either the embryos or adults of terrestrial vertebrates, they remain as necessary primordia for a great variety of other structures. For example, the first arch and its pouch (the space between adjacent arches) form the upper and lower jaws and inner ear of higher vertebrates. The second, third, and fourth arches contribute to the tongue, tonsils, parathyroid gland, and thymus. We can understand then why gill arches and other fishlike structures appear in early mammalian embryos. Their original function has been abandoned, but the structures are retained for new purposes. It is the great conservatism of early embryonic development that has so conveniently provided us with a telescoped evolutionary history.

Derivatives of mesoderm: support, movement, and beating heart. The intermediate germ layer, the mesoderm, forms the vertebrate skeletal, muscular, and circulatory structures and the kidney. As vertebrates have increased in size and complexity, the mesodermally derived supportive, movement, and transport structures make up an ever greater proportion of the body bulk. Although the primitive mesoderm appears after the ectoderm and endoderm, it gives rise to the first functional organ, the embryonic heart. Guided by the underlying endoderm, clusters of precardiac mesodermal cells move ameboid-like into a central position between the underlying primitive gut and the overlying neural tube. Here the heart is established, first as a single, thin tube. Even while the cells group together, the first twitchings are evident. In the chick embryo, a favorite and nearly ideal animal for experimental embryology studies, the primitive heart begins to beat on the second day of the 21-day incubation period—begins beating before any blood vessels have formed and before there is any blood to pump. As the ventricle primordium develops, the spontaneous cellular twitchings become coordinated into a feeble, but rhythmic, beat. Then, as the atrium develops behind the ventricle, followed by the sinus venosus behind the atrium, the heart rate quickens. Each new heart chamber has an intrinsic beat faster than its predecessor. Finally, a specialized area of heart muscle called the **sinoatrial** node develops in the sinus venosus and takes command of the entire heart beat. This becomes the heart's **pacemaker.** As the heart builds up a strong and efficient beat, vascular channels open within the embryo and across the yolk. Within the vessels are the first primitive blood cells suspended in plasma. The early development of the heart and circulation is crucial to continued embryonic development because without a circulation the embryo could not obtain materials for growth. Food is absorbed from the yolk and carried to the embryonic body; oxygen is delivered to all the tissues, and carbon dioxide and other wastes are carried away. The embryo is totally dependent on these and other extraembryonic support systems, and the circulation is the vital link between them.

EMBRYONIC SUPPORT SYSTEMS

Extraembryonic membranes and amniotic egg. As rapidly growing living organisms, embryos have the same basic animal requirements as adults—food, oxygen, and disposal of wastes. For the embryos of marine invertebrates, living a contact existence with their environment, gas exchange is a simple matter of direct diffusion. Food can be acquired as soon as the embryo develops a mouth and begins feeding on plankton. All eggs of aquatic animals are provided with just enough stored yolk to allow growth to this critical stage. Beyond this point, the embryo (now called a free-swimming **larva**) is on its own. Yolk enclosed in a membranous **yolk sac** is a conspicuous feature of all fish embryos (Fig. 19-6). The yolk is gradually used up as the embryo grows; the yolk sac shrinks and finally is enclosed within the body of the embryo. The mass of yolk is an **extraembryonic** structure, since it is not really a part of the embryo proper, and the yolk sac is an **extraembryonic membrane.** Bird and reptile eggs are also provided with large amounts of yolk to support early development. In birds the yolk reaches relatively massive proportions, since it must build a baby bird in a much more advanced stage of growth at hatching than a larval fish.

In abandoning an aquatic life for a land existence the first terrestrial animals had to evolve a sophisticated egg containing a complete set of life-support systems. Thus appeared the **amniotic egg,** equipped to protect and support the growth of embryos on dry land. In addition to the yolk sac containing the nourishing yolk are three other membranous sacs—amnion, chorion, and allantois. All are referred to as extraembryonic membranes because, again, they are accessory structures that develop beyond the embryonic body and are discarded when the embryo hatches.

The **amnion** is a fluid-filled bag that encloses the embryo and provides a private aquarium for development (Fig. 19-7). Floating freely in this aquatic environment, the embryo is fully protected from shocks and adhesions. The evolution of this structure, from which the amniotic egg takes its name, was crucial to the successful habitation of land. The **allantois,** another component in the support system for embryos of land animals, is a bag that grows out of the hindgut of the embryo (Fig. 19-7). It collects the wastes of metabolism (mostly uric acid). At hatching, the young animal breaks its connection with the allantois and leaves it and its refuse behind in the shell. The **chorion** (also called **serosa**) is an outermost extraembryonic membrane that completely encloses the rest of the embryonic sys-

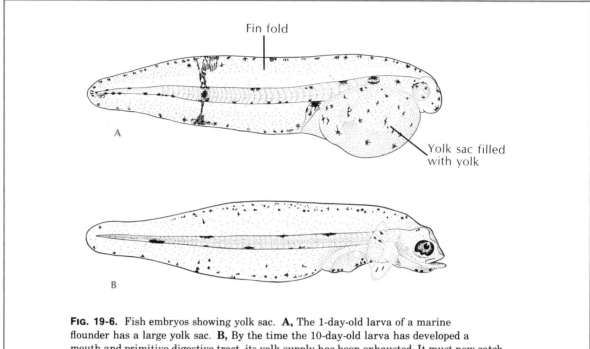

FIG. 19-6. Fish embryos showing yolk sac. **A,** The 1-day-old larva of a marine flounder has a large yolk sac. **B,** By the time the 10-day-old larva has developed a mouth and primitive digestive tract, its yolk supply has been exhausted. It must now catch its own food to survive and continue growing. (From Hickman, C. P., Jr.: The larval development of the sand sole, *Psettichthys melanostictus,* Washington State Fisheries Research Papers **2:**38-47, 1959.)

tem. It lies just beneath the shell (Fig. 19-7). As the embryo grows and its need for oxygen increases, the allantois and chorion fuse together to form a **chorioallantoic membrane.** This double membrane is provided with a rich vascular network, connected to the embryonic circulation. Lying just beneath the porous shell, the vascular chorioallantoic membrane serves as a kind of lung across which oxygen and carbon dioxide can freely exchange. And although nature did not plan for it, the chorioallantoic membrane of the chicken egg has been used extensively by generations of experimental embryologists as a grafting site for small explants of young chick embryos in order to easily observe their development.

The great importance of the amniotic egg to the establishment of a land existence cannot be overemphasized. Amphibians must return to water to lay their eggs. But the reptiles, even before they took to land, developed the amniotic egg with its self-contained aquatic environment enclosed by a tough outer shell. Protected from drying out and provided with yolk for nourishment, such eggs could be laid on dry land, far from water. Reptiles were thus freed from aquatic life and could become the first true terrestrial tetrapods.

Incidentally, the sexual act itself comes from the requirement that the egg be fertilized **before** the egg shell is wrapped around it, if it is to develop. Thus the male must introduce the sperm into the female tract so that the sperm can reach the egg before it passes to that part of the oviduct where the shell is secreted. Hence, as one biologist puts it, it is the egg shell, and not the devil, that deserves the blame for the happy event we know as sex.

Mammalian placenta. The amniotic egg, for all its virtues, has one basic flaw: placed neatly in a nest, it makes fine food for other animals. It was left

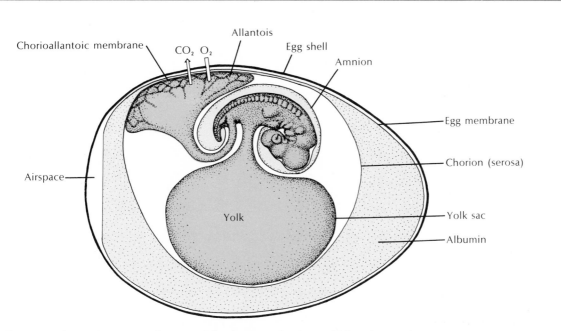

FIG. 19-7. Amniotic egg at early stage of development showing a chick embryo and its extraembryonic membranes. Porous shell allows gaseous exchange of oxygen and carbon dioxide. Circulatory channels from embryo's body to allantois and yolk sac are not shown.

for the mammals to evolve the best solution for early development: allow the embryo to grow within the protective confines of the mother's body. This has resulted in important modifications in mammalian development as compared with other vertebrates. The earliest mammals, descended from early reptiles, were egg layers. Even today the most primitive mammals, the monotremes (for example, duck-billed platypus, spiny anteater), lay large yolky eggs that closely resemble bird eggs. In the marsupials (pouched mammals such as the opossum and kangàroo), the embryos develop for a time within the mother's uterus. But the embryo does not "take root" in the uterine wall, as do the embryos of the more advanced placental mammals, and consequently it receives little nourishment from the mother. The young are therefore born immature and are sheltered and nourished in a pouch of the abdominal wall.

All other mammals, the placentalians, nourish their young in the uterus by means of a remarkable fetal-maternal structure, the **placenta.** The placenta is essentially a device wherein the developing embryo can draw upon the uterine circulation of the mother. Mammalian eggs are virtually yolkless. In the absence of this supply of raw food, the mammalian embryo must derive **everything** it needs for growth from the maternal circulation—food, oxygen, and antibodies. And it must be able to discharge its carbon dioxide and nitrogenous wastes into the maternal circulation. All of these functions take place across the placenta.

Since the mammalian embryo is protected and nourished by the mother's placenta, what becomes of the various embryonic membranes of the amniotic egg whose functions are no longer required? Surprisingly, perhaps, all of these special membranes are still present, although they may be serving a new function. The yolk sac is retained, empty and

275

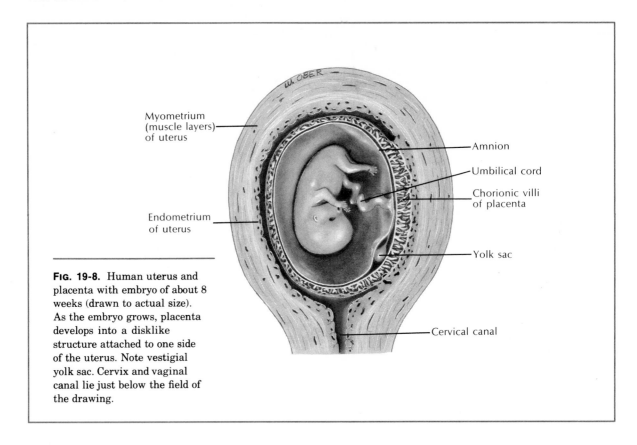

Myometrium
(muscle layers)
of uterus

Amnion

Umbilical cord

Chorionic villi
of placenta

Endometrium
of uterus

Yolk sac

Cervical canal

FIG. 19-8. Human uterus and placenta with embryo of about 8 weeks (drawn to actual size). As the embryo grows, placenta develops into a disklike structure attached to one side of the uterus. Note vestigial yolk sac. Cervix and vaginal canal lie just below the field of the drawing.

purposeless, a vestige of our distant past (Fig. 19-8). Perhaps evolution has not had enough time to discard it. The amnion remains unchanged, a protective water jacket in which the embryo weightlessly floats. The remaining two extraembryonic membranes, the allantois and chorion, have been totally redesigned. The allantois is no longer needed as a urinary bladder. Instead it becomes the stalk, or **umbilical cord,** that links the embryo physically and functionally with the placenta. The chorion, the outermost membrane, forms most of the placenta itself. The placenta is a marvel of biologic engineering. Serving as a provisional lung, intestine, and kidney for the embryo, it performs all these activities without ever allowing the mother's and the embryo's bloods to intermix. The two circulations are physically separated at the placenta by an exceedingly thin membrane only 2 μ thick across which materials are transferred by diffusive interchange. The transfer occurs across thousands of tiny finger-

like projections, called **chorionic villi,** which develop from the original chorion membrane (Fig. 19-8). These projections sink like roots into the uterine endometrium after the embryo implants. As development proceeds and embryonic demands for food and gas exchange increase, the great proliferation of villi in the placenta vastly increases its total surface area. Although the human placenta at term measures only about 7 inches across, its total absorbing surface is about 140 square feet—fifty times the surface area of the skin of the newborn infant.

One of the most intriguing questions the placenta presents is why it is not rejected by the mother's tissues. The placenta is a uniquely successful foreign transplant, or **allograft.** Since the placenta is an embryonic structure, containing both paternal and maternal antigens, we should expect it to be rejected by the uterine tissues, just as a piece of a child's skin will be rejected by the child's mother should a surgeon attempt a grafting transplant. The placenta in some way circumvents the normal rejection phenomenon, a matter of the greatest interest to immunologists seeking ways to successfully transplant tissues and organs.

CONTROL OF DEVELOPMENT

At the beginning of this chapter we asked what it was that enabled the fertilized egg to unfold into billions of differentiated cells, organized into a functional animal. Although this remains one of the great unanswered riddles of biology, much of the mist is being cleared away.

First of all, we know that a fertilized egg contains a full complement of maternal and paternal genes. The DNA contained within the egg nucleus carries all the information needed to form not just another egg cell, or a heart cell, or a liver cell, but a complete animal. What happens to this genetic information as the egg cell divides again and again? Does each daughter cell receive a full set of genes, or are the original genes, some 90,000 in the human, parcelled out among the cleaving cells, according to the ultimate fate of these cells? Simple arithmetic tells us that if genes were parcelled out—the first two daughter cells each receiving half the genes, and these being split again in half by the next division—that all the genes would be exhausted long before a billion-cell embryo could form. But there is direct proof, as well, that every cell in the body contains all the genes necessary to grow a complete organism. Recently, two Oxford scientists J. B. Gurdon and R. Laskey were able to grow a normal adult frog from an unfertilized egg containing the nucleus of a differentiated intestinal cell of frog tadpoles. The technique they used was developed by R. Briggs and T. J. King at Indiana University in 1952. Using minute glass tools, they were able to pluck out the nucleus from an egg and replace it with a nucleus from a fully differentiated cell. These experiments are of great significance because they demonstrate that a cell, or actually the nucleus from a cell, can be forced backward from its specialized state, and once again make available all its genetic information.

The basic problem then is **gene expression.** As cells differentiate, they use only a part of the genetic instructions their nuclei contain. The unneeded genes are in some way switched off, but not destroyed. But what switches off genes (or turns them on) at the right moment in development? What determines that a particular blastomere of say, a 100-cell embryo, will differentiate into muscle or skin or thyroid gland? Presumably a gene, or set of genes, responsible for the development of a thyroid gland will be set in motion by the chemical environment found **only** in the region of the future thyroid gland. But how can such a unique chemical environment be created unless some **previous** genetic action made the thyroid region different from the rest of the body? And even this earlier genic action must have been expressed in a unique chemical environment, or else thyroid glands would grow all over the body.

It is easy to see that this kind of argument quickly takes one back to the fertilized egg itself. If genes are the same in all nuclei of the early embryo, then the only way differences can develop is through some interaction between these nuclei and the surrounding cytoplasm. This is the basis for the great present-day interest in nucleocytoplasmic interactions.

Role of the cytoplasm. The importance of the egg cytoplasm in differentiation was demonstrated many years ago by Hans Spemann, a German embryologist. Spemann put ligatures of human hair around newt eggs (amphibian eggs similar to frog eggs) just as they were about to divide, constricting them until they were almost, but not quite, separated into two halves (Fig. 19-9). The nucleus lay in one half of the partially divided egg; the other side was anucleate, containing only cytoplasm. The egg then completed its first cleavage division on the side containing the nucleus; the anucleate side remained undivided. Eventually, when the nucleated side had divided into about 16 cells, one of the cleavage nuclei would wander across the narrow cytoplasmic bridge to the anucleate side. Immediately this side began to divide. Now with both halves of the embryo containing nuclei, Spemann drew the ligature tight, separating the two halves of the embryo. He then watched their development. Usually two complete embryos resulted. Although the one embryo possessed only 1/16 the original nuclear material, and the other contained 15/16, they both developed

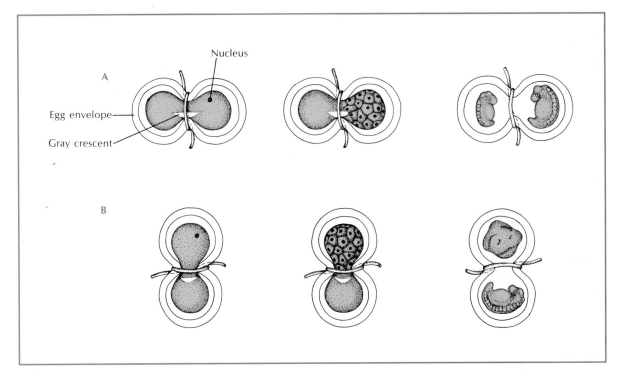

FIG. 19-9. Spemann's delayed nucleation experiments. Two kinds of experiments were performed. **A,** Hair ligature was used to partly divide an uncleaved fertilized newt egg. Both sides contained part of the gray crescent. Nucleated side alone cleaved until a descendent nucleus crossed over the cytoplasmic bridge. Then both sides completed cleavage and formed two complete embryos. **B,** Hair ligature was placed so that the nucleus and gray crescent were completely separated. Side lacking the gray crescent became an unorganized piece of belly tissue; other side developed normally.

normally. The 1/16 embryo was initially smaller, but caught up by about 140 days. This proves that every nucleus of the 16-cell embryo contains a complete set of genes; all are equivalent.

Sometimes, however, Spemann observed that the nucleated half of the embryo developed only into an abnormal ball of "belly" tissue, although the half that received the delayed nucleus developed normally. This was odd. Why should the more generously endowed 15/16 embryo fail to develop and the small 1/16 embryo live? The explanation, Spemann discovered, depended on the position of the **gray crescent,** a crescent-shaped, pigment-free area on the egg surface. In amphibian eggs the gray crescent forms at the moment of fertilization and determines the plane of bilateral symmetry of the future animal. If one half of the constricted embryo lacked any part of the gray crescent, it would not develop.

Obviously, then, there must be cytoplasmic inequalities involved. The gray crescent cytoplasm contains substances that are essential for normal development. Since all the nuclei of the 16-cell embryo are equivalent, each capable of supporting full development, it is clear that the cytoplasmic environment is crucial to nuclear expression. The nuclei are all alike, but the cytoplasm throughout the embryo is not all alike. In some way chemically different regions of the egg, created during the early growth of the egg (oogenesis), are segregated out into specific cells during early cleavage. Thus, although all nuclei have the same information content, cytoplasmic substances surrounding the nucleus determine what part of the genome will be expressed and when.

Differential gene action. Attention is now turned to the nuclear DNA, from which genes are made. It is clear that not all genes are active all the time. In fact, for any cell at any given time, only a very small part of the genetic information present is being used.

Let us briefly summarize what is presently known of the transmission of genetic information. Genetic information is coded in the sequence of nucleotides in DNA molecules. DNA serves as a template for the synthesis of messenger RNA in the nucleus. Messenger RNA then migrates out through nuclear pores into the cytoplasm, where it attaches to a ribosome. Here the messenger RNA serves as a template for the synthesis of specific proteins. In this way cytoplasmic proteins are formed that may be specific for that cell.

Evidence to date suggests that at the beginning of development most, if not all, nuclear genes are inactivated, or "repressed." No cellular proteins are formed during this early embryonic period (early cleavage), that is, no messenger RNA is being produced on the DNA templates. Then, as development proceeds, new cytoplasmic proteins appear, indicating that some genic DNA is producing messenger RNA. Evidently different genes are activated (or "derepressed") in different parts of the young em-

bryo, and this differential gene activity is responsible for embryonic differentiation.

What is the mechanism by which genes are repressed and then derepressed at specific times during development? We presently do not know. However, one appealing possibility is the kind of control described for the bacterium *Escherichia coli* by two French biologists F. Jacob and J. Monod. This model, discussed in more detail on p. 291, proposes that the synthesis of specific cellular proteins is governed by a set of structural genes, called an **operon.** The operon is in turn controlled by other genes (regulator and operator genes) that are responsive to specific chemical materials in the surrounding cytoplasm. We must stress that this model has been developed from research with bacteria only. Nevertheless, it offers a plausible mechanism for explaining the control of gene action in higher animals. Whatever the mechanism is, it seems certain that the kinds of cytoplasm present in different cells determine what genes will come into action. Nucleocytoplasmic interactions form the basis of the organized differentiation of tissues that characterizes animal development.

20

Genetics (heredity)

MEANING OF HEREDITY

Heredity is one of the great stabilizing agencies in nature. Although offspring and parents in a particular generation may look different, there is nonetheless a basic sameness that runs from generation to generation for any species of plant or animal. But sometimes large variations appear; some of these are inherited and others disappear with the generation in which they arise. These inherited characteristics, which may be like or unlike those of the parents, are now known to be due to the segregation of hereditary factors. Those not inherited are caused by environmental conditions.

The inheritance of any characteristic depends on the interaction of many genes. There is a germinal basis for every characteristic that appears in the development of the organism, but environmental factors of food, disease, etc. may greatly affect the physical expression of the characteristics so that their potentialities are never fully realized.

In a unisexual organism only one parent is involved, as in binary fission or sporulation. Unless there is mutation or other genetic variation, the offspring and parent are genetically alike. But in bisexual reproduction the offspring shares genetic potentialities of both parents as well as mutation variations.

CYTOLOGIC BACKGROUND OF HEREDITY

Heredity is a protoplasmic continuity between parents and offspring. In bisexual animals the gametes are responsible for establishing this continuity. Scientific explanation of genetic principles required a study of germ cells and their behavior, which meant working backward from certain visible results of inheritance to the mechanism responsible for such results. The nuclei of sex cells were early suspected of furnishing the real answer to the mechanism. This applied especially to the chromosomes, for they appeared to be the only entities passed on in equal quantities from parents to offspring.

When the discovery of Mendel's laws was announced in 1900, the parallelism that existed between these fundamental laws of inheritance and the cytologic behavior of the chromosomes was worked out. In a series of experiments by Boveri, Sutton, McClung, and Wilson, the mechanism of heredity was definitely assigned to the chromosomes. The next problem was to find out how chromosomes affected the hereditary pattern. The work of T. H. Morgan and his colleagues on the fruit fly *(Drosophila)* led to mapping of the genes on the chromosomes. This work led to the development of cytogenetics.

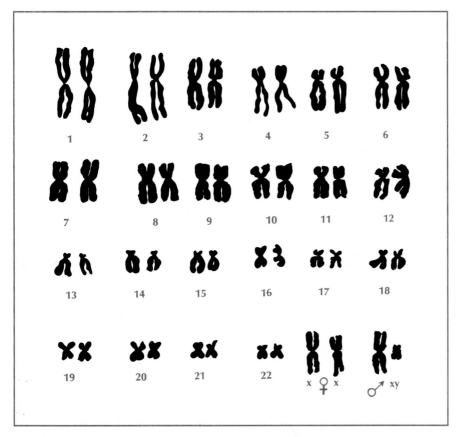

FIG. 20-1. Human chromosomes showing both male and female sex chromosomes. Diploid number is 46. Chromosomes are arranged according to standard pattern (karyotype) of homologous pairs. Chromosomes differ in size, shape, and position of centromeres; arrangement is based on these characteristics, two members of homologous pair (one from each parent) being identical except in case of XY sex chromosomes. Techniques for preparation of human cells for chromosome counting are based upon tissue cultures, biopsy material, and bone marrow studies. Procedures involve tissue exposure to trypsin and special growth media. During part of culture period, colchicine (which arrests cell division at metaphase) is added, and treatment with hypotonic salt solution swells and disperses chromosomes. Squash preparations are usually stained with acetocarmine or Feulgen reagent. (Modified from several sources.)

NATURE OF CHROMOSOMES

Chromosomes for a particular organism have in general a definite size and shape in each stage of their cycle. Each chromosome is made up of a central spiral thread (chromonema) that bears bead-like enlargements (chromomeres). These may be the location of the genes. The various chromomeres appear in regular and constant pattern in a particular chromosome.

The exact structure of chromosomes is not yet fully understood. At present, chromosomes are known to contain deoxyribonucleic acid (DNA), ribonucleic acid (RNA), histones or protamines, and some large complex proteins (residual proteins). The histone or protamine is wound around the DNA double helix (p. 285), and the basic residues of the protein are bound to the phosphoric acid residues of DNA. The nucleoprotein molecules are about 4,000 Å long and 40 Å thick and are lined up end-to-end to form long fibers.

CHROMOSOMES CONSTANT FOR A SPECIES

Each somatic cell in a given organism contains the same number of chromosomes. Somatic chromosomes of diploid organisms are found in pairs, the members of each pair (except sex chromosomes) being alike in size, in position of spindle attachment, and in bearing genes relating to the same hereditary characters. In each homologous pair of chromosomes, one has come from the father and the other from the mother.

Some chromosomes are as small as 0.25 μ in length, and (with the exception of salivary chromosomes that are very long) some are as long as 50 μ. Within the haploid set of chromosomes of most species there are considerable differences in size and shape. In man, most chromosomes are 4 to 6 μ in length.

Mature germ cells have only one half as many chromosomes as somatic cells. Only one member of each homologous pair of chromosomes is found in a mature, functional gamete.

CHROMOSOMES OF MAN

Until 1956 the diploid number in man was supposed to be 48. Newer and improved techniques, however, have resulted in a more accurate count of 46 chromosomes. A standard arrangement of the chromosomes according to size and position of centromeres has been adopted (Fig. 20-1). In a somatic cell, such as a white blood corpuscle, there is a pair of each type of chromosomes, including 2 X chromosomes (female) and a pair of each type, with the exception of the X and Y chromosomes in the male. The Y chromosome is one of the smallest chromosomes, although there are some autosomes almost as small.

SIGNIFICANCE OF REDUCTION DIVISION

Genes are in paired, homologous chromosomes. Therefore, when these chromosomes separate at the reduction division of meiosis, the homologous genes must also separate, one gene going to each of the two germ cells produced. Thus at the end of the maturation process each mature gamete (egg or sperm) contains one gene of every pair or a single set of every kind of gene (haploid number) instead of two genes for each character as in somatic cells (diploid number). A particular germ cell does not necessarily contain all the chromosomes from one or the other parent of the individual producing that germ cell. As we discovered earlier (p. 263), it is a matter of chance in the reduction division whether the paternal chromosome of a homologous pair goes to one daughter cell or the other, and the same is true of the maternal chromosome.

The real significance of the reduction division for explaining the principles of heredity lies mainly in the segregation of the chromosomes and consequently the genes that the chromosomes carry. Of course, when the zygote is formed at fertilization, the homologous pairs of chromosomes (and genes) will be restored.

SALIVARY GLAND CHROMOSOMES

Details about the structure of chromosomes have been difficult to obtain because of their small size in most animals. About 1934 Professor Painter of the University of Texas and some German investigators independently discovered in the salivary glands of the larvae of *Drosophila* and other flies chromosomes many times larger than their ordinary somatic or germinal chromosomes. This discovery marked a new era in the development of cytogenetics, for it afforded much new information about the structure and nature of chromosomes.

The salivary glands of the larval flies are a pair of club-shaped bodies attached to the pharynx. Each consists of only about 100 unusually large cells. Salivary tissue grows by an increase in cell size and not by an increase in cell number. When the cell membrane is disintegrated, the chromosomes are scattered out and can be easily studied. The giant chromosomes are elongated, ribbonlike bodies about one hundred to two hundred times longer than the ordinary chromosome (Fig. 20-2). In some flies they lie separated from each other; in *Drosophila* they are attached to a dark mass called the **chromocenter.**

What are these chromosomes and what do they show? The chromosomes are somatic prophase chromosomes with the homologous chromosomes closely paired throughout their length. One of their most

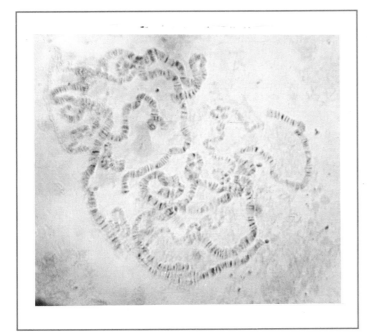

FIG. 20-2. Chromosomes from salivary gland of larval fruit fly *Drosophila*. These are among largest chromosomes found in animal cells. Bands of nucleoproteins may be loci of genes. Such chromosomes are sometimes called polytene because they appear to be made up of many chromonemata. These chromosomes are not confined to salivary glands but are also known to occur in other organs, such as gut and malpighian tubules of most dipteran insects. Technique for their study is simply to crush salivary glands between cover glass and slide in drop of acetocarmine so that chromosomes are set free from nuclei and are spread out as shown in photograph.

FIG. 20-3. Puffing in one of the bands of a salivary gland chromosome of a midge larva *(Chironomus)*. Swelling, or puff, indicates activity in a region where protein and RNA (and perhaps some DNA) are being produced, and may include single bands or adjacent ones. Puffs always include same bands that occur in a definite sequence during development of larva. (From several sources.)

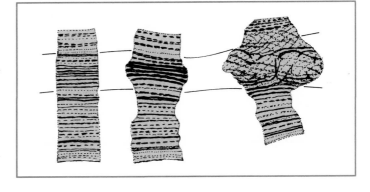

striking characteristics is the transverse bands with which they are made. Another feature is the number of chromonemata (central threads) they possess. In the ordinary somatic chromosome there may be only one or two of these gene strings, but in the salivary gland chromosomes there may be between 512 and 1,024 *(Drosophila)*. This indicates that the chromonemata may have divided many times without being accompanied by the division of the whole chromosome; hence, they are often called **polytene** chromosomes. A polytene chromosome is a typical mitotic chromosome that has uncoiled and undergone many repeated duplications

that have remained together in the same nucleus.

The transverse bands are made up of chromatic granules, the chromomeres. These bands result from the lateral apposition of the chromomeres on the adjacent fibrils or chromonemata. More than 6,000 of these bands have been found on the three large chromosomes of *Drosophila*. The bands contain much DNA and each may be considered the equivalent of the conceptual gene. In the regions between the bands there is little DNA. Another aspect of giant chromosomes is the so-called "puffs," which are local and reversible enlargements in the bands (Fig. 20-3). Each "puff" may be due to

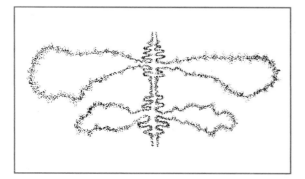

FIG. 20-4. Small portion of a lampbrush chromosome showing two pairs of loops. These chromosomes are found in germinal vesicles (nuclei) of oocytes during diplotene phase of first meiotic division and may indicate synthesis of yolk. They appear to be largest in certain salamanders. Loops represent lateral extensions of chromatids, or half chromosomes. RNA is being transcribed along loop and, with protein formed there, gives a fuzzy appearance to loop. Central axis with closely coiled chromomeres is made up of DNA. Exact relation of loop to gene is not yet known. (Modified from J. G. Gall, 1956.)

the unfolding or uncoiling of the chromosomes in a band. The puffing may be large (Balbiani's rings) or small. In addition to the DNA of the band, the puff contains a great deal of RNA. The size of the puff is an indication of gene activity. Evidence seems to indicate that messenger RNA is produced at the puff, where it makes a complementary copy of a DNA strand. The RNA messenger is then carried to the ribosome, where it serves for the synthesis of proteins. (See discussion of genetic code, pp. 286 to 289.)

The even larger lampbrush chromosomes found in the oocytes of many vertebrates and some invertebrates are characterized by loops extending laterally that give them the appearance of a brush (Fig. 20-4). These chromosomes appear to be composed of two chromatids that form loops (gene loci) when they are active but are coiled up within a chromomere when at rest.

GENE THEORY

The term **gene** was given by W. Johannsen in 1909 to the hereditary factors of Mendel (1865). Genes

are the chemical entities responsible for the hereditary pattern of an organism. No one perhaps has seen a definite gene and we still have much to learn about the nature of genes. Yet, by long, patient genetic experiments, their relative positions (loci) on the chromosomes have been mapped in many cases. Evidence indicates that they are arranged in linear order on the chromosome-like beads on a string. In some cases, as in salivary gland chromosomes, genes are assigned to definite bands. Since chromosomes are few and genes are many in number, each chromosome must contain many genes (linkage group). An organism may have 10,000 (*Drosophila*) to 90,000 (man) genes.

Each zygote of sexual reproduction has two sets (diploid) of homologous chromosomes, one set from each parent; in other words, there are two of each kind of chromosomes (except in sex determination), one from the father and the other from the mother. We saw earlier that when the gametes are formed in meiosis, disjunction of the homologous chromosomes occurs so that each germ cell receives one or the other of the pair at random. Since the genes are a part of the chromosomes, their distribution will parallel that of the chromosomes. All the genes of a particular homologous chromosome, or **linkage group,** will go at meiosis into one gamete and all the genes of the other homologous mate will go into another gamete.

The two members of a homologous pair of chromosomes often exchange corresponding segments or blocks of genes. This is called **crossing-over.** There is visible evidence of this physical exchange, for at the beginning of the first meiotic division the two members of each pair of chromosomes come into side-by-side contact (synapsis) and become twisted. When they separate they have exchanged parts. Naturally, the genes on the traded portions will be exchanged also. The new combinations so formed are as stable as the original ones. Linkage groups are also altered by such rearrangements as the linear reversal of genes sequence in the **reversal of a chromosome segment,** by the shifting of a chromosome segment to another part of the same chromosome (**translocation**), by **polyploid changes** in chromosome number, etc.

Although genes can reproduce themselves exactly for many generations, they do occasionally undergo abrupt changes called **mutations.** A mutation is a sudden chemical rearrangement of a gene. The changed structure and action of a gene may change the inherited characteristic controlled by that gene. The mutant gene now faithfully reproduces itself in the new mutant form.

There is evidence that some genetic variability is the result of self-duplicating, hereditary units in the cytoplasm. Such units are called **plasmagenes** and are apparently transmitted only by the cytoplasm. Their exact role is still obscure.

We now know that genes, like chromosomes, are made up chiefly of nucleoproteins that consist of nucleic acids and proteins (histones and protamines). Life as we know it began with the first formation of nucleoproteins because they have the properties of self-duplication and specificity. Nucleoproteins are the only molecules with this power so far as we know.

The nucleic acids are each chemically made up of a purine or pyrimidine base, a sugar, and phosphoric acid. On the basis of the kind of sugar (deoxyribose or ribose), the nucleic acids are divided into two main groups: **deoxyribonucleic acid (DNA)** and **ribonucleic acid (RNA).** DNA occurs only in the nucleus, where it is the major structural component of genes; RNA is found throughout the cell, being especially abundant in nucleoli and in the cytoplasm. The nucleic acids may be broken down chemically or enzymatically into **nucleotides.** Thus a nucleic acid molecule is made up of many nucleotides joined to form long chains. Each nucleotide consists of phosphoric acid, either deoxyribose or ribose sugar, and a pyrimidine or purine base (Fig. 2-7, p. 22). The purine units are adenine and guanine; pyrimidines are cytosine, thymine, and uracil. Five kinds of nucleotides are recognized on the basis of these purines and pyrimidines: (1) adenine-sugar-phosphate, (2) guanine-sugar-phosphate, (3) cytosine-sugar-phosphate, (4) thymine-sugar-phosphate, and (5) uracil-sugar-phosphate. The DNA molecule has the first four of these nucleotides (Fig. 20-5); the RNA has the first three and the last one. Although the phosphate-sugar part of the long chain of nucleotides is regular, the base attached to the sugar is not always the same. The order of these bases is irregular and varies from one section to another of the nucleic acid molecule. Depending on the propor-

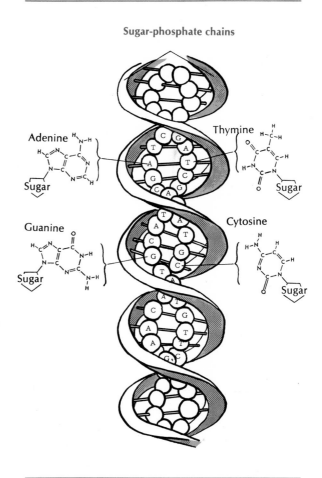

Sugar-phosphate chains

Adenine

Thymine

Guanine

Cytosine

Sugar

FIG. 20-5. Structure of DNA (deoxyribonucleic acid) molecule. Evidence indicates that molecule is formed by two interlocking helixes or chains of nucleotides. Each nucleotide is made up of deoxyribose sugar attached to phosphoric acid molecule on one side and to nitrogenous base on another side. Bases are of four kinds: adenine (**A**), cytosine (**C**), guanine (**G**), and thymine (**T**). Adenine is always paired with thymine and guanine with cytosine because in every pair only a large base and a small one will fill the space available between parallel sugar-phosphate chains. Apparently nucleotides occur in great variety of sequences so that long double chains contain many combinations that confer specificity on given DNA molecule, just as many words are formed out of a few letters of alphabet. Since each spiral thread is a complement of the other, this structure of DNA molecule provides basis for duplication, for either spiral chain can serve as a template on which missing part of the other spiral can reconstruct itself. Various sections (genes?) of long molecule may also serve as pattern codes for enzymes, proteins, and other kinds of molecules.

tion and sequence of the nucleotides, there is an almost unlimited variety of nucleic acids.

THE GENETIC CODE
THE WATSON-CRICK MODEL
OF STRUCTURE OF DNA MOLECULE

In 1953 J. D. Watson and F. H. Crick, with the aid of x-ray diffraction studies of M. H. F. Wilkins, proposed a model of the structure of the DNA molecule that has been widely accepted (Fig. 20-5). The model could not have been made earlier, for its pattern depended on experimental evidence of many investigators over several years. Their model had to suggest plausible answers to such problems as (1) how specific directions are transmitted from one generation to another, (2) how DNA could control protein synthesis, and (3) how the DNA molecule could duplicate itself. Classical genetics and cytology had shown how a cell divides to form two cells and how each cell receives a set of chromosomes with their genes identical in structure to the preexisting set. But nothing was known about how a chemical substance could carry out the specifications required by the genetic substance of a gene. The elegance of the Watson-Crick hypothesis lies in the perfect manner it fits the data and in the way it can be tested.

Wilkins succeeded in getting very sharp x-ray diffraction patterns that revealed three major periodic spacings in crystalline DNA. These periodicities of 3.4 Å, 20 Å, and 34 Å, were interpreted by Watson and Crick as the space distance between successive nucleotides in the DNA chains, the width of the chain, and the distance between successive turns of the helix, respectively. The x-ray diffraction photograph, with certain limitations, also gave indications of the spatial arrangement of some of the atoms within the large molecule. These investigators came up with the idea that the molecule is bipartite, with an overall helical configuration. Accordingly, their model showed that the DNA molecule consists of two complementary polynucleotide chains helically wound around a central axis with the sugar-phosphate backbones of each chain or strand on the outside of the molecular helix and the purines and pyrimidines on the inside of the helix. The two strands are held together by hydrogen bonds between specific pairs of purines and pyrimidines (Fig. 20-5). The two strands are complementary in that the sequence of bases along one strand specifies the sequence of bases along the other strand. It will be noted that the sequence of bases travels in opposite directions on the two strands or, in other words, are of opposite polarity. If the sequence of bases on one strand is known, it is possible to identify the base sequence on the other chain. Each separate strand could then serve as a template for the production of its complement. According to the base-pairing rule, adenine at one point on a strand must have thymine at the corresponding point on the other strand, and likewise guanine with cytosine. There are no restrictions on the sequence of nucleotides in the DNA double helix, for the base sequence along one strand specifies the base sequence along the other strand. Because each strand has complete genetic information, the presence of a complementary strand can serve as a template in the repair of a damaged one. The purine bases adenine and guanine (A and G) are large, whereas the pyrimidine bases thymine and cytosine (T and C) are small. To fit into the structure of the DNA molecule, each pair must consist of one large and one small base; thus the small bases T and C, T and T, or C and C could not bond in pairs as shown because they would not meet in the middle. The big bonds (A and G) could not bond in any combination because there is no room for them. Neither can A and C nor G and T (although in these pairs one is large and one is small) form bond mates because their hydrogen bonds would not pair off properly. Thus there is only one correct way for the bases to bond: A with T and G with C (Fig. 20-5).

Although all DNA molecules have the same general pattern, each one is unique because of the varying sequence of bases attached to the backbone. This sequence spells out the genetic instructions that determine each inherited characteristic. This genetic message is written in a language of only four symbols (adenine, guanine, thymine, and cytosine), and since any sequence of nucleotides is possible, almost limitless variations are possible in the genetic instructions. The nucleotides within the molecule are so arranged that the sugar of 1 molecule is always attached to the phosphate group of the next nucleotide in the sequence. The sugar of DNA is deoxyribose, a pentose sugar with 5 carbons, and to it is attached either a purine or pyrimidine base in each nucleotide.

No one knows how many nucleotides are possible in a DNA molecule; there may be hundreds of thousands, but the number naturally varies. DNA molecules are of enormous length. It is estimated that the total DNA in the 46 chromosomes of a cell (man) contains something like a billion base pairs with a length of 3 feet, and all of it has to unwind during each act of replication.

Building new DNA molecules. Everytime the cell divides, the structure of DNA must be carefully copied in the daughter cells. The double chains, or strands, of the DNA helix explain how this may be done. The two strands of the helix could unwind, and each separate strand could then serve as a template for the production of its complement or to guide the synthesis of a new companion chain. It is thought that four kinds of building blocks are used in building the new strand. They are found in the nuclear environment surrounding the new strand. Each of the building blocks consists of one of the four DNA bases, one sugar unit, and three phosphate groups. Only one of the phosphates is used in making the backbone of the new strand; the other two provide the energy for the synthesis. A large protein catalyst, **DNA polymerase,** acts to accelerate the reaction. This catalyst, or enzyme, does not determine which of the four bases (adenine, thymine, guanine, or cytosine) will be added to the new chain, but it does hold the reacting molecules steady. The selection of the base is made by the complementary base on the old, or template, chain. The copying of the sequence of the four bases is always done by the same catalyst. The sugar-phosphate linkages between successive nucleotides would result in a double-stranded DNA molecule in which the new strand has been specified by the old template strand. In this way each separate strand serves as a template for the formation of its complement, and the two daughter molecules would be identical to the parental molecule of the cell before it divides. Every daughter molecule is half old and half new; only one strand has been synthesized, whereas the other has been conserved.

Mutations from mistakes in the DNA molecule. All of the DNA bases occasionally occur in tautomeric forms in which certain bonds within the molecules are rearranged. Watson and Crick suggested that these rare tautomeric alternatives for each of the bases provide a mechanism for mutation during the replication of DNA. In some cases a tautomeric form of adenine can pair with cytosine instead of with thymine, or a tautomeric form of thymine can pair better with guanine than with adenine. To do this, the hydrogen atoms may assume new positions and their shift may be responsible for abnormal pairing. This incorrect choice of bases made during the copying process may cause a change in the base sequence. Such a mistake is preserved in the DNA each time it is copied and may be reflected in a physical alteration in the organism carrying the changed DNA. Although most mutations are disadvantageous, some may be improvements. In this case favorable mutations will be preserved by natural selection and will spread through a population. This may be one of the principal mechanisms of evolution. A genic mutation, therefore, is an alteration in the DNA that changes the information content of the molecules so that new alleles are produced. Nucleotides may be deleted from or added to the sequence, or a nucleotide may be exchanged for a different one. A codon (see below), for instance, having normally the base composition of CGC, might be changed to CAC. Such a codon would code for a different amino acid, and such a type of mutation is called a base substitution. Base substitutions may cause nonsense triplets that may reduce or destroy the activity of the enzyme coded by the gene. Base substitutions that result in the replacement of a single amino acid may reduce the activity of an enzyme, but rarely cause its complete inactivity.

A variety of types of radiations—x-rays, gamma rays, ultraviolet rays, cosmic rays, or by-products of atomic rays—are mutagenic and can lead to changes in base pairs. Many chemical substances are also known to produce mutagenic changes in DNA, such as mustard gas, peroxides, nitrites, certain purines, and pyrimidines. In some cases an insertion may be made at one point in a genetic message, turning it into nonsense, but when a deletion is made at a later time, the message comes back into step again and once more has meaning.

DNA coding by base sequence. The Watson-Crick model suggested how new DNA may be made from old. The coding problem indicated that there must be some relation between the sequence of the four bases of DNA and the sequence of the twenty amino acids of proteins. The coding hypothesis had to account for the way these four bases (adenine, thymine, cytosine, guanine) must arrange themselves so that each permutation is the code for an amino acid. In the coding procedure it is obvious that there cannot be a 1:1 correlation between four bases and twenty amino acids. If the coding unit (often called a word, or **codon**) consists of two bases, only sixteen words can be formed, which cannot account for twenty amino acids. Therefore, the protein code must consist of at least three bases or three letters because sixty-four possible words can be formed by four bases when taken as triplets. DNA must then be considered a language written in a four-letter alphabet. The particular composition or sequence of amino acids in a given protein are thus specified by the particular sequence of nucleotide pairs in a specific DNA molecule.

Messenger RNA. Information is coded in DNA of the nucleus, whereas protein synthesis occurs in the cytoplasm. An intermediary of some kind between the two regions is necessary. This intermediary appears to be a special kind of RNA called **messenger RNA.** (It will be recalled that RNA differs from DNA in having thymine [T] replaced by uracil [U] and having a sugar residue of ribose instead of deoxyribose.) Messenger RNA is thought to be transcribed directly from DNA in the nucleus, each of the many messengers RNA being determined by a gene or a particular segment of DNA. In this process of making a complementary copy of one strand or gene of DNA in the formation of messenger RNA, an enzyme, RNA-DNA polymerase, is needed. The messenger RNA contains a sequence of bases that complements the bases in one of the two DNA strands just as the DNA strands complement each other. Thus, A in the coding DNA strand is replaced by U in messenger RNA; C is replaced by G; G is replaced by C; and T is replaced by A. It appears that only one of the two chains is used as the template for RNA synthesis, although either one could be so used. The reason why only one strand of the double-stranded DNA is a "coding strand" is that messenger RNA otherwise would always be formed in complementary pairs and enzymes also would be synthesized in complementary pairs. In other words, two different enzymes would be produced for every DNA coding sequence instead of one. This would certainly lead to metabolic chaos.

The messenger RNA when formed is separated from the DNA and migrates through nuclear pores (Fig. 20-6) into the cytoplasm of the cell, where it becomes attached to a granular **ribosome.** Ribosomes are submicroscopic structures comprised of protein and a second kind of nonspecific RNA. Ribosomes carry no information but just serve as attachment points for messenger RNA. Most messenger RNA molecules are so long that they bind a cluster of ribosomes together. On this ribosome cluster, or polysome (Fig. 20-7), the messenger RNA molecule serves as a template against which amino acids are lined up in a sequence according to the coded instructions in messenger RNA. (These amino acids are either obtained in the food supply or synthesized by the organism.)

FIG. 20-6. Electron micrograph of smooth muscle cell of frog showing part of nucleus with nuclear pores and nuclear envelope. Two nuclear pores are seen. (×120,000.) (Courtesy G. E. Palade and National Academy of Sciences, Washington, D. C.)

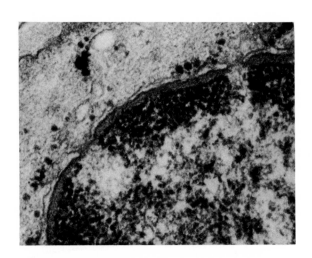

Role of transfer RNA. While this is taking place, various amino acids are activated by the energy of ATP and enzymatically attached to a third type of RNA called **transfer RNA,** a small molecule of about 80 ribonucleotides and folded back on itself to form a double helix. This RNA molecule, like messenger RNA and ribosome RNA, is also synthesized on a DNA template. In a specific region of each transfer RNA, there is a coding sequence of three bases that have a complementary sequence on messenger RNA. Thus the triplet code UUU (three uracil bases) on the messenger would furnish the complementary site for the coding sequence of AAA (three adenine bases) on a transfer RNA. A different transfer RNA molecule corresponds to each triplet code on messenger RNA. Each transfer RNA is specific for a particular amino acid (Table 20-1). The coding sequence of three unpaired nucleotides is found in the region where the chain of the transfer RNA turns back on itself and, at a different place, a recognition site where the amino acid is attached. There must be 20 recognition sites, one for each amino acid. At the end to which the amino acid is attached, the base triplet ACC is always the same on all kinds of transfer RNA. The sequence of three nucleotides at the other end of transfer RNA represents a code that determines where transfer RNA fits into the template. More than one kind of transfer RNA is found for certain amino acids. By stepwise addition, the amino acids are guided by the coding sequence on transfer RNA, to which they are attached, and arranged in the correct order along messenger RNA to form a protein molecule (Fig. 20-7). Each gene codes for about 500 amino acids, which is the average of a polypeptide chain, and since the code triplets, or codons, are in groups of three nucleotides, there would be 1,500 nucleotide pairs in a single gene. These figures naturally will vary with the protein or enzyme being coded.

The general scheme of the code may be abbreviated thus:

TABLE 20-1. Proposed codons (code triplets) between messenger RNA and specific amino acids

Codons	Amino acid
GCU, GCC, GCA, GCG	Alanine
CGU, CGC, CGA, CGG, AGA	Arginine
AAU, AAC	Asparagine
GAU, GAC	Aspartic acid
UGU, UGC	Cysteine
GAA, CAG	Glutamic acid
CAA, CAG	Glutamine
GGU, GGC, GGA, GGG	Glycine
CAU, CAC	Histidine
AUU, AUC, AUA	Isoleucine
CUU, CUC, CUA, CUG, UUA, UUG	Leucine
AAA, AAG	Lysine
AUG	Methionine
UUU, UUC	Phenylalanine
CCU, CCC, CCA, CCG	Proline
AGU, AGC, UCU, UCC, UCA, UCG	Serine
ACU, ACC, ACA, ACG	Threonine
UGG	Tryptophan
UAU, UAC	Tyrosine
GUU, GUC, GUA, GUG	Valine

SYNTHESIS OF ARTIFICIAL GENE

The synthesis of an artificial gene by H. G. Khorana and his colleagues was accomplished in 1970. Dr. Khorana made use of a short molecule of DNA that codes for the production of a certain transfer RNA in yeast cells. By knowing the order of the nucleotides in the transfer RNA (previously determined by R. Holley), he linked the bases (adenine, thymine, guanine, and cytosine) to sugar and phosphoric acid in the same order in which they occur in the yeast gene that made the molecule of transfer RNA. By knowing the sequence of the 77 nucleotides that specifies the synthesis in yeast of the molecule alenine transfer RNA; he was able to link together commercial nucleotides to form short single strands of DNA. Complementary, or opposite, strands were

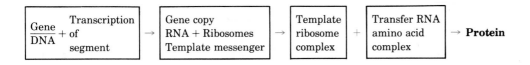

| $\dfrac{\text{Gene}}{\text{DNA}}$ + Transcription of segment | → | Gene copy RNA + Ribosomes Template messenger | → | Template ribosome complex | + | Transfer RNA amino acid complex | → **Protein** |

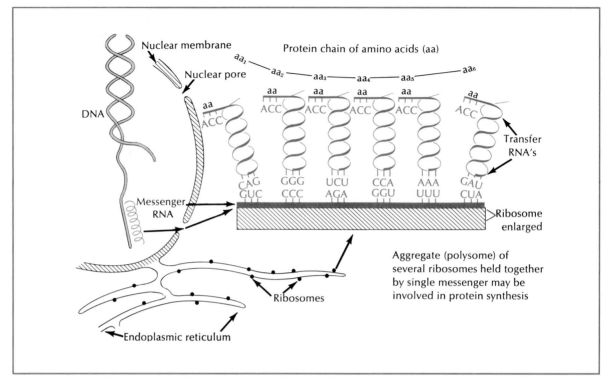

FIG. 20-7. Genetic code. Illustrates how genetic information may be passed from DNA molecule in nucleus by means of messenger RNA to ribosomes of endoplasmic reticulum where amino acids are arranged in proper sequence to form specific enzyme. Messenger RNA is thought to be synthesized from particular segment (gene?) of one of DNA chains serving as template. Messenger RNA (shown as red coil), with its specific code from DNA, passes through nuclear pore to endoplasmic reticulum where it becomes attached to ribosome or ribosomes. Amino acids, obtained preformed in food supply or else synthesized by organism, are probably activated by specific enzyme for each of twenty or so amino acids. High-energy ATP may play a part in activation. Activated amino acid is then attached to specific transfer RNA molecule at recognition site on folded double helix of molecule. Transfer RNA bearing specific amino acid is now lined up and brought into correct position by base pairing of triplet-code between transfer RNA and messenger RNA (C always pairs with G and A with U). Diagram shows some transfer RNA molecules, with their amino acids being transferred to form peptide or protein chain (aa$_1$, aa$_2$, etc.) as they are "read off" on template of messenger RNA. Each specific transfer RNA bears base triplet ACC at amino acid acceptor end.

also formed by base pairing, using the enzyme DNA ligase for tying the double strands together.

PRESENT CONCEPT OF THE GENE

The classic gene is no longer regarded as the indivisible minimal unit of heredity. It consists instead of smaller functional subunits. For example, alleles (all of the genes that may be situated at a particular chromosome locus) may differ only slightly in phenotypic expression. This may be due to slight chemical differences brought about by small changes in the sequences of the nucleotides within a particular region of a DNA molecule. New alleles for a particular locus are produced by gene mutations. Thus, if the gene represents some section of a DNA molecule, it is conceivable that a mutation may affect only a small part of the nucleotides within a gene. A **muton,** then, is the smallest segment of a gene that can produce an altered trait by mutation. The

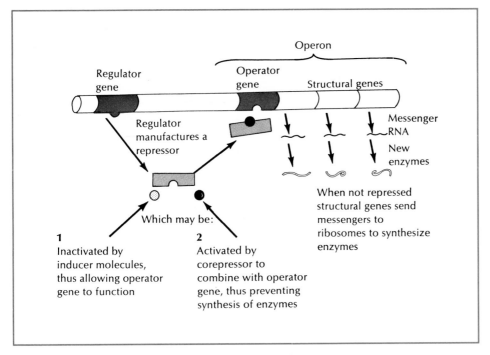

FIG. 20-8. Operon hypothesis. Regulator gene acts by way of a repressor on the operator gene. These regulator genes control rate of information sent to operator gene, either to induce more of a particular enzyme or to cut off (repress) additional amounts of unneeded enzymes. Repressor is a cytoplasmic factor, probably a macromolecule. When operator is "turned on," entire operon is active in synthesis of enzymes; when it is "off," operon is inactive. Inducer molecules modify regulator substance to prevent it from switching off operator; repressor molecules react with regulator substance and cause it to switch off operator (and genes that it controls). (From several authors.)

smallest segment within a gene that is interchangeable but not divided by genetic recombination is called a **recon** and corresponds to the distance between adjacent nucleotides in the DNA chain. A recon may contain no more than two nucleotide pairs. A larger subunit of the gene is the **cistron,** which refers to the smallest number of mutons or recons of a gene that must remain together on one chromosome to perform a biochemical or genetic function. If a series of consecutive mutons, for instance, are necessary for the synthesis of a certain protein and are located on the same chromosome (**cis** position), the whole gene functions normally. But if one part of the mutons are on one chromosome and the other part on the other chromosome (**trans** position), the protein synthesis will occur normally only if the two groups of mutons complement each other. Only a few cistrons are found in a gene and each must be made up of many nucleotides. It has been suggested that recons and mutons may control the synthesis of individual amino acids, whereas cistrons control the formation of polypeptide (chains of amino acids).

OPERON CONCEPT

The genetic code as given in the foregoing description simply explains how the code carried on the DNA molecules of the nucleus is transcribed into a definite protein or enzyme synthesized in the cytoplasm. It does not explain how genes are turned off and on as their products are needed by the cell. It does not explain why certain enzymes are not formed when they are not needed. Obviously, enzyme-forming systems require control because they produce different amounts of the same enzyme at different times. It is also apparent that this control must have two components: (1) the genetic apparatus of the code and ribosome transcription and (2) factors from the environment such as the amount of products accumulated. Thus there must be mechanisms in the cell for repressing the synthesis of enzymes when they are not needed and for inducing them when they are needed.

In 1960 the two French scientists, F. Jacob and J. Monod, proposed the **operon hypothesis,** or model, for explaining how repressions and inductions of protein synthesis might occur (Fig. 20-8). This important hypothesis is based entirely on work with bacteria; it remains to be seen whether their hypothesis also applies to higher living forms. The gist of their hypothesis, for which they were awarded the Nobel Prize in 1965, may be stated in the following way:

1. There are two types of genes, **structural genes** and **regulator genes.** The structural genes contain the coded formulas for the synthesis of the primary structure of a protein, or enzyme, that are useful in cellular metabolism. The regulatory genes control the function of the structural genes.

2. There are two kinds of regulator genes. One is the **operator** gene that determines whether or not the formula, or code, in structural genes adjacent to it are to be transcribed into an enzyme. The **regulator** gene codes for the structure of a cytoplasmic factor (the **repressor**), which turns the operator on and off.

3. The **operon** is that portion of a chromosome that regulates all the steps in the synthesis of an enzyme, or protein. Some operons may contain only one gene; others contain several. An operon consists of an operator gene and the segment of DNA it controls. The operator may control either a single structural gene to which it is adjacent or several structural genes of related function. Thus all the nine enzymes in the histidine pathway are controlled by a single operator.

4. The regulator genes produce a substance called the **repressor** that blocks the operator genes and thus prevents the structural gene from functioning

normally. Repression occurs when the repressor substance combines with the operator gene and prevents the formation of messenger RNA along the segment of DNA controlled by the operator gene. The operator is that part of the operon that is the receptor site for the repressor.

5. If the repressor substance reacts with an appropriate cytoplasmic substance, it **derepresses** the repressor substance and permits the operator gene to act. This is called the **inducible system.** In other words, it renders the repressor incapable of turning the operator off.

6. In this way the two antagonistic systems, the inducible system and the repressor system, maintain a refinement in the amount and kind of enzymes necessary for the steady states of the cell. For instance, if there is a high concentration of a particular enzyme in the cell, this high concentration can act as a "feedback" through the repressor system to block the action of the operator gene so that the structural gene can no longer produce the enzymes. The repressor may be changed to an inactive form by a lower-than-normal concentration of the enzyme, or by a specific substance synthesized in the cytoplasm or from the environment (ions and amino acids) so that the operator gene is turned on to produce more of the enzyme. In this way the genes influence the cytoplasm and the cytoplasm exerts a "feedback" influence on the genes for turning on or off their action.

The operon model is still a hypothesis but many believe that this or a similar system must operate in animals.

MENDEL'S INVESTIGATIONS

The first man to formulate the cardinal principles of heredity was Gregor Johann Mendel (1822-1884), who was connected with the Augustinian monastery at Bruenn, Moravia, then a part of Austria, later a part of Czechoslovakia. In the conduction of experiments from 1856 to 1864 he examined with great care and accuracy many thousands of plants. His classic observations were made on the garden pea because gardeners over a long period by careful selection had produced pure strains. For example, some varieties were definitely dwarf and others were tall. A second reason for selecting peas was

Experiments on which Mendel based his postulates

Results of monohybrid crosses for first and second generations

Round-wrinkled seeds
F_1 all round
F_2 5474 round
1850 wrinkled
Ratio 2.96:1

Colored-white flowers
F_1 all colored
F_2 705 colored
224 white
Ratio 3.15:1

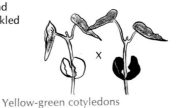

Yellow-green cotyledons
F_1 all yellow
F_2 6022 yellow
2001 green
Ratio 3.01:1

Green-yellow pods
F_1 all green
F_2 428 green
152 yellow
Ratio 2.82:1

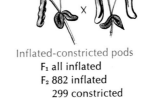

Inflated-constricted pods
F_1 all inflated
F_2 882 inflated
299 constricted
Ratio 2.95:1

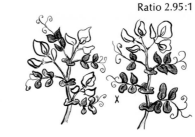

Long-short stems
F_1 all long
F_2 787 long
277 short
Ratio 2.84:1

Axial-terminal flowers
F_1 all axial
F_2 651 axial
207 terminal
Ratio 3.14:1

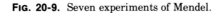

FIG. 20-9. Seven experiments of Mendel.

that they are self-fertilizing, but are also capable of cross-fertilization. To simplify his problem he chose single characters and those that were sharply contrasted. Mere quantitative and intermediate characters he carefully avoided. Mendel selected seven pairs of these contrasting characters. These were tall plants and dwarf plants, smooth seeds and wrinkled seeds, green cotyledons and yellow cotyledons, inflated pods and constricted pods, yellow pods and green pods, axial position of flowers and terminal position of flowers, and transparent seed coats and brown seed coats (Fig. 20-9). Mendel crossed a plant having one of these characters with one having the contrasting character. He did this by removing the stamens from a flower to prevent self-fertilization, and then by placing on the stigma of this flower pollen from the flower of the plant that had the contrasting character. He then prevented the experimental flowers from being pollinated from other sources, such as wind and insects. When the cross-fertilized flower bore seeds, he noted the kind of plants (hybrids) they produced when planted. His next step was to cross these hybrids among themselves and to see what happened.

MENDEL'S LAWS

Mendel knew nothing about the cytologic background of heredity, for chromosomes and genes were unknown to him. Instead of using the term "genes" as we do today, he called his inheritance units **factors.** He reasoned that the factors for tallness and dwarfness were units that did not blend when they were together. The F_1 generation (the first generation of hybrids, or first filial generation) contained both these units or factors, but when these plants formed their germ cells, the factors separated out so that each germ cell had only one factor. In a pure plant both factors were alike; in a hybrid they were different. He concluded that individual germ cells are always pure with respect to a pair of contrasting factors, even though the germs cells are formed from hybrids in which the contrasting characters were mixed. This idea formed the basis for his first principle, the **law of segregation,** which states that whenever two factors are brought together in a hybrid, when that hybrid forms its germ cells, the factors segregate into separate gametes and each germ cell is pure with respect to that character.

In the crosses involving the factors for tallness and dwarfness, in which the resulting hybrids were tall, Mendel called the tall factor **dominant** and the short **recessive.** Similarly, the other pairs of characters that he studied showed dominance and recessiveness. Thus when plants with yellow unripe pods were crossed with green unripe pods, the hybrids all contained yellow pods. In the F_2 generation (second filial generation) the expected ratio of 3 yellow to 1 green was obtained. Whenever a dominant factor (gene) is present, the recessive one cannot produce an effect. The recessive factor will show up only when both factors are recessive, or, in other words, a pure condition. When a tall plant with the yellow type of pod was crossed with a dwarf plant bearing green pods, the F_1 generation was all tall and yellow, for these factors are dominant. When the F_1 hybrids were crossed with each other, the result was 9 tall and yellow, 3 tall and green, 3 dwarf and yellow, and 1 dwarf and green. In this experiment each factor separated independently of the other and showed up in new combinations. This is Mendel's second law, or the **law of independent assortment,** which states that, whenever two or more pairs of contrasting characters are brought together in a hybrid, the factors of different pairs segregate independently of one another. Rarely do two organisms differ in only one pair of contrasting characters; nearly always they differ in many. The second law of Mendel therefore deals with two or more pairs of contrasting characters.

It happened that all seven pairs of characters Mendel worked with were on different pairs of chromosomes, but since his laws became known, many pairs of characters have been found on the same chromosome, which alters the original mendelian ratios. Since his time, too, the phenomena of linkage and crossing-over have necessitated a modification in his second law. This modification does not detract, however, from the basic significance of his great laws.

EXPLANATION OF MENDELIAN RATIOS

In representing his crosses Mendel used letters as symbols. For dominant characters he employed capitals, and for recessives, corresponding small letters. Thus the factors, or genes, for pure tall plants might be represented by TT, the pure recessive by tt, and the hybrid of the two plants by Tt. In diagram form, one of Mendel's original crosses (tall plant and dwarf plant) could be represented in this manner:

	(tall)		(dwarf)	
Parents	TT	×	tt	
Gametes	all T		all t	
F_1		Tt		
		(hybrid tall)		
Crossing hybrids	Tt	×	Tt	
Gametes	T,t		T,t	
F_2	TT	Tt	tT	tt
		(3 tall to 1 dwarf)		

It is convenient in most mendelian crosses to use the checkerboard method devised by Punnett for representing the various combinations resulting

from a cross. Thus in the previous F_2 cross, the following scheme would apply:

		Eggs	
		T	t
Sperm	T	**TT** (pure tall)	**Tt** (hybrid tall)
	t	**Tt** (hybrid tall)	**tt** (pure dwarf)

Ratio: 3 tall to 1 dwarf.

Mendel's experiment involving two pairs of contrasting characters instead of one pair may be demonstrated in the diagram shown below.

	(tall, yellow)		(dwarf, green)
Parents	TTYY	×	ttyy
Gametes	all TY		all ty
F_1		TtYy	
		(hybrid tall, hybrid yellow)	
Crossing hybrids	TtYy	×	TtYy
Gametes	TY, Ty, tY, ty		TY, Ty, tY, ty
F_2		(see checkerboard)	

	TY	**Ty**	**tY**	**ty**
TY	**TTYY** ① pure tall pure yellow	**TTYy** ① pure tall hybrid yellow	**TtYY** ① hybrid tall pure yellow	**TtYy** ① hybrid tall hybrid yellow
Ty	**TTYy** ① pure tall hybrid yellow	**TTyy** ② pure tall pure green	**TtYy** ① hybrid tall hybrid yellow	**Ttyy** ② hybrid tall pure green
tY	**TtYY** ① hybrid tall pure yellow	**TtYy** ① hybrid tall hybrid yellow	**ttYY** ③ pure dwarf pure yellow	**ttYy** ③ pure dwarf hybrid yellow
ty	**TtYy** ① hybrid tall hybrid yellow	**Ttyy** ② hybrid tall pure green	**ttYy** ③ pure dwarf hybrid yellow	**ttyy** ④ pure green pure green

Ratio: 9 tall yellow to 3 tall green; 3 dwarf yellow to 1 dwarf green.

In the cross between tall and dwarf it will be noted that there are two types of visible characters—**tall** and **dwarf.** These are called **phenotypes.** On the basis of genetic formulas there are three hereditary types—TT, Tt, and tt. These are called **genotypes.** In the cross involving two pairs of contrasting characters (**tall yellow** and **dwarf green**) there are in the F_2 generation four phenotypes: **tall yellow, tall green, dwarf yellow,** and **dwarf green.** The genotypes are nine in number: TTYY, TTYy, TtYY, TtYy, TTyy, Ttyy, ttYY, ttYy, and ttyy. The F_2 ratios in any cross involving more than one pair of contrasting pairs can be found by combining the ratios in the cross of one pair of factors. Thus the genotypes will be $(3)^n$ and the phenotypes $(3:1)^n$. To illustrate, in a cross of two pairs of factors the phenotypes will be in the ratio of $(3:1)^2 = 9:3:3:1$. The genotypes in such a cross will be $(3)^2 = 9$. If three pairs of characters are involved, the phenotypes will be $(3:1)^3 = 27:9:9:9:3:3:3:1$. The genotypes will be $(3)^3 = 27$. Thus it is seen that the numerical ratio of the various phenotypes is a power of the binomial $(3 + 1)^n$ whose exponent (n) equals the number of pairs of heterozygous genes in F_2. This is true only when one member of each pair of genes is dominant. By experience, then, one may determine the ratios of phenotypes in a cross without using the checkerboard. In a dihybrid (9:3:3:1 ratio), for instance, it will be seen that those phenotypes that make up the dominants of each pair will be $9/16$ of the whole F_2; each of the $3/16$ phenotypes will consist of one dominant and one recessive; and the $1/16$ phenotype will consist of two recessives.

LAWS OF PROBABILITY

When Mendel worked out the ratios for his various crosses, they were approximations and not certainties. In his 3 to 1 ratio of tall and short plants, for instance, the resulting phenotypes did not come out exactly 3 tall to 1 short. All genetic experiments are based on probability; that is, the outcome of the events is uncertain and there is an element of chance in the final results. Probabilities are expressed in fractions, or it is always a number between 0 and 1. This probability number (p) is found by dividing the number (m) of favorable cases (for

example, a certain event) by the total number (n) of possible outcomes:

$$p = \frac{m}{n}$$

When there are two possible outcomes, such as in tossing a coin, the chance of getting heads is $p = 1/2$, or 1 chance in 2.

The more often a particular event occurs, the more closely will the number of favorable cases approach the number predicted by the p value.

The probability of independent events occurring together involves the **product rule,** which is simply the product of their individual probabilities. When two coins are tossed together, the probability of getting two heads is $1/2 \times 1/2 = 1/4$, or 1 chance in 4. Here, again, this prediction is most likely to occur if the coins are tossed a sufficient number of times.

The ratios of inheritance in a monohybrid cross of dominant and recessive genes can be explained by the product rule. In the gametes of the hybrids the sperm may carry either the dominant or the recessive gene; the same applies to the eggs. The probability that the sperm carries the dominant is $1/2$ and the probability of an egg carrying the dominant is also $1/2$. The probability of a zygote obtaining two dominant genes is $1/2 \times 1/2$, or $1/4$. Thus 25% of the offspring will probably be pure dominants. The same principle applies to the recessive gene, which will be pure for 25% of the offspring. The heterozygous gene combinations will be found by the sum of the two possible combinations—a sperm with a dominant gene and an egg with a recessive gene, and a sperm with a recessive gene and an egg with a dominant gene—which yields 50% heterozygotes. Thus we have the 1:2:1 ratio.

TERMINOLOGY OF GENETICS

Genetics, in common with other branches of science, has built up its own terminology. Some of the terms first proposed by Mendel have been replaced by those that seem more suitable in the light of present-day knowledge. These terms are all important to the student of heredity because they are essential in understanding the analyses of genetic problems. Whenever a cross involves only one pair of contrasting characters, it is called a **monohybrid;** when the cross has two pairs, it is a **dihybrid;** when the cross has three pairs, it is a **trihybrid;** and when it has more than three pairs, it is a **polyhybrid.** Characters that show in the F_1 are **dominant;** those that are hidden are **recessive.** When a dominant always shows up in the phenotype, it is said to have **complete dominance;** when it sometimes fails to manifest itself it is called **incomplete dominance.** When two characters form a contrasting pair, they are called **alleles** or **allelomorphs.** The term **factor** that Mendel used is replaced by **gene.** A **zygote** is the union of two gametes; whenever the two members of a pair of genes are alike in a zygote, the latter is **homozygous** for that particular character; when the genes are unlike for a given character, the zygote is **heterozygous.** A **hybrid,** for instance, is a heterozygote, and a **pure** character is a homozygote.

TESTCROSS

The dominant characters in the offspring of a cross are all of the same phenotypes whether they are homozygous or heterozygous. For instance, in Mendel's experiment of tall and dwarf characters, it is impossible to determine the genetic constitution of the tall plants of the F_2 generation by mere inspection of the tall plants. Three fourths of this generation are tall, but which of them are heterozygous recessive dwarf? The test is to cross the F_2 generation (dominant hybrids) with pure recessives. If the tall plant is homozygous, all the plants in such a testcross will be tall, thus:

TT (tall) × tt (dwarf)
Tt (hybrid tall)

If, on the other hand, the tall plant is heterozygous, the offspring will be half tall and half dwarf, thus:

Tt × tt
Tt (tall) or tt (dwarf)

The testcross is often used in modern genetics for the analysis of the genetic constitution of the offspring as well as for a quick way to make homozygous desirable stocks of animals and plants.

INCOMPLETE DOMINANCE

A cross that always shows the heterozygotes as distinguished from the pure dominants is afforded by the four-o'clock flower *(Mirabilis)* (Fig. 20-10), discovered since Mendel's time. Whenever a red-flowered variety is crossed with a white-flowered variety, the hybrid (F_1), instead of being red or white according to whichever is dominant, is actually intermediate between the two and is pink. Thus the homozygotes are either red or white, but the heterozygotes are pink. The testcross is therefore unnecessary to determine the nature of the genotype.

In the F_2 generation, when pink flowers are crossed with pink flowers, one fourth will be red, one half pink and one fourth white.

This cross may be represented in this fashion:

	(red flower)		(white flower)	
Parents	RR	×	rr	
Gametes	R,R		r,r	
F_1		Rr		
		(all pink)		
Crossing hybrids	Rr	×	Rr	
Gametes	R,r		R,r	
F_2	RR	Rr	rR	rr
	(red)	(pink)	(pink)	(white)

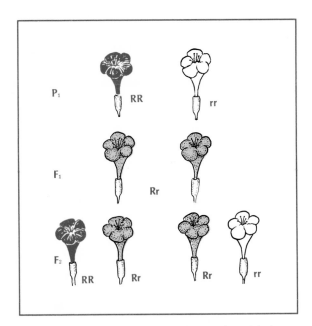

FIG. 20-10. Cross between red and white four-o'clock flowers. Red and white are homozygous; pink is heterozygous.

PENETRANCE AND EXPRESSIVITY

Penetrance refers to the percentage frequency with which a gene manifests phenotypic effect. If a dominant gene or a recessive gene in a homozygous state always produces a detectable effect, it is said to have **complete penetrance.** If dominant or homozygous recessive genes fail to show phenotypic expression in every case, it is called **incomplete** or **reduced penetrance.** All of Mendel's experiments apparently had 100% penetrance.

The phenotypic variation in the expression of a gene is known as **expressivity.** Environmental factors may cause different degrees in the appearance of a phenotype. For example, lower temperatures permit expression of the genes for a black color in certain regions of the Siamese cat.

SOME SPECIAL FORMS OF HEREDITY

The types of crosses already described are simple in that the characters involved are due to the action of a single gene, but many cases are known in which the characters are the result of two or more genes. Mendel probably did not appreciate the real significance of the genotype as contrasted with the visible character—the phenotype. We now know that many different genotypes may be expressed as a single phenotype.

It is also known that many genes have more than a single effect. A gene for eye color, for instance, may be the ultimate cause for eye color, yet at the same time it may be responsible for influencing the development of other characters as well. Also, many unlike genes may occupy the same locus on a chromosome, but not, of course, all at one time. Thus more than two alternative characters may effect the same character. Such genes are called **multiple alleles,** or factors. In the fruit fly (*Drosophila*) there are 18 alleles for eye color alone. Not more than two of these genes can be in any one individual and only one in a gamete. The remaining 16 alleles are present in other individuals of the population. All genes can mutate in several different ways if given time and thus can give rise to several alternative conditions. In this way, many alleles for a particular locus on a chromosome may have evolved and added to the genetic pool of a population. It is thought that all genes present in an organism are mutants. In some cases dominance is lacking between two members of a set of multiple alleles. In *Drosophila* the gene for red eye color (wild type) is dominant over all other alleles of the eye color series; the gene for white eye is recessive to all the others.

Some of these unusual cases of inheritance are described in the following discussions on supplementary, complementary, cumulative, and lethal factors.

Supplementary factors. The variety of comb forms found in chickens illustrates the action of supplementary genes (Fig. 20-11). The common forms of comb are **rose, pea, walnut,** and **single.** Of these, the pea comb and the rose comb are dominant to the single comb. For example, when a pea comb is crossed with a single comb, all the F₁ are pea and the F₂ show a ratio of 3 pea to 1 single. When the two dominants, pea and rose, are crossed with each other, an entirely new kind of comb, walnut, is found in the F₁ generation. Each of these genes supplements the other in the production of a kind of comb different from each of the dominants. In the F₂ generation the ratio is 9 walnut, 3 rose, 3 pea, and 1 single. The walnut comb cannot thus be considered a unit character, but is merely the

FIG. 20-11. Heredity of comb forms in chickens.

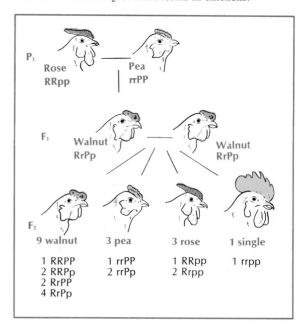

phenotype's expression of pea and rose when they act together.

Inspection of the ratio reveals that two pairs of genes are involved. If P represents the gene for the pea comb and p its recessive allelomorph and if R represents the gene for the rose comb and r its recessive allelomorph, then the pea comb formula would be rrPP; and the one for rose comb, RRpp. Any individual having both dominant genes has a walnut comb. When no dominant gene is present, the comb is single. The cross may be diagrammed as follows:

Parents	rrPP	×	RRpp	
	(pea comb)		(rose comb)	
Gametes	all rP		all Rp	
F_1		RrPp	×	RrPp
		(walnut)		(walnut)
Gametes		RP, rP, Rp, rp	×	RP, rP, Rp, rp

By the checkerboard method, the F_2 will show 9 walnut, 3 pea, 3 rose, and 1 single. It will be seen that genotypes with the combinations of PR will give walnut phenotypes; those with P, pea; those with R, rose; and those lacking in both P and R, single.

Complementary factors. When two genes produce a visible effect together, but each alone will show no visible effect, they are referred to as **complementary genes.** Some varieties of sweet peas can be used to illustrate this kind of cross. When two white-flowered varieties of these are crossed, the F_1 will show all colored (reddish or purplish) flowers. When these F_1's are self-fertilized, the F_2 will show a phenotypic ratio of 9 colored to 7 white flowers. This is really a ratio of 9:3:3:1 because the last three groups cannot be distinguished phenotypically. The explanation lies in the fact that in one of the white varieties of flowers there is a gene (C) for a colorless color base (chromogen) and in the other white variety a gene (E) for an enzyme that can change chromogen into a color. Only when chromogen and the enzyme are brought together is a colored flower produced. The cross may be diagrammed in this way:

Parents	CCee	×	ccEE	
	(white)		(white)	
Gametes	all Ce		all cE	
F_1		CcEe	×	CcEe
		(colored)		(colored)
Gametes		CE, Ce, cE, ce		CE, Ce, cE, ce

By the checkerboard method, the F_2 phenotypes will be as follows:

9 colored (CCEE, CCEe, CcEE, or CcEe)—both chromogen and enzyme

7 { 3 white (CCee, or Ccee)—only chromogen
3 white (ccEE, or ccEe)—only enzyme
1 white (ccee)—no chromogen or enzyme

Cumulative factors. Whenever several sets of alleles produce a cumulative effect on the same character, they are called **multiple genes,** or factors. Several characteristics in man are influenced by multiple genes. In such cases the characters, instead of being sharply marked off show continuous variation between two extremes. This is called **blending,** or **quantitative inheritance.** In this kind of inheritance the children are more or less intermediate between the two parents. The best illustration of such a type is the degree of pigmentation in crosses between the black and the white race. The cumulative genes in such crosses have a quantitative expression. A pure-blooded Negro has two pairs of genes on separate chromosomes for pigmentation (AABB). On the other hand, a pure-blooded white will have the genes (aabb) for nonblack. In a mating between a homozygous Negro and a homozygous white, the mulatto (AaBb) will have a skin color intermediate between the black parent and the white. The genes for pigmentation in the cross show incomplete dominance. In a mating between two mulattoes (F_2), the children will show a variety of skin color, depending on the number of genes for pigmentation they inherit. Their skin color will range all the way from black (AABB) through dark brown (AABb or AaBB), half-colored (AAbb or AaBb or aaBB), light brown (Aabb or aaBb) to white (aabb). In the F_2 there will be the possibility of a child resembling the skin color of either grandparent, and the others will show intermediate grades. It is thus possible for parents heterozygous for skin color to produce children with darker or lighter colors than themselves.

The relationships can be seen in the following diagram:

Parents	AABB	×	aabb	
	(black)		(white)	
Gametes	AB		ab	
F_1		AaBb	×	AaBb
		(mulatto)		(mulatto)
Gametes		AB, Ab, aB, ab		AB, Ab, aB, ab

By the checkerboard method, the F$_2$ will show this ratio:

1 black (AABB)
4 dark brown (AABb or aABB)
6 half-colored mulattoes (AaBb, AAbb, or aaBB)
4 light brown (Aabb, aaBb)
1 white (aabb)

The student should realize that when the terms "white" and "black" are used in a cross involving mulattoes, they refer to skin color (in this instance) and not to other characteristics, for other racial characteristics are inherited independently. Thus in such a cross an individual may have white color (no genes for black) but could have other Negro characteristics.

When there are many genes involved in the production of traits, the latter often take the form of distribution curves. One such trait is stature in man. Where between a few extremely short and tall individuals, there are many in between these extremes.

Lethal factors. A lethal gene is one that, when present in a homozygous condition, will cause the death of the offspring. It has been known for a long time that the yellow race of the house mouse *(Mus musculus)* is heterozygous. Whenever two yellow mice are bred together, the progeny are always 2 yellow to 1 nonyellow. In such a case the expected ratio should be 1 pure yellow, 2 hybrid yellow, and 1 pure nonyellow. Examination of the pregnant yellow females shows that the homozygous yellow always dies as an embryo, which accounts for the unusual ratio of 2:1.

SEX DETERMINATION

The first really scientific clue to the cause of sex was discovered in 1902 by C. McClung, who found that in some species of bugs (Hemiptera) two kinds of sperm were formed in equal numbers. One kind contained among its regular set of chromosomes a so-called accessory chromosome that was lacking in the other kind of sperm. Since all the eggs of these species had the same number of haploid chromosomes, half the sperm would have the same number of chromosomes as the eggs and half of them would have one chromosome less. When an egg is fertilized by a spermatozoan carrying the accessory (sex) chromosome, the

resulting offspring is a female; when fertilized by the spermatozoan without an accessory chromosome, the offspring is a male. There are, therefore, two kinds of chromosomes in every cell; X chromosomes determine sex (and sex-linked traits), and **autosomes** determine the other bodily traits. The particular type of sex determination just described is often called the XX-XO type, which indicates that the females have 2 X chromosomes and the male only 1 X chromosome (the O stands for its absence).

Later, other types of sex determination were discovered. In man and many other forms there are the same number of chromosomes in each sex, but the sex chromosomes (XX) are alike in the female but unlike (XY) in the male. Hence the human egg contains 22 autosomes + 1 X chromosome; the sperm are of two kinds: half will carry 22 autosomes + 1 X and half will bear 22 autosomes + 1 Y. The Y chromosome is much smaller than the X. At fertilization, when 2 X chromosomes come together, the offspring will be a girl; when XY, it will be a boy.

A third type of sex determination is found in birds, moths, and butterflies in which the male has 2 X (or sometimes called ZZ) chromosomes and the female an X and Y (or ZW). In this latter case the male is homozygous for sex and the female is heterozygous.

Sex ratios in certain forms can also be influenced by environmental forces. When toad eggs are partially dried out, the proportion of females over males is thereby increased. Variations in diet are also known to upset the ratio between the sexes of some animals.

MICROSCOPIC DETERMINATION OF SEX

Sexual dimorphism in the nuclei was first discovered in 1949 by M. Barr and E. G. Bertram. These investigators discovered a chromatin mass in female nuclei that they identified as the heterochromatic parts of 2 X chromosomes in the interphase stage. Such a body, which is often called the nucleolar satellite, or the Barr body (Fig. 20-12, *A*), is found lying against the nuclear membrane. It is not found in the nuclei of the male because the male has only 1 X. Although first found in the cat, the sex chromatin has been found in other organisms including

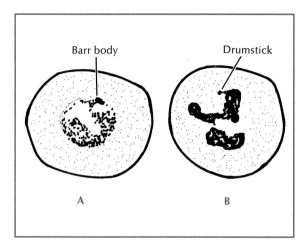

Barr body   Drumstick

A   B

FIG. 20-12. Cellular determination of sex. **A,** Squamous epithelial cell of human female showing Barr body, which is absent in male. **B,** White blood cell of human female showing accessory nuclear lobule ("drumstick"), which is mostly lacking in male.

man. A smear of the oral epithelium is one of the simplest places in man for demonstrating the Barr body.

Another type of sexual dimorphism is found in the polymorphonuclear leukocyte of blood smears. An accessory nuclear lobule called the "drumstick" (Fig. 20-12, *B*) is found in leukocytes from females but is lacking or else is very diminutive in males. Both drumsticks and Barr bodies indicate the presence of more than 1 X chromosome. In those abnormal sex chromosome cases, such as the XXX constitution, there may be two Barr bodies and two "drumsticks."

SEX-LINKED INHERITANCE

Sex-linked inheritance refers to the carrying of genes by the X chromosomes for body characters that have nothing to do with sex. The X chromosome is known to contain many such genes, the Y chromosome only a few because of its small size. Such sex-linked traits are not always limited to one sex but may be transmitted from the mother to her male offspring or from the father to his female offspring. One of the examples of a sex-linked charac-

ter was discovered by Morgan in *Drosophila.* The normal eye color of this fly is red, but mutations for white eyes do occur. The genes for eye color are known to be carried in the X chromosome. If a white-eyed male and a red-eyed female are crossed, all the F_1's are red eyed, because this trait is dominant (Fig. 20-13). If these F_1's are interbred, all the females of F_2 will have red eyes and half the males will have red eyes and the other half white eyes. No white-eyed females are found in this generation; only the males have the recessive character (white eyes). The gene for white-eyed, being recessive, should appear in a homozygous condition. However, since the male has only 1 X chromosome (the Y does not carry a gene for eye color), white eyes will appear whenever the X chromosome carries the gene for this trait. If the reciprocal cross is made in which the females are white eyed and the males red eyed, all the F_1 females are red eyed and all the males are white eyed (Fig. 20-14). This is called **crisscross inheritance.** If these F_1's are interbred, the F_2 will show equal numbers of red-eyed and white-eyed males and females.

If the allele for red-eyed is represented by R and white-eyed by r, the following diagrams will show how this eye color inheritance works:

Parents	RR (red ♀)		×		rY (white ♂)
F_1	Rr (red ♀)		×		RY (red ♂)
Gametes	R,r				R,Y
F_2	RR	RY		rY	rR
	red ♀	red ♂		white ♂	red ♀

RECIPROCAL CROSS

Parents	rr (white ♀)		×		RY (red ♂)
F_1	rR (red ♀)		×		rY (white ♂)
Gametes	r,R				r,Y
F_2	rr	rY		Rr	RY
	white ♀	white ♂		red ♀	red ♂

Many sex-linked genes are known in man, such as bleeder's disease (hemophilia), night blindness, and color blindness. The latter is often used as one of the most striking cases of sex-linked inheritance in man. The particular form of color blindness involved is called daltonism, or the inability to distinguish between red and green. The defect is recessive and requires both genes with the defect in the female but only one defective gene in the male to acquire a visible defect. If the defective allele is represented

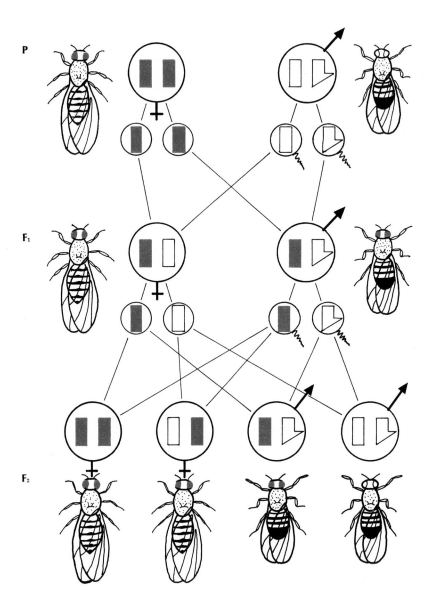

Fig. 20-13. Sex determination and sex-linked inheritance of eye color in fruit fly
(Drosophila). Normal red eye color is dominant to white eye color. If a homozygous red-eyed
female and a white-eyed male are mated, all F_1 flies are red eyed. When F_1 flies are
intercrossed, F_2 yields approximately 1 homozygous red-eyed female and 1 heterozygous
red-eyed female to 1 red-eyed male and 1 white-eyed male. Genes for red eyes and
white eyes are carried by sex (X) chromosomes; Y carries no genes for eye color.

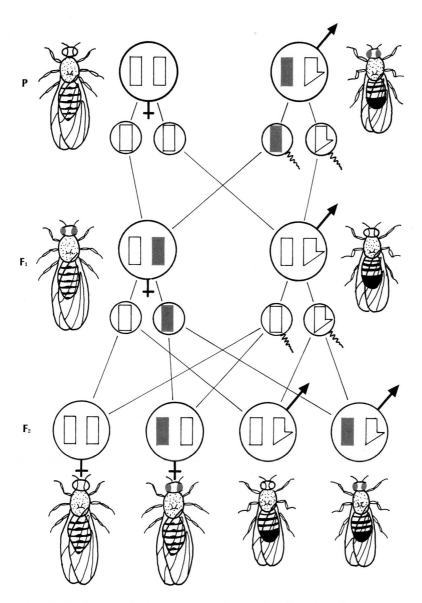

FIG. 20-14. In cross of a homozygous white-eyed female and a heterozygous red-eyed male (reciprocal cross of Fig. 20-13), F_1 consists of white-eyed males and red-eyed females. In the F_2, there are equal numbers of red-eyed and white-eyed females and red-eyed and white-eyed males.

by an asterisk (*), the following diagram will show the inheritance pattern:

CROSS BETWEEN COLOR-BLIND MALE (X*Y)
AND HOMOZYGOUS NORMAL FEMALE (XX)

	X*Y	×	XX
F$_1$	X*X	×	XY (all normal)
Gametes	X*,X		X,Y
F$_2$	X*X, X*Y, XX, XY		
	(X*Y, color-blind individual)		

It will be seen from this cross that the color-blind father will transmit the defect to his daughters (who do not show it because each has only one defective gene), but these daughters transmit the defect to one half their sons (who show it because a sex-linked recessive gene in the male has a visible effect).

LINKAGE AND CROSSING-OVER

Since the number of chromosomes in any animal is relatively few compared to the number of traits, it is evident that each chromosome bears many genes. All the genes contained in a chromosome tend to be inherited together and are therefore said to be **linked.** The sex-linked phenomena described in the previous section are examples of linkage; genes borne on chromosomes other than sex chromosomes form **autosomal linkage.** Genes, therefore, occur in linkage groups and there should be as many linkage groups as there are pairs of chromosomes. In *Drosophila,* in which this principle has been worked out most extensively, there are four linkage groups that correspond to the four pairs of chromosomes found in these fruit flies. Small chromosomes have small linkage groups and large chromosomes have large groups. Five hundred genes have been mapped in the fruit fly and all these are distributed among the four pairs.

How the mendelian ratios can be altered by linkage can best be illustrated by one of Morgan's experiments on *Drosophila.* When a wild-type fly with gray body and long wings is crossed with a fly bearing two recessive mutant characters of black body and vestigial wings, the dihybrids (F$_1$) all have gray bodies and long wings. If a male of one of the F$_1$'s is testcrossed with a female with a black body and vestigial wings, the flies are all gray-long and black-vestigial. If there had been free assortment, that is, if the various characters had been carried on different chromosomes, the expected offspring would have been represented by four types of flies: gray-long, gray-vestigial, black-long, and black-vestigial. However, in this case gray-long and black-vestigial had entered the dihybrid cross together and stayed together, or linked.

Linkage, however, is usually only partial, for it is broken up frequently by what is known as **crossing-over.** In this phenomenon the characters usually separate with a certain frequency. How often two genes break their linkage, or their percentage of crossing-over, varies with different genes. In some cases this percentage of crossing-over is only 1% or less; with others it may be nearly 50%. The explanation for crossing-over lies in the synapsis of homologous chromosomes during the maturation division when two nonsister chromatids sometimes become intertwined and, before separating from each other, exchange homologous portions of the chromatids (and the genes they bear) (Fig. 20-15).

Crossing-over makes possible the construction of chromosome maps and proof that the genes lie in a linear order on the chromosomes. To illustrate how this is done one may take a hypothetical case of three genes (A, B, C) on the same chromosome. In the determination of their comparative linear position on the chromosome, it will first be necessary to find the crossing-over value between any two of these genes. If A and B have a crossing-over of 2% and B and C of 8%, then the crossing-over percentage between A and C should be either the sum (2 + 8) or the difference (8 − 2). If it is 10%, B lies between A and C; if 6%, A is between B and C. By laborious genetic experiments over many years, the famed chromosome maps in *Drosophila* were worked out in this manner (Fig. 20-16). Cytologic investigations on the giant chromosomes since these maps were made tend to prove the correctness of the linear order, if not the actual position, of the genes on the chromosomes. There is no evidence of crossing-over occurring in the giant chromosomes themselves, and it is also absent in the male *Drosophila.*

HYBRID VIGOR

Hybrid vigor, or heterosis, refers to the greater vitality and vigor manifested by the hybrids produced by crossing individuals of two pure races that differ from each other in a number of genes. Such

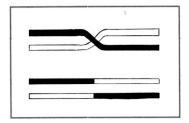

FIG. 20-15. Simple case of crossing-over between two chromatids of homologous chromosomes. Block of genes shown in black is thus transferred to white and vice versa.

hybrids are often bigger and more vigorous than either parent. What actually happens is that such crosses may bring together dominant genes for vigor in the hybrid, provided each of the pure line races carries a vigor gene lacking in the other. In this way it is conceivable that the hybrid may contain homozygous dominant vigor genes that would account for its more desirable traits. There is also the possibility that some genes in the heterozygous condition in the hybrid might produce more vigor

X	II	III
0.0 — Yellow body	0.0 — Net veins (wings)	0.0 — Roughoid eye
1.5 — White eye		
7.5 — Ruby eye	13.0 — Dumpy wings	
	16.5 — Clot eye	
20.0 — Cut wings		19.2 — Javelin bristles
27.7 — Lozenge eye		26.0 — Sepia eye
		26.5 — Hairy body
33.0 — Vermilion eye	48.5 — Black body	
36.1 — Miniature wings		
	57.5 — Cinnabar eye	
43.0 — Sable wings		41.4 — Glued eye
44.4 — Garnet eye		
	67.0 — Vestigial wings	48.0 — Pink eye
57.0 — Bar eye		50.0 — Curled wings
	72.0 — Lobe eye	
62.5 — Carnation eye	75.5 — Curved wings	58.5 — Stubble bristles
		63.1 — Glass eye
66.0 — Bobbed bristles		
		70.7 — Ebony body
	93.3 — Humpy body	75.7 — Cardinal eye
	100.5 — Plexus veins (wings)	91.1 — Rough eye
IV		100.7 — Claret eye
0.0 — Bent body	107.0 — Speck wings	
0.2 — Eyeless		104.3 — Brevis bristles

FIG. 20-16. Chromosome maps of certain representative genes in *Drosophila*. One chromosome of each pair is shown with figures, indicating relative positions of genes in crossover units from end of chromosome.

than genes in the homozygous state found in the inbred race. Hybrid vigor, however, tends to be lost in succeeding generations when hybrids are crossed because the desirable dominant genes may segregate out again and the undesirable recessives would again make their effects visible.

Examples of hybrid vigor in plants and animals are shown by hybrid corn and by the mule. The mule is the hybrid between the jackass and the mare and is stronger and sturdier than either parent.

CYTOPLASMIC INHERITANCE

The genetic behavior so far considered has stressed the nuclear elements (chromosomes and genes) as the bearers of heredity. Does the cytoplasm bear any hereditary factors? Evidence for cytoplasmic inheritance is often sought in reciprocal cross differences (for example, whether or not the genetic type is introduced into a cross by the father or the mother) because the egg (maternal contribution) contains most of the cytoplasm of the new organism, and the phenotype of the latter should follow that of the mother. A few cases of cytoplasmic inheritance seem well established. One of these is the chlorophyll-bearing plastids in the cytoplasm of certain plants. These are self-duplicating bodies and some of them are always maternal in character regardless of the kind introduced by the pollen in the cross. Another case is the cortical pattern of certain paramecia (T. M. Sonneborn) that has been explained by non-chromosomal inheritance. When two paramecia, one with a double cortical pattern and the other with a single cortical pattern, conjugate, it was occasionally found that one conjugant had received a piece of the cortex of the other and became intermediate in appearance between the other two types and thereafter reproduced true to this new intermediate type. The demonstration of plasmagenes (inheritance factors in the egg cytoplasm) bids to shed some light upon the difficult problem of developmental differentiation.

PRACTICAL APPLICATION OF GENETICS

Although genetics as a science has developed within this century, many of the practical applications of heredity extend as far back as the dawn of civiliza-tion. The improvement of plants and animals has always interested man.

Methods of increasing productivity in plants and animals. Plant and animal breeders have two main objectives—to improve strains that give greater productivity and to produce strains that are pure in their pedigree. Desirable hereditary qualities are carefully selected and propagated by breeding. Often the production of pure lines is for the purpose of crossing with other strains of plants to take advantage of hybrid vigor, a practice used by both plant and animal breeders. Practical application of Mendel's laws forms the basis for most plant and animal experiments. The testcross, for instance, is a common procedure for getting homozygous stock.

Hybrid varieties. Pure lines of corn were found to be of low fertility and gave scanty yields. When pure strains were crossed, the hybrids showed a remarkable increase in both fertility and general yield. Only the seed of the first generation following a cross can be used for replanting.

Progeny selection. The breeder may use either phenotypic or progeny selection. The phenotypic appearance of the plant or animals is not necessarily an indication of its genotype. Progeny selection emphasizes the genotype as the chief basis of selection.

Disease-resistant strains. By careful selection and breeding, plant breeders produce varieties that do not become infected by common plant diseases such as wheat rust and corn smut.

Polyploidy. Many polyploids have been experimentally produced, such as tulips, roses, fruit trees, and others, and represent an improvement over the original stock. When they cannot breed true, they may be propagated asexually.

HUMAN HEREDITY

The study of human heredity cannot be treated as that of other animals, so observation is relied on to a great extent. But there is every reason to suppose that man follows the same principles of genetics as other organisms.

GENETIC DISEASES

The work of G. W. Beadle and E. L. Tatum on the mold *Neurospora,* by which it was shown that a single gene controlled the specificity of a particular enzyme, has helped explain many inherited human diseases. An English physician A. E. Garrod contended in 1908 that enzyme deficiencies were to blame for certain disorders that he described as "inborn errors of metabolism." Among such disorders is phenylketonuria (PKU), which is produced by the absence of the enzyme phenylalanine hydroxylase carried by a recessive gene. A person afflicted with this disorder cannot convert the amino acid phenylalanine into tyrosine. Some phenylalanine is converted into phenylpyruvic acid, which produces injury to the nervous system and mental deficiency. The disorder is detected by a simple test with blood or urine and can be controlled by restricting phenylalanine in the diet.

About one person in 20,000 has the condition known as albinism, which is characterized by the lack of the dark pigment melanin in the skin and hair. Albinism is caused by the absence of the enzyme tyrosine, which is necessary for the synthesis of melanin. Albinism is inherited as a recessive gene, and an individual must be homozygous to show it.

Many other such genetic diseases have been uncovered in man in the past decade or so. Almost all traits are produced by a sequence of chemical reactions, each under the control of a specific enzyme.

Some genetic disorders that are caused by recessive genes may show some effect in a heterozygous condition. One of these is the sickle cell mutant gene that produces the abnormal S-hemoglobin. This hemoglobin differs from normal hemoglobin in having a valine amino acid in place of the glutamic acid molecule in a chain of more than 300 amino acids. A person with one gene for normal hemoglobin and one gene for the abnormal S-hemoglobin will usually suffer a mild anemia. However, such persons have a better resistance to falciparum, or malignant malaria, which may account for the high prevalence of the heterozygotes in parts of Africa. A homozygous condition of S-hemoglobin is usually fatal.

Geneticists estimate that one person in every four or five of the population is born with a defective gene, which can bring about serious consequences to the population, depending on whether the genes are dominant or recessive. It is thought that one conception in every five results in a genetic defect that destroys the individual at some time before maturity. In addition, various types of abnormalities, not necessarily lethal, are caused by both dominant and recessive genes. These include sex-linked and nonsex-linked mutant genes as well as those caused by gross chromosomal abnormalities. The noted geneticist H. J. Muller estimates that each individual carries at least eight recessive lethal genes in his sex cells. These will cause death only in the homozygous condition, the possibilities of which can be increased by marriage of relatives.

ROLES OF HEREDITY AND ENVIRONMENT

If the human organism is the product of both heredity and environment, which is the more important factor? The genes determine the general pattern of development, but at every step of the way these genes must interact with environmental factors. Although the genic constitution within the nucleus remains essentially the same in all cells and tissues, the cytoplasm undergoes various changes under the joint influence of genes on the one hand and a constantly changing environment on the other. Both these factors are, therefore, important in the realization of the individual's possibilities.

HAIR AND EYE COLOR

The color of the hair seems to be determined by several genes or several pairs of modifying factors. Often the pigmentation of hair and eye color are correlated. In general, blond hair is recessive to the darker shades, but the presence of varying shades of blond and dark hair indicates a blending effect or the interaction of more than one pair of genes. Red hair is recessive to the other shades of hair. One may be homozygous for genes for red hair and have a darkish shade of red hair color because of the presence of genes for darker hair.

The color of the eyes is due to the presence and location of pigment in the iris. If the pigment is on the back of the iris, the eyes are blue; if the pigment is on both back and front of the iris, the eyes are of the darker shades. If pigment is lacking altogether (albinism), the eyes are pinkish because of the blood vessels. Blue eyes are recessive to the other eye shades, and when the parents are blue eyed, the children are normally blue eyed. Blue-eyed children may appear also if parents are heterozygous for the darker eye colors. The manner in which eye color is inherited indicates that there are many kinds of genes, all variant forms of the gene for dark color, which behave in a simple mendelian way in their hereditary expression.

BLOOD GROUP INHERITANCE

The inheritance of blood groups follows Mendel's laws. This inheritance is based on three genes or allelomorphs, I^a, I^b, and i. I^a and I^b are antigens and are dominant, and they never appear in a child's blood unless present in at least one of the parents; i represents no antigen and is recessive to the antigens. Neither I^a nor I^b is dominant to each other, but each is dominant to i. The relationships of the blood groups and genotypes are as follows:

Blood groups	O	A	B	AB
Genotypes	ii	$I^a I^a$ or $I^a i$	$I^b I^b$ or $I^b i$	$I^a I^b$

If both parents belong to group O, a child must also belong to group O. On the other hand, if one parent belongs to group A and the other to B, the child could belong in any one of the four groups. The various possibilities of inheritance are shown in Table 20-2.

This pattern of heredity has some practical applications in medicolegal cases involving disputed parentage. If a child in question has group A and the supposed parents group O, it is obvious that the child could not belong to them. On the other hand, if the parents belong to groups A and B, the blood tests would prove nothing.

The frequencies of the blood groups O, A, B, and AB have been tabulated by investigators for most races all over the world, although data are scanty and incomplete in many instances. Among the interesting facts revealed by these studies is the absence of the allele B in the American Indians and Australian aborigines, the high frequency of B in Asia and India and its decline in Western Europe, and the high frequency of group O in Ireland and Iceland. The distribution of the Rh factor also shows a varied pattern. Reasons for these varied distributions are obscure, but some explanations have been advanced, such as genetic drift involving small populations, natural selection, and migration and mixing of races.

INHERITANCE OF MENTAL CHARACTERISTICS

Mental traits, both good and bad, are known to be inherited, although because of environmental influences it is not always possible to appraise them genetically. Heredity is in general responsible for the basic patterns of intelligence and mental deficiencies, although environmental factors can and do influence the development of intellectual capacities. Intelligence is not a single hereditary unit; apparently many different genes are responsible for its expression. The field of inheritance affords many examples of both outstanding abilities and mental defects being handed down through long family histories. The best kind of environment in the world cannot make a superior intelligence out of one who has inherited a moron potentiality, and conversely, superior inherited mental potential may always remain undeveloped unless stimulated by a favorable environment. Identical twins, for instance, with exactly the same genes for intelligence may show considerable difference in their I.Q.s when reared under different advantages of education and culture.

Of the several million feebleminded persons in the United States, only a small percent have come

TABLE 20-2. Inheritance of blood groups

Parent groups	Possible children groups	Impossible children groups
O × O	O	A, B, AB
O × A	O, A	B, AB
O × B	O, B	A, AB
A × A	O, A	B, AB
A × B	O, A, B, AB	None
B × B	O, B	A, AB
O × AB	A, B	O, AB
A × AB	A, B, AB	O
B × AB	A, B, AB	O
AB × AB	A, B, AB	O

from feebleminded parents. The great majority of them are born of parents heterozygous for this trait. Such parents may be in all particulars normal for intelligence and carry the recessive gene for feeblemindedness. Recessive genes are hidden and may remain so until they become homozygous.

The inheritance of special abilities, such as that of music, has been thoroughly investigated, and the evidence indicates that talents for music may have a hereditary basis. The hereditary pattern, however, is very complex and probably involves a number of genes. It is not always possible to distinguish between the influences of heredity and environment in musical families.

INHERITANCE OF TWINNING

Fraternal twins, which are four or five times as common as identical ones, result from the independent fertilization of separate ova by separate sperm. They are simply conceived and born together, and genetically they have the same likenesses and differences as ordinary brothers and sisters. They may be of the same or opposite sex. **Identical** twins come from a single zygote, the result of the segregation of cells in the inner cell mass. They have the same genetic constitution and are always of the same sex. Identical twins show a remarkable similarity in their general characters, both physical and mental. When the halves of the zygote fail to separate completely, Siamese twins are the result.

INHERITANCE OF CERTAIN PHYSICAL TRAITS IN MAN

It is impossible to deal with man's heredity in a simple mendelian ratio. Information about his heredity must be acquired by inspection and analysis of family life histories, or pedigrees. Such a plan involves the formation of hypotheses to explain hereditary expression and careful checking of these hypotheses to determine whether they apply to the data obtained. The inheritance of many abnormal characters has been stressed in family pedigrees because they are easily followed.

Following are some of the more common traits, but their dominance or recessiveness is not always clear cut.

Dominant	Recessive
Curly hair	Straight hair
Dark skin color	Light skin color
Hairy body	Normal hair
Skin pigmentation	Albinism (no pigment)
Near or farsightedness	Normal vision
Hereditary cataract	Normal vision
Astigmatism	Normal vision
Glaucoma	Normal vision
Normal hearing	Deaf-mutism
Normal color vision	Color blindness
Normal blood clotting	Hemophilia
Broad lips	Thin lips
Large eyes	Small eyes
Long eyelashes	Short eyelashes
Short stature	Tall stature
Polydactylism (extra fingers or toes)	Normal number of digits
Brachydactylism (short digits)	Normal length of digits
Syndactylism (webbed digits)	Normal digits
Normal muscles	Progressive muscular atrophy
Hypertension	Normal blood pressure
Diabetes insipidus	Normal excretory system
Tasters (of certain substances)	Nontasters
Huntington's chorea	Normal
Normal mentality	Schizophrenia
Nervous temperament	Phlegmatic temperament
Average intellect	Very great or very small
Normal intellect	Feeblemindedness

INFLUENCE OF RADIATION ON HUMAN HEREDITY

Radiation and radioactive substances such as x-rays, radium rays, and ultraviolet light greatly increase the mutation rate of the gene. Most of these gene mutations are lethal and either destroy or produce abnormalities of various types in the offspring of animals exposed under certain conditions to irradiation.

In the light of the fierce controversy that is raging over the possible effects of radiation in an atomic age, it may suffice to summarize certain generalizations that the data seem to substantiate:

1. All life is constantly exposed to high-energy radiations. Some of this radiation comes from natural sources, such as cosmic rays from outer space, radioactive elements (radium, thorium, radioactive isotopes of potassium), and atomic disintegration

within the organism; some comes from the technologic use of radioactive substances in medicine and industry (x-rays, radium treatment, mustard gas) and some from the fallout of the explosion of atomic bombs.

2. High-energy radiations are definitely known to increase the rate of mutation in every organism tested.

3. Most mutations are harmful to the organism whether they are natural mutants or artificial ones induced by man. This harm is produced chiefly by the ionization effect by which electrons are removed from atoms and attached to other atoms so that positive and negative ion pairs are produced. Protoplasm is greatly injured by ionization because the molecular organizations of chromosomes especially is disrupted and mutations (mostly harmful) of the code result.

4. Mutations do not as a rule occur in more than one gene at a time, for a mutation is highly localized and may involve only one gene locus when the latter is struck by a quantum of radiation.

5. Radiations may affect any cell in the body, but only those changes that are produced in sex cells can be transmitted to the offspring.

6. Sensitivity to radiation effects vary from species to species. Mice are far more sensitive than fruit flies.

7. An atomic fallout occurs when atom bombs are exploded. Unstable and radioactive isotopes of many elements are hurled high into the air and carried about the earth by the winds, eventually settling down to earth on a large or small area, depending on circumstances. Some radioactive elements may settle out quickly and others may remain aloft for years.

8. Genetic radiations from whatever source are insidious because their effects seem to be cumulative, although this is denied by some authorities. Small exposures add up, and what really counts in the long run is the total amount of radiation one is exposed to during one's reproductive life. The rate of delivery of the radiation is of no consequence; the genetic effect is the same for low or high rate.

9. Some of the radioactive elements released in an atom bomb explosion decompose very slowly and have half-lives of many years. For example, strontium 90 has a half-life of 28 years. This means that

the body may be exposed to these isotopes for a long time.

10. The danger of radioactive substances is far greater to future generations than to present ones because of the genetic implications referred to.

11. In addition to genetic effects of radiation, there are also physiologic or somatic effects such as excessive burns. Some somatic effects are leukemia, cancer, and a syndrome of radiation illness, depending on the amount of exposure. Body tissues of rapidly dividing cells are especially prone to damage. If the exposure has not been excessive, many somatic effects may be healed by therapeutic measures, but healing does not apply to genetic damages.

HUMAN CHROMOSOMAL ABNORMALITIES

Instead of the normal number of 46 chromosomes, a number of individuals have been reported who had one more or one less than the normal number. In mongolism, which is characterized by mental and physical retardation together with a mongolian type of eyelid fold, the individual has 47 chromosomes. This extra chromosome is thought to be due to nondisjunction of a pair of chromosomes (autosomes) during meiosis in the maternal ovum so that some of the eggs carry 24 chromosomes. The occurrence of mongolism varies with the age of the mother. With mothers under 30 years of age, its frequency is about 15 in 10,000; in those 45 years of age or older, its frequency may be increased fiftyfold. Few mongolians live to maturity, but those who have done so and had offspring produced mongolians and normal children in about equal proportions. The predisposition to have a mongolian child can be detected by an examination of tissue culture cells made from certain blood cells of the woman, where there is a likelihood that such a condition could occur.

Certain other conditions are associated with abnormalities of the sex chromosomes. Klinefelter's syndrome is produced by the presence of two X chromosomes plus a Y, or an XXY complex. Such an individual is a sterile male with undeveloped testes and a tendency toward female breasts (gynecomastia).

The XYY pattern also occurs and appears to be

associated with antisocial and criminal tendencies.

Turner's syndrome (45 chromosomes) is a condition in which the individual has only one X chromosome with no Y chromosome as a mate instead of the normal female (XX) or male (XY) state. In Turner's disease the XO (O = absence) constitution produces an external appearance of femaleness, although the person may lack ovaries or else have imperfect ones and so is sterile.

Both Klinefelter's and Turner's syndromes are caused by meiotic nondisjunction in the paternal or maternal germ lines. Since hormones are involved in regulative processes of sex, abnormalities in primary and secondary sex characters just mentioned may be due to hormonal imbalance produced by the faulty chromosome behavior.

21

The evolutionary concept

MEANING OF EVOLUTION

It is evident that animals and plants, as we see them, present an immense variety of different forms, ranging all the way from those of small size and low degree of complexity to those of large size and complicated structure. Moreover, organisms are found in nearly every kind of habitat that will support life at all and manifest every conceivable kind of adaptation to their surroundings. The principle of evolution, which attempts to explain why there is such tremendous variety in plants and animals, is the framework of biology.

Evolution is the doctrine that modern organisms have attained their diversity of form and behavior through hereditary modifications of preexisting lines of common ancestors. It means that all organisms are related to each other because of common descent. Basically, evolution is the change in the relative frequency of genes. Organic evolution is only one aspect of the larger view that the entire earth has undergone an amazing evolution of its own. The grand concept of evolution arouses in all thoughtful individuals a feeling of awe and wonder, and one of inspiration, at the great drama that has been and is unfolding.

Evolution is a continuous process and is taking place today. Almost any group of animals one may select is undergoing the process of evolution. Any diversity a group may possess, now or past, has been the product of prior evolution. The variability in structural and physiologic features of any form is the basis of evolution still to come. It is a very slow process but some striking evolutionary changes have taken place within historic times. Once it occurs, evolution seems to be irreversible except for small, minor reversals.

Although the evolution of every great group of organisms is more or less unique in its details, three basic kinds of organic evolution have been followed in producing the great diversity of life we know today. There is first a buildup within a population of a **gene pool** of specific characters of structure, function, or habit. The term gene pool refers collectively to all of the alleles of all of the genes in a population. From one generation to the next there is usually a change in the gene pool so that there is a gradual alteration in the range or kind of these characters. Variations in the environment may account for such changes because the genetic composition will develop in a different way under variable environmental conditions, but changes in the genetic composition may also occur by mutation. Through successive generations, there will be a succession of different gene pools that will be expressed in different characters. This results in a

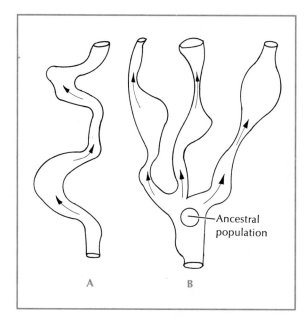

FIG. 21-1. Two basic patterns of evolution. **A,** Phyletic or sequential pattern. Phyletic evolution involves progressive, directional shift of average genotypes and phenotypes of population from one generation to another. Although population changes, there is no splitting up of population as it continues within zone of varying breadth, depending on fluctuations of characters at particular times. Differences may be great enough, however, for origin of new species. **B,** Divergent or cladogenesis pattern. This type involves splitting up of population into two or more separate lines, each of which may give rise to taxa of different ranks. It is well represented by adaptive radiation characteristics of most major groups. Both these patterns are usually found in any population over long periods of time.

change of the population as a whole and is not a splitting up of the population. This kind of evolution is usually restricted to a fairly stable organism-environment complex and is called **sequential** or **phyletic evolution** (Fig. 21-1, *A*). Such evolutionary changes may be slight if the environment is stable, but shifts of gene combinations in adaptive types may occur in response to a changing environment, thus producing distinguishably different forms (species).

A rapid shift of a population into another and different adaptive zone where it is necessary for the organisms to adapt quickly results in a type of rapid

evolution called **quantum evolution.** Quantum evolution is promoted by excessive mutation, relaxation of selective pressure, and such short cuts as paedomorphosis. It may produce taxa of all ranks but probably applies best to such taxonomic units as families, orders, and classes (Simpson).

Divergent evolution or **cladogenesis** (Fig. 21-1, *B*) involves a splitting process in which new populations originate from old populations. It is the differentiation of two or more groups within a widespread population. It involves taking advantage of adaptations to spread out into various available ecologic niches (opportunism). Within each of the adaptive lines, more or less phyletic evolution occurs, with alternate stages of advancement and stability. In some cases there is enough diversification within a line for the formation of a new species.

Probably the most important factor contributing to the evolutionary processes are mutation rate, natural selection, chance or random genetic drift, and isolation and population structure.

Mutation rate. Mutation rate is the chance, random change of hereditary mechanisms. This and the subsequent reshuffling of the genes into various new combinations by bisexual reproduction make possible structural differences in populations.

Natural selection. Natural selection is the basic external factor that molds the development of a species. Selection determines which mutant genes survive and which ones are eliminated. It molds the genes into a coordinated whole, and although it cannot originate new characters, it can determine what sets of genes can be of immediate biologic usefulness to the species. It can produce rapid evolutionary transformation or it can stabilize the evolutionary trend.

Chance or random genetic drift. In small populations some genes may be entirely lost because certain mutant genes may not be included in gametes in the process of meiosis, or they may not be present at all. Others may be present with a greater frequency than they were in the original larger population. Close interbreeding in such populations would tend to make heterozygous genic pairs homozygous.

Isolation and population structure. An isolated population would not share its mutant genes with other populations and would develop its own unique evolution. This process could lead to the formation of a new species in each of the isolated groups.

How these four basic factors operate in the evolutionary process may be made clear by selecting a hypothetical species population of great extent and following it through its possible evolutionary fate. Suppose this population is found in a wide geographic area and exhibits the characteristics of a **cline;** that is, there is a gradual, continuous, gradient change in the members of the population because of adjustments to local conditions that show considerable variations in different parts of the cline. Thus in certain parts of the cline, climate conditions may be hot, in others cold; weather may vary from extreme moisture in some parts to very dry in others, etc.

Within a cline the species will be divided into smaller units of population (**demes**) that are more or less isolated from each other in the different habitats found within the range of a cline. The members of a deme may breed freely with each other but usually not with the members of other demes. Each deme more or less can develop an evolution of its own, for small hereditary differences (mutations) that occur in demes may be, to some extent, unique, and thus each deme in time becomes different from other demes. Natural selection will operate to select the better-adapted characters in each case and their possessors will increase in frequency. Some of the mutations may confer only small advantages at first, but reshuffling of the genes at meiosis and further recombinations may increase their effects. Recessive mutants will spread slowly, dominant ones more rapidly. If the environmental conditions remain fairly stable, there is a slow successional evolution within the deme; readaptation to changing conditions, however, speeds up the evolutionary rate.

Some of the demes may come together (if interbreeding can occur) and fuse; others may differentiate far enough to prevent interbreeding and form true species. Whenever two demes fuse, each contributes its pool of genes to the future offspring, which thus acquire advantageous genes of both demes. Still other demes may become extinct in a large cline. A common pattern of demes consists of many subspecies formed by divergent evolution in partially isolated demes. The time factor may cause these subspecies to differentiate into true species. Because of the smallness of some of the demes, genetic drift can operate to produce a marked differential in gene frequency. It will thus be seen that whenever a population is subdivided into small subunits there is a strong possibility that there may be a rapid evolutionary process of a divergent nature, with results favorable to the formation of new species.

INHERITANCE OF ACQUIRED CHARACTERS

One of the earliest evolutionary theories proposed was the one by Jean Baptiste de Lamarck (1744-1829), a French biologist. This theory is named after him and is commonly known as **lamarckianism.** According to Lamarck, new organs arose in response to the demands of the environment. The size of an organ depended on its use or disuse. He would explain the limbless condition of the snake by the handicap of legs in crawling through dense vegetation, and the long neck of the giraffe by its habit of reaching up into trees to browse. There is little or no evidence for Lamarck's theory of inheritance of acquired characters.

The effects of use and disuse are restricted mainly to somatic tissues but genetics does not give support to the transmission of somatic characters. There can be no permanent change unless the germ cells themselves are altered.

NATURAL SELECTION THEORY OF DARWIN

The first real mechanical explanation of evolution was the one proposed by Charles Darwin in his work *The Origin of Species* published in 1859. Another English scientist, Alfred Russell Wallace (1823-1913), should receive some credit for this theory of **natural selection.** Both Darwin and Wallace had arrived at the main conclusions of this theory independently, but the publication of *The Origin of Species* the year after Wallace had announced his conclusions in a brief essay really clinched Darwin's position and prestige.

As a young man Darwin had spent five years (1831-1836) as a naturalist on board the *Beagle,* a vessel that had been commissioned to make oceanographic charts for the British admiralty. This vessel in the course of its voyage around the world spent much time in the harbors and coastal waters of South America and adjacent regions, and Darwin had ample time to make extensive collections and studies of the flora and fauna of those regions. He kept a detailed journal of his observations on the animal and plant life, an account that he later published. After the idea of natural selection dawned on him, he spent the next twenty years accumulating data from all fields of biology to prove or disprove his theory. So much evidence was brought forth and so forcible were the arguments advanced for natural selection by Darwin in his book, that one may safely say that a new era in thinking not only in organic evolution but also in many related fields, dates from the publication of the book.

The essential steps in the theory of natural selection as advanced by Darwin include nature of variation, great rate of increase among offspring, struggle for survival, natural selection and variation, survival of the fittest, and formation of new species.

Nature of variation. No two individuals are exactly alike. There are variations in size, coloration, physiology, habits, and other characteristics. Darwin did not know the causes of these variations and was not always able to distinguish between those that were heritable and those that were not. Only the heritable ones are important in evolution, since those caused by environmental factors of temperature, food, etc., are not passed on to succeeding generations. He laid stress on artificial selection in domestic animals and plants and the role it played in the production of breeds and races of livestock and plants. Darwin thought that this idea of selective breeding could also be produced by agencies operating in the wild state.

Great rate of increase among offspring. In every generation the young are far more numerous than the parents. Even in a slow-breeding form, such as the elephant, if all the offspring lived and produced offspring in turn, in a few hundred years the earth could not hold the elephants. Most of the offspring perish, and the population of most species remains fairly constant under natural conditions. Usually natural checks of food, enemies, diseases, etc. keep the populations within bounds.

Struggle for survival. If more individuals are born than can survive, there must be a severe struggle for existence among them. This competition for food, shelter, breeding places, and other environmental factors results in the elimination of those that are not favorably suited to meet these requirements.

Natural selection and variation. Individuals of the same species tend to be different through variations. Some of these variations make it easier for their possessors to survive in this struggle for existence; other variations are a handicap and result in the elimination of the unfit.

Survival of fittest. Out of the struggle for existence there results the survival of the fittest. Natural selection therefore determines that those individuals that have favorable variations will survive and will have a chance to breed and transmit their characteristics to their offspring. The less fit will naturally die without reproducing themselves. This process will operate anew on each succeeding generation so that the organisms will gradually become better and better adapted to their environment.

Formation of new species. How does this result in new species? According to Darwin, whenever two parts of an animal or plant population are each faced with slightly different environmental conditions, they would diverge from each other and in the course of time become different enough from each other to form separate species. In this way two or more species may arise from a single ancestral species. Also a group of animals, through adaptation to a changed environment, may become different enough from their ancestors to be a separate species. Variations that are neither useful nor harmful will not be affected by natural selection and may be transmitted to succeeding generations as fluctuating variations. This explains many variations that have no significance from an evolutionary viewpoint.

Appraisal of theory of natural selection. Certain weaknesses in Darwin's theory centered around the concepts of variation and inheritance. Darwin made little distinction between variations induced by the environment and those involving alterations of the germ plasm. It is now known that many of the types of variation Darwin stressed are noninheritable and can have no significance in evolution. Only variations arising from changes in the genes (mutations) can furnish material on which natural selection can act. Darwin referred to sudden and radical variations as "sports" and considered them of little importance. However, modern biology now considers mutations as the cornerstone of evolutionary change.

Darwin thought that natural selection could operate indefinitely in promoting the development of a desirable variation. We now know that when the population becomes homozygous for the genes of a particular trait, natural selection can no longer operate. Additional genic changes (mutations) must occur before selection can continue.

Darwin did not appreciate the real nature of isolation in the differentiation of new species. There are really three types of isolation involved—geographic, genetic, and ecologic. Darwin noticed that groups of animals cut off from each other by natural barriers, such as mountains, deserts, and water, were often quite different from each other. This **geographic isolation** accounts for the existence of many different species and subspecies of animals in rather small regions that are broken up by natural barriers. When groups of individuals of a given species are in slightly different environments, mutations may arise within each group, enabling each to diverge from the others. If separated long enough, they may become distinct species. **Genetic isolation** is as effective in isolating separate species as geographic isolation. When two groups of animals become infertile with each other because of genetic changes, they become distinct species. **Ecologic isolation** may be due to differences in habitats and ecologic niches that tend to separate animals during the breeding season. Or, two groups of animals may actually be able to interbreed, but because of a seasonal variation in their breeding habits, they are effectively isolated from each other.

Natural selection seems to play its most important role in the later stages of the evolutionary process and may determine which mutations have survival merit. A particular trait in a successful species may be nonadaptive (have no value), yet not sufficiently injurious to jeopardize the survival of the species. However, if the environment changes or if the animal moves into a new environment, this trait may prove to be advantageous. Thus **preadaptation** may have played an important role in the evolutionary process and may account for the remarkable way in which some animals have flourished in new surroundings.

ORTHOGENESIS

Orthogenesis refers to the tendency for a group of animals to continue to change in a definite direction. This is also called **straight-line evolution.** The idea involved is that certain structural changes once started may continue without deviation for an indefinite period unless checked by extinction. Paleontology furnishes many examples, as for instance the horse, where certain structural trends such as increased size, reduction in number of toes, and increased differentiation of teeth are marked. The theory does not explain the underlying causes for the evolutionary change along a particular line.

This theory also involves the idea of the irreversibility of evolution, or the view that the trend of change does not return toward a former condition. The sequence of the adaptive patterns in orthogenesis gives the impression of direction and orientation, but paleontologists now recognize that most examples do not have direct, unbranched lines. Instead there have been many side branches which have been eliminated by natural selection.

MUTATION

Little was known in Darwin's time about the behavior of chromosomes and their bearing on heredity. Soon after the principles of mendelism were rediscovered in 1900, the parallelism between the chromosomal behavior and mendelian segregation was worked out. Hugo de Vries, a Dutch botanist, had stressed the importance of **mutations** in the evolutionary process. Working with the evening primrose, he had found certain types of this plant differing materially from the original wild plant and, more important, he found that these aberrant forms bred true thereafter. de Vries explained these mutant forms mainly on the basis of a recombination of chromosomes, but since his time, stress has been laid on genetic transformations. The experimental production of mutations through radiation, x-rays, etc. and the discovery of the giant chromosomes in the salivary glands of certain larval insects have added to this understanding.

Before a new species can evolve, there must be both mutation and natural selection. Mutations furnish many possibilities, natural selection determines which of them have survival merit, and the environment imposes a screening process that passes the fit and eliminates the unfit, but elimination may be slow.

Nature of mutations. Mutations may be harmful, beneficial, or neutral in their action. Perhaps the majority are harmful because most animals are already adapted and any new change would likely be disadvantageous. In *Drosophila* it has been found that the same mutation occurs with a certain frequency. Thus the red-eyed wild type can be expected to mutate to the white-eyed type every so often. Moreover, the white-eyed type undergoes mutation, either changing back to the original red eyes or else to one of the other eye color allelomorphs. Mutant characters tend to be recessive in their hereditary patterns and may show up as a phenotype only when they are homozygous; when paired with normal or original allelomorphs, they usually show no effect. Some mutant characters are lethal, but this action is expressed, as a usual thing, only in the homozygous condition.

Mutations can be divided into those that produce small changes (**micromutations**) and those that produce large changes (**macromutations**). Evolutionary changes produced by the action of the former are referred to as **microevolution;** those produced by the latter, as **macroevolution.** The great majority of evolutionists now favor small mutations as the more important, believing they explain many intermediate forms (races, subspecies, etc.) between the parent species and the new one. Those who hold to macroevolution explain such intermediate forms as being only geographic varieties of basic species.

Types of mutations. There are two main types of mutations—gene mutations and chromosome mutations. A **gene mutation** is a chemicophysical change of a gene resulting in a visible alteration of the original character. Although the actual change in the gene is largely unknown, it is thought to be a rearrangement of the nucleotides within a region of the DNA molecule. Such changes cannot be detected under the microscope, for there are no visible alterations in the chromosomes bearing the genes in question. Many of this type of mutation have been found in *Drosophila* and other forms. Most gene mutations are point mutations, that is, the physical or chemical change of one gene.

Chromosome mutations involve either chromosome rearrangement during meiosis or an alteration in the number of chromosomes. The rearrangement of the chromosomes may involve inversion of the linear order of the genes, the deletion of blocks of genes, or translocation of portions of chromosomes in which a part of one chromosome becomes attached to another nonhomologous chromosome. Such changes are usually detectable and often produce phenotypic changes that are inherited in the regular mendelian manner. Increase in number of chromosomes usually involves the formation of extra sets of chromosomes, sometimes a doubling or tripling of the diploid set, resulting in polyploidy, a condition more common in plants than in animals. Polyploids are usually characterized by a larger size than is the parent stock. Among roses there are species with 14, 28, 42, and 56 diploid chromosomes, although the basic parent rose is thought to have 14 chromosomes. Such duplication of chromosomes is due to the omission of the reduction divisions so that germ cells are formed with the diploid number rather than the usual haploid number. This variability gives natural selection something to work on in the production of a new species, for it can eliminate the unfit and allow the adaptable characters to survive.

Causes and frequency of mutations. Different genes possess different frequencies of mutation rates because some genes are more stable than others. Mutation is also a reversible process. Gene **A,** for instance, may mutate to gene **a,** and gene **a** may mutate back to gene **A.** This reversibility must be taken into account in mutation equilibrium, and the difference between the mutation rate in one direction and the mutation rate in the reverse direction constitutes **mutation pressure,** which is usually of a low magnitude. Some plants may be expected to produce 10% of their offspring with at least one mutant gene. The widely known *Drosophila* is supposed to produce at least one new mutation in every 200 or so flies. In higher vertebrates the average mutation rate per individual may be between 1 in

50,000 to 1 in 200,000 (Mayr). Many mutations are not detected and estimates of their frequency are in many cases only guesses.

Little is known about the causes of natural mutations. Many mutations that were first noted in laboratory stocks are now known to occur also in the wild state. Both gene and chromosome mutations can be produced artificially by the influence of x-rays, ultraviolet rays, chemicals, temperature, and other agencies. It has been suggested that cosmic rays may be responsible for the appearance of mutations in wild populations.

RATES OF EVOLUTION

Evolution has not always proceeded at the same rate among different types of organisms or within different geologic periods. Brachiopods have undergone relatively little change since early geologic times. The primates have evolved rapidly. The nature of geologic periods has had an important bearing on the rate of evolution. When there are geologic changes, only animals with suitable variations can adapt to the changes. On the other hand, periods of geologic uniformity have usually been periods of slow evolutionary progress. Vertebrates have tended to evolve faster than invertebrates, but there are exceptions. The opossum has changed little since the late Cretaceous period. Evolution seems to advance by spurts rather than in a steady, uniform manner.

It has been estimated that it takes 500,000 to 1 million years for a new species of bird or mammal to evolve and 20,000 to 50,000 years for a subspecies in the same group. The horse required 45 to 50 million years to evolve to its present state and passed through some 8 genera. Thus about 6 million years on the average were required for each genus.

EVOLUTION AND ADAPTATION

The major aim of evolution is the adaptation of the organism to its environment. All organisms share in common about the same biochemical compounds, the same kinds of biosynthesis and energy transfer, the same structural features of tissues, and the same metabolic mechanisms of growth, respiration, digestion, etc. These primordial processes of mutual

adaptations were fashioned somewhere in the long evolutionary process and no explanation is at present adequate to account for them. All individual adaptations are shaped by evolution because maladjusted organisms simply do not survive to reproduce. Fitness to the environment must be found at all levels, from cellular ecology to that of populations. Adaptation, then, is fitting a biologic system to harmonize with the environmental factors of its existence. The adaptive responses of any organism are determined by and restricted to the range of environmental conditions the organism has had in its long evolutionary and phylogenetic experience. Only the gross aspects of adaptation are predetermined; the individual must fill in the details to adjust to the conditions it meets in its existence. The limits of its adjustment are fixed by heredity.

No animal is perfectly adaptive to its environment; adaptation is relative because the adaptive endowment of an organism is never able to anticipate all the constantly occurring variations in its environment. The successful adaptation of an organism is determined by the sum total of all its adaptations. This means that, although some of its adaptations may be highly favorable and others unfavorable, the animal survives. This is why many so-called adaptations, such as protective coloration and sexual selection, are now viewed with a critical eye.

In **protective coloration** the organism blends into its environment so that it can escape detection by enemies. A common example is the dark dorsal and lighter ventral side of fish and many birds. Their color blends with the sky to a potential enemy viewing them from underneath, and viewed from above, the darker shades blend with the darker shades and regions below. Many types of concealment blend the animal into its background; frogs and lizards have the added advantage of changing their color to suit their background. **Mimicry** adaptation is common among many forms in which harmless animals have found survival value by imitating other species that are well equipped with defensive or offensive weapons. Thus harmless snakes may have a resemblance to venomous ones, and some flies look very much like bees. There are also many examples of **warning** coloration, which advertises the presence of a well-protected animal, such as the white stripes of the skunk and the brilliant colors of the poisonous coral snake.

These adaptive resemblances, whether of protection, warning, or mimicry, have been subjected to experiments with the idea of determining their selective value. Conclusions from these experiments are often conflicting. Some biologists have gone as far as to deny the selective value of color altogether. The extensive investigations of McAtee on the stomach contents of birds indicated that protectively as well as nonprotectively colored insects were eaten by birds without discrimination, but the relative abundance of the protected and unprotected forms in nature must be considered in such experiments. Other experiments seem to indicate that color patterns do play some useful role in survival selection. It has been suggested that some animals that are not protectively colored to the human eye are so protected when viewed by the type of vision of their enemies. Many experiments have also been performed in which predators had an opportunity to choose among prey bearing concealing, warning, or other color devices, and the results indicated that these color patterns do have selective value.

MODERN SYNTHESIS OF EVOLUTION (NEO-DARWINISM)

In the first thirty years of the present century there was gradually accumulated a great factual amount of information about the chromosomal and genic theory of heredity, the way mendelian heredity operated, and a more fundamental understanding of the mutation theory. All these branches of investigation had become more or less unified into what we now call cytogenetics. Under the influence of a brilliant group of biologic thinkers, such as J. S. Huxley, R. A. Fisher, and J. B. S. Haldane in England and Sewall Wright, H. J. Muller, and T. Dobzhansky in America, there has been a fruitful attempt to unify all the various theories and ideas of evolution into one underlying mechanism of organic evolution. The new outlook on evolutionary causes has pinpointed the genotype and its behavior in the organism as the focal point for understanding how evolution operates. Evolution has thus been found to be mainly a sequence of genic changes.

POPULATION GENETICS AND EVOLUTIONARY PROCESSES

Evolution implies changes in the hereditary characteristics, and it is generally agreed that the best conditions for evolutionary changes occur in large populations that are broken up into small subdivisions. The population must be considered the natural form of existence of all species. Moreover, the materials with which evolution works are the genetic variations produced by mutation and recombination of genes. This is especially true of biparental populations in which the normal mechanisms, aside from mutations, of recombinations can operate to produce great variation. The gene pool of large populations must be enormous, for at observed mutation rates many mutant alleles can be expected at all gene loci. In some cases more than 40 alleles of the same gene have been demonstrated. Suppose there are two alleles present, A and a. Among the individuals of the population there will be three possible genotypes: AA, Aa, aa. When there are three alleles present, there are six possible genotypes. Increasing the number of alleles increases the possible genotypes. The reshuffling of genes at the reduction division of meiosis makes possible combinations of a gene at one locus, with any of several others at other loci.

Changes in uniparental populations occur by the addition and elimination of a mutation; in biparental populations the mutant gene may combine with all existing combinations and thus double the types. With only 10 alleles at each of 100 loci, the number of mating combinations would be 10^{100}. Many genotypes and phenotypes can be produced when only a few pairs of genes are involved, but since organisms have thousands of pairs the number of possible combinations is staggering. Even though many genes are found together on a single chromosome and tend to stay together in inheritance, this linkage is often broken by crossing-over. If no new mutations occurred, the shuffling of the old genes would produce an inconceivably great number of combinations. But this is not the whole story because genes exert different influences in the presence of other genes. Gene A may act differently in the presence of gene B than it does in the presence of gene C. If this diversity is possible in a single population, suppose two different populations with different genes should mix by interbreeding. It is easy to see that many more combinations of genes and their phenotypic expression would occur. All this means that populations have enormous possibilities for variation. Genetic variation produced in whatever manner is the material on which natural selection works to produce evolution. It is a slow process, but on a geologic time scale it brings about striking evolutionary changes.

Why most mutations are recessive. If mutations represent the material for evolutionary change, would not dominant mutants be the most important in evolutionary processes? Why are most mutations recessive? Not all of the answers are known. We do know that the character of each organ in an organism is controlled by many genes, not by a single gene, and that this interaction of genes or mutants may be complex. One may inhibit the action of another or one may have no effect without the other, etc. Dominance and recessiveness can be altered in this manner also. Ordinarily, a recessive gene would express itself only in a homozygous condition, but there is some evidence that new mutations may not be completely recessive and may make their presence felt in a heterozygous condition.

Why recessive genes are not lost from the population. In an interbreeding population why does not the dominant gene gradually supplant the recessive one? It is a common belief that a character dependent on a dominant gene will increase in proportion because of its dominance. This, however, is not the case, for there is a tendency for genes to remain in equilibrium generation after generation. In this way a dominant gene will not change in frequency with respect to its allele. This important principle is based on a basic law of population genetics called the Hardy-Weinberg equilibrium. According to this law, gene frequencies and genotype ratios in large biparental populations will remain constant thereafter unless disturbed by new mutations, by natural selection, or by genetic drift (chance). The rule does not operate in small populations. A rare gene, according to this principle, will not disappear merely because it is rare. That is why certain rare traits, such as albinism, persist for endless generations. It is thus seen that variation is retained even though evolutionary processes are not in active operation. Whatever changes occur

in a population—gene flow from other populations, mutations, and natural selection—involve the establishment of a new equilibrium with respect to the gene pool, and this new balance will be maintained until upset by disturbing factors.

The Hardy-Weinberg formula is a logical consequence of Mendel's first law of segregation and is really the tendency toward equilibrium inherent in mendelian heredity. Select a pair of alleles such as T and t. Represent the proportion of T genes by p and the proportion of t genes by q. Therefore, $p + q = 1$, since the genes must be either T or t. By knowing either p or q, it is possible to calculate the other. Of the male gametes formed, p will contain T and q will contain t, and the same will apply to the female gametes. (See checkerboard in Chapter 20.) As we know from Mendel's law, there will be three possible genotypic individuals, TT, Tt, and tt, in the population. By expanding to the second power, the algebraic formula $p + q$ will be $(p + q)^2 = p^2 + 2pq + q^2$, in which the proportion of TT genotypes will be represented by p^2, Tt by $2pq$, and tt by q^2. Recall the 1:2:1 ratio of a mendelian monohybrid. The homozygotes TT or tt will produce only T or t gametes, whereas the heterozygotes Tt will produce equal numbers of T and t gametes. In the gene pool the frequencies of the T and t gametes will be as follows:

$$T = p^2 + \tfrac{1}{2}(2pq) = p^2 + pq = p(p + q) = p$$
$$t = q^2 + \tfrac{1}{2}(2pq) = q^2 + pq = q(q + p) = q$$

In all random mating the gene frequencies of p and q will remain constant in sexually reproducing populations (subject to sampling errors). It will be seen that the formula $p^2 + 2pq + q^2$ is the algebraic formula of the checkerboard diagram, and thus the formula can be used for calculating expectations without the aid of the checkerboard.

To illustrate how the Hardy-Weinberg formula applies, suppose a gene pool of a population consisted of 60% T genes and 40% t genes. Thus:

$$p = \text{frequency of T } (60\% \text{ or } 0.6)$$
$$q = \text{frequency of t } (40\% \text{ or } 0.4)$$

Substituting numerical values of gene frequency in the following,

$$p^2 + 2pq + q^2$$
$$(0.36 + 0.48 + 0.16)$$
$$TT \quad Tt \quad tt$$

the proportions of the various genotypes will be 36% pure dominants, 48% heterozygotes, and 16% pure recessives. The phenotypes, however, will be 84% (36 + 48) dominants and 16% recessives.

On the other hand, suppose 4% of a population is made up of a certain recessive trait, then:

$$q^2 = 4\% \text{ or } 0.04$$
$$q = \sqrt{0.04} = 0.2 \text{ or } 20\%$$

Thus 20% of the genes are recessive. Even though a recessive trait (the phenotypic expression of a gene) may be quite rare, it is amazing how common a recessive gene may be in a population. Only 1 person in 20,000 is an albino (a recessive trait); yet by the above formula, it is found that 1 person in every 70 carries the gene or is heterozygous for albinism.

How chance operates to upset the equilibrium of genes in a population. The Hardy-Weinberg equilibrium can be disturbed, as already stated, by mutation, by selection, and by chance or genetic drift. The term **genetic drift** (Wright) refers to changes in gene frequency resulting from purely random sampling fluctuations. By such means a new mutant gene may be able to spread through a small population until it becomes homozygous in all the organisms of a population (random fixation) or it may be lost altogether from a population (random extinction). Such a condition naturally would upset the gene frequency equilibrium.

How does the principle apply? Suppose a few individuals at random became isolated from a large general population. This could happen by some freakish accident of physical conditions, such as a flood carrying a small group of field mice to a remote habitat or a disease epidemic wiping out most of a population. Suppose that in the general population individuals would be represented by both homozygotes, TT, for example, and heterozygotes, Tt. It might be possible for the small, isolated group to be made up only of TT individuals and the t gene would be lost altogether. Also, when only a small number of offspring are produced, certain genes may, by sampling errors, be included in the germ cells and others not represented. It is possible in this way for heterozygous genes to become homozygous. The new group may in time have gene pools quite different from the ancestral population.

We should note also that most breeding popula-

tions of animals are usually small. A natural barrier, such as a stream, may be effective in separating two breeding populations. Thus, chance could lead to the presence or absence of genes without being directed at first by natural selection.

Genetic drift has been assigned as the cause of the frequency of certain human traits such as blood groups. Among some American Indian tribes it is known that group B, for instance, is far rarer than it is among other races and may be due to small isolated mating units.

There are many who deny the importance of genetic drift, but it is generally agreed that in bisexually reproducing species evolution proceeds more rapidly when a population is broken up into isolated or partially isolated breeding communities, and the smaller the population the greater will be the importance of genetic drift.

Examples of speciation. Speciation refers to the splitting of one species into two or more other species. Although there may be several patterns that result in the diversification of organisms, it is thought that the most effective sequence is that of geographic isolation, morphologic differences (mutations, etc.), and conditions that prevent future fusion of groups. Evolution seems to be most rapid when a group of organisms are presented with a new environment with many unoccupied ecologic niches. By adaptive radiation many new types can evolve to fill these niches.

A striking illustration of isolation and adaptive radiation is afforded by Darwin's finches of the Galápagos Islands. Darwin was struck by the unusual animal and plant life on these islands as contrasted with that on the nearby mainland of South America. These volcanic islands have never had land connections with the mainland and such life as is found there had to arrive by accidental immigration. Darwin was particularly interested in the bird life of the islands, especially the finches.

In 1947 the English ecologist D. Lack published a book on these finches that has served to renew interest in their evolutionary development. All the different species of finches (family Geospizidae) found on these islands are believed to have descended from a South American finch that reached the island. Their most interesting adaptive structures are the different beak modifications that permit them to exploit the food resources of the various ecologic habitats. The ancestral finch was probably a ground feeder, but as competition increased in this habitat, finches evolved that were adapted (or radiated out) to utilize food in other ecologic niches unavailable to the ancestral form. The ground feeders have thick, conical beaks for seed, the cactus ground finch has a long, curved beak for the nectar of cactus flowers, the vegetarian tree finch has a parrotlike beak for buds and fruit, and the woodpecker finch has a long, stout beak (and uses an improvised spine) for probing into bark. Some of the finches are vegetarian and some are insectivorous. Altogether, some 14 species (and some subgenera) of finches have evolved from the ancestral finch that first reached the islands.

An example of speciation in the making is the common leopard frog *Rana pipiens*. For years herpetologists have puzzled over the status of this frog, which ranges from northern Canada to Panama and from the Atlantic coast to the edge of the Pacific states. Individuals from adjacent localities, such as those from New England and New Jersey, or those from Florida and Louisiana, can be crossed successfully and yield normal and viable embryos. Also, a frog of this species from New Jersey can be crossed with one from Louisiana and will produce normal embryos. But crosses between individuals from widely separated localities, such as a cross between one from Wisconsin and one from Florida, will produce abnormal and nonviable embryos. Evidently the genetic differences between the northern and southern forms are great enough to prevent normal hybrid development. Selection has produced different developmental physiologies in these frogs. Those in the north are adapted for rapid growth at low temperatures; those in the south, for slow growth at high temperatures. This genetic difference is sufficient, if one ignores the individuals from adjacent localities, to produce at least two distinct species—the northern form and the southern form. They certainly meet the chief criterion of separate species in not being able to interbreed.

It is very doubtful that hybridization plays a major role in the evolutionary process in animals. When two different species are crossed, especially those with different chromosome numbers, the offspring are generally sterile.

CONCEPT OF POLYMORPHISM

Many species of plants and animals are represented in nature by two or more clearly distinguishable kinds of individuals. Such a condition is called **polymorphism.** Polymorphic variation may involve not only color but many other characters, physiologic as well as structural. However, the term does not refer to seasonal and some other variations, such as the winter and summer pelage and plumage of certain mammals and birds. Polymorphism always refers to variability within a population. The meaning of polymorphism is well exemplified by the ladybird beetle, *Adalia bipunctat,* studied by N. W. Timofeeff-Ressovsky. In this genus some of the individuals are red with black spots, whereas others are black with red spots. The black color behaves as a mendelian dominant and red as a recessive. The black and red forms live side by side and interbreed freely. Studies show that the black form is predominant from spring to autumn and the red form from autumn to spring. It is thought that the changes are produced by natural selection, which favors the black form during summer and the red during winter. For instance, more black forms, produced by dominant genes, die out during winter, while the recessive red form survives. It is thus possible for the recessive gene, at least during part of the seasonal cycle, to be more common than its dominant allele.

Polymorphism has adaptive value in that it adapts the species to different environmental conditions. Polymorphic populations are thus better able to adjust themselves to environmental changes and exploit more niches and habitats. Without a doubt, the potentialities of polymorphism are much greater than its realization because every population must have many allelic series that are never expressed in the visible phenotype.

Polymorphism is widespread throughout the animal kingdom, and often different forms have been mistaken for separate species. Some examples are the right-handed and left-handed coils of snails of the same species, the blood types of man and other animals (a biochemical distinction), sickle cell anemia in man, albinism in many animals, silver foxes in litters of gray foxes, and rufous and gray phases of screech owls in the same brood.

IS IT POSSIBLE TO OBSERVE EVOLUTION IN ACTION TODAY?

Darwin in his time could not point to a single visible example of evolution in action. However, in England during his lifetime, a striking case of evolution was actually taking place in nature—industrial melanism in moths. Within the last century, certain moths, such as the peppered moth *(Biston)*, has undergone a coloration change from a light *(B. betu-*

laria) to a dark melanic form *(B. carbonaria)*. This moth is active at night and rests on the trunks of trees during the day. The light form is especially well adapted to rest on the background of lichen-encrusted trees where it is largely invisible. In the industrial regions of England, however, the fallout of smoke particles kills the light-colored lichens and blackens the vegetation. Against a dark background, light-colored moths are conspicuous and fall prey to predator birds. Natural selection would

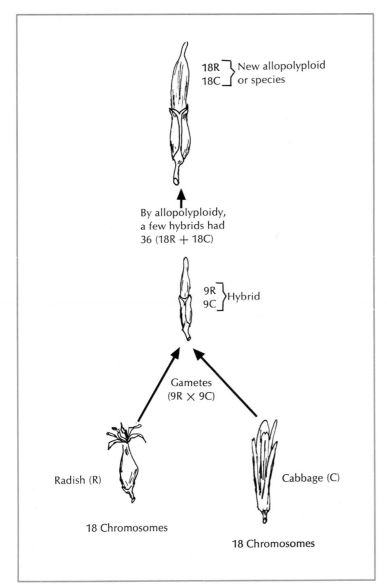

18R ⎤ New allopolyploid
18C ⎦ or species

By allopolyploidy,
a few hybrids had
36 (18R + 18C)

9R ⎤
9C ⎦ Hybrid

Gametes
(9R × 9C)

Radish (R)

Cabbage (C)

18 Chromosomes

18 Chromosomes

FIG. 21-2. Karpechenko's classic experiment of producing artificial species. Allopolyploidy is that form of polyploidy in which there is doubling of chromosomes in hybrid— in this case, result of cross between cabbage gamete of 9 chromosomes, C, and radish gamete of 9 chromosomes, R. (Modified from Karpechenko.)

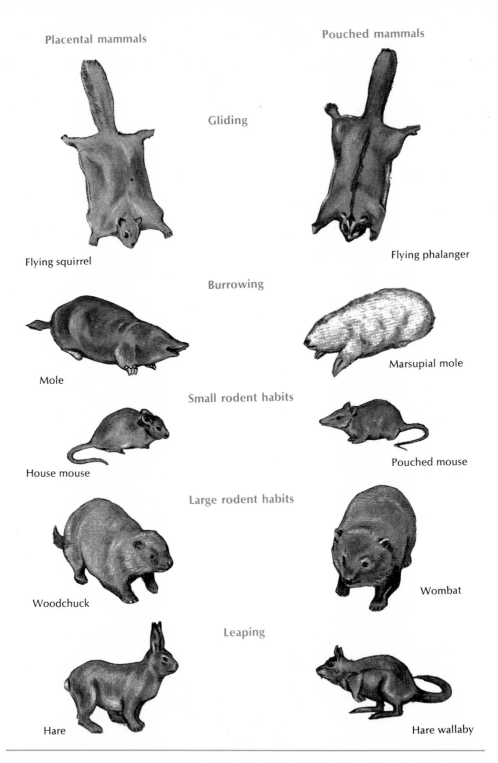

Placental mammals   Pouched mammals

Gliding

Flying squirrel   Flying phalanger

Burrowing

Mole   Marsupial mole

Small rodent habits

House mouse   Pouched mouse

Large rodent habits

Woodchuck   Wombat

Leaping

Hare   Hare wallaby

FIG. 21-3. Convergent evolution among mammals. Two groups of mammals not closely related (placentals and marsupials) have independently evolved similar ways of life and occupy similar ecologic niches. It will be noted that for every member of the ecologic niche in one group there is a counterpart in other group. This correspondence is not restricted to similarity of habit but also includes morphologic features. (From many sources.)

therefore, in the course of time, largely eliminate the moth. However, by mutation the light-colored moth has given rise to a dark-colored melanic form that has a much better survival rate under such surroundings. Within a period of years in polluted areas, the dark species *(B. carbonaria)* has, to a great extent, replaced the light species *(B. betularia)*. The mutation for industrial melanism appears to be controlled by a single dominant gene that is also known to occur in natural environments (not due to pollution) in which a dark color is a distinct advantage.

EXPERIMENTAL PRODUCTION OF NEW SPECIES

In 1924 G. D. Karpechenko was able to produce a new species by hybridization. He crossed the radish *(Raphanus sativus)* with the cabbage *(Brassica oleracea)*, each of which has 9 haploid chromosomes. In most cases the hybrids of two different species of unlike chromosomes (as these were) are unable to produce fertile offspring because unlike chromosomes cannot pair (synapsis) properly in the meiosis of their gametes. The 18-chromosome hydrids (9 chromosomes from each parent) in this cross were entirely sterile and could not produce offspring. However, a few of the hybrids by spontaneous allopolyploidy, or doubling of the chromosomes, had 36 chromosomes (18 radish chromosomes and 18 cabbage chromosomes). When these hybrids underwent meiosis to form their gametes, the homologous radish chromosomes could pair with each other and the homologous cabbage chromosomes could pair with each other so that each gamete had 9 radish and 9 cabbage chromosomes. The union of such eggs and sperm could produce a fertile plant of 36 chromosomes. This new plant, which had charac-

teristics of both the radish and the cabbage, could not be crossed with either one of the original parents and was reproductively isolated. The new synthetic genus was called *Raphanobrassica* and was the first recorded instance of the artificial creation of a new species (Fig. 21-2).

Allopolyploidy has perhaps played an important role in plant evolution but very little in that of animals.

CONVERGENCE AND PARALLELISM

Two groups of animals not closely related may have evolved similar structures because similar habitats have afforded them the same evolutionary opportunity. The two groups may be quite dissimilar in the beginning, or their ancestors may be entirely lacking in the common structures under consideration. The radiations of the two groups have been along lines of similar opportunities. Similar mutations are favored because of the possibilities of the environment. The ways of life are common to both groups and each has taken advantage of similar ecologic niches. Good examples of convergence are seen in the likenesses between the placental and marsupial types of animals (Fig. 21-3) and by the resemblance between the body forms of aquatic mammals and fishes.

When two groups are rather closely related to begin with, the evolutionary phenomenon is called parallelism. In this case, mutations are more likely to be similar in each group, and each will undergo evolutionary changes along similar lines, although the two are never completely identical. For instance, the enamel-crowned molars of the horse, rhinoceros, and elephant came from ancestral forms that lacked this characteristic.

22

Origin of life (biopoiesis) and evidences for evolution

Most biologists agree that life at its beginning arose naturally from nonliving matter. Many disciplines (biochemistry, physical chemistry, astronomy, microbiology, geology, etc.) have been used to learn about the early history of the earth and the chemical properties of living matter. Although the whole problem must always remain speculative, scientists have devised experiments demonstrating the conditions by which evolutionary events may have occurred. There is no evidence that living organisms are now formed from nonliving matter. The primitive conditions under which life first arose are not thought to be available at present; any of the simple organisms that first appeared would now be quickly destroyed by saprophytic bacteria and would have no chance to start an evolutionary pattern.

The historic fossil record is incomplete and gives us no idea of what the earliest life was like. The early organisms up to the Cambrian period were probably small and soft bodied and did not preserve as fossils. Precambrian animals were primarily motile; calcareous skeletons and hard parts appeared only when animals adopted a sessile or sluggish mode of existence (Brooke-Raymond theory). Some primitive fossils (algae) have been found in rocks at least 2.7 billion years old.

Charles Darwin thought that when life originated the conditions in the world may have been quite different from what they are at present. He proposed that amino acids could have survived outside the living organism and that mixed with phosphoric acid salts, ammonia, light, heat, etc., the amino acids might link together to form proteins. In the present century J. B. S. Haldane, the British biologist, proposed that the gases of the early atmosphere of the earth consisted of water, carbon dioxide, and ammonia. When ultraviolet light shines on such a gas mixture, many organic substances, such as sugars and possibly amino acids, are formed. Ultraviolet light must have been very intense before the appearance of oxygen (from plants) to form ozone (the 3-atom form of oxygen), which serves at present as a blanket to prevent ultraviolet rays from reaching the earth's surface. Haldane believes that the early formative substances could accumulate in the early oceans where synthesis of sugars, fats, proteins, and nucleic acids might occur.

However, one proposal has attracted the attention of scientists in recent years—that life arose spontaneously but gradually in the primitive ocean under the impact of certain favorable conditions that no longer exist. This theory stresses the presence of large quantities of organic compounds similar to those that are now found in living organisms.

Forceful arguments in its favor have been advanced by Oparin, Haldane, Urey, and many others in a recent renewal of the hypothesis. A. I. Oparin, the Russian biochemist, through his book *The Origin of Life* (English translation, 1938) has been mainly responsible for this renewed interest. He argued that if the primitive ocean contained large quantities of organic compounds, in time these compounds would react with each other to form structures of increasing complexity. Eventually such a structure would reach a stage that could be called living. His belief was that a living system was gradually synthesized from nonbiologic compounds by logical and probable steps. Energy sources for the creation of the first organic compounds would come from ultraviolet light, electric discharges, localized areas of high temperature, such as volcanoes, and to some extent radioactivity. Since the necessary metabolites were already present in the environment to be used by the primitive organisms, their metabolism would have been heterotrophic. Other conditions necessary for such a scheme of spontaneous generation would be a sterile environment to prevent the destruction of primitive organisms by microorganisms such as exist today and a reducing atmosphere of such precursor compounds as water, methane (CH_4), ammonia (NH_3), and hydrogen instead of the present oxidizing atmosphere of carbon dioxide (CO_2), nitrogen, and oxygen.

Urey thinks that at low temperatures such substances as methane and ammonia would have been formed in the earth's early atmosphere and would have existed for some time during the early origin of life. As evidence for his theory, it has been discovered by spectroscopic analysis that the remote and large planets of Jupiter and Saturn have atmospheres (frozen) of methane and ammonia. This could have been the original atmosphere of the planet Earth.

ORIGIN OF THE EARTH

Of the many theories to account for the origin of our planet, the one that is most seriously considered at present states that the sun and the planets were formed together from a spherical cloud of cosmic dust, which by rotation and gravitation developed a sun at the center and a swirling belt of gas around it. In time the belt broke up into smaller clouds that condensed by gravity to form the planets. Free hydrogen atoms were the most abundant elements in the gas cloud and gravitated toward its center to create the sun, which is largely composed of hydrogen. If the early surface temperature of the earth was high, it should have cooled rapidly by convection and radiation and reached its present condition in the relatively short time of 25,000 years.

While the earth was in a more or less gaseous condition, the various atoms became sorted out according to weight, with the lighter elements (hydrogen, oxygen, carbon, and nitrogen) in the surface gas and with the heavier ones (silicon, aluminum, nickel, and iron) toward the center. At first many gases, such as hydrogen, helium, methane, water, and ammonia, escaped from the earth, but when the gaseous materials became dense enough, the gravitational field tended to prevent these gases from escaping into outer space. As the earth cooled, stable bonds could be formed and free atoms began to disappear as molecules appeared. Since the lighter elements (hydrogen, oxygen, nitrogen, carbon) were the most abundant atoms on the earth's surface, these elements reacted to form the first molecules.

The early atmosphere of the earth was a reducing one rather than an oxidizing one as we have now. Only simple organic molecules would be formed in the reducing atmosphere of the primitive earth. Its carbon would be in the form of methane or carbon monoxide (perhaps a little carbon dioxide), its nitrogen in the form of ammonia, most of the oxygen in the form of water, and there would be considerable hydrogen. These primitive molecules would, then, begin with the following:

H | H—O	O=C=O	H | H—C—H | H	H | N—H | H	H | H
WATER (H_2O)	**CARBON DIOXIDE** (CO_2)	**METHANE** (CH_4)	**AMMONIA** (NH_3)	**HYDROGEN** (H_2)

From these the complex biologic molecules would be formed later. Temperature conditions of the earth were such that these primitive compounds existed as gases and formed the early atmosphere. The key element in the formation of the molecules of the

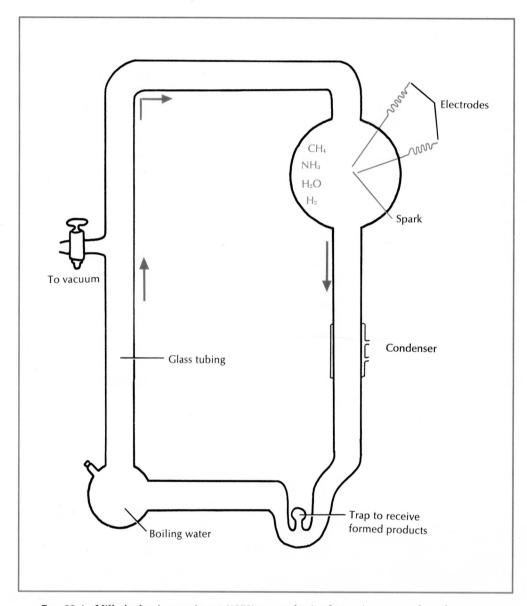

FIG. 22-1. Miller's classic experiment (1953) on synthesis of organic compounds under assumed primitive conditions of earth's early atmosphere. Electric discharge through mixture of gases produced several amino acids and other organic substances. Variations of this experiment, making use of HCN (hydrocyanic acid), adenine, ribose, ethyl metaphosphate, etc. exposed to ultraviolet irradiation, have produced on RNA nucleotide, high-energy ATP, and a host of other organic molecules (aldehydes, purines, pyrimidines, fatty acids, etc.). ATP may have been abundant in early primordial oceans of reacting molecules and may have provided a ready source of energy for chemical reactions.

primitive atmosphere was undoubtedly hydrogen, for when it was present in large amounts, its great reactivity led to the formation of methane, ammonia, and water vapor.

STEPS IN CHEMICAL EVOLUTION

In any living system many molecules are commonly referred to as biomolecules. They may be divided into two groups: (1) molecules of relatively simple structure such as the sugars (disaccharides and polysaccharides), neutral fats, phospholipids, amino acids, and nucleotides and (2) macromolecules such as proteins, nucleic acids, nucleoproteins, and viruses. In a series of chemical reactions free atoms formed simple molecules, which in turn formed larger and larger molecules, leading eventually to chance synthesis of macromolecules with the molecular patterns characteristic of the living organism.

If the early atmosphere contained methane, ammonia, water, and hydrogen, a mixture of these compounds would react only when energy was supplied in some form. Ordinarily, organic compounds are formed by organisms, but organisms were nonexistent in the early origin of life. As biochemists know today, the synthesis of organic compounds involves enzymatic action at every step. But enzymes are complex proteins and in any scheme for explaining early life their formation must be considered, not at the beginning, but well along in the process. Simpler compounds from early primordia must have come along before proteins. However, enzymes only hasten the reaction rate; reactions can occur slowly without them. At the present time the source of free energy to produce all energy directly or indirectly is the sun.

Some light on this problem was added by the now classic experiment of S. L. Miller (1953) (Fig. 22-1). By circulating a mixture of water vapor, methane, ammonia, and hydrogen continuously for a week over an electric spark, he was able to get certain amino acids and some other products. It is possible, therefore, to have had various sources of energy in the early formation of organic compounds—ultraviolet light, electric discharges, radioactivity, etc. In Miller's experiment the mechanism of amino acid formation involved first the appearance of certain intermediate products such as aldehydes and

hydrogen cyanide. The reaction of these compounds led to the synthesis of amino acids.

More recently other workers have varied Miller's experiments by irradiating gaseous mixtures of H_2O, H_2, and NH_3 with electrons, by heating aqueous solutions of hydrocyanic acid (HCN), and by exposing solutions of HCN and formaldehyde to ultraviolet irradiation. From such experiments many different molecules were obtained, such as several amino acids, formic acid, aldehydes, purine and pyrimidine bases, some complex polymers of amino acids and sugars, ribose, and deoxyribose. The high-energy adenosine triphosphate (ATP) has also been produced by the ultraviolet-irradiated solutions of adenine, ribose, and ethyl metaphosphate, using the Miller technique with an oxygen-free, artificial atmosphere.

The formation of water deposits or oceans was no doubt due to the condensation of water vapor from the atmosphere, although another theory accounts for the origin of oceans by the escaping of water from the earth's interior during its cooling stage. Dissolved in ocean water were salts and minerals washed down from the continents as well as atmospheric ammonia and methane. Volcanoes may have also contributed some constituents to the ocean waters.

The presence of water and these early dissolved substances may be considered the crucial conditions that led eventually to the origin of life. The main bulk of organic compounds accumulated in the oceans. In this broth of dissolved organic compounds of great variety, the conversion of multimolecular systems into living systems took place. Methane (CH_4) was probably one of the first molecules to react, for the 4 H atoms could be replaced by other kinds of atoms, and thus many kinds of carbon-containing molecules could be formed. The versatility of the carbon atom, with its bonding capacity of 4, made possible an enormous complexity and variety of molecular structures.

Logical steps in chemical evolution would suggest that some of the earliest organic compounds formed would be the simple sugars, amino acids, fatty acids, pyrimidines, and purines (Fig. 23-2). Each of these compounds represents a chain or ring of carbon atoms attached to various combinations of hydrogen, oxygen, and nitrogen. The sugars, or carbohydrates,

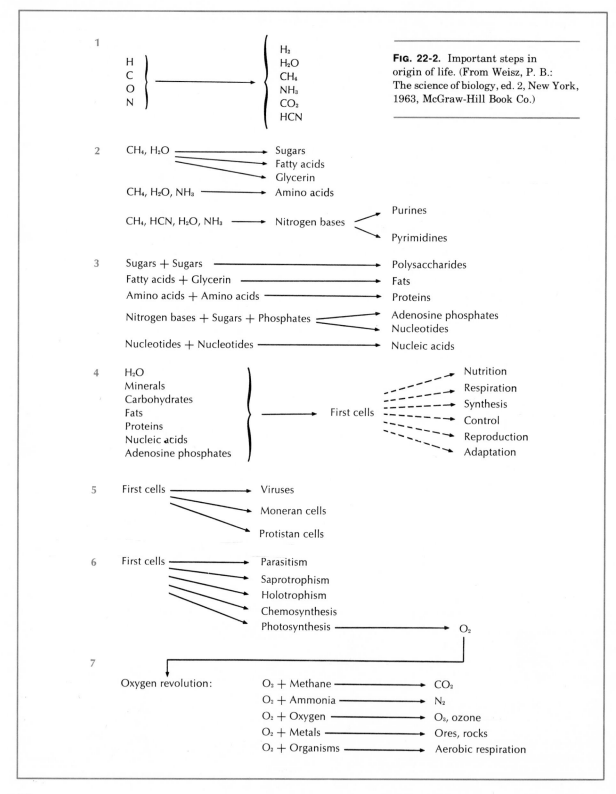

FIG. 22-2. Important steps in origin of life. (From Weisz, P. B.: The science of biology, ed. 2, New York, 1963, McGraw-Hill Book Co.)

and the fatty acids are formed entirely of carbon, hydrogen, and oxygen. The amino acids have, in addition to these elements, nitrogen, which occurs in an amino group (NH_2). Nitrogen could have been obtained in a reaction involving ammonia (NH_3) in which one of the hydrogen atoms was removed.

In the various reactions of these early compounds with each other, simple sugars could form polysaccharides, and fatty acids could combine to produce fats. The appearance of proteins was important because they are absolutely essential to living things. Proteins are made up of amino acids that are linked together in gigantic patterns to form complex and varied protein molecules. However, a mixture of amino acids would not form a protein unless other factors were involved.

How are proteins synthesized from amino acids? The experiment of Miller showed that it is rather easy to account for the amino acids, but the formation of these into the complex polypeptide chains is far more difficult. In 1963 M. Calvin and others subjected a mixture of methane, ammonia, and water to a bombardment of electrons from the Berkeley cyclotron, and after an hour got adenine, 1 of the 4 bases of the DNA molecule. About a year later, S. W. Fox proposed that heat may have played an important part in the synthesis of organic compounds and heated a mixture of methane, ammonia, and water to 1,800° F. Of the many amino acids Fox obtained by this method, he found 14 of them involved in what is now known as the genetic code. He later demonstrated, by using more modest heat ranges (150° to 180° C.), that the 18 kinds of amino acids so treated together formed chains of polypeptides that could be digested by enzymes. By adding polyphosphate to his reaction mixture, polypeptide chains could be formed at only 70° to 80° C. The high temperatures used by Fox could have been caused by hot meteorites in the early history of our planet. The most available energy source could have been ultraviolet light in the synthesis of the early organic compounds.

Another important aspect of protein development was their role in acting as **enzymes.** The first catalysts were probably metal ions and were somewhat weak in their action, but the development of proteins as enzymes greatly accelerated reactions without increasing the temperature.

Other key organic molecules that developed from the purines and pyrimidines were the **nucleotides.** These contain a purine or pyrimidine base combined with a sugar and a phosphate. Combinations of many different nucleotide molecules produced the supermolecules known as nucleic acids. Some of the nucleotides are synthetically produced from ammonium cyanide, which in turn can be formed from methane and ammonia by electric discharges. The polymerization of ribonucleic and deoxyribonucleic acids could have been effected by enzymes or by mineral surfaces. The combination of nucleic acid with a protein produced the nucleoproteins, which are the principal components of the cell nucleus and are intimately associated with the life process.

FIRST ORGANISMS

It is very doubtful that the transition from the nonliving to the living was a sharp or abrupt process. The simplest system subject to evolution would be the replication of a nucleic acid, such as DNA, that would require complex nucleotide triphosphate for its synthesis. For making exact duplicate copies, it is necessary to have in abundance the component parts of nucleoproteins such as sugars, amino acids, purines, pyrimidines, and phosphorus. The first nucleoproteins served as models for the formation of more nucleoproteins. The Watson-Crick model of the structure of deoxyribonucleic acid shows how a molecule can act as a pattern to synthesize another molecule complementary to itself. The first living organism could have been strips of DNA or RNA that, with the necessary enzymes, could duplicate themselves. The sequence of the process may have occurred something like this: (1) A single strand of nucleic acid was formed, (2) nucleotides complementary to the bases in the first chain lined up, (3) the polymerization of the nucleotides to form the complementary strand occurred, (4) the original strand separated from the newly formed complementary strand, and (5) the final stage was the addition of cytoplasm and a membrane to the self-duplicating polynucleotides.

An alternative theory of the origin of the first organism is the coacervate theory of Oparin. According to this theory, coacervate aggregates, a special form of a colloidal solution in which one of

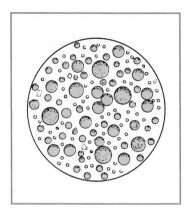

FIG. 22-3. Coacervate droplets in hydrophilic colloid. According to this theory of origin of first organisms, complex coacervates of different proteins made up a type of dilute "broth" in the ocean, and by intermolecular attraction large complex molecules were formed into colloidal aggregates. These colloidal aggregates could selectively concentrate materials from ocean "broth," and by unknown favorable internal organization, some aggregates could pick up molecules better than others and would become dominant. (After J. Keosian, 1964.)

the liquids of a hydrophilic sol appears as viscous drops (coacervates) instead of forming a continuous liquid phase, would absorb proteins and other materials from the environment, increase in size, and then divide (Fig. 22-3). Protoplasm, as now organized, has the structure of a complex coacervate.

There are logical reasons for believing that the early organisms were heterotrophic; that is, they got their energy from sources outside of themselves. The primitive oceans must have accumulated large quantities of compounds, and the first organisms used this reservoir of compounds for their evolution and expansion. When this reservoir of resources was used up, spontaneous generation could no longer occur and biogenesis became the only possible method of origin of organisms. By mutations, certain early organisms acquired the capacity to synthesize organic compounds from simpler ones. Natural selection could operate here to favor those organisms that could synthesize all the essential complex compounds, and competition became a rule of existence. Less successful forms would become extinct.

All organisms require free energy for their chemical reactions and for the synthesis of their body parts. If the early organisms were heterotrophic and anaerobic, the source of their energy must have come from fermentation processes. Many microorganisms get their energy this way. Lactic acid bacteria get free energy from the breaking down of glucose into lactic acid, each molecule of glucose producing 2 molecules of the high-energy ATP. The yeast organisms produce ethyl alcohol and carbon dioxide instead of lactic acid. Most animals at present obtain their free energy from the oxidation of organic molecules by oxygen, and plants get their energy from light in the photosynthetic process (autotrophic). When early organisms used up the fermentable compounds of the ocean "broth," other sources of energy had to be utilized. About this time photosynthesis evolved and was made possible by the accumulation of atmospheric carbon dioxide produced by the metabolism of the early heterotrophs.

Another form of autotrophic nutrition is **chemosynthesis,** in which energy for forming carbohydrates is obtained from metallic or nonmetallic materials such as iron, sulfur, and nitrogen. By using bond energy from chemical reactions it was possible to combine CO_2 and water into carbohydrates. With the appearance of oxygen, an ozone (O_3) layer was formed in the higher atmosphere, and it absorbed most of the sun's ultraviolet rays that had made life possible only in water. It is thought that all the present oxygen of the air can be renewed by photosynthetic processes every 2,000 years and that all the CO_2 molecules pass through photosynthesis every 300 years. Autotrophic forms, or those that could synthesize complex organic compounds from simple renewable resources, were now established. Mutational descendants of these autotrophs produced secondary heterotrophs that could now live on autotrophs. Thus the present scheme of living systems was initiated.

In the overall picture (Fig. 22-2) there were three major evolutionary directions in the formation of organisms. These directions are based on the three methods of nutrition found in the biologic world: (1) the photosynthetic processes, or producers of organic compounds, represented by plants; (2) the ingestion of producers, or consumers, represented by animals; and (3) the reduction, or decomposition, of the dead remains of both producers and consumers to an absorbable state represented by saprophytes such as fungi and certain bacteria. Some organisms may fit into more than one of these nutritional categories.

PALEONTOLOGY: THE HISTORIC RECORD

The strongest and most direct evidence for evolution is the fossil record of the past. The study of paleontology, or the science of ancient life, shows how the ancestors of present-day forms lived in the past and how they became diversified. Incomplete as the record is—and many groups have left few or no fossils—biologists rely on the discoveries of new fossils for the interpretation of the phylogeny and relationships of both plant and animal life. The documentary evidence for evolution as a general process, the progressive changes in life from one geologic era to another, the links between the one grade of taxon to another, the past distribution of lands and seas, and the environmental conditions of the past (paleoecology) are all dependent on what fossils teach us.

A fossil may be defined as any evidence of past life. It refers not only to complete remains (mammoths and amber insects), actual hard parts (teeth and bones), petrified skeletal parts that are infiltrated with silica or other minerals by water seepage (ostracoderms and mollusks), but also to molds, casts, impressions, and fossil excrement (coprolite). Skeletal parts are perhaps the most common of all and paleontologists have been very skillful in reconstructing the whole animal from only a few parts. Vertebrate animals and invertebrates with shells or other hard structures have left the best record (Fig. 22-4). But now and then a rare, chance discovery, such as the Burgess shale deposits of British Columbia and the Precambrian fossil bed of South Australia, reveal an enormous amount of information about soft-bodied organisms.

Fossils may be found in any part of the world, although certain regions, because of ideal conditions for their formation, may have a greater abundance than others, such as the tar pits of Rancho La Brea in Hancock Park, Los Angeles; the great dinosaur beds of Alberta, Canada, and Jensen, Utah; the Olduvai Gorge of South Africa; and many others.

A common method of fossil formation is the burial of animals under the sediment deposited by large bodies of water. Climatic conditions must have also been a great factor, as well as the nature of the deposits. Many fossils are found in regions where there are excellent conditions for preservation, such as asphalt tar pits and cold places for refrigeration.

Most fossils are laid down in deposits that become stratified and, if undisturbed, the older strata are the deeper ones. The five major rock strata were mainly formed by the accumulation of sand and mud at the bottoms of seas or lakes. However, strata are not always in the regular sequence by which they were laid down, for in many regions they have buckled and arched under pressure so that older strata may be shifted over more recent ones.

One of the most useful methods for determining the age of geologic formations and fossils is radioactivity. Radioactive elements are transformed into other elements at certain rates, independent of pressure and temperature. Uranium 238, for example, is slowly changed into lead 206 at the rate of 0.5 gram of lead for each gram of uranium in a period of 4.5 billion years, or the half-life of uranium.

The ratio of lead 206 to the amount of uranium 238 in a sample of rock formation provides an estimate of the age of the stratum from which the specimen is taken. The potassium-argon method is now considered to be even more precise. It is found that the radioactive isotope K^{40} decays into Ca^{40} (88%) and A^{40} (12%) at the rate of a half-life of 1.3 billion years. By knowing the amount of argon emitted (calcium is unreliable) from each unit of potassium in a unit of time, it has been possible to date the age of the rock and that of the fossil laid down in this rock. For fossils not over 50,000 years old, the radiocarbon method of Libby is very accurate. Radioactive C^{14} has a half-life of 5,568 years and is slowly transformed into N^{14}. Its ratio to C^{12} in the living organisms is the same as that in the atmosphere. There is no exchange of carbon atoms after death, and by knowing the $C^{14}:C^{12}$ ratio in a fossil, its age can be estimated.

In recent years the science of **paleoecology** has attracted widespread interest. The major aim of this discipline is to throw light on the kinds of environment under which sedimentary rocks of the

FIG. 22-4. Representative fossils. **A,** Fossil arthropod *Eurypterus,* which was abundant during upper Silurian period; related to modern scorpions. **B,** Some worm tubes. **C,** Bryozoan. **D,** Trilobite; related to king crab of today and one of most abundant of arthropod fossils. **E,** Cephalopod; chambered nautilus of today is little changed from this ancient fossil. **F,** Coral. **G,** Gastropod. **H,** Coelenterate strobila.

past were accumulated. In this way it is possible to derive information about the physical environments in which fossil organisms lived and their relations to each other. Assemblages of fossils may show the nature of sea floors, the chemicophysical nature of water, and climatic conditions at the time of their existence. The problems of the paleoecologist are far more difficult than those of the present-day ecologist. The fossil record is so incomplete and the almost entire absence of certain groups as fossils makes it impossible to determine with any degree of accuracy the nature of populations, communities, and other important ecologic concepts. Whatever is found, however, has been of great importance not only to an understanding of the biologic conditions of the past but also to geologic aspects of the ancient distribution of lands and seas.

The major characteristics of the geologic eras are indicated in Table 22-1. As far as the fossil record is concerned, the recorded history of life begins about the base of the Cambrian period of the Paleozoic era. The Cryptozoic eon, which includes the Archeozoic and Proterozoic eras, has been a great puzzle because of the lack of fossils. The chief evidences of life during this period were mostly the burrows of worms, sponge spicules, algae, and a few others. However, the Precambrian fossil deposits of South Australia, with many invertebrate forms, indicate that life had already evolved to a marked extent for perhaps as long as a billion years before the Cambrian period. There is also a great deal of carbon as a residue of organic matter in the sedimentary rocks of this time. There may have been a great diversity of life in the Precambrian seas, but it was not preserved because of the lack of shells or other hard parts.

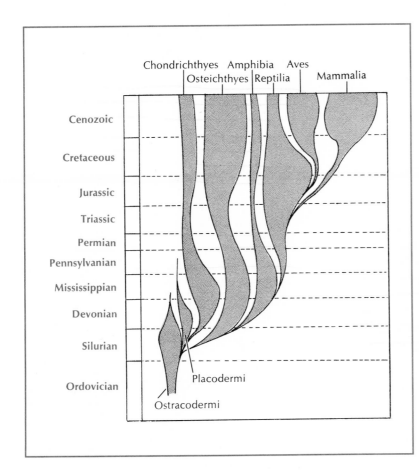

FIG. 22-5. Fossil record of vertebrates. Relative abundance of groups, as indicated by width of shaded areas, is based upon numbers of genera that have been identified. Note expansion and decline of reptiles. Osteichthyes, Aves, and Mammalia are dominant groups at present.

TABLE 22-1. Geologic time scale (younger ages toward top of chart; older ages toward bottom)

Era	Period	Epoch	Time at beginning of each period (millions of years ago)	Geologic events and climate	Biologic characteristics
Cenozoic (Age of Mammals)	Quaternary	Recent	0.025	End of fourth ice age; climate warmer	Dominance of modern man; modern species of animals and plants
		Pleistocene	0.6 to 1	Four ice ages with valley and sheet glaciers covering much of North America and Eurasia; continents in high relief; cold and mild climates	Modern species; extinction of giant mammals and many plants; development of man
	Tertiary	Pliocene	12	Continental elevation; volcanic activity; dry and cool climate	Modern genera of mammals; emergence of man from man-apes; peak of mammals; invertebrates similar to modern kinds
		Miocene	25	Development of plains and grasslands; moderate climates; Sierra mountains renewed	Modern subfamilies rise; development of grazing mammals; first man-apes; temperate kind of plants; saber-toothed cat
		Oligocene	34	Mountain building; mild climates	Primitive apes and monkeys; whales; rise of most mammal families; temperate kind of plants; archaic mammals extinct
		Eocene	55	Land connection between North America and Europe during part of epoch; mountain erosion; heavy rainfall	Modern orders of mammals; adaptive radiation of placental mammals; subtropical forests; first horses
		Paleocene	75	Mountain building; temperate to subtropical climates	Dominance of archaic mammals; modern birds; dinosaurs all extinct; placental mammals; subtropical plants; first tarsiers and lemurs
Mesozoic (Age of Reptiles)	Cretaceous		130	Spread of inland seas and swamps; mountains (Andes, Himalayas, Rocky, etc.) formed; mild to cool climate	Extinction of giant land marine reptiles; pouched and placental mammals rise; flowering plants; gymnosperms decline

TABLE 22-1. Geologic time scale—cont'd

Era	Period	Epoch	Time at beginning of each period (millions of years ago)	Geologic events and climate	Biologic characteristics
	Jurassic		180	Continents with shallow seas; Sierra Nevada Mountains	Giant dinosaurs; reptiles dominant; first mammals; first toothed birds
	Triassic		230	Continents elevated; widespread deserts; red beds	First dinosaurs; marine reptiles; mammallike reptiles; conifers dominant
Paleozoic (Age of Amphibians)	Permian		260	Rise of continents; widespread mountains; Appalachians formed; cold, dry, and moist climate; glaciation; red beds	Adaptive radiation of reptiles which displace amphibians; many marine invertebrates extinct; modern insects; evergreens appear
	Pennsylvanian*		310	Shallow inland seas; glaciation in Southern Hemisphere; warm, moist climate; cool swamp-forests	Origin of reptiles; diversification in amphibians; gigantic insects
	Mississippian*		350	Inland seas; mountain formation; warm climates; hot swamp lands	Amphibian radiation; insects with wings; sharks and bony fish; crinoids
(Age of Fishes)	Devonian		400	Small inland seas; mountain formation; arid land; heavy rainfall	First amphibians; mostly freshwater fish; lungfish and sharks; forests and land plants; brachiopods; wingless insects; bryozoans and corals
	Silurian		425 to 430	Continental seas; relatively flat continents; mild climates into higher latitudes	Eurypterids; fish with lower jaws; brachiopods; graptolites; invasions of land by arthropods and plants
(Age of Invertebrates)	Ordovician		475	Oceans greatly enlarge; submergence of land; warm mild climates into higher latitudes	Ostracoderms (first vertebrates); brachiopods; cephalopods; trilobites abundant; land plants; graptolites

*The Pennsylvanian (upper) and Mississippian (lower) are often referred to as the Carboniferous period. This is the period between the Devonian and Permian periods.

Continued.

TABLE 22-1. Geologic time scale—cont'd

Era	Period	Epoch	Time at beginning of each period (millions of years ago)	Geologic events and climate	Biologic characteristics
Paleozoic (Age of Invertebrates) —cont'd	Cambrian		550	Lowlands; mild climates	Marine invertebrates and algae; all invertebrate phyla and many classes; abundant fossils; trilobites dominant
Proterozoic (Precambrian)			2,000	Volcanic activity; very old sedimentary rocks; mountain building; glaciations; erosions; climate warm moist to dry cold	Fossil algae 2.6 billion years old; sponge spicules; worm burrows; soft-bodied animals; autotrophism established
Archeozoic			4,000 to 4,500	Lava flows; granite formation; sedimentary deposition; erosion	Origin of life; heterotrophism established

Since the Cambrian period when all the major phyla (with one or two exceptions) were well established, it has been a matter of replacing primitive lines with better-adapted ones. Many of the vertebrate classes and orders appeared first in the Ordovician to Devonian periods. By the end of the Paleozoic era, some dominant groups became extinct and were replaced by the expansion of other groups (Fig. 22-5).

EVOLUTION OF ELEPHANT

Paleontology has afforded evidences for tracing the phylogeny of many groups of animals. Two classic examples are those of the elephant and the horse. Elephants have stressed two morphologic features in particular—a prehensile proboscis and teeth. They originated in Africa (in the late Eocene epoch) and gradually spread to other continents. They appear to have evolved from types similar to *Moeritherium* and were about the size of a pig (Fig. 22-6). The second incisors of their upper and lower jaws were a little enlarged. By the Oligocene epoch,

four tusks had developed from these incisors (*Palaeomastodon*) and the size had increased to that of a steer. From this type the larger Miocene and Pleistocene forms, lacking the lower tusks, emerged. Later, during the Pliocene epoch came elephants as large or larger than the modern types such as the wooly mammoths (Fig. 22-7). Of the many types of elephants that have appeared, only the African (*Loxodonta*) and Asiatic form (*Elephas*) exist now.

EVOLUTION OF HORSE

The fossil record affords us no more complete evolutionary line than that of the horse. The evolution of this form extends back to the Eocene epoch, and much of it took place in North America. This record would at first seem to indicate a straight-line evolution, but actually the history of the horse family is made up of many lineages. The phylogeny of the horse is extensively branched, with most of the branches now extinct. There were millions of years when little change occurred; there were other eras when changes took place relatively rapidly. No real

FIG. 22-6. Restoration of heads of fossil elephant-like animals. **1,** *Moeritherium.* **2.** *Palaeomastodon.* **3.** *Trilophodon.* **4,** *Dinotherium.* **5,** *Mastodon.* **6,** *Elephas.* (Courtesy Ward's Natural Science Establishment, Inc., Rochester, N. Y.)

FIG. 22-7. Mammoths. These have been found frozen in Siberia in a good state of preservation. (Courtesy Chicago Natural History Museum.)

change in the feet occurred during the Eocene epoch, but at least three types of feet developed later and were found in different groups during the late Cenozoic. Only one of these three types is found today.

In the evolution of the horse the morphologic changes of the limbs and teeth were of primary importance, along with a progressive increase in size of most of the types in the direct line of descent.

The first member of the horse phylogeny is considered to be *Hyracotherium*, about the size of a small dog. Its forefeet had four digits and a splint (Fig. 22-8); the hind limb had three toes and two splints that represented the first and fifth toes. The teeth had short crowns and long roots, and the teeth in the cheek were specialized to some extent for grinding. This form lived in and grazed on forest underbrush.

The middle Eocene was represented by *Orohippus*, which had a further development of molarlike teeth.

Mesohippus flourished in the Oligocene epoch. It was taller and had three digits on each foot. The middle toe was larger and better developed than the others. *Miohippus*, also found in the Oligocene, was larger but of the three-toed type. These horses were also browsing forms.

The Miocene epoch was represented by *Parahippus* and *Merychippus*. *Merychippus* is considered the direct ancestor of the later horses. They were three toed, but the lateral toes were high above the ground. Thus the weight of the body was thrown upon the middle toe. The teeth were high crowned, and the molar pattern was adapted for grinding, with sharp ridges of enamel. The evidence is that *Merychippus* fed on grasses. It had a larger skull

FIG. 22-8. Evolution of forefoot of horse as revealed by fossil record. (Courtesy Ward's Natural Science Establishment, Inc., Rochester, N. Y.)

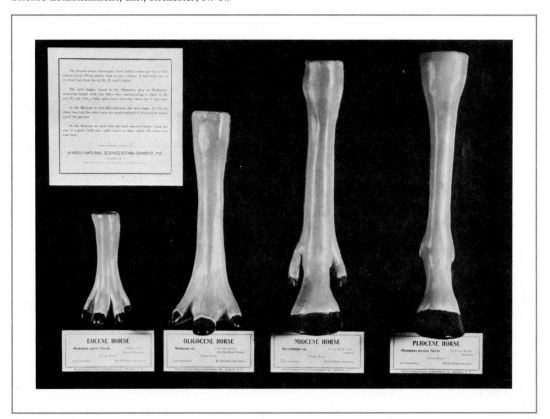

and heavier lower jaws than earlier forms and was between 3 and 4 feet high at the shoulders. *Merychippus* gave rise to a number of horse types, most of which became extinct by the end of the Tertiary period. One that persisted into the Pleistocene was *Pliohippus,* the first one-toed horse.

From *Pliohippus* the genus *Equus,* or modern horse, arose probably in the Pleistocene epoch. It arose in North America and spread to most of the other continents. It has one toe on each foot, but the two splint bones are evidences of the former lateral toes. By the end of the Pleistocene epoch the horse had become extinct in North America, but migrant forms persisted in Eurasia to become the ancestors of the present-day horse. After the discovery of America by Columbus, the horse was reintroduced by the early Spanish colonists.

The development of this great animal from a small foxlike form was closely associated with the geologic development from a hilly, forested country to the great plains of the west. Thus the horse in its evolution represents a close parallelism between the development of an adaptive structural pattern on the one hand and the geologic development of the earth's surface on the other.

OTHER EVIDENCES FOR EVOLUTION

Aside from the evidence of paleontology, other evidence that evolution has occurred rests chiefly upon the same grounds that Darwin and other supporters have advanced, that is, embryology, comparative anatomy (Fig. 22-9), geographic distribution, and taxonomy. Recent additional evidence stresses the biochemical similarities of metabolism and chemical systems in organisms. In general, the more closely organisms are related, the greater number of like functions and structures they have. Genetics

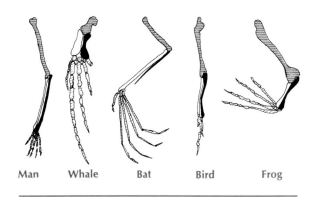

Man   Whale   Bat   Bird   Frog

FIG. 22-9. Forelimbs of five vertebrates to show skeletal homologies. Red = humerus; white = radius; black = ulna; gray = wrist and phalanges. Most generalized or primitive limb is that of man—the feature that has been primary factor in man's evolution because of its wide adaptability. Various types of limbs have been structurally modified for adaptations to particular functions.

has proved to be the key to the mechanical basis of evolution. From previous discussions it can be seen that mutations in genes must be accepted as a fundamental concept in evolution. This viewpoint is all the stronger in the light of the revealing genetic code and the alterations it can produce in the traits of organisms.

Within recent years, one of the most striking evidences for evolution is the rapid, man-influenced evolution of certain viruses, bacteria, and insects. Confronted with antibiotics and insecticides, these forms evolve within short periods a resistance to these drugs. These organisms have short generations, amazing reproduction, and high mutation rates. Such evolution is observable and its mechanism can be verified.

23

Man's evolution

About the beginning of the Miocene age 25 million years ago, the hominoid (resembling man) line divided into two main subbranches, the pongids, or apes, and the hominids (of the family of mankind). Man is therefore a product of evolution from primeval apes. Man's advantage over the apes, the animals he most resembles in structure, has been due chiefly to his brain, which is capable of a self-conscious appraisal of current conditions. Other biologic differences are minor, although they have contributed to his evolutionary developments. His arboreal habits, universal arm sockets, reduction of the tail, swiveling of the hips, and broad chests are all shared with the higher apes. After the pongid and hominid lines diverged from the common hominoid stock, each developed an adaptive radiation of its own. Some authorities think that for a time the hominids, together with the gorilla and chimpanzee, shared a common ancestor after their ancestry had become distinct from that of the other living apes (orangutan and gibbon).

Tropical conditions do not favor fossil formation, but in the last few decades many prehuman and definite human types have been found. Early fossil men are often placed in the same species, *Homo erectus,* which had a brain intermediate between earlier human types and modern man.

Early divergence of man's ancestry from the higher apes is marked especially by bipedalism and the erect posture. Such changes may have been induced by environmental conditions when forested regions were replaced by grasslands. This forced man to change from an arboreal animal to one that hunted and lived upon those that ate grass. Selective pressure would put an evolutionary premium on the erect position, communication, and the ability to use tools. Brachiation (long, grasping arms) may have arisen in apes after they diverged from the common hominoid ancestor, so that man may never have had this characteristic.

MAN'S CLOSEST RELATIVES—PRIMATES

Man belongs to the order of mammals called Primates. This order is commonly divided into the Prosimii and Anthropoidea. The prosimians include the tree shrews, lemurs, lorises, and tarsiers. The anthropoids are the more advanced primates and are further divided into the Platyrrhini, or New World primates, and Catarrhini, or Old World monkeys, apes, and man. The catarrhines are likewise separated into the Cercopithecoidea, or anthropoids with tails, and Hominoidea, or the great apes and man. The hominoids include three families—Hylobatidae (gibbons), Pongidae (orangutans, chimpanzees, and gorillas, and Hominidae (man).

Evolution of this mammalian order has been guided by the arboreal habits of the group. Man abandoned the brachiating habit (using the arms for swinging) and took up a terrestrial existence. But certain anatomic traits of an arboreal life remain. The free rotation of limbs in their sockets, the movable digits on all four limbs, and the opposable thumbs are all modifications for grasping branches and swinging through trees. Omnivorous food habits have produced a characteristic dentition. Better vision and development of other sense organs, as well as the proper coordination of limb and finger muscles, have meant an enlargement of appropriate regions of the brain.

Early in the Cenozoic era the basic radiation of primitive placental mammals began. This continued until more than a score of orders arose and took over the niches formerly occupied by reptiles (Fig. 35-1). The most primitive order was the Insectivora, familiar examples of which are the present terrestrial moles and shrews. But some members of this order, the tree shrews, took to living in trees and gave rise to the great order of primates.

These early primates (**prosimians**) were small nimble animals adapted for living in trees. Although they possessed good muscular coordination and sense organs, their intelligence was low; intelligence among modern primates is a later development in their evolution. Some modern prosimians have many primitive characteristics, such as eyes on the side, long snouts, and long tails (lemurs); but in the tarsiers the large eyes have moved forward and the muzzle is short. Their larger brain and better-developed placenta also place the tarsiers nearer the anthropoids than the lemurs. Certain basic adaptations were developed by primates for a tree-dwelling existence. It was not only necessary to have grasping ability, such as the opposability of the thumb and great toe, but also stereoscopic vision and the ability to judge distance in swinging precariously from limb to limb.

The primitive, **hominoid apes** evolved independently from prosimian ancestors. They were small monkey-sized animals. Many of their fossils have been found in the Miocene and Pliocene deposits. Although these fossils are definitely apelike, they lack, in general, the specializations of the modern apes. Rather, they show many primitive characteristics similar to man. They lack especially the brachiating arms and simian shelf found in living apes. Although modern apes are more specialized than man for arboreal life and in other ways, the fossil record discovered in the past decade or so indicates that man and the higher apes have a very close affinity. The evidence for a common ancestry of the two groups is more and more convincing as new fossils are found.

Apes today are represented by the gibbons, orangutans, chimpanzees, and gorillas (Fig. 23-1, B). Apes have longer arms than legs, long trunk compared to lower limbs, curved legs with the knees turned outward, large canines, laterally compressed dental arch (not rounded), long protruding face, and a brain about one third as large as man's. Some modern apes have largely abandoned the arboreal way of life. Chimpanzees and gorillas are quite at home on the ground. A more or less erect posture is characteristic of many of them, although they may use their long arms for support while walking. More abundant food on the ground may have induced them to come down out of the trees.

When the hominids separated from the higher apes, they underwent a radiation of their own. Giving up arboreal life was a prelude to their amazing evolution. Climatic changes during certain geologic ages may have reduced the forests and forced a ground existence. Increased body size may have been a factor also. Because of predators, selection pressure may have brought about evolution of feet and strong running muscles. The fossil record is incomplete between the rather abundant ape fossils of the Miocene epoch and the hominid fossils of the Pleistocene epoch. In this period of more than 20 million years, little is known about the hominid line. The initial phases are largely unknown. The fossil record indicates that the hominid radiation produced various lines of descent. From its dentofacial features, *Ramapithecus* appears to be the most appropriate ancestral taxon. One of the chief problems of paleontologists is to find fossils that will close the gap between man and the higher apes. The exciting discoveries found in Africa in the past few years have added much to our knowledge of man's emergence.

343

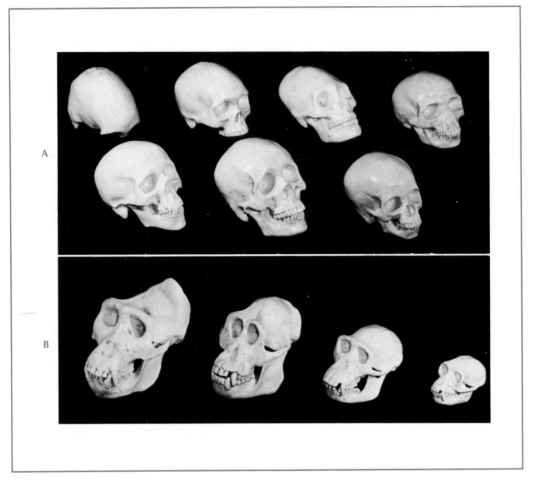

FIG. 23-1. Comparison of skulls of living races of man, **A,** with those of living simians, **B.** (Courtesy Ward's Natural Science Establishment, Inc., Rochester, N. Y.)

FOSSIL RECORD OF MAN

The hominids appear to date back to the early Pleistocene epoch and maybe earlier. Authorities are not agreed on the age of many fossils nor on the interpretations that are made about them. It is generally agreed that man is a polytypic species, that is, composed of several races, each with a different gene pool. The newer idea of speciation would consider the hominid fossils as samples of populations wherein there are many variations in type and form.

During the Pleistocene epoch there are three well-marked stages when man was evolving from apelike to manlike individuals. The transitions that occurred during this period may have been influenced by the climatic changes induced by the advances and recessions of four north polar ice caps. The three stages were (1) the *Australopithecus* group of the basal Pleistocene in Africa, (2) the *Homo erectus* or *Pithecanthropus* forms of the middle Pleistocene, and (3) the *Homo sapiens* types of the late Pleistocene. Students may be interested in some of the famous fossil finds and evidences of their position in the phylogeny of the human race.

SEPARATION OF HOMINID LINEAGE FROM APE LINEAGE

Darwin probably knew only two types of important fossils for apes and man. One was *Dryopithecus* (fossil ape) and the other the Neanderthal (fossil man). Darwin in his *Descent of Man* stated his belief that the lineage of apes and monkey had separated by the late Miocene epoch, which began about 25 million years ago. From this time on apes and monkeys were common throughout Europe, Asia, and Africa. Some think that apes and monkeys separated in the Oligocene epoch several million years earlier. Since the earliest apes were ancestral to man, the divergence of the two groups must have occurred about the same time, probably not later than the Miocene epoch.

AUSTRALOPITHECUS (SOUTHERN APE)

In 1925 the fossil brain cast of an immature anthropoid was discovered in South Africa by Professor R. Dart. This specimen had a mixture of both human and ape characters. Additional fossils of this and related types, including skeletons that were almost complete, have been found by a number of investigators since that time.

Two distinct groups were discovered, *Australopithecus (Paranthropus) robustus* and *Australopithecus africanus.* Both belonged to the human family. They showed some apelike characteristics along with more human ones. They are dated about 2 million years ago. Recently, an *Australopithecus* jawbone found near Lake Rudolf in Kenya, Africa (1967), has been dated at 5.5 million years ago. The two groups coexisted for many years, but finally *A. robustus* became extinct, and *A. africanus* evolved into *Homo.* Similar fossils have been found as far away as Java. The volume of their brain casts varied from about 450 to 600 ml. and overlapped the range of the chimpanzee and gorilla. They walked in an erect or semierect position, as shown by the shape of the leg and foot bones. Their dentition was more human than apelike. From the evidences of primitive stone tools, these so-called man-apes were toolmakers and tool users. They may be regarded as the earliest type with distinct manlike characteristics, and they come close to the anatomical features expected in a "missing link."

Their habitat was mainly terrestrial, which may have some significance.

ZINJANTHROPUS

This fossil was discovered in 1959 by L. S. B. Leakey in the famous Olduvai gorge of Tanganyika, Africa, where so many significant paleontologic specimens have been found. It probably belongs to the *Paranthropus,* one of the genera of the australopithecines. Its age has been calculated by the potassium-argon method at 1,750,000 years. Although evidence of tools was found with the fossil, these tools could have been left by more recent members of the *Homo* type. In some ways this form seems to be closer to the Hominidae than any others of the australopithecines. They were generally less than 5 feet tall, walked almost erect, and had small brains. Some of their skull features were similar to man's. Leakey has also found a tool-using hominid, which he called *Homo habilis,* in the same locality. This one seems to be even closer to *Homo.*

PITHECANTHROPUS (JAVA MAN)

This famous fossil was discovered in eastern Java in 1891 by E. Dubois, a Dutch anatomist. Only a skullcap and thigh bone were first found, but better specimens have since been discovered (Fig. 23-2). This taxon is considered to be almost 500,000 years old (middle Pleistocene). Because so many of the anatomic features were so close to modern man's, Java man has been called *Homo erectus.* They walked fully erect, were over 5 feet tall, and had heavy projecting brow ridges. They may have had a spoken language and used stone tools. The brain capacity was at least 900 ml. This species was found in Asia and Africa, as well as in Java. The Peking man (*Sinanthropus pekinensis* or *Homo erectus pekinensis*) is a northern race of the same species.

HEIDELBERG MAN (HOMO HEIDELBERGENSIS)

The evidence for the existence of the Heidelberg man rests on an almost perfect jaw discovered in 1907 near Heidelberg, Germany, in a deposit of bones from the lower Pleistocene. Its general aspect is human, although it combines a very massive jaw

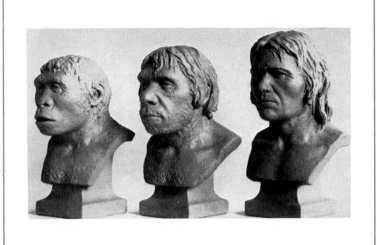

FIG. 23-2. Restoration of prehistoric man. Left to right: Java man, Neanderthal man, and Cro-Magnon man. (Courtesy J. H. McGregor.)

with small humanlike teeth. Some authorities classify this man under the australopithecines, but it differs from the Asiatic type of that group.

SWANSCOMBE AND STEINHELM SKULLS

Between the *Pithecanthropus* group and the establishment of the modern polytypic species of *Homo sapiens,* the hominid evolution has taken a complex course. The earliest fossils that show clear-cut *Homo sapiens* features were the Swanscombe man in England and the Steinhelm man in Germany. These may have evolved from the pithecanthropines about 300,000 years ago. They had prominent eyebrow ridges and a brain capacity not far from that of modern *Homo.* They may be the earliest *Homo* that cannot clearly be differentiated from *Homo sapiens.* Leakey's recent (1961) discovery of the Chellean man at Olduvai gorge in East Africa has been dated at 250,000 years. This new find seems to mark a morphologic transition from *Pithecanthropus* to *Homo,* but further study is awaited. However, more recent fossils than those of the Olduvai gorge are known to occur. It appears that primitive australopithecines, near ape-men, and members of the first true men *(Homo)* were undergoing an evolution, with considerable overlapping of the different types in the period of the middle Pleistocene about 300,000 to 1 million years ago.

EMERGENCE OF MODERN MAN

Modern man, *Homo sapiens,* first appeared in the fossil record about 75,000 to 100,000 years ago in the form of Neanderthals. This race has left many fossils in West Central Europe (and some elsewhere) in glacial deposits. Selective pressure of this severe climatic condition may have been responsible for the development of the race. The first specimen of *Homo sapiens neanderthalensis* (Fig. 23-2) was discovered near Düsseldorf, Germany, in 1856—the first hominid fossil to receive attention by competent scholars. He was little more than 5 feet in height; his brain capacity was similar to that of modern man, and he had developed a crude form of paleolithic culture. He differed from modern *Homo sapiens* in having a flattened braincase, projecting jaws, recessive chin, and strong mandibles. Their populations dominated the scene in late Pleistocene times, from Europe to Africa and Asia. The race varied from place to place in response to local conditions or the intermixing of the different types. Of the various types of *Homo sapiens* that have arisen, there may be mentioned the Solo man (Japan), Rhodesian man (South Africa), and the Mt. Carmel man (Palestine).

About 30,000 to 40,000 years ago, the Neanderthal race was replaced in Europe rather suddenly by the Cro-Magnon race, which emerged from an unknown source (Fig. 23-2). They may have been responsible for the extermination of the Neanderthals. The Cro-Magnon was rather a mixture of people that showed considerable physical variations in different localities. They had a superior culture (Perigordian) and left artistic paintings and carvings in their caves. They are considered ancestors of modern man and represent the modern type of man. They were about 6 feet tall, had a high forehead but no supraorbital ridges, a rather prominent chin, and a brain capacity as large as (or larger than) present-day man. Their physical characteristics are matched today by the Basques in northern Spain and certain Swedes in southern Sweden. Attempts to discern the characteristics of present-day races in early populations of *Homo sapiens* have not been successful. It is not known whether they were white, black, or brown.

The evolutionary course of modern man in his 30,000 to 40,000 years has exhibited the same pattern of divergence and extinction demonstrated by his forbears and that of other organisms. The essential characters of human phylogeny, such as man's superior brain and wide adaptability, can all be attributed to the strictly quantitative effects of mutations that could have happened at any evolutionary level. In other words, man has been the outcome of the basic factors of evolution. Mutation, selection, population factors, genetic drift, and isolation—all the general processes of evolutionary progress have operated for man the same as for other animals.

Of the many types of man mentioned in the foregoing account, it is doubtful if there were more than two species coexisting at one time, and in most cases perhaps only one. A polytypic species such as man could have many types in a widespread population.

MAN'S UNIQUE POSITION

Man has what no other animal has—a psychosocial evolution, or a directional cultural pattern that involves a constant feedback between past and future experience. Although human evolution has become increasingly cultural as opposed to genetic, man is still subjected to the same biologic forces and principles that regulate other animals.

When one compares man with other animals, he finds a broad gap. First, he is the only animal that knows how to make and use tools effectively. This more than any other factor has been responsible in giving man his dominant position. Another unique characteristic of man is his capacity for conceptual thought. Man has a symbolic language of wide and specific expression. With words, he can carve concepts out of experience. This has resulted in cumulative experience that can be transmitted from one generation to another. In other animals, transmission never spans more than one generation. Man owes much to his arboreal ancestry. It promoted his binocular vision, a fine visual-tactile discrimination, and manipulative skills in the use of his hands. If a horse (with one toe) had man's intellect and culture, could it accomplish what man has done?

Man's population is commonly divided into races or populations that are genetically distinguished from others. A so-called race has certain genes or gene combinations that may be more or less unique, although races grade into each other and do not have definite boundaries. Pure races are nonexistent, and there is no fixed number of races. Races are adaptations to local conditions. It is thought that as primitive men spread over geographic areas, they became adapted to certain regions, and natural selection stamped on them certain distinguishing features. Races at present are losing their biologic significance because human adaptation to environment is becoming largely cultural. Rapid mobility and quick communication and intercourse has shrunk the size of our planet so that isolation of races rarely exists. Genes can shift through human populations now with amazing speed, and racial intermixtures are far more common than formerly. Since all races have more or less unique potentialities, hybrid vigor could operate within racial interbreeding just as it does for other animals.

REFERENCES FOR PART V

Balinsky, B. I. 1970. An introduction to embryology, ed. 3. Philadelphia, W. B. Saunders Co.

Barnett, L., and editors of Life Magazine. 1955. The world we live in. New York, Time, Inc.

Barnett, L., and editors of Life Magazine. 1960. The wonders of life on earth. New York, Time, Inc.

Bates, M., and P. S. Humphrey (editors). 1956. The Darwin reader. New York, Charles Scribner's Sons.

Blum, H. F. 1955. Time's arrow and evolution. New York, Harper & Row, Publishers.

Bulmer, M. G. 1970. The biology of twinning in man. New York, Oxford University Press, Inc.

Colbert, E. H. 1969. Evolution of the vertebrates, ed. 2. New York, John Wiley & Sons, Inc.

Crow, J. F. 1966. Genetic notes, ed. 6. Minneapolis, Burgess Publishing Co.

Darwin, C. R. 1859. On the origin of species by means of natural selection. Cambridge, Mass., Harvard University Press.

De Beer, G. R. 1951. Embryos and ancestors. London, Oxford University Press.

Dobzhansky, T. 1951. Genetics and the origin of species, ed. 3. New York, Columbia University Press.

Fox, S. W. 1964. Experiments in molecular evolution and criteria of extraterrestrial life. Bioscience 14(12):13-21.

Kerkut, G. A. 1965. Implications of evolution. Elmsford, N. Y., Pergamon Press, Inc.

Kettlewell, H. A. D. 1959. Darwin's missing evidence. Sci. Amer. 200:48-53 (March).

Leakey, L. S. B., and V. M. Goodall. 1969. Unveiling man's origins. Ten decades of thought about human evolution. Cambridge, Mass., Schenkman Publishing Co., Inc.

Leakey, L. S. B., and H. Van Lawick. 1963. Adventures in the search for man. Nat. Geogr. Mag. 123:132-152 (Jan.).

Mayr, E. 1963. Animal species and their evolution. Cambridge, Mass., Harvard University Press.

Moore, J. A. 1963. Heredity and development. New York, Oxford University Press, Inc.

Oppenheimer, J. M. 1967. Essays in the history of embryology and biology. Cambridge, Mass., M. I. T. Press.

Scheinfeld, A. 1965. Your heredity and environment. Philadelphia, J. B. Lippincott Co.

Simpson, G. G. 1964. This view of life: the world of an evolutionist. New York, Harcourt Brace & World, Inc.

Sonneborn, T. M. (editor). 1965. The control of human heredity and evolution. New York, The Macmillan Co.

Tax, S. (editor). 1960. Evolution after Darwin. Chicago, University of Chicago Press, 3 vol.

Teilhard de Chardin, P. 1965. The phenomenon of man, ed. 2. New York, Harper & Row, Publishers.

Torrey, T. W. 1967. Morphogenesis of the vertebrates, ed. 2. New York, John Wiley & Sons, Inc.

Young, L. B. (editor). 1970. Evolution of man. New York, Oxford University Press, Inc.

part six

The diversity of life

The enormous diversity of animal life is seen in the 1.5 million species that have already been named and the many more that are still to be named. Two important principles may be gleaned from this great variety of life. One is that all the varieties of animals appear to be related to each other, some varieties closely and others distantly related. The other principle is that by progressive adaptation to diversified environmental opportunities, animals have expanded into practically every environmental niche that can support life.

Adaptations are brought about chiefly by heredity and evolution. Heredity tends to stabilize the reproductive process from generation to generation so that evolutionary innovation for adaptation is usually slow. Thus it has been possible to trace evolutionary relationship as the basis of natural classification. No contemporary type of animal is the ancestor of any other contemporary animal. All existing animals are too specialized to have given rise to other existing animals.

It is commonly assumed that the first animals arose from one-celled forms and increased gradually in complexity of structure and function. Metazoan animals may have arisen from protozoan ancestors or they may have come from early protists (unicellular forms with both plant and animal characteristics). Metazoans may have a polyphyletic origin and could have come from several sources. Two structural features must have been emphasized from the first—alimentation and motility. Because animals are heterotrophic and require organic food materials that have originated in other plants or animals, motility, or the ability to get to food, is a basic characteristic of most animals.

As we study the various animal groups from the simplest to the highest, we can see that there is a gradual progression from generalized traits to specialized ones and that general traits are shared by more animals than are specialized ones. This vast assemblage of diversified forms becomes more meaningful and easier to study and understand when divided into groups of more modest numbers. Such grouping has been done on the basis of similarities in morphologic organization, embryologic development, chemical constituents, and in many other ways. Other bases of separation are types of symmetry, grades of organization, and the presence or absence of a body cavity.

24

The concept of phylogeny

The diversity of living things is both the delight and the despair of those who would study and try to understand living things. It becomes a practical necessity to classify them into some sort of order. A phylum approach groups organisms on the basis of similarities in morphologic organization, embryologic development, basic adaptive features, and other criteria of fundamental importance. To understand an animal it is vital to know where it belongs and how it relates to other animals.

The phylum approach may give the false impression that evolution has been linear and that the higher existing phyla have evolved from lower existing phyla. Although the so-called higher forms are more complex than lower forms, there is no basis for assuming that characteristics have been added by succeeding phyla. Many characteristics shared by groups of animals have been acquired independently (convergent evolution). Environmental factors may foster and promote certain structural patterns in evolutionary development, but these are not necessarily evidences of relationship.

CONCEPT OF THE SPECIES

A species is often defined as a collection of individuals or populations sharing the same gene pool. They are capable in nature of interbreeding with one another to produce fertile offspring. Within the pool a free flow of genes occurs, but between two such pools a genetic flow of genes rarely if ever happens. We must make some reservations here because fertile offspring sometimes occur when genes are exchanged between members of different species. But usually different species cannot mate because of some isolating mechanism (geographic, ecologic, structural, physiologic, or genetic). Members of a fertile group may become scattered geographically into isolated groups that begin to differ in evolutionary development. When this happens, gene exchange will not occur when the members are exposed to each other. In this way a species may become fragmented into a number of subspecies, which have evolved genetic differences.

The term "species characters" usually refers to those attributes that distinguish one species from all other species. Criteria commonly used are morphologic characters, range of population, ecologic segregation, genetic composition, and isolating mechanism. Species are usually genetically closed systems, whereas races within a species are open systems and can exchange genes.

In defining a species most biologists have shunned purely morphologic distinctions. In contrast, genetic implications are stated or implied in the definition. A species therefore may be defined as **a group of** organisms of interbreeding natural populations that are reproductively isolated from other groups and that share in common gene pools (E. Mayr and T. Dobzhansky).

EARLY HISTORY OF TAXONOMY

Although Aristotle, the great Greek philosopher and student of zoology, attempted to classify animals on the basis of their structural similarities, little was done about the grouping of animals until the English naturalist John Ray (1627-1705) brought forth his system of classification. He employed structural likenesses as the basis of his classification and worked out a number of groups. Carolus Linnaeus (1707-1778), Swedish botanist connected with the University of Uppsala, gave us the modern scheme of classification. He worked out a fairly extensive system of classification for both plants and animals that was published in his classic work *Systema Naturae,* which had gone through ten editions by 1758. Linnaeus emphasized structural features, and he believed strongly in the permanence of species. He divided the animal kingdom down to species, and according to his scheme each species was given a distinctive name. He recognized four classes of vertebrates and two classes of invertebrates. He divided these classes into orders, the orders into genera, and the genera into species. Since his knowledge of animals was limited, much of his classification has been drastically altered, yet the basic principle of his scheme is followed at the present time.

Linnaeus recognized four units, or taxa, in classification—class, order, genus, and species. Since his time other major units have been added, such as the phylum and the family, so that the units now used are **phylum, class, order, family, genus,** and **species.** The major units can be subdivided into finer distinctions, such as subphylum, subclass, suborder, subfamily, subgenus, and subspecies.

BINOMIAL NOMENCLATURE AND THE NAMING OF ANIMALS

Linnaeus early adopted the use of two names for each species: the genus name and the species name. These words are from Latin or in Latinized form be-

cause Latin was the language of scholars and universally understood. The generic name is usually a noun and the specific name an adjective. For instance, the scientific name of the common robin is *Turdus migratorius* (L. *turdus,* thrush; *migratorius,* of the migratory habit). This usage of two names to designate a species is called **binomial nomenclature.** There are times when a **trinomial nomenclature** is employed. Thus to distinguish the southern form of the robin from the eastern robin, the scientific term *Turdus migratorius achrustera* (duller color) is employed for the southern type.

Taxonomy aims to apply a name tag to every species of animal in the animal kingdom. Since each species has a universal specific name, students of all languages know what animal is meant when the scientific name is designated. Common names vary with the different languages or even in different parts of one country, but the scientific ones are universal. The woodpecker *Colaptes auratus luteus,* for instance, is called the golden-winged woodpecker, the flicker, the highhole, etc., depending on the part of the United States in which it is found. But it has only one valid scientific name wherever it is found.

SOME EXAMPLES OF SCIENTIFIC NOMENCLATURE

The examples listed in Table 24-1 will give you some idea of how animals are classified on the basis of relationship and likeness. Of all animals the anthropoid apes are generally agreed to be nearest man in relationship and structural features. In contrast to man and the gorilla are the frog, also a vertebrate like the others but diverging from them much earlier, and the little katydid, which is not a chordate but belongs to a lower phylum.

THREE KINGDOMS OF LIVING THINGS

One may divide all living things into three great kingdoms—plants, protists, and animals.

The **Protista** are chiefly single-celled organisms. Some are considered plantlike, such as the blue-green algae, bacteria, algae, and fungi. The animal-like protists are the protozoans. Slime molds are also placed among the protists. Some protists are

TABLE 24-1. Examples of classification of animals

	Man	**Gorilla**	**Grass frog**	**Katydid**
Phylum	Chordata	Chordata	Chordata	Arthropoda
Subphylum	Vertebrata	Vertebrata	Vertebrata	
Class	Mammalia	Mammalia	Amphibia	Insecta
Subclass	Eutheria	Eutheria		
Order	Primates	Primates	Salientia	Orthoptera
Suborder	Anthropoidea	Anthropoidea		
Family	Hominidae	Simiidae	Ranidae	Tettigoniidae
Subfamily			Raninae	
Genus	*Homo*	*Gorilla*	*Rana*	*Scudderia*
Species	*sapiens*	*gorilla*	*pipiens*	*furcata*
Subspecies			*pipiens*	*Brunner*

unicellular and others colonial. Some may have only one nucleus or some may be multinucleated.

Although rigid distinctions between plants and animals are not always possible, the chief differences are as follows:

Plants	**Animals**
Synthesize food by photosynthesis	Food mostly organic substances, ultimately supplied by plants
Have functional chlorophyll	Chlorophyll absent except as pigment
Starch as principal food reserve	Glycogen or fat as principal food reserve
Cellulose cell walls (rigid)	Plasma cell membrane
Mostly sessile	Active movements
Indefinite growth and shape	Usually fixed size and shape

SUBKINGDOMS, BRANCHES, AND GRADES

Although the phylum is often considered to be the largest and most distinctive taxonomic unit, biologists often find it convenient to combine phyla under a few large groups because of certain common embryologic and anatomic features. Such large divisions may have a logical basis, for the members of some of these arbitrary groups are not only united by common traits, but evidence also indicates some relationship in phylogenetic descent. The scheme proposed below for some of these larger groupings should give the student a more comprehensive view of the classification of the animal kingdom.

Subkingdom Protozoa, acellular animals—phylum Protozoa
Subkingdom Metazoa, cellular animals—all other phyla
  Branch A, Mesozoa—phylum Mesozoa
  Branch B, Parazoa—phylum Porifera
  Branch C, Eumetazoa—all other phyla
    Grade I, Radiata, the radiate animals—phyla Coelenterata and Ctenophora
    Grade II, Bilateria, bilaterally symmetrical animals—all other phyla
    Acoelomata, animals without a coelom—phyla Platyhelminthes, Rhynchocoela
    Pseudocoelomata, with a body cavity, but not lined with peritoneum—phyla Acanthocephala, Aschelminthes, Entoprocta
    Eucoelomata, with a true coelom—all other phyla

PROTOSTOME AND DEUTEROSTOME DIVISIONS OF BILATERAL ANIMALS

The Bilateria may be arranged into two major divisions on the basis of their embryonic development—Protostomia and Deuterostomia. These divisions form the two main lines of evolutionary ascent in the animal kingdom and are often referred to as the diphyletic theory of phylogeny. The characteristics and phyla of each of these divisions are as follows:

Protostomia
Mouth usually formed from blastopore
Schizocoelous formation of body cavity (coelom)
Mostly spiral cleavage
Determinate or mosaic pattern of egg cleavage
Ciliated larva (when present) a trochophore or trochosphere type

353

Protostomia—cont'd
  Includes the acoelomates, the pseudocoelomates, and, among the eucoelomates, phyla Annelida, Mollusca, Arthropoda, and a number of minor phyla
Deuterostomia
  Anus formed from blastopore
  Enterocoelous formation of coelom
  Mostly radial cleavage
  Indeterminate pattern of egg cleavage
  Ciliated larva (when present) a pluteus type
  Includes phyla Echinodermata, Hemichordata, Pogonophora, Chaetognatha, and Chordata

THE PHYLOGENETIC TREE

Exact relationships of the members of the animal kingdom are often vague or nonexistent according to our present knowledge. This is especially true of the major groups (phyla). The sequence schemes in which biologists present the various groups do not imply that each group has arisen directly from the one preceding it. Most common ancestors are sufficiently generalized to give rise to many divergent groups, but such ancestors have either undergone evolutionary change or else have become extinct. The closing of the gaps in relationships is much like supplying the missing parts of a jigsaw puzzle.

Although multicellular animals may have evolved from protozoans, it seems to be doubtful that metazoans have had a monophyletic origin, that is, have arisen from a single common stock. A diphyletic or even polyphyletic origin appears more likely.

It is customary among biologists to express evolutionary relationships by a schematic diagram, or phylogenetic tree, showing a genealogy of living things on the basis of characteristics of both present and fossil forms (Fig. 24-1). All such schemes are of a necessity based on available information, which changes from time to time. The fossil record (if available), similarities of structure and development, presence of common traits, ecologic distribution, etc. are usually the lines of evidence stressed in the formation of a common scheme. If two different organisms share many common traits, biologists assume that there is a basic relationship for the similarity and that it has not been due to convergent or coincidental evolution. Perhaps no phylogenetic tree should be stressed too much, and certainly not dogmatically, but there is some ad-

vantage to the student in tying the groups together in some sort of evolutionary blueprint. A better way, perhaps, to represent phylogeny is by a fan-shaped scheme because there is little difference in the age of most phyla from a geologic viewpoint (Simpson).

THE FIRST ANIMALS

The origin of life from nonliving beginnings may have taken more than 2 billion years. The nature of the first life is purely speculative. It is generally agreed that such primitive forms could not now survive in the face of stronger competitors such as bacteria and fungi. The earliest form of life must have had the basic characteristics of a living form, such as self-duplication and some capacity for metabolism and adaptive adjustments to its environment.

Since plants and animals cannot be separated by biotic criteria, it is a moot question whether the earliest organisms were unicellular plants or animals. They were undoubtedly simpler than the bacteria, and they were able to live an independent existence with the simplest of requirements for duplication, metabolism, and environmental adjustments.

ORIGIN OF METAZOA

It is generally believed that both plant and animal traits evolved gradually from the Protista (Colonial theory). Some of the primitive flagellates have both ameboid and flagellate methods of locomotion, and many flagellates have a tendency to form colonies of few to many cells. It is possible to pick out a progressive series of such aggregations of gradually increasing complexity (Fig. 24-2). For example, *Chlamydomonas* is single celled, *Gonium* is a colony made up of 4 to 16 cells, *Pandorina* has 16 cells, *Eudorina* 32 cells, *Pleodorina* 32 to 128 cells (depending upon the species), and *Volvox* many thousands. The cells, or zooids (individuals), may be loosely connected or may be held together in a mucilaginous jelly. In *Pleodorina* the cells are differentiated into a few somatic cells that are sterile and other cells that are capable of reproduction. In *Volvox* the majority of cells are sterile and only a few

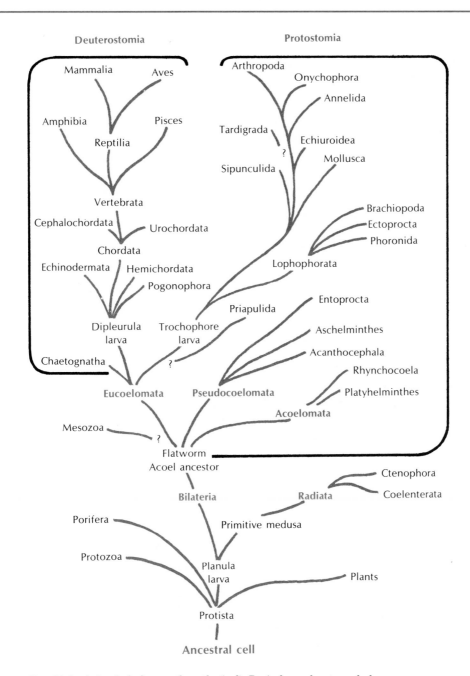

FIG. 24-1. Animal phylogeny (hypothetical). Basis for such a tree phylogeny is evolutionary relationship, or common ancestry, but such information is scanty. In general, time is represented by vertical levels — the higher the branch, the more recent the taxon — but this is also uncertain because fossil record is incomplete. Interpretations of relationship are based on common characters, similarity of embryologic development, basic structural patterns, fossils, etc.

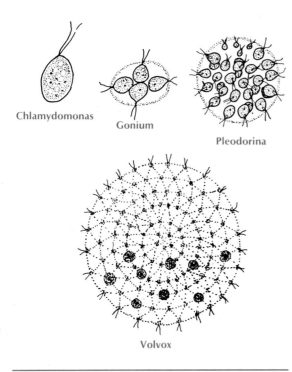

Chlamydomonas

Gonium

Pleodorina

Volvox

FIG. 24-2. Some flagellates live in clumps or colonies, joined at their outer surfaces or enclosed in gelatinous envelope. In *Gonium* all cells are alike and each may divide to form a new colony. Some colonies have division of labor, possessing both somatic and reproductive cells. *Pleodorina illinoisensis* has 4 somatic and 28 reproductive cells; *Volvox* has hundreds of somatic and only a few reproductive cells. *Chlamydomonas* is solitary and does not form colonies.

are reproductive. In the other examples given each zooid is capable of both asexual and sexual reproduction. These examples, however, do not afford evidence that one form gave rise to another. Some zoologists believe it unlikely that flagellates evolved beyond the colony.

The essential difference between a metazoan and a colony of protozoans is polarization, or the establishment of an axis along which there is both morphologic and functional differentiation. Such a polarization is manifested by a capability for direct

forward movement with one pole forward. This is lacking in protozoan colonies.

The metazoans are often considered a diphyletic group, with sponges (Parazoa) as one separate line and the Eumetazoa as the other. Sponges probably arose from flagellated protozoans and represent complex flagellate colonies at a level of differentiated cells and with a tendency toward a tissue level of construction. Their digestive collar cells are similar to protozoan collared flagellates. Sponges, of course, have other differences in construction and embryonic development.

There is also a theory of metazoan origin from syncytial ciliates. The syncytial theory differs from the colonial theory in that the ancestral animal, a ciliate, was at first a syncytium (that is, it was multinucleated) and later it became multicellular by the development of internal cell boundaries. According to this theory, the first true metazoans were flatworms. They resembled the acoel flatworms that are considered to be the most primitive of the living bilateral animals. One advantage of the theory is that some syncytial ciliates have an established bilateral symmetry and an anteroposterior axis. However, the theory implies that coelenterates with radial symmetry are derived from flatworms rather than the reverse, as is more commonly accepted.

Another theory favors the origin of metazoans from plants (Metaphyta). Simple plantlike protists absorb nutriments from all sides equally. Early metazoans, after they had evolved from such primitive plants, no doubt also absorbed nutriments from the environment, but they may, as a result of a shortage of available phosphates or nitrates in the environment, have begun to capture and feed upon small organisms.

Many other theories have been proposed for the origin of metazoans. All are purely speculative. Metazoans may have had a polyphyletic origin, that is, the major groups may have evolved independently of each other and in different ways. Relationships may have played little or no role in the process. The only thing we can be certain of is that we do not yet know what relationship, if any, actually existed. We can only theorize.

25

Protozoans and sponges

Protozoans and sponges are the most primitive animals. Protozoans are often considered to be unicellular, but since each animal is a complete organism, it is better to call them acellular. They obviously do not correspond to the cells of metazoan (many-celled) animals.

Sponges are an aberrant group out of the direct line of animal evolution. They are multicellular, but differ so much from other multicellular animals that they are more correctly called parazoans than metazoans. A sponge is in many ways a group of independent but physically associated cells that coordinate their activities to the benefit of the whole —not far advanced beyond the protozoan colony.

PROTOZOANS (PHYLUM PROTOZOA*)

Most protozoans are free living in water or damp soil, although some of the more notorious are parasitic in other forms. The number of named species of Protozoa lies somewhere between 15,000 and 50,000, but even this upper figure probably represents only a fraction of the total number of species. Most protozoans are small or microscopic, usually from 3 to 300 mm. long; certain amebas may be 4 to 5 mm. in diameter.

*Pro"to-zo′ a (Gr. *protos*, first, + *zoon*, animal).

A protozoan is an organism that is not divided into cells, but carries on all the life processes within the limits of a single plasma membrane. They represent a protoplasmic level of organization. Many biologists place them close to the common ancestor of the many-celled forms. Some protozoans are quite close to the plants and may be considered as connecting links between animals and plants.

Some protozoans are found in colonies in which each individual carries on its functions independently of the others. However, in a few colonies there is a small amount of "division of labor," that is, some cells serve for locomotion and nutrition, others for reproduction (Fig. 24-2). The distinction between colonial Protozoa and Metazoa lies mainly in the degree of division of labor. If the cells are completely dependent on each other for such functions as nutrition, movement, excretion, and reproduction, the colony belongs properly to the Metazoa; if only certain cells are for reproduction and the rest can perform all other bodily functions, they are considered a protozoan colony.

From a practical standpoint protozoans have played an important role in building up soil and forming earth deposits; they have important roles in food chains; they form important symbiotic relationships; they have been responsible for much of the contamination of water; and many of them as parasites have been responsible for serious diseases in man and other animals.

CLASSIFICATION

Class Mastigophora (mas″ti-gof′o-ra). Move by flagella or by pseudopodia. Examples: *Euglena, Volvox, Trypanosoma.*

Class Sarcodina (sar″ko-di′na). Locomotion by pseudopodia; no definite pellicle; free-living or parasitic; uninucleate or multinucleate; mostly holozoic. Example: *Amoeba.*

Class Sporozoa (spor″o-zo′a). No locomotor organelles; asexual and sexual phases; saprozoic; parasitic. Examples: *Plasmodium, Nosema.*

Class Ciliata. Cilia present at some or all stages; nuclei of two kinds. Examples: *Paramecium, Vorticella, Stentor.*

CHARACTERISTICS

1. **Acellular** (or one cell), some colonial
2. **Mostly microscopic,** although some large enough to be seen with the unaided eye
3. All symmetries represented in the group; shape variable or constant (oval, spherical, etc.)
4. **No germ layer present**
5. No organs or tissues, but **specialized organelles** found; nucleus single or multiple
6. Free living, mutualism, commensalism, parasitism all represented in the group
7. Locomotion by **pseudopodia, flagella, cilia,** and direct cell movements; some sessile
8. Some provided with a **simple protective exoskeleton,** but mostly naked
9. Nutrition includes all types: holozoic (feeds on other organisms), holophytic (makes own food by photosynthesis), saprozoic (absorbs simple organic materials), and saprophytic (on dissolved substances)
10. Habitat: aquatic, terrestrial, or parasitic
11. Reproduction asexually by fission, budding, and cysts and sexually by conjugation of gametes

SYMBIOTIC RELATIONSHIPS

The term **symbiosis** refers to the intimate interrelationships between two organisms of different species for the purpose of deriving energy or for some other benefit. This special relationship may be beneficial to both (**mutualism**) or beneficial to only one but not harmful to the other species (**commensalism**), or the relationship may be forced so that one receives benefit and the other furnishes all the energy and may actually be harmed (**parasitism**). Such relationships are not limited to protozoans, but protozoans are represented by all three major types of symbiosis.

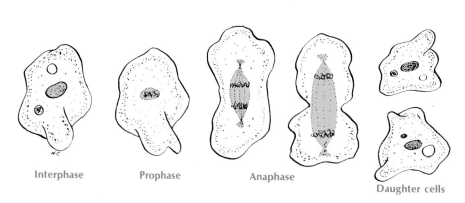

Interphase   Prophase   Anaphase   Daughter cells

FIG. 25-1. Mitosis in nucleus of *Amoeba*. There are many mitotic patterns among protozoans. In most cases nuclear membrane persists throughout mitosis, and division bodies of centrioles, centrosphere, and spindle are of nuclear rather than of cytoplasmic origin. Sometimes one of these division bodies may be absent. In many cases, chromosomes behave as in metazoans. In others, chromatin mass splits and passes to poles without forming chromosomes. Amitosis in protozoans is restricted mainly to macronucleus of ciliates.

REPRODUCTION

Reproduction in most protozoans is by cell division (asexual), but the method of reproduction varies. **Binary fission,** the most common process, involves the division of the organism, both nucleus and cytoplasm, into two essentially equal daughter organisms (Fig. 25-1). The nucleus divides by mitosis. **Budding** involves unequal cell division. The parent organism retains its identity while forming one or more small cells, each of which assumes the parent form after it becomes free. In **multiple division (sporulation)** the nucleus divides a number of times, followed by the division of the organism into as many parts as there are nuclei. It is a method of rapid multiplication characteristic of parasitic forms.

Protozoan **colonies** are formed when the daughter zooids remain associated together instead of moving apart and living a separate existence. The shape of a colony may be linear, spherical, discoid,

FIG. 25-2. Examples of Sarcodina showing wide variety of forms.

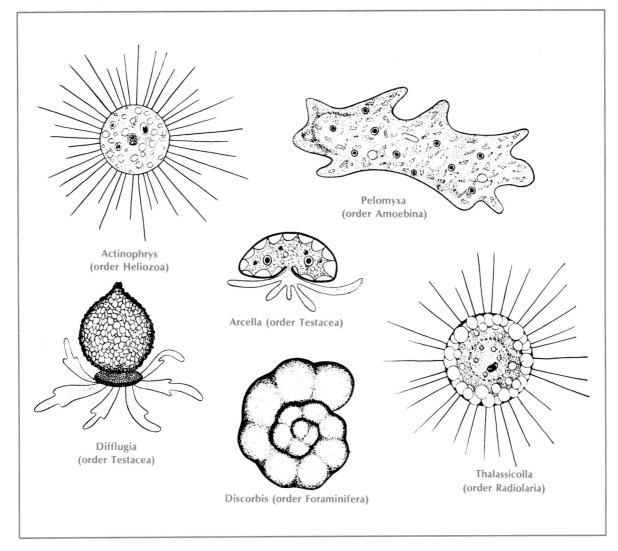

Actinophrys
(order Heliozoa)

Pelomyxa
(order Amoebina)

Arcella (order Testacea)

Difflugia
(order Testacea)

Discorbis (order Foraminifera)

Thalassicolla
(order Radiolaria)

or branched. The individuals are usually similar, although there may be a differentiation of reproductive and somatic zooids.

Sexual reproduction is found in certain protozoans. It may involve the formation of male and female gametes that unite to form a zygote, or there may be the complete union of two mature sexual individuals that merge their cytoplasm and nuclei together to form a zygote, or conjugation may occur during which the mates, or conjugants, exchange nuclear materials.

PROTOZOAN FAUNA OF PLANKTON

Plankton is a general term for those organisms that passively float and drift with the wind, tides, and currents of both fresh water and marine water. It is comprised mostly of microscopic animals and plants, of which protozoans form an important part. Plankton is important as food for a wealth of other marine animals—from tiny crustaceans to the giant whale bone whale. As the animals of the surface plankton die, they sink and serve as food for animals at lower levels.

AMEBOID PROTOZOANS (CLASS SARCODINA)

Sarcodina is a class (Fig. 25-2) characterized chiefly by ameboid movement. This movement involves the formation of cytoplasmic extensions called **pseudopodia,** which are also used for intake of food. Sarcodinians are found in moist soils and in both fresh and salt water, where some are planktonic forms and others prefer a substratum.

These forms are usually covered by a very thin membrane, the plasmalemma (Fig. 25-3), so that

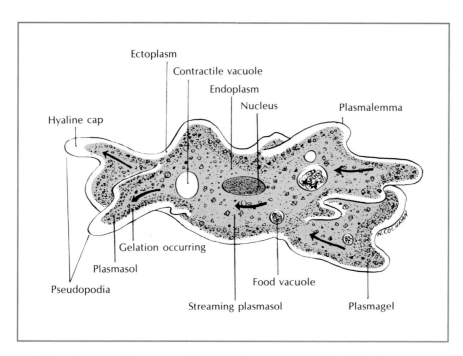

FIG. 25-3. Structure of *Amoeba* in active locomotion. Arrows indicate direction of streaming plasmasol. There is no entirely satisfactory theory of ameboid movement. First sign of formation of new pseudopodium is thickening of ectoplasm to form clear hyaline cap. Into this hyaline region flow granules from fluid endoplasm (plasmasol), forming a type of tube with walls of plasmagel and core of plasmasol. As plasmasol flows forward it is converted into plasmagel, which may involve contraction that squeezes pseudopodium in a definite direction. Substratum is necessary for ameboid movement, but only tips of pseudopodia touch it.

the body is plastic unless enclosed in a skeleton of some sort. The outer part of the cytoplasm underlying the cell membrane is a gelatinous and nongranular **ectoplasm.** The internal cytoplasm is more fluid and called **endoplasm.** Ectoplasm and endoplasm are merely gel and sol states of colloidal protoplasm that are reversible.

Amoeba proteus (Fig. 25-3) is one of the naked amebas, having no pellicle or skeleton. Locomotion may involve one or more pseudopodia. A necessary feature of ameboid movement is the attachment of the tip of the pseudopodium to a substratum during the act of moving.

Nutrition in amebas and other sarcodinians is **holozoic,** that is, they ingest and digest solid particles of food, living largely on algae, bacteria, protozoans, and other microscopic forms. A pseudopod flows around and encloses a food object, forming a **food vacuole** (Fig. 25-3), which may also contain some of the environmental water. Digestion occurs within the food vacuole by enzymatic action (**intracellular digestion).**

Respiration is simply a matter of diffusion of gases through the plasmalemma. Excess water and liquid wastes accumulate in a **contractile vacuole** (Fig. 25-3), which, when full, contracts, expelling its contents to the outside.

Asexual reproduction occurs by **binary fission,** which is ordinary mitosis (Fig. 25-1). An ameba can divide in about 30 minutes. Certain shelled amebas undergo division of the shell as well as the cytoplasm, each daughter cell receiving a part of the shell.

Shelled sarcodinians

The foraminiferans (order Foraminifera) have many types of shells, mostly of calcium carbonate, and the shells are many chambered (Fig. 25-2). Slender pseudopodia extend through openings in the shell, or test, then branch and run together to form a net to ensnare their prey. Their life cycles are complex. The radiolarians (order Radiolaria) have intricate and beautiful skeletons of silica or strontium sulfate (Fig. 25-2), usually with a radial arrangement of spines that extend through a capsule from the center of the body.

The foraminiferans and radiolarians have left excellent fossil records. Many extinct species are identical to present ones, but some fossils were among the largest of protozoans. Living forms range from 0.02 to 50 mm. in diameter, whereas some fossil forms may measure up to 100 mm. or more in diameter.

For untold millions of years the shells of dead foraminiferans have been sinking to the bottom of the ocean, building up a characteristic ooze rich in lime and silica. About one third of all sea bottom (50 million square miles) is covered with the ooze of a single species, *Globigerina.* As many as 50,000 shells of foraminiferans may be found in a single gram of sediment.

The radiolarians, with their less soluble siliceous shells (Fig. 25-2), usually are found at greater depths (15,000 to 20,000 feet), mainly in the Pacific and Indian Oceans, and probably cover 2 to 3 million square miles. These deep-sea sediments are estimated to be from 2,000 to 12,000 feet deep.

Limestone and chalk deposits on land were laid down when a deep sea covered the continents. Later, through geologic changes, this sedimentary rock emerged as dry land. The pyramids of Egypt were made from limestone beds formed from the tests of a very large foraminiferan that flourished many millions of years ago.

Parasitic sarcodinians

Many of the amebas are parasitic. *Entamoeba histolytica* is responsible for amebic dysentery in man. It lives in the intestinal wall, often produces lesions and abscesses, and may even spread to other parts of the body. These amebas are spread chiefly through contaminated water and food containing cysts discharged through the feces.

FLAGELLATES (CLASS MASTIGOPHORA)

Class Mastigophora includes protozoans that bear at some stage, usually as adults, one or more flagella. They are quite varied in size range, life cycles, and types of reproduction and nutrition (Fig. 25-4). There are both free-living and parasitic forms. They are the most abundant members of both freshwater and marine plankton.

Mastigophorans have two main groups—the

phytoflagellates, or plantlike forms, and the **zooflagellates,** or animallike forms; they differ largely in their manner of nutrition and reproduction. Phytoflagellates usually have **chloroplasts** (Fig. 25-5), which contain the chlorophyll necessary for photosynthesis. Their nutrition is almost entirely **holophytic.** Zooflagellates lack chloroplasts and their nutrition is **holozoic** (ingesting liquids and solid particles of food) or **saprozoic** (absorbing dissolved salts, simple organic materials, or decayed animal matter from their environment) or both.

The structure of a **flagellum** is similar to that of a cilium (p. 92), whether found in protozoans or in higher animals. The movement of the flagellum varies in different flagellates, turning in some with a spiral propellar-like motion that pulls the animal forward, but moving in others with a rowing stroke that pushes the animal forward. Flagellates travel from a few tenths to 1 mm. per second, according to their size. *Euglena,* a solitary form (Fig. 25-5), and *Volvox,* a spherical colony (Fig. 24-2), are common pondwater forms.

Many flagellates are parasitic, such as the various species of *Trypanosoma,* a blood parasite that causes African sleeping sickness (Fig. 25-4). It is carried by the tsetse fly.

The outbreaks of "red tides" that occur occasion-

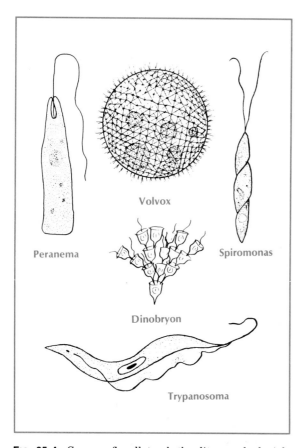

FIG. 25-4. Common flagellates, both solitary and colonial.

FIG. 25-5. General structure of a flagellate protozoan, *Euglena.* Features shown are combination of those visible in living and stained preparations.

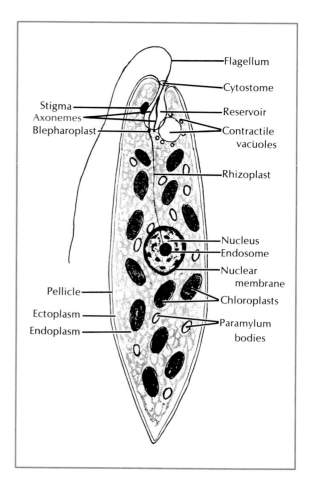

ally along our Florida and California coasts, killing large numbers of fish and other marine life, are caused by enormous swarms of certain species of dinoflagellates.

SPOROZOANS (CLASS SPOROZOA)

The class Sporozoa is made up entirely of endoparasites. They lack locomotor organelles and reproduce by spore formation, which may involve both asexual and sexual methods.

Plasmodium

Malaria is caused by the sporozoan parasite *Plasmodium* (Fig. 25-6). The carriers, or vectors, of the parasites are female mosquitoes, which introduce the parasites from their salivary glands into the

blood in the form of **sporozoites.** The sporozoites first enter the cells of the liver where they pass through a process of multiple division (**schizogony**), the products of which enter the red corpuscles. During this incubation period when the parasites are in the liver, antimalarial drugs have little effect upon them. In the red blood corpuscles they again undergo multiple fission; the many daughter cells break out to enter other red corpuscles and repeat

FIG. 25-6. Life cycle of *Plasmodium vivax,* protozoan (class Sporozoa) that causes malaria in man. **A,** Sexual cycle produces sporozoites in body of mosquito. **B,** Sporozoites infect man and reproduce asexually, first in the cells of liver sinusoids and finally in red blood cells. Malaria is spread by the mosquito, which sucks up gametocytes along with the blood of human and later, when biting another victim, leaves sporozoites in the new wound.

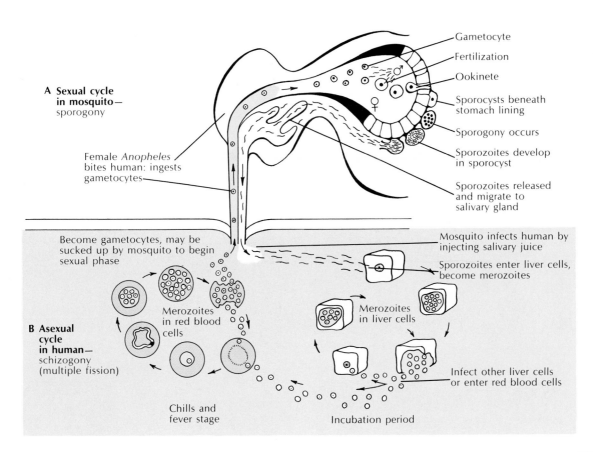

A Sexual cycle in mosquito—sporogony

Gametocyte
Fertilization
Ookinete
Sporocysts beneath stomach lining
Sporogony occurs
Sporozoites develop in sporocyst
Sporozoites released and migrate to salivary gland

Female *Anopheles* bites human: ingests gametocytes

Become gametocytes, may be sucked up by mosquito to begin sexual phase

Merozoites in red blood cells

B Asexual cycle in human—schizogony (multiple fission)

Chills and fever stage

Mosquito infects human by injecting salivary juice

Sporozoites enter liver cells, become merozoites

Merozoites in liver cells

Infect other liver cells or enter red blood cells

Incubation period

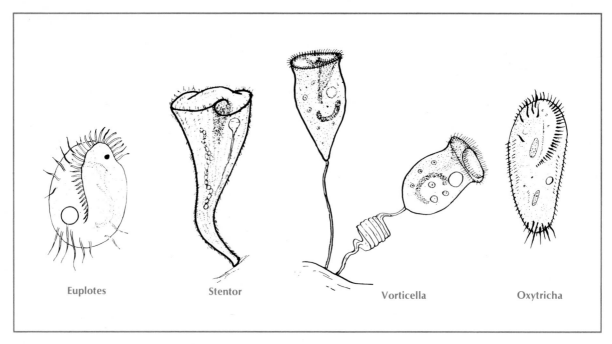

Fig. 25-7. Some representative ciliates. *Euplotes* and *Oxytricha* can use stiff cirri for crawling about. Contractile myonemes in ectoplasm of *Stentor* and in stalks of *Vorticella* allow them to expand and contract.

Fig. 25-8. General structure of *Paramecium caudatum*. (Shown cut into three sections.)

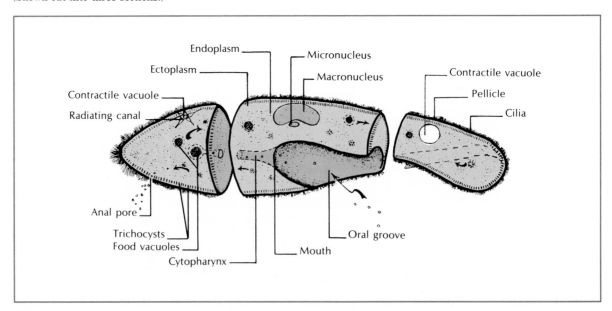

the asexual cycle. In a few days the number of parasites is so great that the characteristic chills and fever occur; these symptoms are caused mainly by the toxins released by the parasites.

After a period of asexual reproduction, some of the cells become sexual forms (gametes). When these are sucked up into the stomach of the mosquito, the gametes unite to form a zygote. From the zygotes oocysts develop, which in a few days divide into thousands of sporozoites that migrate to the salivary glands, whence they are transferred to man by the bite of the mosquito. The developmental cycle in the mosquito requires from 7 to 18 days. After being inoculated by the mosquito, man usually manifests the symptoms of the disease 10 to 14 days later.

Other species of *Plasmodium* parastize birds, reptiles, and mammals. Those of birds are transmitted chiefly by the *Culex* mosquito, those of the human by the *Anopheles* mosquito.

Coccidia

The **coccidians** are sporozoan parasites that infect epithelial tissues in both invertebrates (annelids, arthropods, mollusks) and vertebrates. The symptoms of coccidiosis are severe diarrhea or dysentery, and infection is by the ingestion of oocysts and sometimes by separate sporozoites.

CILIATES (CLASS CILIATA)

The ciliates differ from other protozoans by the presence, at some stage, of **cilia** for locomotion and food getting, and by the presence of two kinds of nuclei—a large **macronucleus** and one or more smaller **micronuclei.** Ciliates vary in shape, and they range in size from 10 to 3,000 μ (Fig. 25-7). Most of them are free living and solitary, but some are sessile and attached, such as *Vorticella;* some ciliates are colonial, and some are commensal.

Ciliates are usually abundant in fresh water that contains a great deal of decaying organic matter. *Paramecium caudatum,* a common ciliate, is from 150 to 300 μ in length, somewhat slipper shaped, and with an asymmetric appearance because of the **oral groove** that runs obliquely backward, ending just behind the middle of the body (Fig. 25-8). The flexible pellicle is covered over with fine cilia.

Ciliates are **holozoic,** using the ciliary action to sweep food particles—bacteria, algae, and other small organisms—into the gullet (cytopharynx). At the end of the gullet the particles collect into food vacuoles that move about the cytoplasm while digestion takes place. Fecal material is discharged through a temporary **anal pore** back of the oral groove.

Cilia can beat either forward or backward so that the animal can swim in either direction. The cilia in paramecia beat obliquely, causing the animal to rotate on its long axis. Oral cilia are longer and beat more vigorously so that the anterior end swerves aborally. These factors cause paramecia to follow a spiral path when moving forward.

Paramecia reproduce only by transverse **binary fission** (Fig. 25-9). However, they do frequently go

FIG. 25-9. Binary fission in paramecium. (Photograph of stained slide.)

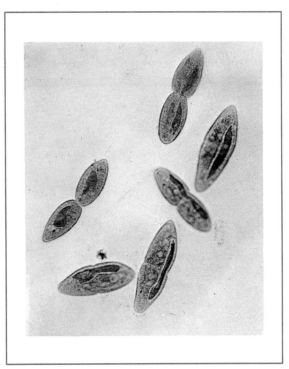

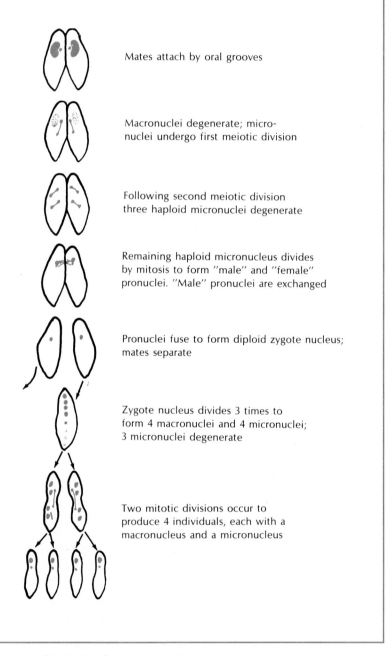

Mates attach by oral grooves

Macronuclei degenerate; micronuclei undergo first meiotic division

Following second meiotic division
three haploid micronuclei degenerate

Remaining haploid micronucleus divides
by mitosis to form "male" and "female"
pronuclei. "Male" pronuclei are exchanged

Pronuclei fuse to form diploid zygote nucleus;
mates separate

Zygote nucleus divides 3 times to
form 4 macronuclei and 4 micronuclei;
3 micronuclei degenerate

Two mitotic divisions occur to
produce 4 individuals, each with a
macronucleus and a micronucleus

FIG. 25-10. Conjugation in *Paramecium caudatum*.

through a process of micronuclear exchange called **conjugation** (Fig. 25-10). Conjugation is a temporary union of two individuals that forms a common protoplasmic bridge through which chromosomal material is exchanged. Conjugation is not the same as the union of eggs and sperm (zygote formation) for the animals continue to divide by fission. It does achieve about the same result, however, since a conjugate ends up with the hereditary material of two individuals.

SPONGES (PHYLUM PORIFERA*)

Sponges belong to phylum Porifera, which means "to bear pores." Most biologists consider the group to be aberrant, that is, an offshoot from standard patterns and not in the direct line of evolution of other animals. More than one hundred years ago sponges were thought to be plants. Because they vary so much from other metazoans they are often called Parazoa, which means "beside the animals." The characteristic collar cells, or choanocytes, of the sponges resemble the collar cells of certain flagellate colonies. There is no evidence that any higher metazoans arose from the sponges.

Organization in sponges goes little further than the cellular level, although there is some indication of coordination of cells to form tissues. Sponges are said to have a primitive nervous system, although evidence is lacking on this point.

Sponges are found in both fresh water and seawater. They are abundant in the sea, from the shallow water of the shore to the abyssal depths. All adult sponges are sessile and are attached to rocks and other solid objects.

Most sponges form large colonies. They vary greatly in color, ranging from dull gray and brown to brilliant scarlet and orange. The best commercial bath sponges are found in the warm, shallow waters of the Mediterranean Sea, the Gulf of Mexico, the West Indies, and off the coast of Florida. Sponges are often cultured by cutting out pieces of the individual animals, fastening them to concrete or rocks, and dropping them into the proper water conditions. It takes many years for sponges to grow to market size.

*Po-rif'e-ra (L. *porus*, pore, + *ferre*, to bear).

CLASSIFICATION

There are three classes of sponges, classified mainly by the kinds of skeletons they possess.

Class Calcispongiae (kal"si-spon'ji-e). Spicules of carbonate of lime that often form a fringe around the osculum; spicules single or three- or four-branched; all three types of canal systems represented; all marine. Example: *Scypha*.

Class Hyalospongiae (hy"a-lo-spon'ji-e). Six-rayed siliceous spicules in three dimensions; often cylindric or funnel shaped; habitat mostly in deep water; all marine. Example: the glass sponges, such as Venus's flower basket.

Class Demospongiae (de"mo-spon'ji-e). Siliceous spicules, spongin, or both; one family found in fresh water; all others marine. Example: bath sponges.

CHARACTERISTICS

1. All aquatic; mostly marine
2. All sponges attached and with a variety of body forms; radial symmetry or none
3. **Multicellular;** body a loose aggregation of cells of mesenchymal origin; body surface simply a colloid with freely movable epithelial cells; mesenchyme with skeletal spicules or horny fibers and free ameboid cells
4. Body with many **pores (ostia), canals,** and **chambers** that serve for the passage of water
5. Most of the inner chambers and interior surfaces lined with **choanocytes, or flagellate collar cells**
6. No organs or definite tissues
7. Digestion intracellular and no excretory or respiratory organs
8. **Skeleton usually of calcareous** or **siliceous crystalline spicules** or of **protein spongin**
9. Asexual reproduction by **buds or gemmules** and sexual reproduction by **eggs** and **sperm;** freeswimming, ciliated larva

STRUCTURE

Most dry sponges we see consist only of skeletal framework. In the living condition many of them appear as slimy gelatinous masses. They are characterized by a complex system of pores, canals, and chambers and depend on the maintenance of water currents through this system for their supply of food and oxygen.

The surface of sponges possesses many small pores **(ostia)** for the inflow of water (Fig. 25-11). The ostia open into canals, simple or complex, which run into flagellated chambers and then into a central cavity, the **spongocoel.** The opening of the spongocoel to the outside is known as the **osculum.** Colonial sponges have many oscula. The sponge has no mouth and no organs.

In simple sponges the sponogocoel is lined with characteristic flagella-bearing cells, the **choano-cytes,** commonly called "collar cells" because each has a little collar around the base of the flagellum. It is these flagellated cells that keep the water currents moving.

The canal structure of most sponges fall into one of the three principal types (Fig. 25-11).

1. **Asconoid type.** The canals pass directly from the ostia to the spongocoel, which is lined with collar cells. These are in general small, inconspicuous sponges. *Leucosolenia* is an example.

2. **Syconoid type.** The incurrent canals (from the outside) lie alongside the radial canals that empty into the spongocoel. The radial canals are lined with collar cells. Both types of canals end blindly in the body wall but are connected by minute pores. *Scypha* is a common syconid sponge.

3. **Leuconoid type.** The canals of this type are much branched and complex, with numerous chambers lined with collar cells. The larger sponges, including the bath sponge, are of this type.

These three types of canals are correlated with the evolution of sponges from the simple to the complex forms. It has been mainly a matter of increasing the surface in proportion to the volume, so that there are enough collar cells to meet food demands. This problem has been met by the out-pushing of the spongocoel of a simple sponge such as the asconoid type to form radial canals (lined

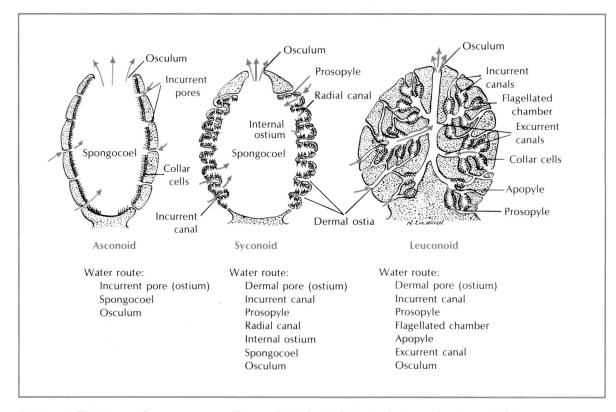

Water route:
    Incurrent pore (ostium)
    Spongocoel
    Osculum

Water route:
    Dermal pore (ostium)
    Incurrent canal
    Prosopyle
    Radial canal
    Internal ostium
    Spongocoel
    Osculum

Water route:
    Dermal pore (ostium)
    Incurrent canal
    Prosopyle
    Flagellated chamber
    Apopyle
    Excurrent canal
    Osculum

Fig. 25-11. Three types of sponge structure. Degree of complexity from simple asconoid type to complex leuconoid type has involved mainly the water and skeletal systems, accompanied by outfolding and branching of collar cell layer. Leuconoid type considered major plan for sponges, for it permits greater size and more efficient water circulation.

with choanocytes) of the syconoid type. The formation of incurrent canals between the blind outer ends of the radial canals completes this type. Further increase in the body wall foldings produces the complex canals and chambers (with collar cells) of the leuconoid type.

SKELETONS

The skeleton gives support to the sponge, preventing collapse of the canals and chambers. It also serves as the basis for classifying sponges (Fig. 25-12). In many small marine sponges the skeleton consists of spicules of calcium carbonate; glass

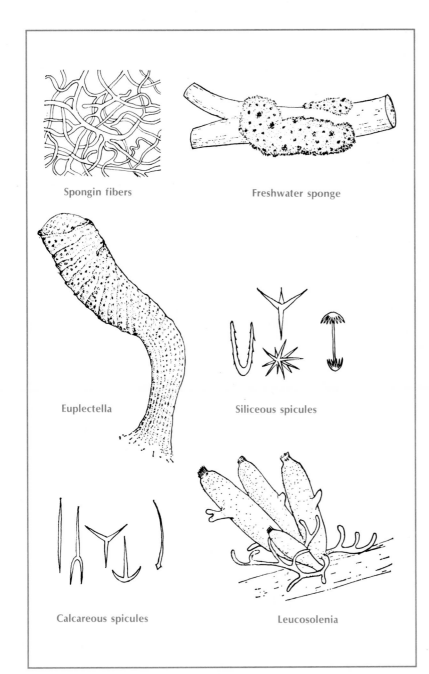

Spongin fibers

Freshwater sponge

Euplectella

Siliceous spicules

Calcareous spicules

Leucosolenia

FIG. 25-12. Types of skeletal structure found in sponges, with example of each. Amazing diversity, complexity, and beauty of form exists among the many types of spicules.

sponge spicules are formed of siliceous material (Fig. 25-12). These spicules are of many different forms and shapes. They are secreted by special cells called scleroblasts.

Sponges, such as the bath sponge and freshwater sponges, contain spongin (Fig. 25-12), a protein-like substance. This type of skeleton is a branching, fibrous network that supports the soft, living cells of the sponge. Special cells called spongioblasts form this type of skeleton.

METABOLISM

Metabolism in sponges is mainly a matter of individual cellular function. Their food consists of small organic substances, both plant and animal, which are drawn into the ostia and through the canal system by currents of water induced by the waving of the flagella on the choanocytes. Particles of food adhere to the outer surface of the collars, and later pass into the cytoplasm, where food vacuoles are formed. Digestion is therefore intracellular in sponges. Undigested food is ejected by the amebocytes into outgoing currents.

Excretion and respiration are by simple diffusion. Contractile vacuoles have been found in the amebocytes and choanocytes of freshwater sponges (*Spongilla* and *Ephydatia*). The movement of the flagellated collar cells moves the water through, but the rate of flow is different in the various parts of the passageway. A large sponge with many oscula was found to filter more than 1,500 L. of water a day. It is to the advantage of the sponge to discharge the current of water from its oscula as far away as possible to prevent reusing water containing its own waste. The water flow is regulated by contraction and relaxation of the pores and oscula.

REPRODUCTION

Sponges reproduce both asexually and sexually. Asexual reproduction is mainly a matter of bud formation. After reaching a certain size, these buds may become detached, or they may remain to form colonies. Internal buds, or gemmules, are formed in freshwater and some marine sponges. The gemmules survive during periods of drought or freezing, and later the cells in the gemmules escape and develop into new sponges.

In sexual reproduction ova and sperm develop; the ova are fertilized in the mesenchyme, develop there, and finally break out into the spongocoel and out through the osculum. During development the zygote undergoes cleavage and differentiation in the mesenchyme, and finally a flagellated larva emerges.

Some sponges are monoecious (having both male and female sex organs in one individual), and some are dioecious (having separate sexes).

26

Coelenterates
and
comb jellies

These two groups of animals, coelenterates and comb jellies, are distinguished by their primary radial or biradial symmetry. They are the most primitive of the true metazoans.

COELENTERATES
(PHYLUM COELENTERATA*)

The phylum Coelenterata contains a large and interesting group of animals (Fig. 26-1). The name "coelenteron" means "hollow intestine" and refers to the large body cavity that serves as an intestine.

Most coelenterates are marine, although a few freshwater forms, such as the hydra, are found attached to the underside of aquatic plants in quiet streams, lakes, and ponds. Colonial polyp coelenterates are found along the coast, where they may be attached to mollusk shells, rocks, and wharves, although some hydroids have been found at great ocean depths. Corals are found in reefs along the shallow waters of the southern seas.

The free-swimming coelenterates (medusae) are found in the open sea and lakes. Floating colonies, such as the Portuguese man-of-war, are kept afloat by large floats or sails called pneumatophores.

CLASSIFICATION

Class Hydrozoa (hy"dro-zo'a). Solitary or colonial; asexual polyps and sexual medusae, although one type may be suppressed; both fresh water and marine. Examples: *Hydra, Obelia, Physalia.*
Class Scyphozoa (si"fo-zo'a). Solitary; polyp stage reduced or absent; bell-shaped medusae without velum; gelatinous mesoglea much enlarged; all marine. Examples: *Aurelia, Cassiopeia.*
Class Anthozoa (an"tho-zo'a). All polyps; no medusae; solitary or colonial; enteron subdivided by at least eight mesenteries or septa with nematocysts; gonads endodermal; all marine. Examples: sea anemones, corals, sea pens.

CHARACTERISTICS

1. **Radial** or **biradial symmetry** around a longitudinal axis with **oral** and **aboral ends;** no definite head
2. Two types of individuals—**attached polyps** and **free medusae**
3. Entirely aquatic, mostly marine
4. Body with two layers, epidermis and gastrodermis, with mesoglea between
5. Exoskeleton (perisarc) of chitin or lime in some

---

*Se-len"te-ra'ta (Gr. *koilos,* hollow, + *enteron,* gut, + *ata,* pl. suffix meaning characterized by).

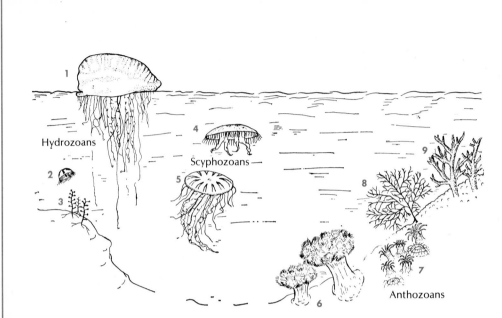

Fig. 26-1. Representative coelenterates. Class Hydrozoa: **1,** *Physalia;* **2,** *Gonionemus;* **3,** *Obelia.* Class Scyphozoa: **4,** *Aurelia;* **5,** *Chrysaora.* Class Anthozoa: **6,** *Metridium;* **7,** *Astrangia;* **8,** *Gorgonia;* **9,** staghorn coral *(Acropora).* (Not sized to scale.)

6. **Gastrovascular cavity** (often branched or divided with septa) with a single opening that serves as both mouth and anus; extensible tentacles often encircling the mouth or oral region
7. Special stinging cell organoids called **nematocysts** in epidermis or gastrodermis or both
8. **Nerve net** of synaptic and nonsynaptic patterns; with some sensory organs; diffuse conduction
9. Muscular system (epitheliomuscular type) of an outer layer of longitudinal fibers at base of epidermis and an inner one of circular fibers at base of gastrodermis
10. Reproduction by asexual budding (in polyps) or sexual reproduction by gametes (in all medusae and some polyps). Sexual forms monoecious or dioecious; **planula larva;** holoblastic cleavage; mouth from blastopore
11. No excretory or respiratory system; no coelomic cavity

Dimorphism: two types of individuals. One of the most basic characteristics of coelenterates — and often responsible for confusion among students

— is that two morphologic types of individuals are recognized in the group — **polyps** and **medusae.**

Polyps have tubular bodies with a mouth surrounded by tentacles at one end. The other end is blind and usually attached by a pedal disk or other device to a substratum. Medusae, or free-swimming jellyfish, have umbrella-shaped bodies, with a mouth centrally located on a projection of the concave side. Around the margin of the umbrella are the tentacles, which are provided with stinging cells.

Some species have both types of individuals in their life history (*Obelia* and other hydroids); others have only the polyp stage (hydra and Anthozoa); and still others have only the jellyfish, or medusa, stage (certain Scyphozoa). Many of the colonial hydrozoans are polymorphic, having medusae and two or more types of polyps in their life cycle.

Although they look very different from each other,

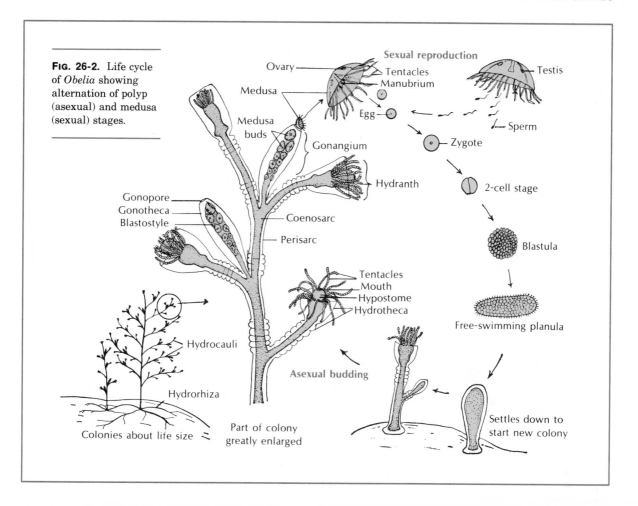

FIG. 26-2. Life cycle of *Obelia* showing alternation of polyp (asexual) and medusa (sexual) stages.

polyps and medusae are structurally similar. If a polyp form, such as that of the hydra, were inverted and broadened out laterally to shorten the oral-aboral axis, the hypostome lengthened to form a manubrium, and the mesoglea greatly increased, the result would be a structure similar to a medusa, or jellyfish.

Stinging cells. Nematocysts, or stinging cells, are one of the most characteristic structures in the entire coelenterate group. The nematocyst is a tiny capsule containing a coiled tubular "thread" or filament, which is continuous with the narrowed end of the capsule. This end of the capsule is covered by a little lid, or operculum. The nematocyst is found in a cell called a **cnidoblast,** which is provided with a projecting triggerlike **cnidocil.** When the cnidocil is stimulated by food, prey, or enemies,

the coiled thread turns inside out with explosive force, the barbs unfolding to the outside as the tube everts. Nematocysts may occur singly, in batteries, or in rings and are especially abundant on the tentacles. One type of nematocyst is long and thread-like, with spines and thorns. When discharged it can pierce the body of the prey and secrete a paralyzing toxin. Another type has a long, barbed thread that coils when discharged and produces an adhesive secretion used in locomotion and attachment. A third type of nematocyst has a short thread that, when released, can coil around some portion of the prey.

Nematocysts of most coelenterates are not harmful to man, but the stings of the Portuguese man-of-war and certain large jellyfish such as *Cyanea* are quite painful.

373

HYDROIDS (CLASS HYDROZOA)

Most members of this class are polymorphic and have both a **medusa** and one or more types of **polyp** in their life cycles; or they may be wholly polypoid or wholly medusoid; or they may have one of the stages predominate and the other inconspicuous. Many of the hydroids are colonial forms, which by asexual budding produce medusae (jellyfish). The medusae produce eggs and sperm that result in, not medusae, but polyps.

Obelia is often selected to represent the hydroid and medusa types because the hydroid stage of *Obelia* is large enough to study readily. *Obelia* is a little colonial hydroid whose colonies may grow to as much as an inch in height (Fig. 26-2). In one species the colony grows to nearly a foot long. It is found attached to stones, shells, and other objects by its **hydrorhiza** (Fig. 26-2). Most of the polyps are **nutritive hydranths** that do the feeding for the colony. They look much like miniature hydras and eat tiny crustaceans, worms, or insect larvae. Digestion is both extracellular and intracellular.

The reproductive polyps, **gonangia,** produce young medusae by asexual budding. The medusae reproduce sexually and are dioecious.

Since the *Obelia* medusae stage is small and inconspicuous, we frequently study another hydroid in the laboratory, *Gonionemus,* which has a large medusae stage (Fig. 26-3). The *Gonionemus* medusa is bell shaped and about 1/2 to 1 inch in diameter. Around the margin of the bell are a score or more of **tentacles,** each one bearing an **adhesive pad.** Inside the margin is the thin muscular **velum,** which partly closes the open side of the bell and is used in swimming. Muscular pulsations that alternately fill and empty the cavity of the bell propel

FIG. 26-3. Medusa of *Gonionemus,* partly cut away to show internal structure.

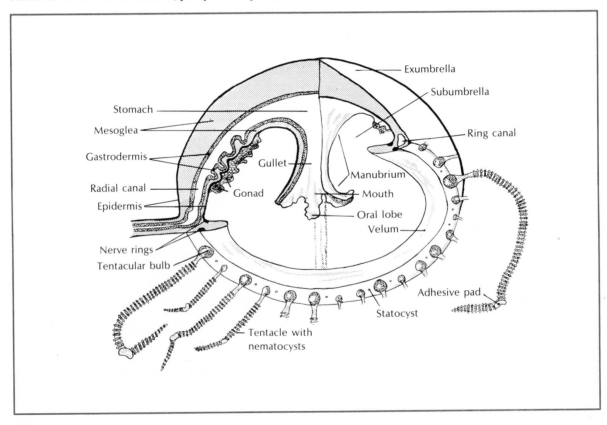

the animal forward, aboral side first, with a type of "jet propulsion."

The mouth is at the end of a suspended **manubrium** (Fig. 26-3). A stomach at the base of the manubrium is connected by four **radial canals** to a **ring canal** around the margin, which in turn connects with all the tentacles. Consequently the **gastrovascular cavity** is continuous from the mouth to the tips of the tentacles.

A **gonad** is suspended beneath each radial canal, and fertilization is external. The zygote develops into a ciliated **planula larva** that swims about for a time, then settles down to some substratum and develops into a minute polyp. The cycle begins again with the young polyp budding off additional polyps that finally produce tiny medusae by asexual means (Fig. 26-2).

FIG. 26-4. General structure of hydra. Although specimen shown is hermaphoroditic, most species are dioecious.

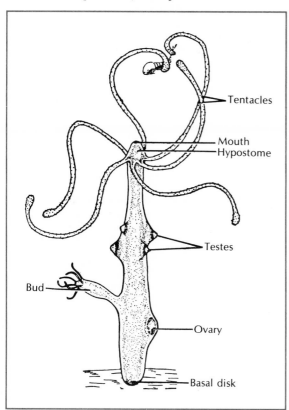

Hydra

The common freshwater hydra (Fig. 26-4) is a solitary polyp and one of the few coelenterates found in fresh water. Its normal habitat is the underside of aquatic leaves and lily pads in cool, clean, fresh water throughout the world. Common species are the green hydra *(Chlorohydra viridissima),* which owes its color to symbiotic algae in its cells, and the brown hydra *(Pelmatohydra oligactis).*

The body of the hydra is a cylindric tube with the lower (aboral) end drawn out into a slender **stalk.** The mouth is encircled by six to ten hollow tentacles (Fig. 26-4). The hydra feeds upon a variety of small crustaceans, insect larvae, and annelid worms. A hungry hydra waits for its food to come to it. It may, if necessary, shift to a more favorable location, but once attached to its chosen substratum, it waits motionless, its tentacles fully extended. When the tentacles are attached, the prey is moved toward the hydra's mouth, the mouth slowly opens, and the prey slides in (Fig. 26-5).

A hydra may move by gliding on the basal disk, aided by mucous secretions; or it may use a "measuring-worm" type of movement. It may even do handsprings, attaching alternately by its basal disk and its tentacles. It often forms a gas bubble on its basal disk and floats up in water.

The hydra uses both sexual and asexual methods of reproduction. **Asexual reproduction** is by budding, in which outpocketings of the body wall are produced by proliferation of cells (Fig. 26-4). Eventually the buds constrict and detach themselves. In **sexual reproduction** most species are **dioecious,** but some are monoecious (hermaphroditic). The formation of temporary gonads occurs in autumn (Fig. 26-6). Reduction of water temperature and reduced aeration in stagnant water may stimulate gonad formation. The medusa stage is lacking in freshwater hydras.

Physalia

Some of the hydrozoans are floating colonies made up of modified medusa and polyp types. *Physalia,* the "Portuguese man-of-war," is a colony with a rainbow-hued float that carries it along on the surface waters of the southern seas (Fig. 26-7). Bathers frequently find that the long, graceful tentacles

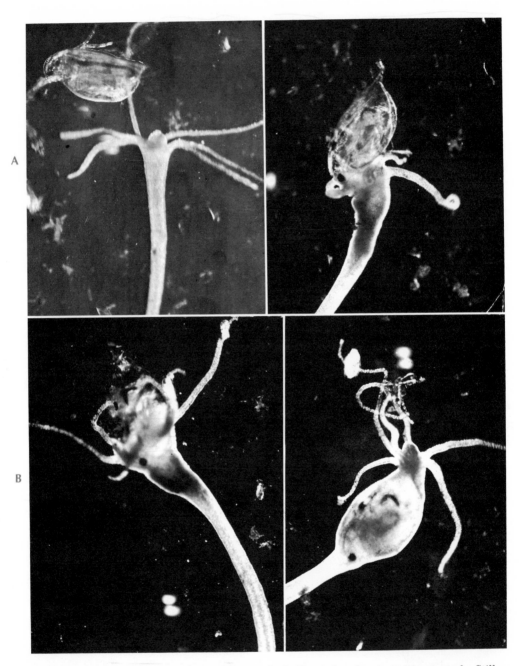

FIG. 26-5. Lunch for hungry hydra. **A,** Unwary daphnid gets too close to waiting tentacle. Still kicking (right), he is drawn into widening mouth of captor. **B,** Swallowing process is slow but odds are against daphnid. Mission accomplished (right), hydra settles down to digest lunch.

FIG. 26-6. **A,** Living hydras, mostly males with testes. **B,** Living hydras, mostly females with ovaries and eggs. (Courtesy General Biological Supply House, Inc., Chicago.)

(sometimes as much as 60 feet long) are laden with nematocysts and are capable of painful and sometimes dangerous stings.

The float of *Physalia* contains an air sac filled with a gas similar to air and acts as a carrier for the generations of individuals that bud from it and hang suspended in the water. There are several types of individuals, including feeding polyps, reproductive polyps, long stinging tentacles, and the so-called jelly polyps.

THE TRUE JELLYFISH (CLASS SCYPHOZOA)

Class Scyphozoa contains most of the large jellyfish of the oceans. They can be distinguished from the jellyfish of class Hydrozoa by the absence of a velum and the presence of a notched margin of the umbrella. Most of them have polyp and medusa stages, but the polyp stage is very small. Some jellyfish are several feet in diameter, with tentacles more than 75 feet long *(Cyanea)* (Fig. 26-8). Other

FIG. 26-7. Portuguese man-of-war *Physalia physalis* (order Siphonophora, class Hydrozoa) eating fish. This colony of medusa and polyp types is integrated to act as one individual. As many as a thousand zooids may be found in one colony. Although a drifter, the colony has restricted directional movement. Their stinging organoids secrete a powerful neurotoxin. Colonies often drift onto southern ocean beaches, where they are a hazard to bathers. (Courtesy New York Zoological Society.)

FIG. 26-8. Large jellyfish *Cyanea*. Some may attain diameter of more than 6 feet in Arctic waters, but are smaller in warmer waters. Its many hundred tentacles may reach length of 75 feet or more. It is one of the most striking of all jellyfish.

jellyfish, however, are quite small. Most are found floating in the open sea, although some are attached.

SEA ANEMONES, CORALS, AND RELATIVES (CLASS ANTHOZOA)

Class Anthozoa includes the sea anemones, the stony corals, the horny, black, and soft corals, the sea pens, sea pansies, sea feathers, and others. Some of these anthozoans are provided with an external or internal skeleton, and they may be solitary or colonial. No medusa stages are found.

Metridium, one of the **sea anemones,** is 2 to 3 inches long and is commonly found on wharves and rocky bottoms along the North Atlantic coast. It is cylindric in form, with a crown of hollow tentacles arranged in circlets around the mouth (Fig. 26-9).

Anemones can glide along slowly on their pedal disks. They can expand and stretch out their ten-

tacles in search of small vertebrates and invertebrates, which they overpower with tentacles and nematocysts and carry to their mouth. When disturbed, they contract and draw in their tentacles and oral disks.

Most **corals** belong to the class Anthozoa, although one or two are from the class Hydrozoa. The coral organism is a small polyp (Fig. 26-10) that looks

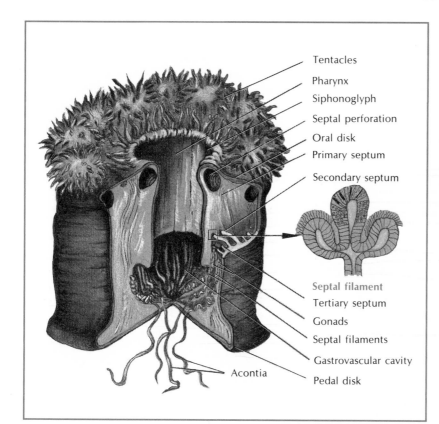

Tentacles
Pharynx
Siphonoglyph
Septal perforation
Oral disk
Primary septum
Secondary septum

Septal filament
Tertiary septum
Gonads
Septal filaments
Gastrovascular cavity
Acontia
Pedal disk

FIG. 26-9. Structure of *Metridium* cut to show pharynx and septal arrangement. **Inset,** Section through septal filament. (From Hickman, C. P.: Biology of the invertebrates, 1967, The C. V. Mosby Co.)

FIG. 26-10. Group of coral polyps, *Astrangia danae,* protruding from their shallow cups. This is a common coral of Atlantic coast. (Courtesy General Biological Supply House, Inc., Chicago.)

FIG. 26-11. Stony coral *Oculina* showing cups in which polyps once lived.

like a miniature sea anemone living in a stony cup with radial ridges (Fig. 26-11). The cup is a limy exoskeleton secreted by the epidermis around the polyp and between its mesenteries. Living polyps can withdraw into these cavities when not feeding. Most of them are in colonies, which assume a great variety of forms. Living polyps are found only on the surface layers of coral masses.

Reef-building corals are mainly restricted to shallow water of tropic seas. Three kinds of coral reefs are commonly recognized, depending on how they are formed. The **fringing reef** may extend out to a distance of a quarter mile from the shore, with the most active zone of coral growth facing the sea. A **barrier reef** differs from a fringing reef in being separated from the shoreland by a lagoon of varying width and depth. The Great Barrier Reef off the northeast coast of Australia is more than 1,200 miles long and up to 90 miles from the shore. An **atoll** is a reef that encircles a lagoon but not an island.

Most reefs grow at the rate of 10 to 200 mm. each year. Most of the existing reefs could have been formed in 15,000 to 30,000 years. Coral reefs have great economic importance, for they serve as habi-

tats for a large variety of organisms, such as sponges, worms, echinoderms, mollusks, many kinds of fish, and man.

Besides the stony corals there are many other types of corals such as the sea fans, sea plumes, sea pansies, and sea pens. Some of these have horny skeletons instead of calcareous ones. Horny skeletons consist chiefly of rods that pass through stems and branches, with the polyps arranged around the rods.

COMB JELLIES (PHYLUM CTENOPHORA*)

The Ctenophora comprise a small group of fewer than 100 species and take their name from the eight comblike plates they bear for locomotion. Common names for them are "sea walnuts" and "comb jellies" (Fig. 26-12). Coelenterates and ctenophores represent the only two phyla with basic radial symmetry. Nematocysts are lacking in ctenophores, except in one species.

There is no convincing evidence that ctenophores were derived directly from coelenterates, although

*Te-nof' o-ra (Gr. *ktenos*, comb, + *phoros*, bearing).

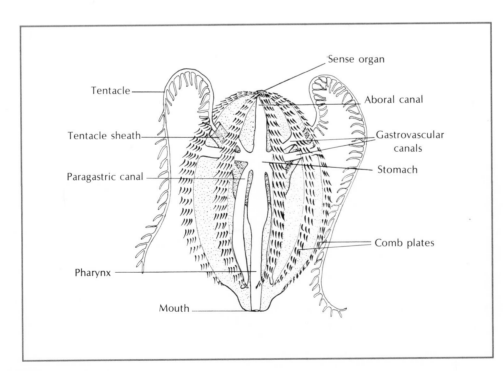

FIG. 26-12. Structure of the comb jelly *Pleurobrachia*.

Labels: Sense organ, Tentacle, Aboral canal, Tentacle sheath, Gastrovascular canals, Paragastric canal, Stomach, Pharynx, Comb plates, Mouth

there may be some kinship. Ctenophores have some resemblance to some of the hydrozoan medusae and may have diverged very early from that coelenterate stem.

The ctenophores are strictly marine, and all are free swimming. They are feeble swimmers and are carried by tides and currents. Storms may drive them in concentrated numbers onto beaches and sea bays and inlets. Although ctenophores are more common in surface waters, they also occur at great depths in the sea.

One of the most striking ctenophores is *Beroë*, which may be more than 100 mm. in length and 50 mm. in breadth. It is conical or ovoid, pink colored, and provided with a large mouth but no tentacles. Venus's girdle *(Cestum)* is compressed and bandlike, may be more than a yard long, and presents a graceful appearance as it swims. A common ctenophore along the Atlantic coast is *Mnemiopsis*, which has a laterally compressed body with two large oral lobes and unsheathed tentacles. The little sea walnut, *Pleurobrachia* (Fig. 26-12), is a common ctenophore. Nearly all ctenophores give off flashes of luminescence at night. The vivid flashes of light seen at night in southern seas are often due to members of this phylum.

27

Flatworms and ribbon worms

These two groups of worms, flatworms and ribbon worms, represent the most primitive bilateral animals. They have only one internal space, the digestive cavity; they lack the true body cavity, or **coelom**, which appears in all the more advanced animals. Consequently, these two groups are called **acoelomate** animals. Both flatworms and ribbon worms are closely related, although the ribbon worms are considered more advanced, having a complete digestive system, a vascular system, and a more highly organized nervous system.

The term "worm" has been loosely applied to elongated invertebrate animals without appendages and with bilateral symmetry. At one time zoologists placed all the worms together in a phylum called Vermes. Modern classification has broken up this group into phyla and reclassified them. By tradition, however, zoologists still refer to these animals as "flatworms," "ribbon worms," "cavity worms," "segmented worms," etc.

FLATWORMS (PHYLUM PLATYHELMINTHES*)

This large group of animals includes some of the notorious animal undesirables—the parasitic flukes

*Plat″y-hel-min′thes (Gr. *platys*, flat, + *helmins*, worm).

and tapeworms. There are, however, a large number of free-living forms, as well.

CLASSIFICATION

Class Turbellaria (tur″bel-la′re-a)—**turbellarians.** Usually free-living forms with soft flattened bodies; mouth usually on ventral surface. Example: *Dugesia* (planaria).

Class Trematoda (trem″a-to′da)—**flukes.** Body covered with thick living cuticle without cilia; leaflike or cylindric in shape; suckers and sometimes hooks for attachment; all parasitic. Examples: *Fasciola, Opisthorchis, Schistosoma.*

Class Cestoda (ses-to′da)—**tapeworms.** Body covered with thick, nonciliated, living cuticle; scolex with suckers or hooks and sometimes both for attachment; body divided into series of proglottids; general form of body tapelike; all parasitic, usually with alternate hosts. Examples: *Diphyllobothrium, Taenia, Echinococcus.*

CHARACTERISTICS

1. Three germ layers (**triploblastic**)
2. **Bilateral symmetry;** definite polarity of anterior and posterior ends
3. **Body flattened dorsoventrally;** oral and genital apertures mostly on ventral surface
4. Body segmented in one class (Cestoda)
5. No definite coelom (**acoelomate**)
6. Digestive system incomplete (gastrovascular type); absent in cestodes

7. **Nervous system consisting of a pair of anterior ganglia with longitudinal nerve cords connected by transverse nerves;** simple sense organs; eyespots in some
8. Excretory system of two lateral canals with branches bearing **flame cells (protonephridia);** lacking in some primitive forms
9. Respiratory, circulatory, and skeletal systems lacking
10. Most forms monoecious; reproductive system complex with well-developed gonads, ducts, and accessory organs; internal fertilization; internal parasites may have complicated life cycle often involving several hosts

TURBELLARIANS (CLASS TURBELLARIA)

A well-known representative of the turbellarians is the common freshwater planarian *Dugesia* (Fig. 27-1). Several species of this genus are found on the underside of rocks and debris in ponds, brooks, or springs of cold running water. The head region is triangular, with two **eyespots** on the dorsal side of the head near the midventral line. Near the center of the ventral side is the mouth, through which the muscular **pharynx** (proboscis) can be extended. The skin is ciliated epidermis containing rod-shaped **rhabdites** that, when discharged into water, swell and form a protective gelatinous sheath around the body.

Freshwater planarians move by gliding over a slime tract secreted by adhesive glands. The beating of the epidermal cilia in the slime tract and the rhythmic waves of muscular movement drive the animal along. In crawling the worm lengthens, anchors its anterior end with adhesive mucus, and contracts to pull up the rest of its body.

Planarians are mainly carnivorous and feed upon intact, injured, or dead prey, such as small crustaceans, nematodes, rotifers, and insects. The planarian grips its prey with its anterior end, sucks up minute bits of food into the intestine, where the

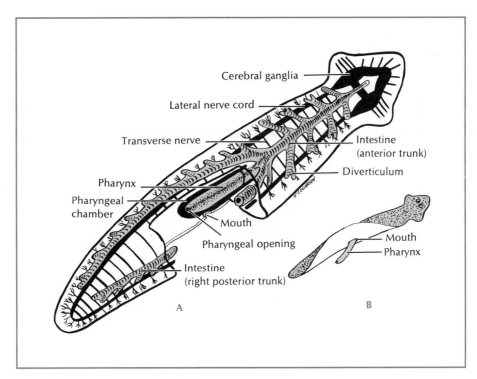

FIG. 27-1. A, Digestive system and ladder-type nervous system of planaria. Cut section shows relation of pharynx, in resting position, to digestive system (in red) and mouth on ventral surface. **B,** Pharynx extended through ventral mouth.

phagocytic cells of the gastrodermis complete the digestion **(intracellular).** The **gastrovascular cavity** ramifies to most parts of the body (Fig. 27-1), and food is absorbed through its walls into body cells. Since these forms have no anus, undigested food is egested through the pharynx. Planarians can go a long time without feeding by drawing nutrients from the parenchyma cells back into the intestinal cells where it is digested.

Water regulation and some nitrogenous excretion are carried out by an extensive system of tubules that end in **flame cells (protonephridia)** (Fig. 12-5, p. 148). Exchange of gases takes place through the body surface.

There is a beginning of a central nervous system. Two **cerebral ganglia** beneath the eyespots serve as the brain. Two ventral longitudinal **nerve cords** that extend from the brain and transverse nerves that connect them form a "ladder-type" nervous system (Fig. 27-1). The **eyespots** are made up of pigment cups (Fig. 27-2). Retinal cells extend from the brain to dip into the pigment cups, with the photosensitive ends of the cells inside the cup. They are sensitive to light intensities and direction, but can form no images. The eyespots enable planarians to avoid strong light and seek out dimly lighted regions. The auricular lobes appear to be sensitive to both water currents and chemical stimuli. Animal behaviorists consider flatworms to be the lowest group to show capacity for learning in response to simple conditioned changes.

Turbellarians reproduce both sexually and asexually. Asexually the animal merely constricts behind the pharyngeal region and separates into two animals, and each new animal regenerates its missing parts. Sexually the worm is monoecious; each individual is provided with both male and female organs.

FLUKES (CLASS TREMATODA)

Most flukes are almost exclusively endoparasitic, having from two to four hosts in the life cycle. The

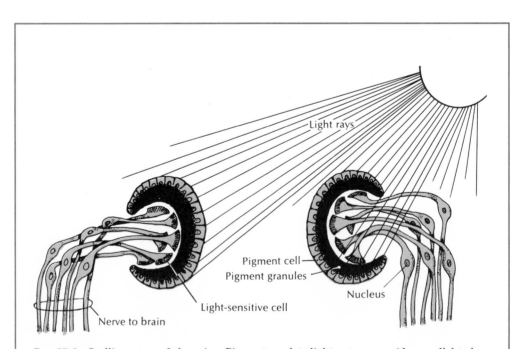

FIG. 27-2. Ocelli, or eyes, of planarian. Pigment cup lets light enter open side, parallel to long axis of retinal (light-sensitive) cells. Planarian determines light direction from stimulation of light-sensitive cells.

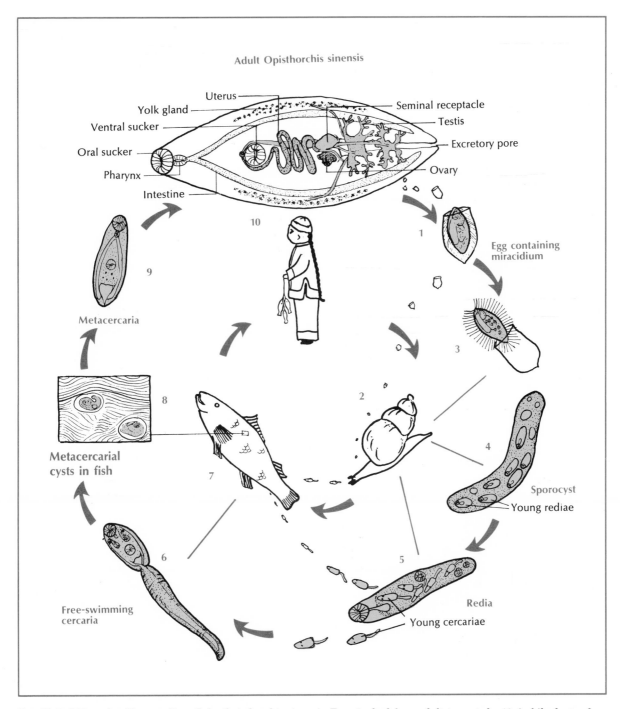

FIG. 27-3. Life cycle of human liver fluke *Opisthorchis sinensis*. Egg, **1**, shed from adult trematode, **10**, in bile ducts of man, is carried out of body in feces and is ingested by snail *(Bythinia)*, **2**, in which miracidium, **3**, hatches and becomes mother sporocyst. **4**, Young rediae are produced in sporocyst, grow, **5**, and in turn produce young cercariae. Cercariae now leave snail, **6**, find a fish host, **7**, and burrow under scales to encyst in muscle. **8**, When raw or improperly cooked fish containing cysts is eaten by man, metacercaria is released, **9**, and enters bile duct, where it matures, **10**, to shed eggs into feces, **1**, thus starting another cycle.

adults are found mainly in terrestrial, freshwater, and marine vertebrates and are usually specific for organs, such as the intestines, bile passages, lungs, kidneys, urinary bladder, etc. Flukes have the most complicated life histories in the animal kingdom. They usually have four larval stages (Fig. 27-3)—**miracidium, sporocyst, redia,** and **cercaria.** Another stage, the **metacercaria,** is considered a juvenile fluke. The miracidium nearly always enters a mollusk (bivalve or snail) as the first intermediate host.

Opisthorchis (Clonorchis) sinensis— liver fluke of man

Opisthorchis (Fig. 27-3) is the most important liver fluke of man and is common in many regions of the Orient, especially in China, southern Asia, and Japan. This fluke has two intermediate hosts for the larval stages and a final host for the adult.

The normal habitat of the adults is in the bile passageways of man. The eggs, each containing a miracidium, are shed into the water with the feces

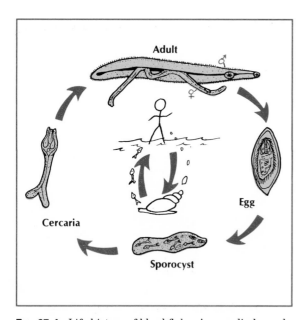

FIG. 27-4. Life history of blood fluke. An egg discharged in human feces passes into water and hatches into miracidium (shown enclosed in egg), which enters a snail and transforms into a sporocyst. Cercariae escape from snail and penetrate exposed skin of man to enter bloodstream.

but do not hatch until they are ingested by the snail *Bythinia* or related genera. In the snail the miracidium enters the tissues and is transformed into the sporocyst, which produces one generation of rediae. The redia is elongated, with an alimentary canal, a nervous system, and many germ cells in the process of development. The rediae pass into the liver of the snail, where, by a process of internal budding, they give rise to the tadpolelike cercariae. The cercariae escape into the water, swim about until they meet with fish of the family Cyprinidae (for example, carp, goldfish, minnows) and then bore into the muscles or under the scales, and encyst as metacercariae. If man eats raw infested fish, the cysts dissolve in the intestine, and the metacercariae migrate up the bile duct, where they become adults. Here the flukes may live for fifteen to thirty years. A heavy infestation may cause a marked cirrhosis of the liver and result in death. Cases are diagnosed through fecal examination. To avoid infection, all fish used as food should be thoroughly cooked. Destruction of the snails that carry larval stages is a method of control.

Schistosoma—blood flukes

Infection by blood flukes is called schistosomiasis, a disorder very common in Africa, China, southern Asia, and parts of South America. The blood flukes are dioecious, and the male is usually broader and encloses the very slender female in a ventral fold on the body (Fig. 27-4).

The plan of the life history of blood flukes is similar in all species (Fig. 27-4). Eggs are discharged in human feces or urine; if they get into water, they hatch out as ciliated miracidia that must contact a certain kind of snail within 24 hours to survive. In the snail they transform into sporocysts, which develop cercariae directly. These cercariae escape from the snail and swim about until they come in contact with the bare skin of a human being. They penetrate through the skin into blood vessels and there develop into adults without a metacercaria stage. In the adult stage when the eggs have been fertilized, the female leaves the male canal and goes into small blood vessels where she lays the eggs. By the time these ova reach the exterior, they have within them fully formed miracidia.

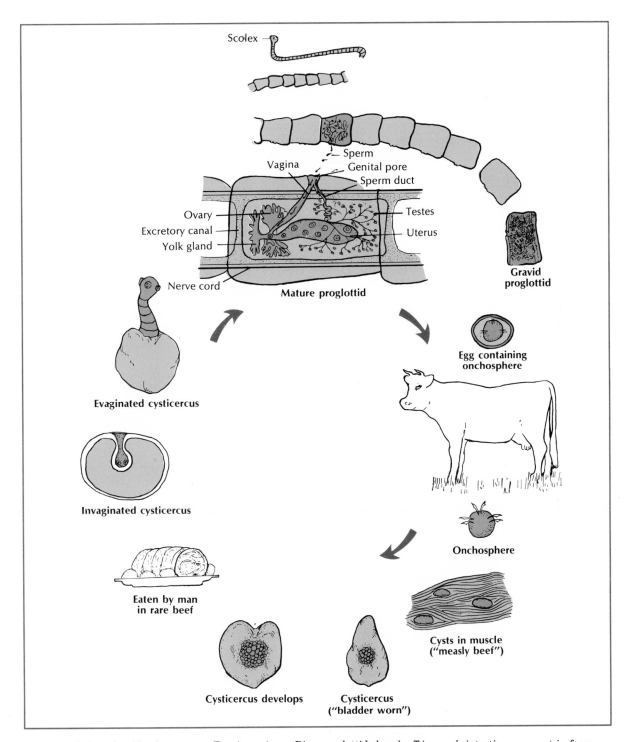

Scolex

Sperm

Vagina
Genital pore
Sperm duct

Ovary
Testes
Excretory canal
Uterus
Yolk gland

Nerve cord

Mature proglottid

Gravid proglottid

Evaginated cysticercus

Egg containing onchosphere

Invaginated cysticercus

Onchosphere

Eaten by man in rare beef

Cysts in muscle ("measly beef")

Cysticercus develops

Cysticercus ("bladder worn")

FIG. 27-5. Life cycle of beef tapeworm *Taenia saginata*. Ripe proglottids break off in man's intestine, pass out in feces, and are ingested by cows. Eggs hatch in cow's intestine, freeing onchospheres which penetrate into muscles and encyst, developing into "bladder worms." Man eats infested rare beef and cysticercus is freed in intestine where it develops, forms a scolex, attaches to intestine wall, and matures.

Schistosoma haematobium is found chiefly in Africa and is one of the most dangerous of the blood flukes. The adults live in the blood vessels of the bladder and urinary tract. *Schistosoma mansoni,* common in the West Indies, parts of South America, and Africa, lives in branches of the portal and mesenteric veins and cause severe dysentery and anemia.

Schistosoma dermatitis causes a skin irritation (swimmer's itch) by the penetration of the cercariae into the skin. Bathers in our northern lakes often feel the effects soon after leaving the water. Several species of cercariae are known to cause this irritation. Man is an accidental host and a "dead end" for these flukes, for the cercariae do not survive in the skin. Aquatic birds are the normal vertebrate hosts of these flukes, and snails are intermediate hosts.

TAPEWORMS (CLASS CESTODA)

The Cestoda, or tapeworms, differs in many respects from the preceding classes: their long flat bodies are usually made up of many sections, or **proglottids,** and there is a complete lack of a digestive system. One of their most specialized structures is the **scolex,** or holdfast, which is the organ of attachment (Fig. 27-5). It is provided with a varying number of suckers and, in some cases, also with hooks. As long as the scolex is present, it is impossible to get rid of a tapeworm, for new proglottids will be formed as the old ones are shed.

All members of this class are endoparasites, and all, with a few exceptions, involve at least two hosts. The adults are always found in vertebrates; other stages may be found in either vertebrates or invertebrates.

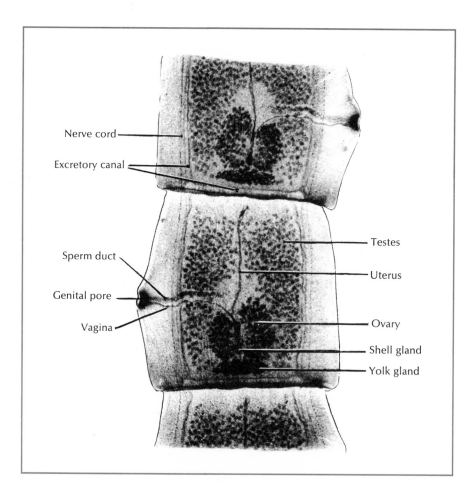

Nerve cord

Excretory canal

Sperm duct

Genital pore

Vagina

Testes

Uterus

Ovary

Shell gland

Yolk gland

FIG. 27-6. Photomicrograph of mature proglottids of *Taenia pisiformis,* dog tapeworm. (Courtesy General Biological Supply House, Inc., Chicago.)

Taenia saginata—beef tapeworm

Structure. The beef tapeworm lives as an adult in the alimentary canal of man, whereas the larval form is found primarily in the intermuscular tissue of cattle (Fig. 27-5). The mature adult may reach a length of 30 feet or more. Its scolex has four suckers for attachment to the intestinal wall, but no hooks. Back of the suckers a short neck connects the scolex to the body; the body may be made up of as many as 2,000 proglottids. As they move backward, the proglottids increase in size so that the proglottids are narrow near the scolex and broader and larger toward the posterior end, where they are finally detached and shed in the feces. The youngest proglottid is therefore nearest the scolex and the oldest one is at the posterior end.

Each mature proglottid contains a complete set of male and female organs (Fig. 27-6). Ova may be fertilized by the sperm of the same proglottid, from other proglottids of the same individual, or from other tapeworms if more than one should be present. When the terminal proglottids (called **gravid proglottids**) become enlarged with the developing zygotes, they break off and pass out with the feces. The embryos may be scattered on the soil or grass where they may be picked up by grazing cattle.

Life cycle. When cattle swallow the eggs or proglottids, the six-hooked larvae (**oncospheres**) burrow through into the blood or lymph vessels and finally reach voluntary muscle, where they encyst to become **bladder worms (cysticerci)** (Fig. 27-5). In 10 to 20 weeks the larvae develop an invaginated scolex with suckers and remain quiescent until the uncooked muscle is eaten by man or another suitable host. Such infested meat is known as "measly" meat. In the new host the scolex evaginates and becomes attached to the intestinal mucosa, and the new proglottids begin to develop. It takes 2 or 3 weeks for a mature worm to form. When man is infested with one of these tapeworms, many ripe proglottids are expelled daily from his intestine. Man usually becomes infested by eating rare beef.

Other tapeworms

More than 1,000 species of tapeworms are known to parasitologists, and almost all vertebrates are infested. Nearly all tapeworms have an intermediate host and a final host that is infested by preying upon the former. Some of these tapeworms do considerable harm to the host by absorbing nourishment and by secreting toxic substances, but unless present in large numbers, they are rarely fatal.

Taenia solium (pork tapeworm) lives in the small intestine of man, whereas the larvae live in the muscles of the pig. Man becomes infested by eating improperly cooked pork.

Diphyllobothrium latum (fish tapeworm), often called the broad tapeworm of man, is the largest and most destructive of the cestodes that infest man. It sometimes reaches a length of 60 feet and may have more than 3,000 proglottids. Eggs discharged into water by a human host hatch into a ciliated larva that is swallowed by a crustacean *(Cyclops)*. When the crustacean is eaten by the second host, a fish, the larvae encyst in the muscles. When raw or poorly cooked fish is eaten by man or other suitable host, the larva is liberated and matures in the intestine. Broad tapeworm infestations are found all over the world.

RIBBON WORMS (PHYLUM RHYNCHOCOELA*)

The ribbon worms, also known as nemertine worms, number about 600 species. Most species are marine, although some are found in moist soil and fresh water. They may be collected at low tide from under stones, in empty mollusk shells, or in seaweed, or secured by dredging at depths of 15 to 25 feet or deeper.

The general body plan is similar to that of Turbellaria, with ciliated epidermis and with flame cells in the excretory system (Fig. 27-7). Other flatworm characteristics are the presence of bilateral symmetry, mesoderm, and lack of coelom. Ribbon worms, however, show some distinct advancements over the flatworms. The one-opening gastrovascular cavity has given way to a complete mouth-to-anus digestive tract, and there is a retractile proboscis that can be withdrawn into a special proboscis cavity above the digestive tract. Ribbon worms have also evolved a blood vascular system, usually consisting of one dorsal and two lateral blood vessels that propel the colorless blood.

*Ring″ko-se′la (Gr. *rhynchos*, beak, + *koilos*, hollow).

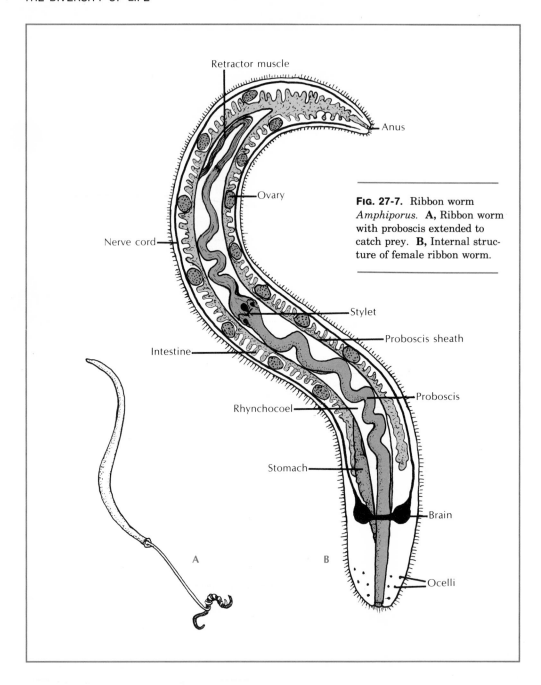

Retractor muscle

Anus

Ovary

Nerve cord

Stylet

Proboscis sheath

Intestine

Proboscis

Rhynchocoel

Stomach

Brain

A

B

Ocelli

FIG. 27-7. Ribbon worm *Amphiporus.* **A,** Ribbon worm with proboscis extended to catch prey. **B,** Internal structure of female ribbon worm.

The ribbon worms are carnivorous and very voracious, eating either dead or living prey. In seizing their prey they thrust out with deadly accuracy the slime-covered proboscis, which quickly ensnares the prey by wrapping around it (Fig. 27-7, *A*). The stylet also pierces and holds the prey. Then by retracting the proboscis, the prey is drawn near the mouth and is engulfed by the esophagus that is thrust out to meet it. Ribbon worms do not hesitate to eat each other when they are confined together.

28

Cavity worms
and
spiny-headed worms

These two groups, cavity and spiny-headed worms, include a heterogeneous array of worms, or wormlike animals, that have a **pseudocoel.** This is a body cavity that performs many of the same functions that the coelom performs in the true coelomate animals. But it develops as a remnant of the blastocoel, and lacks a peritoneal lining.

Affinities between groups are difficult to establish, but there is much evidence that these groups are derived from the flatworms.

CAVITY WORMS (PHYLUM ASCHELMINTHES*)

The animals that make up cavity worms all have some form of pseudocoel, bilateral symmetry, and a wormlike form or modification of it. Most of them are small, although some of the parasitic nematodes may reach a length of more than a meter. The body is often round or cylindric, although it is distinctly flattened in some.

Aschelminths occupy a wide range of habitat distribution. Many are aquatic in both fresh and marine water; others occupy terrestrial habitats. Some aschelminths are among the most common of

*As"kel-min' thes (Gr. *askos,* a leather bag, + *helmins,* worm).

all parasites. All vertebrates and most of the invertebrates are parasitized by one or more kinds of aschelminths. The number of species in this phylum must be very great; large numbers of species still remain to be named and classified.

CLASSIFICATION

Some authorities consider Aschelminthes as a superphylum with each of the following classes being given the rank of phylum.

Class Rotifera (ro-tif' e-ra). Aquatic and microscopic; anterior end with ciliary organ (corona); forked foot with cement gland; separate sexes, males much smaller than females; parthenogenesis and sexual reproduction; about 1,500 species. Example: *Philodina.*

Class Gastrotricha (gas-trot' ri-ka). Aquatic and microscopic; body usually posteriorly forked with adhesive tubes and glands for attachment; only females found in some species, eggs develop parthenogenetically; some species hermaphroditic; about 140 species. Example: *Chaetonotus.*

Class Kinorhyncha (kin"o-ring' ka). Marine, microscopic animals; body of 13 or 14 rings (zonites); sexes separate; metamorphosis of several larval stages; about 100 species. Example: *Echinoderella.*

Class Nematoda (nem"a-to' da). Aquatic, terrestrial, or parasitic worms; body cylindric, unsegmented, and elongated; sexes usually separate, with female generally larger than male; fertilization internal; development usually direct but life history may be intricate; over 12,000 named species, but their number has been estimated to be 500,000 species. Examples: *Ascaris* (intestinal roundworm), *Necator* (hookworm), *Wuchereria* (filarial worm), *Dioctophyma* (giant kidney worm), *Trichinella* (trichina worm), *Enterobius* (pinworm).

Class Nematomorpha (nem″a-to-mor′fa). Long, slender worms with cylindric bodies; size from a few millimeters to a meter in length; larval forms parasitic, adults free living; separate sexes; development mostly direct. Example: *Paragordius* ("horsehair" worm).

CHARACTERISTICS

1. Symmetry bilateral; unsegmented; triploblastic (three germ layers)
2. Size mostly small; some microscopic; a few a meter or more in length
3. Body usually vermiform, cylindric, or flattened; **cilia** mostly absent
4. Muscular layers of the body mostly of **longitudinal fibers,** with few exceptions
5. Body cavity an unlined **pseudocoel**
6. **Digestive system complete** with mouth, enteron, and anus—"a tube-within-a-tube" arrangement; digestive tract usually only an epithelial tube with **no definite muscle layer;** pharynx muscular and well developed.
7. Circulatory and respiratory organs lacking
8. Excretory system of canals and protonephridia (in some)
9. Nervous system of cerebral ganglia or of a circumenteric nerve ring connected to anterior and posterior nerves; sense organs of **ciliated pits,** papillae, bristles, and eyespots (few)
10. Reproductive system of gonads and ducts that may be single or double; sexes nearly always separate, with the male usually smaller than the female
11. Development may be direct or with a complicated life history; cleavage mostly determinate; **cell or nuclear constancy common**

WHEEL ANIMALCULES (CLASS ROTIFERA)

The rotifers, or "wheel animalcules," are microscopic, ranging from 0.5 to 1.5 mm. in length. They have a worldwide distribution; most are freshwater forms.

The body is usually made up of a head, trunk, and foot. The **corona** of the head may form a type of ciliated funnel with its upper edges folded into lobes bearing bristles, or, as in the familiar *Philodina* (Fig. 28-1), the corona may be made up of a pair of

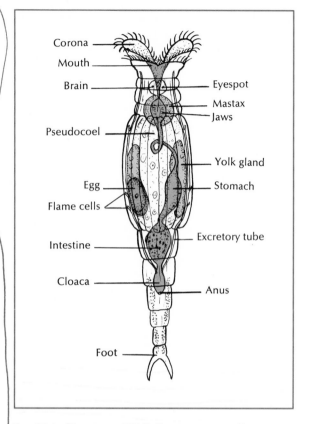

FIG. 28-1. Structure of *Philodina,* common rotifer.

ciliated disks. The cilia cause currents of water to flow toward the mouth, bringing with it small planktonic forms for food.

GASTROTRICHS (CLASS GASTROTRICHA)

Gastrotrichs (Fig. 28-2) have habitats about the same as those of rotifers. They are ventrally flattened, have dorsal spines, and can glide by means of ventral cilia. The head bears cilia and, in some species, long bristles, and some have forked posteriors. Adhesive tubes secrete a substance for attachment to substrata. Only females occur in freshwater, and the eggs develop parthenogenetically. Eggs are laid on some substratum, such as weeds, and hatch in a few days. Species of *Chaetonotus* (Fig. 28-2) are common freshwater gastrotrichs.

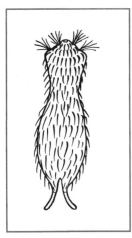

FIG. 28-2. *Chaetonotus,* a common gastrotrich.

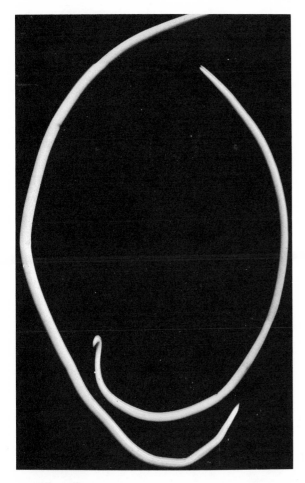

FIG. 28-3. Intestinal roundworm *Ascaris lumbricoides,* male and female. Male (right) is smaller and distinguished by the kink in its tail.

THE ROUNDWORMS (CLASS NEMATODA)

It has been said that if the earth were to disappear leaving only the nematode worms, the general contour of the earth's surface would be outlined by the worms, for they are present in nearly every conceivable kind of ecologic niche. They live both in freshwater and terrestrial habitats and some can withstand unusual extremes of temperature and desiccation. Plant parasitic nematodes do extensive damage to many crops because several generations of worms each year are possible in warm soil. Nematode parasites also cause severe disabilities in man and other animals. Nematode eggs are especially resistant and can be carried by animals and winds.

Not all nematodes are parasites; most are free living. Many feed on plant juices, algae, and bacteria; others feed on small live forms; or some may be scavengers that feed upon dead animals and plants. Those parasitic in plants cause galls or nodular growths on roots and leaves. The vinegar eel *Turbatrix,* often found in cider, is a free-living nematode.

Nearly all vertebrates and many invertebrates are parasitized by nematodes. Some of the more common nematode parasites of man are described below.

Ascaris worm. *Ascaris lumbricoides* is a parasitic roundworm of man (Fig. 28-3). They can be acquired through unsanitary habits in which contaminated food and vegetables containing the ova are conveyed to the mouth.

A large female, which can easily reach a foot in length, may lay 200,000 eggs a day. These are eliminated in the host feces. On the ground under suitable conditions, small worms may develop in the shells within 2 or 3 weeks. If ingested at this stage the worms will mature in the host. When embryonated eggs are swallowed, they hatch into tiny larvae that burrow through the intestinal wall into the veins or lymph vessels. In the blood they are carried through the heart to the lungs, move up the trachea, cross over into the esophagus, and then go down the alimentary canal to the intestine, where they grow to maturity in about 2 months. Here they copulate and the female begins her egg laying. Thus only one host is involved in the life cycle. They do their greatest damage to the host while the juvenile worms are migrating.

393

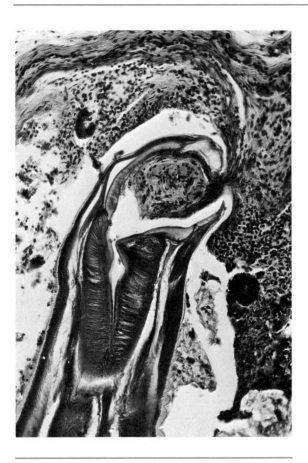

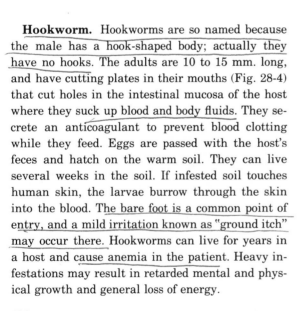

FIG. 28-4. Section through hookworm attached to human intestine. Note cutting plates of mouth pinching off bit of mucosa from which muscular pharynx sucks blood. Mouth secretes anticoagulant. (AFIP No. 33810.)

FIG. 28-5. Muscle infested with trichina worm *Trichinella spiralis.* Larvae may live ten to twenty years in these cysts. If eaten in poorly cooked meat, larvae are liberated in intestine. They quickly mature and release many larvae into blood of host.

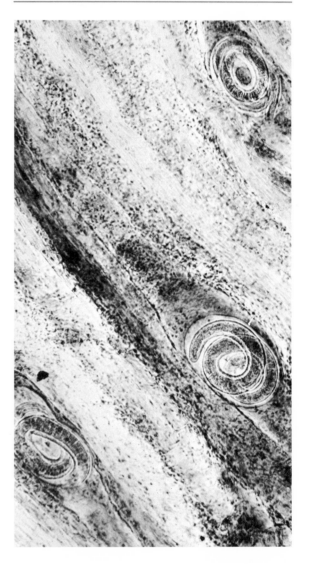

Hookworm. Hookworms are so named because the male has a hook-shaped body; actually they have no hooks. The adults are 10 to 15 mm. long, and have cutting plates in their mouths (Fig. 28-4) that cut holes in the intestinal mucosa of the host where they suck up blood and body fluids. They secrete an anticoagulant to prevent blood clotting while they feed. Eggs are passed with the host's feces and hatch on the warm soil. They can live several weeks in the soil. If infested soil touches human skin, the larvae burrow through the skin into the blood. The bare foot is a common point of entry, and a mild irritation known as "ground itch" may occur there. Hookworms can live for years in a host and cause anemia in the patient. Heavy infestations may result in retarded mental and physical growth and general loss of energy.

FIG. 28-6. Male (left) and female (larger) pinworms, *Enterobius vermicularis.* Infestation in up to 40% of school children has been found in some communities; this worm may be most common and most widely distributed of human helminth parasites. (Courtesy Indiana University School of Medicine, Indianapolis.)

FIG. 28-7. Guinea worm *Dracunculus* is slowly extracted from body of its host by making one turn of stick each day to prevent rupturing worm and causing bacterial infection. If not removed, the worm would die in time and become calcified.

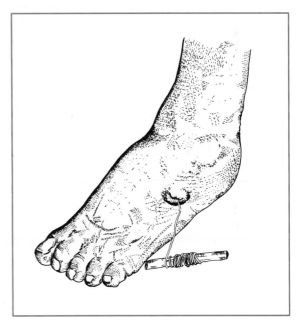

Trichina worm. This is a tiny nematode responsible for the serious disease **trichinosis.** Adults burrow into the mucosa of the small intestine where the female produces living larvae. The larvae penetrate into blood vessels and are carried to the skeletal muscles where they coil up and form cysts that become calcified (Fig. 28-5). The worms may live in the cysts for years if undisturbed. When meat containing live cysts is swallowed, the larvae are liberated, mature, and produce living larvae.

Besides man the worms infest hogs, rats, cats, dogs, etc. Man acquires the parasite by eating improperly cooked pork. Hogs acquire them by eating garbage containing pork scraps with cysts or by eating infested rats. Nearly 75% of all rats are said to be infested.

Pinworms. The pinworm is very common, especially in warm countries. In some communities 40% to 100% of the children are affected. The adults (Fig. 28-6) live in the large intestine. Females with

395

eggs migrate to the anal region at night to lay their eggs. Since this causes irritation, scratching at night is common, and fingers and bedding become contaminated. When ova are swallowed, they hatch in the duodenum and mature in the large intestine. No intermediate host is necessary. Each generation lasts 3 to 4 weeks. If reinfestation does not occur, they will die out.

Filarial worms. Filarial worms are tropical nematodes that live in the lymphatic glands where they often obstruct the flow of lymph. Larvae are passed from person to person by mosquitoes. Repeated infestations cause enormous swelling of tissues, a condition known as elephantiasis.

Guinea worms. This is a tropical worm, 2 to 4 feet long, that lives under the surface of the skin, discharging its living young through skin ulcers. The larvae are carried by an aquatic crustacean.

The time-honored, but dangerous, method of removing a guinea worm is by winding it out on a stick, a little each day (Fig. 28-7).

HORSE-HAIR WORMS (CLASS NEMATOMORPHA)

The popular name for this class is based on an old superstition that they arose from horse hairs that fell into water. They are found in aquatic habitats in both the temperate and tropical zones. Adults are free living and never feed, for their digestive system is degenerate. The larval stages are parasitic in some arthropod hosts. The different species vary in length from a few millimeters to about a meter, and many of them have the habit of coiling themselves into a knot, often around an aquatic plant or other object.

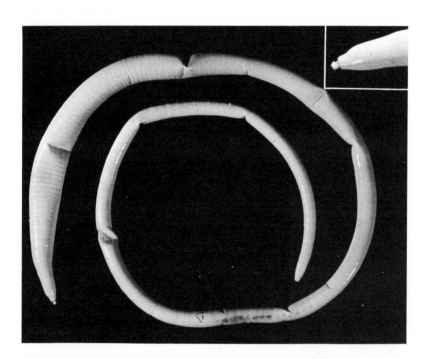

FIG. 28-8. Spiny-headed worm of pigs, *Macracanthorhynchus hirudinaceus* (female), shown about life size. Inset is enlarged view of head showing proboscis.

SPINY-HEADED WORMS (PHYLUM ACANTHOCEPHALA*)

The acanthocephs are all endoparasitic, living as adults in the intestines of vertebrates and as larval forms in arthropods. The spiny-headed worm derives its name from one of its most distinctive characteristics, a cylindric invaginable proboscis bearing rows of recurved spines, by which it attaches itself to the intestine.

Acanthocephs are very harmful parasites, for great damage is done mechanically by the spiny heads. Multiple infestations may do considerable damage to the pig's intestine and perforation may occur.

Acanthocephs resemble the flatworms in their reproductive systems and their method of development. They resemble the aschelminths in possessing a pseudocoel and a syncytial nucleated epidermis. They share with the rotifers, nematodes, and tunicates a tendency to cell or nuclear constancy.

About 500 species of acanthocephs have been named, most of which parasitize fish, birds, and mammals. Their distribution is worldwide. They range from 2 to 650 mm. in length.

CHARACTERISTICS

1. Anterior end with **spiny retractile proboscis** and **sheath**
2. Body cylindric in form, in three sections—proboscis, neck, and trunk
3. **Epidermis syncytial in structure** and covered with cuticle and containing **fluid-filled lacunae;** cell or nuclei constancy pronounced
4. **No digestive tract**
5. No circulatory or respiratory organs
6. Nervous system with a central ganglion on the proboscis sheath and nerves to the proboscis and posterior parts of the body
7. Separate sexes; male organs of paired testes, vas deferens, cement glands, and penis; female organs of paired ovaries formed in a ligament and breaking down into ova; young develop in body cavity of female; special selector apparatus in female system

The best known spiny-headed worm, and the one most frequently used in student laboratories, is the spiny-headed worm of pigs, *Macracanthorhynchus hirudinaceus* (Fig. 28-8).

The body of *Macracanthorhynchus* is cylindric and tapers posteriorly. The female is 10 to 65 cm. long and usually less than 1 cm. thick; the male about one fourth as large. The body is enclosed by a thin cuticle through which food is absorbed from the host. The proboscis, bearing rows of recurved hooks, is attached to the neck region (Fig. 28-9) and may be retracted into a proboscis receptacle.

The eggs, which are discharged in the feces of the vertebrate host, do not hatch until eaten by the intermediate host. In this host, which is the larva, or grub, of the well-known June beetle, the first larva (acanthor) burrows through the intestine and encysts as a juvenile. Pigs become infested by eating the grubs. Man rarely becomes infested.

FIG. 28-9. Longitudinal section through anterior end of acanthoceph with proboscis partially everted. When proboscis is inverted, the spines point forward. (After Hamann, modified from Hyman; from Hickman, C. P.: Biology of invertebrates, 1967, The C. V. Mosby Co.)

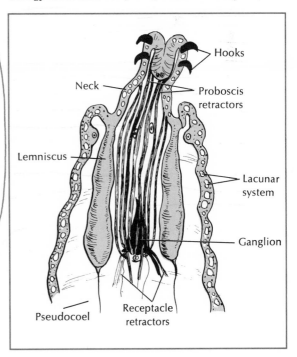

*A-kan"tho-sef'a-la (Gr. *akantha,* spine or thorn, + *kephale,* head).

29

Mollusks
and
segmented worms

The person who first compares a segmented worm, such as an earthworm, with a mollusk, such as a clam or snail, could be forgiven for wondering what the two groups could possibly have in common. Nevertheless many believe that these two groups share a common origin, probably the flatworms. Both groups have a true **coelom,** which reaches a high stage of development. Both have well-developed circulatory systems, and, in general, a high degree of nervous system centralization. Both belong to the protostome branch of the animal kingdom and show spiral and determinate cleavage during embryonic development. Probably the segmented worms can be considered somewhat more advanced than the mollusks because mollusks, lacking segmentation, must have branched off the main evolutionary line at an earlier time. More important, however, is the fact that segmentation is a highly successful innovation retained by all the more advanced groups (for example, arthropods and vertebrates).

MOLLUSKS (PHYLUM MOLLUSCA*)

Next to the arthropods the mollusks have the most named species in the animal kingdom. They range

*Mol-lus′ka (L. *molluscus*, soft).

398

from fairly simple organisms to some of the most complex of invertebrates, and from microscopic to 50 feet long—the largest of invertebrates. Most mollusks are marine; some live in fresh water and some have been successful on land.

The fossil record of the phylum has been continuous since Cambrian times, but it has been somewhat difficult to establish relationships to other groups. The trochophore-like larvae in many marine mollusks and the type of egg cleavage indicate an affinity with the annelids. This view is supported by the recent discovery of living monoplacophorans that are segmented (although their segmentation may be secondary rather than primitive). Some mollusks have a ladderlike nervous system similar to that of the turbellarian flatworms. Both the annelids and the mollusks may have diverged from a common platyhelminth form.

According to fossil evidence the mollusks originated in the sea, and much of their evolution probably took place along the shore line. Only the snails and the bivalves left the sea and invaded the land and fresh water. The filter feeding of bivalves restricted them to water, but many found their way up the rivers. Among the snails a vascular lining of the mantle cavity made possible the development of a lung so the gastropods were able to invade the land as well as the rivers.

Characteristic mollusk adaptations are the mantle, the muscular foot, and the radula. The **mantle** is a sheath of skin that enfolds the soft body and forms a mantle cavity between the body and the mantle. The mantle secretes the shell, and in the various mollusks it is modified to form siphons, gills, or lungs. It is often used for respiration, and in cephalopods and sea hares it is used in locomotion. The muscular **foot** is variously adapted for attachment, locomotion, and burrowing. It has great power of contraction and is aided by an internal hemoskeleton of blood turgidity. The **radula** is a flexible band of tiny chitinous teeth located in the pharynx. It moves back and forth over a cartilage with a rasping motion to scrape off food particles.

CLASSIFICATION

Class Monoplacophora (mon″o-pla-kof′o-ra) — **segmented mollusks.** Body bilaterally symmetric with a broad flat foot; mantle covered with a single limpetlike shell; internal segmentation only; separate sexes. Example: *Neopilina.*

Class Amphineura (am″fi-neu′ra) — **chitons.** Elongated body with reduced head; bilaterally symmetric; row of eight dorsal plates usually; large, flat ventral foot, which is absent in some; sexes usually separate. Example: *Chiton.*

Class Scaphopoda (ska-fop′o-da) — **elephant tusk shells.** Body enclosed in a one-piece tubular shell open at both ends; sexes separate. Example: *Dentalium.*

Class Gastropoda (gas-trop′o-da) — **snails and others.** Body usually asymmetric in a coiled shell (shell absent in some); head well developed; foot large and flat; dioecious or monoecious. Examples: *Littorina, Physa, Helix.*

Class Pelecypoda (pel-e-syp′o-da) — **bivalves.** Body enclosed in a two-lobed mantle; shell of two lateral valves of variable size and form, with dorsal hinge; foot usually wedge shaped; sexes usually separate. Examples: *Anodonta, Teredo, Venus.*

Class Cephalopoda (cef″a-lop′o-da) — **squids and octopuses.** Body with a shell, often reduced or absent; head well developed with eyes; foot modified into arms or tentacles; sexes separate. Examples: *Loligo, Octopus, Sepia.*

CHARACTERISTICS

1. Body unsegmented (except in Monoplacophora) and typically bilaterally symmetric (bilateral asymmetry in some)
2. External body with three typical divisions: an anterior head; a **ventral muscular foot** variously modified but chiefly for locomotion; and a dorsal **mantle** that usually secretes a shell (absent in some)
3. Coelom reduced and represented mainly by the pericardium, gonadal cavity, and kidney
4. Digestive system complete with digestive glands and liver; with a rasping organ **(radula)** usually present
5. Circulatory system of three-chambered heart, pericardial space and blood vessels; blood mostly colorless
6. Respiration by gills, by lungs, or direct
7. Excretion by one or two pairs of nephridia or a single nephridium opening internally into pericardium and externally onto body surface
8. Nervous system of paired **cerebral, pleural, pedal,** and **visceral ganglia,** with nerves; ganglia centralized in ring (Cephalopoda and Gastropoda)
9. Reproduction dioecious or monoecious; fertilization external or internal
10. Spiral and determinate cleavage

SEGMENTED MOLLUSKS (CLASS MONOPLACOPHORA)

Until 1952 this class consisted only of Paleozoic shells. However, in that year a few living specimens of *Neopilina* were dredged up from the ocean bottom near the west coast of Mexico. These mollusks are small and have a low, rounded shell and a creeping foot. Five or six pairs each of gills, nephridia, and gill hearts are arranged segmentally. The mouth bears the characteristic radula. Monoplacophorans are thought to indicate an annelid-mollusk relationship.

CHITONS (CLASS AMPHINEURA)

The **chitons** (Fig. 29-1) have a convex dorsal surface that bears eight overlapping articulating limy plates secreted by the mantle. Most chitons are only 1 or 2 inches long; the largest, *Cryptochiton,* rarely exceeds 8 or 10 inches. A marginal girdle, formed from the mantle, may contain calcareous spines and scales that give it a shaggy appearance. The head on the underside bears the mouth but has no eyes or tentacles.

Chitons creep slowly on the foot or else use it to adhere to rocky surfaces. They live on seaweed or other marine plant life. Chitons are mostly intertidal invertebrates, although some have been taken at great depths.

FIG. 29-1. Dorsal view of *Chiton*.

TOOTH SHELLS (CLASS SCAPHOPODA)

Scaphopods, commonly called tooth shells or tusk shells, have a slender, tubelike shell open at both ends. A burrowing foot protrudes through the larger end of the shell. The mouth, which is near the foot, is provided with a radula and with tentacles that are sensory and prehensile. Respiration takes place through the mantle. Scaphopods are marine, live embedded in sand, and feed upon microscopic animals and plants. *Dentalium* (Fig. 29-2) is a familiar example along our eastern seashore. Most scaphopods are under 3 inches long, but some fossil forms reached a length of 2 feet.

CLASS GASTROPODA

The gastropods, the largest class of mollusks, include the snails, limpets, slugs, whelks, conchs, sea hares, sea butterflies, and sea slugs. These animals are basically bilaterally symmetric, but by torsion the visceral mass in most of them has become asymmetric. The shell, when present, is always single (univalve) and may be coiled or uncoiled. In the sea, gastropods are common both in littoral zones and at great depths. Some are adapted to

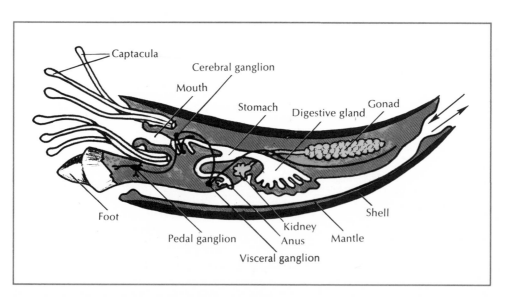

FIG. 29-2. Internal anatomy of a tooth shell, *Dentalium*. In nature it buries its head end (to which are attached numerous tentacles) in sand, leaving posterior narrow end projecting into water. Captacula (L. *captare*, to catch) are ciliated, contractile tentacles, which are prehensile and used in capturing microscopic organisms.

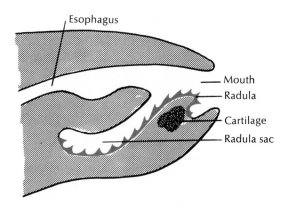

FIG. 29-3. Median section through head and mouth of pond snail showing radula. Muscles work radula back and forth for rasping food. As the forward end wears away, radula is replaced at the caudal end. Radula is usually distinctive for each species and is useful for taxonomic studies.

brackish water and some to fresh water. On land their range of habitats is large. Gastropods range from microscopic forms to giant marine snails that exceed 2 feet; most are from 1/2 inch to 3 inches.

Gastropods are mostly sluggish sedentary animals, but some are specialized for climbing, burrowing, or even swimming. Most gastropods live on plants or plant debris. Many are scavangers; others are carnivorous, preying upon clams, oysters, and worms. Some carnivores drill holes in the shells of other mollusks to eat the soft parts. To obtain and rasp food, snails use the radula (Fig. 29-3).

The larva is at first symmetric, but during development it undergoes **torsion.** In this process the mantle cavity, anus, gills, and other visceral parts rotate a full 180 degrees in a counterclockwise direction so that the anus and mantle cavity come to lie just above the head. In some forms, such as the

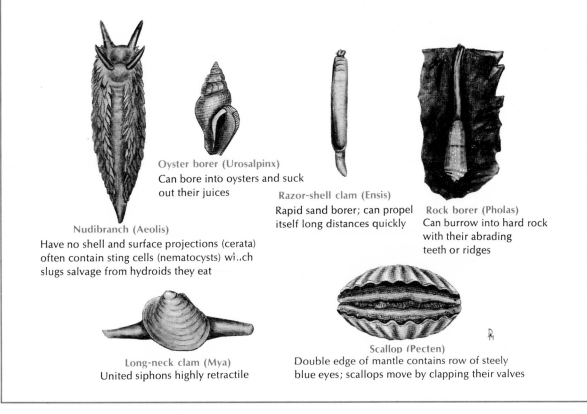

Oyster borer (Urosalpinx)
Can bore into oysters and suck out their juices

Razor-shell clam (Ensis)
Rapid sand borer; can propel itself long distances quickly

Rock borer (Pholas)
Can burrow into hard rock with their abrading teeth or ridges

Nudibranch (Aeolis)
Have no shell and surface projections (cerata) often contain sting cells (nematocysts) which slugs salvage from hydroids they eat

Long-neck clam (Mya)
United siphons highly retractile

Scallop (Pecten)
Double edge of mantle contains row of steely blue eyes; scallops move by clapping their valves

FIG. 29-4. Some mollusks with unusual habits or structures. Nudibranchs, oyster borer, and rock borer belong to class Gastropoda; the others belong to class Pelecypoda.

nudibranchs and sea hares, a process of detorsion may occur in the embryo following the torsion, so that the adult becomes bilateral, or nearly so.

Torsion must be distinguished from the **coiling** or spiral winding of the visceral hump and shell. The coiling of the visceral hump occurs as a result of uneven growth of the hump so that a corkscrew-shaped cone is produced at right angles to the axis of torsion. Coiling does not occur in some snails, or else it is suppressed in some way (as in limpets and abalones). In the nudibranchs a coiled shell is found in the embryo but is absent in the adult. The direction of coiling of the shell—right- or left-handed—is a hereditary characteristic.

There are three major groups of gastropods, which some authorities call subclasses and others call orders.

Prosobranchia. In the prosobranch gastropods the gills are located anteriorly in front of the heart. There is usually an operculum, or horny plate, on the foot used to seal the opening of the shell when the snail withdraws into the shell. This group contains most of the marine snails and a few of the

freshwater ones. Familiar examples are the abalone with an ear-shaped shell, the periwinkles, limpets, whelks, conchs, slipper shells, oyster borers (Fig. 29-4) that bore into oysters and suck out their juices, and rock borers that can burrow into hard rock (Fig. 29-4).

Opisthobranchia. In the opisthobranch gastropods the gill is displaced to the right side or rear of the body. They are all marine, most are shallow-water forms that hide under stones or seaweed, and a few are pelagic. The opisthobranchs include nudibranchs (Fig. 29-4), sea hares, sea butterflies, canoe shells, etc. in all of which the shell is reduced or absent.

Pulmonata. The pulmonates are an extensive group that includes land and freshwater snails and slugs (Fig. 29-5) and a few brackish water forms. They have a lung instead of a gill, one or two pairs of tentacles, and a single gonad.

Common examples of terrestrial snails are the woodland snails and _Helix,_ a garden snail introduced here from Europe. They feed mostly at night on green vegetation, which they rasp off with the radula. When threatened by very dry weather, they can draw into the shell and cover the opening with a protective layer of mucus and limy secretions.

CLASS PELECYPODA

The pelecypods, or bivalves, include over 7,000 species of mussels, clams, scallops, oysters, and shipworms ranging from 1 mm. to 1 meter. Most pelecypods are specialized for a sedentary life and have evolved a filtering mechanism that depends on the gills for assistance in obtaining food by ciliary currents. They have no radula. Most pelecypods are marine, but many also live in brackish and fresh water. Because of their calcareous shells they must live in "hard" water, or water rich in limy salts.

Pelecypods move by extending a slender muscular foot between the valves (Figs. 29-4 and 29-6). Blood swells the foot to anchor it in mud or sand, then muscles contract to shorten the foot and pull the animal forward. In most bivalves the foot is used for burrowing. The scallops swim jerkily by clapping their valves to create a sort of jet propulsion. Some forms are sessile; marine mussels _(Mytilus)_ attach

FIG. 29-5. Pulmonate snails. **A,** Land snail _Anguispira_ extended and traveling. **B,** Common garden slug _Limax._

A

B

themselves by secreting adhesive byssal threads, and oysters cement one valve to a surface.

Bivalves are laterally compressed and covered by a pair of **valves** (shells) secreted by the mantle lobes and hinged together dorsally. An elastic hinge ligament (Fig. 29-6) holds the left and right valves together at the hinge and causes them to gape ventrally. Powerful adductor muscles draw the valves together.

The shell consists of the three layers, with a horny, protective layer on the outside and a layer of nacre, or mother-of-pearl on the inside. The oldest and thickest part of the valve is the **umbo**, a rounded prominence near the hinge (Fig. 29-6), and surrounding it are successive concentric lines of growth.

The production of a pearl is a protective device on the part of a bivalve. When a foreign substance such as a grain of sand or a parasite becomes en-

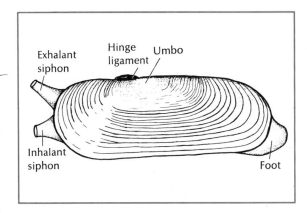

FIG. 29-6. Stubby razor clam *Tagelus gibbus* showing its two siphons (left) and muscular foot (right). It burrows into mud in shallow water and extends its siphons upward to maintain a steady flow of water which bears oxygen and food.

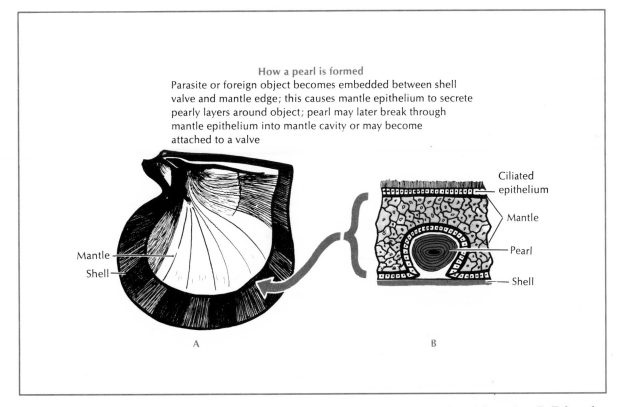

How a pearl is formed
Parasite or foreign object becomes embedded between shell valve and mantle edge; this causes mantle epithelium to secrete pearly layers around object; pearl may later break through mantle epithelium into mantle cavity or may become attached to a valve

FIG. 29-7. Pearl oyster *Margaritifera.* **A,** Interior of valve, with arrow pointing to site of pearl formation. **B,** Enlarged section of mantle, with pearl in position.

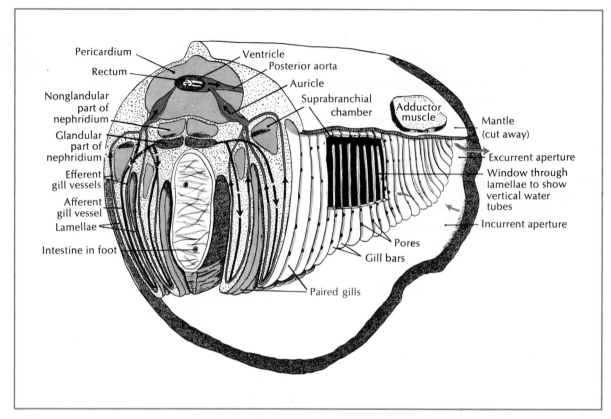

FIG. 29-8. Section through heart and gills of clam to show relationship of circulatory and respiratory systems. Scheme of blood circulation: ventricle pumps blood through aorta to body tissues; blood from body returns to nephridia (kidneys) to give off wastes; then to gills for gaseous exchange; then to auricles and ventricles. Scheme of water circulation: water enters incurrent aperture; through gill pores into water tubes for gaseous exchange; up to suprabranchial chamber; out excurrent aperture, carrying wastes.

closed between the mantle and the shell, the mantle secretes successive layers of nacre around the irritating substance (Fig. 29-7). *Meleagrina* is a pearl oyster used extensively in pearl culture by the Japanese.

The **mantle** is formed by folds of the body wall that hang down, one on each side of the soft body, and adhere to the valves. The space between the mantle lobes is the **mantle cavity.** The body itself is made up of the visceral mass, suspended from the dorsal midline, with a muscular **foot** attached anteroventrally to the visceral mass and a pair of **gills** on each side (Fig. 29-8). Both the outer surface of the gills and the inner surface of the mantle are ciliated. Water drawn in between the posterior ends of the mantle lobes (incurrent aperture) enters tiny pores in the water tubes and is carried up to a common chamber above the gills and out through an excurrent aperture. In some burrowing bivalves, the mantle lobes are modified into long muscular siphons (Fig. 29-6) that can be extended up to the water to bring in food and oxygen.

CLASS CEPHALOPODA

The cephalopods are the most specialized of the mollusks and in some ways are the most advanced of the invertebrates. All are marine. They include the squids (Fig. 29-9), octopuses (Fig. 29-10), nautiluses (Fig. 29-11), devilfish, and cuttlefish. They

FIG. 29-9. School of young squids. As they course back and forth through aquarium, each individual carefully maintains his position and distance with reference to others. If this pattern is disturbed, squids quickly revert to their original formation.

FIG. 29-10. Living octopus *Octopus*. Its eight arms bear powerful suckers for crawling over rocks and seizing prey (mostly crabs). (From film *Marine Life,* Encyclopaedia Britannica Films, Inc.)

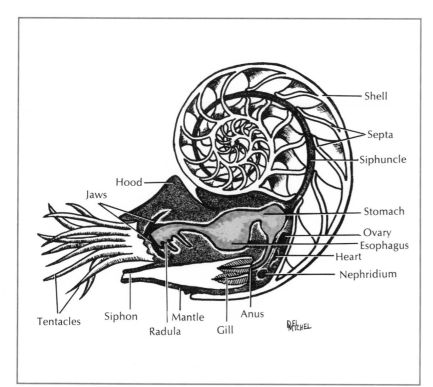

FIG. 29-11. Median section through shell and viscera of *Nautilus* showing internal anatomy (semidiagrammatic). This is only surviving genus of Tetrabranchia, which flourished millions of years ago in tropical seas. It inspired O. W. Holmes' famous poem "The Chambered Nautilus."

are found at various depths of the ocean. Octopuses are often found in shallow intertidal zones, the *Nautilus* is usually taken from the ocean floor (southwestern Pacific) several hundred meters deep, and some squids have been taken at 5,000 feet. The little west coast squid *Rossia* is only 1 1/2 inches long, the giant squid *Architeuthis* may grow to 50 feet long, and the giant octopus of the Pacific has been known to reach an overall diameter of 30 feet. Most cephalopods, however, are of moderate size.

Cephalopods are the "head-footed" mollusks, so-called because of the concentration of the foot in the head region. The edges of the foot are drawn out into arms and tentacles, which bear sucker disks for seizing prey. Part of the foot is modified to form the **funnel** (siphon) for carrying water out of the mantle cavity.

During the Paleozoic and Mesozoic eras there were thousands of species, but today there are only about 600. Although the earliest cephalopods bore heavy external shells, the shells were made buoyant with a series of gas chambers. *Nautilus* (Fig. 29-11) is the only genus of this type remaining today. In later cephalopods the shell was reduced until it was internal or absent and the animals relied on their speed, color changes, and ink screen for protection.

Color changes of cephalopods are produced in the skin by contraction and expansion of special pigment cells called **chromatophores,** manipulated by tiny muscles attached to the cells.

All cephalopods except *Nautilus* have ink sacs from which they expel ink when attacked. The ink contains a black pigment and probably distracts the enemy while the animal makes its escape. The cuttlefish *Sepia* produces a sepia-colored ink, which can be processed into a pigment used by artists.

All cephalopods swim by some form of jet propulsion. They take water into the mantle cavity in the neck region between the mantle and the head; then the collar of the mantle closes and the water is forcibly ejected in a jet through the funnel, or siphon. The siphon may be directed forward or backward to direct the movement. Fins on the mantle are used by squids for steering.

All cephalopods are predaceous and carnivorous, living on small fish, mollusks, crustaceans, and worms. They capture them with their specialized tentacles and use their horny **jaws** to tear or bite the flesh. Both jaws and **radula** are operated by the muscular pharynx.

Cephalopods have complex **eyes** with lenses and retinas that can form images just as vertebrate eyes do, although they are not homologous to vertebrate eyes because the photoreceptors in cephalopods are directed toward the source of light rather than away from the light source as in vertebrates.

Sexes are separate. One of the male arms is modified into a **hectocotylus arm** for transferring sperm to the female. He uses the arm to draw bundles of spermatophores from his funnel and places them in her mantle cavity. The female squid lays her eggs in pencil-shaped masses of jelly and attaches them to some object. The octopus lays eggs in long strings or bunches attached to a rock.

Development differs from that of other mollusks. The larvae at first are quite different from the adults and form a part of the plankton population.

Learning and behavior

The cephalopods have unusual capacities for learning and discrimination. Investigations have shown that octopuses are highly skilled in powers of discrimination in the visual and tactile fields. Early investigators found that octopuses could discriminate between squares of different sizes as well as between squares, circles, and triangles. In recent years J. Z. Young, M. J. Wells, and others have studied their reactions under many sophisticated conditions. Food rewards, electric shocks, and analyses of brain parts have been used to discover what octopuses can and cannot learn. The brain of an octopus has many distinct lobes, each made up of a layer of nerve cell bodies with a mass of nerve fibers. Stimulation of these lobes with electrodes, or removal of them, has given some knowledge of their functions. The well-developed eye indicates that they have marked visual acuity, but there is some limitation regarding the shapes of figures they can distinguish. Although they readily distinguish between horizontal and vertical rectangles, they are unable to do so when the figures are placed in an oblique position. Their tactile sense can distinguish the general roughness or smoothness of a surface, but not a detailed analysis of the contour. They cannot distinguish objects by their weight.

Retention of memory for several weeks has been noted. They can quickly be conditioned to attack or to avoid an object, and such behavior persists for a long time.

SEGMENTED WORMS (PHYLUM ANNELIDA*)

The annelids are worms whose bodies are divided into similar rings or segments. This is commonly called metamerism, and the body divisions are known as segments, somites, or metameres. Circulatory, excretory, nervous, muscular, and reproductive systems all show a segmental arrangement, and there are internal partitions between the somites.

The annelids show some relationships to certain other phyla. Their larval stages are similar to those of the flatworms, and their marine members have a larva that closely resembles that of the mollusks and some of the minor phyla. The adult forms of segmented worms and of flatworms may have come from a common ancestor. Annelids also have some arthropod characteristics, such as metamerism, the hypodermis-secreted cuticle, and the nervous system.

Annelids are found in the soil, in fresh water, and in the sea. Many are free living, but some are parasitic in whole or in part of their life cycles.

CLASSIFICATION

The annelids are classified primarily on the basis of the presence or absence of parapodia, setae, metameres, and other morphologic features.

Class Polychaeta (pol″y-ke′ta). Body of numerous segments with lateral parapodia bearing many setae; head distinct, with eyes and tentacles; clitellum absent; mostly marine. Examples: the clam worm *Neanthes*, the lugworm *Arenicola*.

Class Clitellata (kli″tel-la′ta). Body with clitellum; segmentation conspicuous; segments with or without annuli; parapodia absent; mostly freshwater and terrestrial.

Order Oligochaeta (ol″i-go-ke′ta). Body with conspicuous segmentation; setae few per metamere; head absent; chiefly terrestrial and freshwater. Example: the earthworm *Lumbricus*.

Order Hirudinea (hir″u-din′e-a). Body with definite number of segments (33 or 34) with many annuli; body with anterior and posterior suckers usually;

*An-nel′i-da (L. *annellus*, ring, + *ida*, pl. suffix).

terrestrial, freshwater, and marine. Example: the leech *Hirudo*.

CHARACTERISTICS

1. Body **metamerically segmented;** symmetry bilateral; three germ layers
2. Body wall with outer circular and inner longitudinal muscle layers; transparent moist cuticle covers body
3. **Chitinous setae,** often present on fleshy **parapodia;** absent in some
4. Coelom (schizocoel) well developed in most and usually divided by septa; coelomic fluid for turgidity
5. **Blood system closed** and segmentally arranged
6. Digestive system complete and not metamerically arranged
7. Respiration by skin or **gills**
8. Excretory system typically a **pair of nephridia for each metamere**
9. Nervous system with a double ventral nerve cord and a pair of ganglia with lateral nerves in each metamere; brain a pair of dorsally located cerebral ganglia
10. Hermaphroditic or separate sexes; larvae, if present, are trochophore type; asexual reproduction by budding in some; spiral and determinate cleavage

CLASS POLYCHAETA

The polychaetes are the larger and older of the classes of annelids, with more than 10,000 species; most of them are marine. Some are brightly colored in reds and greens; others are dull or irridescent. Some are picturesque, as the "feather-duster" worms (Fig. 29-12). Ecologically they are divided into two groups. The Errantia are free-moving pelagic forms, active burrowers, crawlers, and tube worms that leave their tubes for feeding or breeding. The Sedentaria are sedentary worms that rarely expose more than the head end from their tubes.

Tube dwellers may secrete parchmentlike tubes, or firm calcareous tubes attached to rocks or other surface (Fig. 29-12), or they may simply cement together grains of sand or bits of shell or seaweed. Many burrowers in sand and mud flats line their burrows with mucus. Most tube and burrow dwellers use mucus in some way to trap their food, whereas others are ciliary feeders.

Polychaetes have a well-differentiated head with sensory appendages, **parapodia** (lateral appendages) with many **setae** (chitinous bristles), a much

407

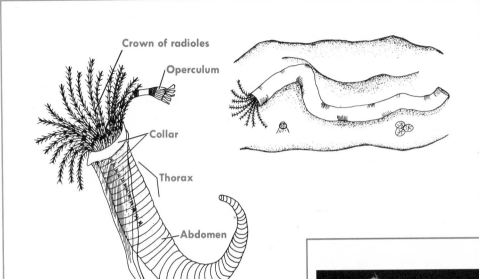

Crown of radioles

Operculum

Collar

Thorax

Abdomen

FIG. 29-12. *Eupomatus dianthus,* a sedentary tubeworm. Glands beneath collar produce calcium carbonate secretions, which collar molds into place at end of tube. Operculum serves as plug in tube when head is withdrawn. The calcareous tube (right) is usually cemented to rock or shell.

FIG. 29-13. Clam worm *Nereis virens.* Note specialized head (at top) and paired parapodia on each segment. (Preserved specimen.)

more marked differentiation of some body somites, and a greater specialization of sensory organs than do the oligochaetes.

The clam worm *Nereis (Neanthes) virens* (sandworm) lives in tidal water in mucus-lined burrows in mud or sand, usually venturing out only at night to search for food. The body, 10 to 15 inches long, may have as many as 200 somites (Fig. 29-13). Along the sides of the body are the two-lobed parapodia supported by chitinous rods (cirri) and bearing numerous setae. Covering the body is a cuticle and epidermis, underneath which are circular and longitudinal muscles.

Sexes are separate but the reproductive organs are not permanent. Sex cells are budded off from the coelomic lining and carried out by ducts or by bursting through the body wall. Only during the breeding season do these worms venture far from their burrows. During this period there may be a radical change in the body form, called **epitoky,** in which there is a differentiation in color and structure between the anterior region (**atoke**) and the posterior region (**epitoke**) of the body. In the epitokal region the segments become swollen with gametes, and the parapodia and setae enlarge and become adapted for swimming. Usually at a particular phase of the moon, the epitokes leave the rest of the worm and rise in great swarms to the surface of the water where they discharge their gametes. In the case of the palola worm *Eunice viridis,* which lives among coral reefs of the South Pacific, this swarming occurs on the first day of the last quarter of the October-November moon. The surface of the sea is usually covered with the epitokes of these worms that burst just as the sun rises. Fertilization occurs at this time. The fertilized eggs develop into free-swimming larvae that later transform into worms.

CLASS CLITELLATA

The Clitellata include those annelids that bear a clitellum. The clitellum is a thickened saddlelike portion of certain segments that is involved in copulation and in the production of cocoons. This group includes the oligochaetes, in which segmentation is usually conspicuous and the setae are present but few in number, and the leeches, Hirudinea, in which the somites lack setae and are marked by transverse grooves called annuli that produce an appearance of more segments than there really are.

Common earthworm (order Oligochaeta)

Earthworms *(Lumbricus terrestris)* are almost worldwide in distribution, preferring moist, rich soil for their burrows. Much of the soil from their burrows passes through the digestive tract and is left as castings at the burrow entrance. At night they emerge to explore their surroundings ("night crawlers"), often keeping the tail in the burrow in order to escape rapidly if disturbed. During dry weather they may coil up in a slime-lined chamber several feet underground and pass into a state of dormancy.

The food of the earthworm is mainly decayed organic matter and bits of vegetation drawn in by the action of the muscular pharynx and liplike prostomium. The calcium from the soil swallowed with food tends to produce a high blood calcium level. To counteract this, calciferous glands along the esophagus secrete calcium ions into the gut to reduce the calcium ion concentration of the blood. Food is stored temporarily in a thin-walled **crop** before being passed on to a thick muscular **gizzard** for grinding up (Fig. 29-14). Digestion and absorption occur in the long intestine. From the intestine food products are absorbed into the blood and carried throughout the body for assimilation.

The **circulatory system** of annelids is a "closed system" of blood vessels and capillaries that ramify to all parts of the body. The dorsal vessel above the alimentary canal (Fig. 29-14) functions as the true heart, pumping the blood forward.

The blood is a liquid plasma containing colorless ameboid corpuscles. Dissolved in the plasma is the pigment hemoglobin, which gives the blood its red color and aids in the transportation of oxygen for respiration.

The excretory organs are the paired **nephridia** found in all but the first three and the last segments (Fig. 12-6, p. 149). **Respiration** is carried on in the moist skin where blood capillaries are numerous.

The **central nervous system** of the earthworm is made up of a brain (Fig. 29-14) above the pharynx, a pair of connectives that run around the pharynx, and a ventral **nerve cord** (really double) that bears a pair of fused ganglia in each somite. The ganglia give off lateral nerves to the body structures.

For rapid escape movements the nerve cord of an earthworm is provided with three giant fibers (neurochords) which run the length of the cord. Each giant fiber is made up of axons contributed by nerve cells in each segment. The axons are fused end-to-end at the intersegmental synapses, allowing rapid transmission of impulses because of the direct one-to-one relation.

The reproductive system of the earthworms is monoecious, that is, each worm has a full set of male and female reproductive organs. The earthworms exchange sperm during copulation, which occurs usually at night. When mating, the worms extend their anterior ends from their burrows and bring their ventral surfaces together (Fig. 29-15). Copulation requires about 2 hours. Later each worm

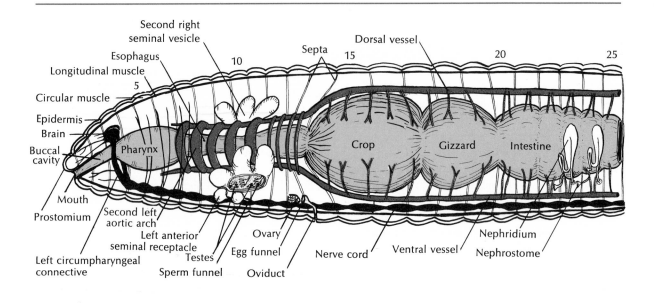

FIG. 29.14. Chief internal features of anterior portion of earthworm, as shown by removal of left body wall. For greater clarity nephridia are shown in only two segments.

FIG. 29-16. Slime tube and cocoon of earthworm after being cast from body. Each copulating earthworm secretes a slime tube around its segments 9 to 36. Sperm are exchanged, and then cocoons are secreted over clitellum within slime tube. Eggs, albumin, and sperm are deposited in cocoon, then worm backs out of slime tube and cocoon. When cocoon is free, its ends close by constriction of slime tubes. In some species many eggs develop in each cocoon; in *Lumbricus,* only one.

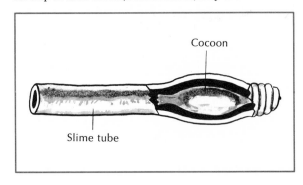

FIG. 29-15. Two earthworms in copulation. Anterior ends point in opposite directions as their ventral surfaces are held together by mucus bands secreted by clitellum. (Courtesy Guy Carter.)

410

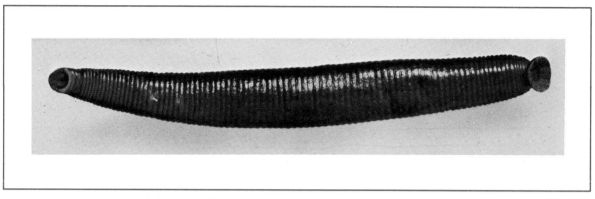

Fig. 29-17. *Hirudo medicinalis,* medicinal leech. This form was once widely used in bloodletting.

secretes a barrel-shaped cocoon about its clitellum within the posterior end of the slime tube. Eggs are passed into the cocoon, then the worm backs out allowing the slime tube and cocoon to be slipped forward over its head, adding the stored sperm to it as it goes. Fertilization occurs within the cocoon. When the cocoon leaves the worm, its ends close. (Fig. 29-16). Cocoons are deposited in the earth or at the entrance to the burrow. In *Lumbricus* cocoons only one of the several fertilized eggs in a cocoon develops into a worm; the others act as nurse cells. The juvenile worm escapes from the cocoon in 2 to 3 weeks.

Leeches (order Hirudinea)

Leeches are well adapted for bloodsucking, for they have anterior and posterior suckers for locomotion and attachment and chitinous jaws for making incisions in the skin to promote the flow of blood from their victims. Leeches are found both on land and in water. They move in a measuring worm fashion, by looping movements of the body.

There are many different species of leeches. Some of them are jawless but have an eversible proboscis

with which they can pierce the body of another animal. Fish and turtles are parasitized by them. Others live mainly on dead animals, and some are predaceous. Tropical countries are plagued by more leeches than are temperate countries.

For centuries the medicinal leech *Hirudo medicinalis* (Fig. 29-17) was used in medical practice for bloodletting in the mistaken belief that most disorders were caused by a "plethora" of blood. Since the medicinal leech was 4 or 5 inches long and could extend to a much greater length when distended with blood, the amount of blood it could suck from a patient was considerable. Leech collecting and leech culture in ponds were practiced in Europe on a commercial basis during the nineteenth century.

When the medicinal leech fastens itself to another animal by means of its suckers, the sharp chitinous teeth make an incision in the skin, and at the same time an anticoagulant substance (hirudin) from the salivary glands is introduced into the wound to make the blood flow freely. Then the blood is sucked up by the muscular pharynx and stored in the large crop. A leech is said to be capable of sucking up three times its own weight in blood, which may take several months to digest.

30

Arthropods

The arthropods are the most extensive phylum in the animal kingdom, with between 700,000 and 800,000 named species and many more yet to be identified. It includes crustaceans, spiders, ticks, millipedes, centipedes, and insects. Many fossil forms go back to Precambrian times.

Arthropods are characterized by having a chitinous exoskeleton and a linear series of somites, each with a pair of jointed appendages. Their body organs and systems are well developed, and they share with the nematodes an almost complete absence of cilia.

The arthropods are more closely related to annelids than to any other group. They probably came from the same ancestors as did segmented worms. They have been guided in their adaptive radiation mainly by the modifications and specializations of their exoskeletons and jointed appendages.

PHYLUM ARTHROPODA

Arthropods belong to the protostome branch of the animal kingdom, along with the mollusks and annelids.

CLASSIFICATION

Subphylum Trilobita (tri″lo-bi′ta) — **trilobites.** All extinct forms; Cambrian to Carboniferous; body divided by two longitudinal furrows into three lobes; head, thorax, abdomen distinct.

Subphylum Chelicerata (ke-lis″e-ra′ta) — **eurypterids, horseshoe crabs, spiders, ticks.** First pair of appendages modified to form chelicerae with claws; pair of pedipalps and four pairs of legs; no antennae; no mandibles; cephalothorax and abdomen usually segmented.

Class Merostomata (mer″o-sto′ma-ta). Cephalothorax; compound lateral eyes; appendages with gills; sharp telson.

Subclass Eurypterida (u″rip-ter′i-da) — **eurypterids.** Extinct; cephalothorax covered by dorsal carapace; abdomen with twelve segments and postanal telson.

Subclass Xiphosurida (zif″o-su′ri-da) — **horseshoe crabs.** Cephalothorax with convex, horseshoe-shaped carapace; abdomen unsegmented and terminated by spine; book gills.

Class Pycnogonida (pik″no-gon′i-da) — **sea spiders.** Small (3 to 4 mm.), but some reach 500 mm.; body chiefly cephalothorax; abdomen tiny; usually eight pairs of long walking legs.

Class Arachnida (ar-ack′ni-da) — **scorpions, spiders, mites, ticks.** Four pairs of legs; abdomen segmented or unsegmented with or without appendages and generally distinct from cephalothorax; respiration by gills, tracheae, or book lungs.

Subphylum Mandibulata (man-dib″u-la′ta). One or two pairs of antennae form first two pairs of appendages, and functional jaws (mandibles) form third pair of cephalic appendages.

Class Crustacea (crus-ta′she-a) — **crustaceans.** Aquatic with gills; body with dorsal carapace; telson at posterior end; hard exoskeleton of chitin; appendages biramous; head of five segments with two pairs of maxillae.

Class Diplopoda (di-plop′ o-da) — **millipedes.** Body subcylindric; head with short antennae and simple eyes; body with variable number of somites; short legs, usually two pairs to a somite.

Class Chilopoda (ki-lop′ o-da) — **centipedes.** Form elongated and dorsoventrally flattened; variable number of somites, each with a pair of walking legs; pair of long antennae, jaws, and pair of maxillae.

Class Pauropoda (pau-rop′ o-da) — **pauropods.** Minute (1 to 1.5 mm.); cylindric body of double segments and bearing nine or ten pairs of legs; no eyes.

Class Symphyla (sym′ fy-la) — **garden centipedes.** Slender (1 to 8 mm.) with long, filiform antennae; body of fifteen to twenty-two segments with ten to twelve pairs of legs; no eyes.

Class Insecta (in-sek′ ta) — **insects.** Body with head, thorax, and abdomen distinct and usually marked constriction between thorax and abdomen; pair of antennae; mouthparts modified for different food habits; head with six somites; thorax with three somites, and abdomen with variable number, usually eleven somites; thorax with two pairs of wings (sometimes one pair or none) and three pairs of jointed legs; metamorphosis gradual or abrupt.

CHARACTERISTICS

1. Symmetry bilateral; body metameric
2. **Appendages jointed,** with one or two pairs to a somite and often modified for specialized functions
3. **Exoskeleton of chitin** secreted by the underlying epidermis and shed at intervals
4. Body often divided into **three regions:** the **head,** usually of six somites, the **thorax,** and the **abdomen,** the latter two divisions having a variable number of somites; head and thorax often united into a cephalothorax
5. True coelom small in adult; most of body cavity a **hemocoel filled with blood**
6. Digestive system complete with mouth, enteron, and anus; **mouthparts modified from somites and adapted for different methods of feeding**
7. Circulatory system open, with dorsal heart, arteries, and mesenchymal blood cavities (sinuses)
8. **Cilia practically absent throughout group**
9. Respiration by body surface, gills, **air tubes (tracheae),** or **book lungs**
10. Excretory system of green glands or of a variable number of **malpighian tubules** opening into the digestive system
11. Nervous system of dorsal brain connected by a ring around the gullet to a double nerve chain of ventral ganglia; sensory organs well developed
12. Sexes nearly always separate, with paired reproductive organs and ducts; fertilization internal; ovip-

arous or ovoviviparous; metamorphosis direct or indirect; parthenogenesis in a few forms

WHY ARTHROPODS HAVE BEEN SO SUCCESSFUL

Some of the criteria for the success of an animal group are number and variety of species, variety of habitats, widespread distribution, ability to defend themselves, variety of food habits, and power to adapt themselves to changing conditions. Arthropods are so diversified that they have met most of these requirements. Some structural and physiologic patterns that have been helpful to them are briefly summarized here.

A highly versatile chitinous covering. Chitin is a nonliving, noncellular protein-carbohydrate compound secreted by the underlying epidermis (Fig. 7-1, *A*, p. 84). It is made up of an outer waxy layer, a middle horny layer, and an inner flexible layer. The chitin may be soft and permeable, or it may form a veritable coat of armor. Between joints it is flexible and thin to permit free movements. In crustaceans it is infiltrated with calcium salts. In general, it is admirably adapted for attachment of muscles, serving as levers and centers of movement; it prevents the entrance and loss of water; and it affords the maximum of protection without sacrificing mobility. It is also used for biting jaws, as grinders in the stomach, as lenses for the eye, in sound protection, sensory organs, and copulatory organs, and for ornamental purposes — in all, chitin is a remarkably versatile material.

A chitinous exoskeleton does, however, limit the size of the animal. In order to grow it must shed its outer shell at intervals and grow a larger one — a process called **ecdysis,** or molting. Arthropods molt from four to seven times before reaching adulthood. Few arthropods exceed 2 feet in length, and most are far below this limit. The largest is the Japanese crab *Macrocheira,* which has about an 11-foot span; the smallest is the parasitic mite *Demodex,* which is less than 0.1 mm. long.

Segmentation and appendages for greater efficiency. Arthropods share with annelids and vertebrates the characteristic of segmentation. Typically each somite is provided with a pair of jointed appendages, but this arrangement is often modified, with both segments and appendages specialized for adaptive functions. This has made for greater efficiency and wider capacity for adjustment to different habitats.

Air piped directly to cells. Aquatic arthropods breathe mainly by some form of gill that is quite efficient; most land arthropods have the highly efficient **tracheal system** of air tubes that delivers oxygen directly to the tissues and cells and makes a high rate of metabolism possible.

Highly developed sensory organs. Sense organs are found in great variety, from the mosaic eye to those simpler senses that have to do with touch, smell, hearing, balancing, chemical reception, etc. Arthropods are keenly alert to what goes on in their environment.

Complex behavior patterns. Arthropods exceed most other invertebrates in the complexity and organization of their activities. Whether learning is involved to any great extent in their reactions to environmental stimuli is still open to question, but no one can deny the complex adaptibility of the group.

ECONOMIC IMPORTANCE

Arthropods serve as an important source of food for man and other animals. Lobsters, crabs, shrimp, and crayfish are highly esteemed all over the world. Plankton contains many crustaceans and is food for fish and other aquatic animals. Insects are important as food for many birds and other land vertebrates.

Some insects are predators that live on other insects, helping to keep the harmful ones in check. Arthropod products include shellac produced by scale insects, cochineal (a dye) from other scale insects, silk spun by silkworm larvae, honey and beeswax from bees, and so on. Bees are essential for pollination of many fruit and cereal crops.

Many arthropods are harmful. Insects destroy millions of dollars worth of food each year. Arthropods carry devastating diseases: a mosquito carries malaria, copepods carry larval stages of guinea worm and fish tapeworm, and mites and ticks carry diseases and also live as ectoparasites. Some spiders and scorpions deliver poisonous bites; barnacles

foul ship bottoms, and sow bugs and other insects damage gardens and greenhouse crops.

TRILOBITES (SUBPHYLUM TRILOBITA)

The trilobites probably had their beginnings millions of years before the Cambrian period in which they were the most numerous inhabitants of the seas. They have been extinct some 200 million years. Their Precambrian ancestor probably also gave rise to all other arthropods as well. Most trilo-

FIG. 30-1. Trilobite (dorsal view) from plaster cast impression. All members of this class are now extinct. Some of the abundant fossils of this group may be the remains of molted exoskeletons.

bites could roll up like pill bugs and were from 2 to 27 inches long (Fig. 30-1).

CHELICERATES (SUBPHYLUM CHELICERATA)

The chelicerate arthropods are an ancient group that includes the euripterids (extinct), king crabs, spiders, ticks, and some others. Their six pairs of appendages include a pair of chelicerae, a pair of pedipalps, and four pairs of walking legs.

KING CRABS (SUBCLASS XIPHOSURIDA)

The king crabs are a very ancient group going back to the Cambrian period. Our common horseshoe or king crab *Limulus* (Fig. 30-2), often called a "living fossil," goes back to the Triassic period. There are only three living genera (five species) today.

Horseshoe crabs have an unsegmented, horseshoe-shaped carapace (hard dorsal shield) and a broad abdomen, which has a long spinelike telson, or tailpiece. They swim awkwardly by means of the abdominal plates and walk on their walking legs. They feed at night on worms and small mollusks. Horseshoe crabs are harmless to man, although pests to clam and oyster fishermen.

SPIDERS, SCORPIONS, AND TICKS (CLASS ARACHNIDA)

The arachnids include the spiders, scorpions, ticks, mites, harvestmen, and some others. They lack antennae and mandibles and have the head and thorax fused into a cephalothorax. The **chelicerae,** or first pair of appendages, have terminal fangs, usually provided with ducts from poison glands. The **pedipalps,** or second pair, are used in chewing and by the males for transfer of sperm. Arachnids have four pairs of walking legs instead of three pairs as in insects. Most of them are free living, and they are highly predaceous, sucking the fluids and soft tissues from the bodies of their prey.

Most arachnids are harmless and are actually beneficial because they destroy huge quantities of harmful insects. A few spiders can inflict dangerous bites, and the sting of the scorpion may be quite painful. Some ticks and mites are vectors of diseases, as well as causing annoying irritations. Some mites are damaging to plants and fruits.

More than 35,000 species of **spiders** are distributed over the world and found in most kinds of habitats. All are predaceous. Some chase their prey, others ambush them, but most of them spin a net or web in which to trap their prey (Fig. 30-3). **Spinnerets** on the ventral side of the abdomen contain openings from the silk glands. Orb weavers use inelastic silk to form a framework, then an elastic sticky silk for the spiral threads that entangle the prey. The spider punctures its prey with its fangs, then alternately injects digestive fluid through the puncture and sucks up the dissolved tissues until the prey has been sucked dry. Although many spiders have poison fangs, very few have poisonous

FIG. 30-2. Ventral view of horseshoe crab *Limulus* (class Merostomata). They grow up to 18 inches long.

FIG. 30-3. Web of orb-weaving spider. Most spiders bear three pairs of spinnerets on posterior ventral surface of abdomen. From silk glands inside abdomen, liquid is forced out of spigots of spinnerets under pressure and hardens on contact with air. Each of four or five different kinds of glands secrete different kinds of silk. Starting with a "bridge thread," spider makes framework of dry silk-glued threads, usually an irregular four- or five-sided figure. Radii of dry silk are next spun and held at center by platform of spiral dry silk. Then, working inward from rim spider spins the functional spiral of sticky threads, to hub, being careful to walk on nonsticky radii while doing so.

bites that are harmful to man. Most spiders will fight and bite only when tormented or when they are defending their young or their egg sacs. The black widow spider *Latrodectus mactans* (Fig. 30-4), however, can give severe or even fatal bites, and the brown recluse spider *Loxosceles* is even more dangerous.

Scorpions (Fig. 30-5) have been reported from at least thirty of our states. Their habitat is in trash piles, around dwellings, and in burrows of desert regions. Most of them will burrow and are most active at night when they seek their prey—insects, spiders, etc. Of the more than forty American species, only two are said to be dangerous to man; both belong to the genus *Centruroides*. Their venum affects the nervous system, whereas the bite of most scorpions produces only a painful swelling. Scorpions bring forth living young that are carried on

FIG. 30-4. This black widow spider *Latrodectus,* suspended on her web, has just eaten larger cockroach. Note "hourglass" marking (orange colored) on ventral side of abdomen.

FIG. 30-5. Tropical blue scorpion *Centrurus,* which is common in Cuba. Note terminal poison stinger. (Preserved specimen.)

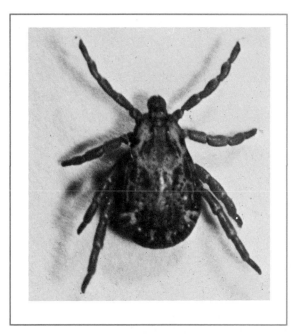

FIG. 30-6. Wood tick *Dermacentor.* One species (*D. andersoni*) transmits Rocky Mountain spotted fever and tularemia, as well as producing tick paralysis.

the back of the mother. The popular superstition that they feed upon her tissues is false.

Mites and **ticks** are arachnids in which the cephalothorax and abdomen are fused into an unsegmented ovoid with eight legs. Mites and ticks are found almost everywhere—in both fresh and salt water, on vegetation, on the ground, and in animals (parasitic). There are about 15,000 known species, a great many of which are of direct importance to humans.

Ticks feed upon the blood of vertebrates. They pierce the skin and suck up blood till enormously distended. Eggs, laid on the ground, hatch into larvae that climb on vegetation to await a host passing by. Texas cattle fever and Rocky Mountain spotted fever are both transmitted by ticks (Fig. 30-6).

The itch mite burrows into the skin where the female lays her eggs. Another species of mite is responsible for mange in dogs and other domestic animals. Chiggers, or red bugs, are the larval forms of red mites, which burrow under the skin to feed.

MANDIBULATES (SUBPHYLUM MANDIBULATA)

The mandibulate arthropods have as head appendages one or two pairs of antennae, a pair of mandibles (jaws), and one or two pairs of maxillae as food handlers. There are no chelicerae or pedipalps. The mandibulates include the crustaceans, millipedes, centipedes, myriapods, and insects.

CRUSTACEANS (CLASS CRUSTACEA)

This class includes some 30,000 species of lobsters, crayfish, shrimp, crabs, water fleas, copepods, sow bugs, wood lice, barnacles, and a few others. Crustaceans are such a highly diversified group that it is difficult to describe them briefly. An idea of their enormous range in body form is gained from Figs. 30-7 and 30-8. Their protective body coverings, chitin impregnated with lime, are much harder than those of most other arthropods. Their segmented nature is usually quite obvious, and the segments usually number between 16 and 20, although they

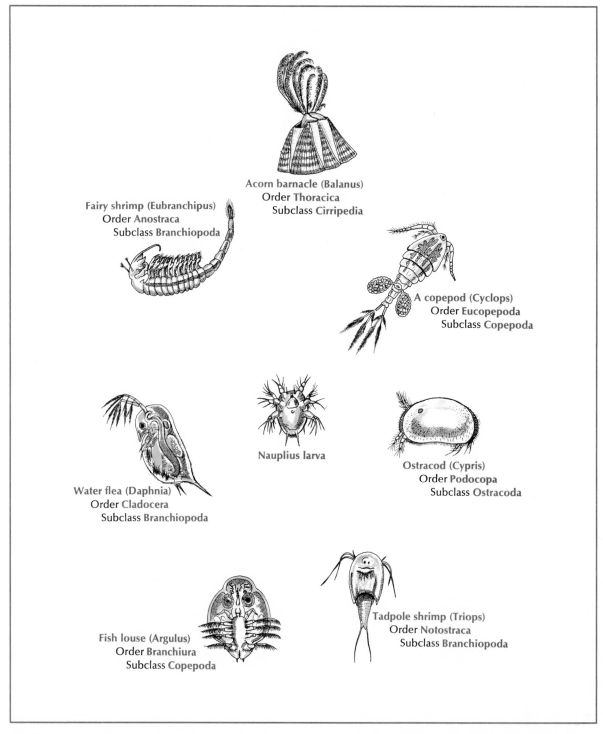

FIG. 30-7. Group of smaller crustaceans. Orders Anostraca and Notostraca live exclusively in fresh water; order Thoracica is exclusively marine; and orders Branchiura, Cladocera, Eucopepoda, and Podocopa are found in both fresh water and marine water. Nauplius larva is common to group.

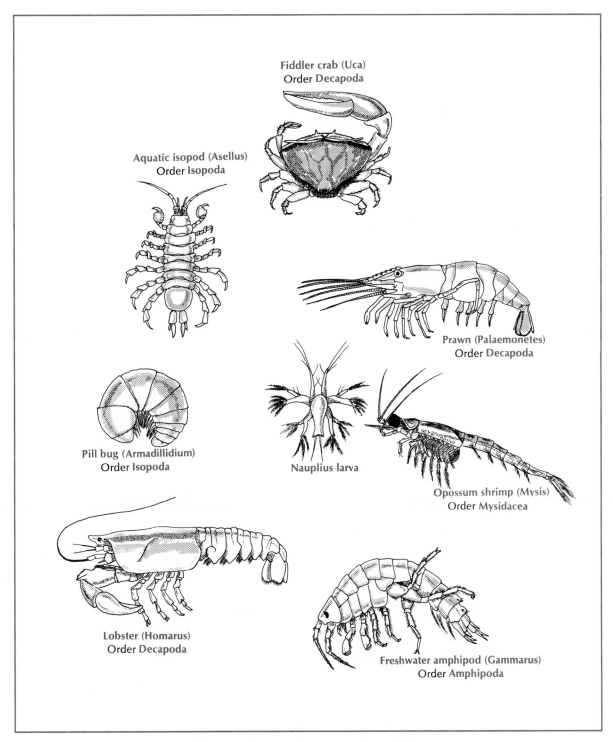

FIG. 30-8. Larger crustaceans (subclass Malacostraca). All members of this subclass have abdominal appendages, gastric mill, eight-segmented thorax, and typical body of nineteen segments. (Not drawn to scale.)

may vary from few to 60 or more. The appendages are often highly specialized and may be adapted for sensory perception, food handling, fighting, walking, swimming, copulation, protection of eggs and young, and so on. Crustaceans have compound eyes similar to those of insects, although they are often borne on movable stalks.

Although the edible crab *Cancer* is immortalized as a constellation and in astrology, most people see so few crustaceans in their normal lives that they are unaware of their abundance and importance.

The small crustaceans, called the microcrustaceans (Fig. 30-7), have been referred to as the "insects of the sea," although they are by no means restricted to the marine environment; they are found in fresh water and moist land habitats as well. But in the sea the countless tiny free-swimming crustaceans are crucial energy-converters of the **plankton.** The position of the microcrustaceans in the food chain is between the minute plant life and the filter-feeding fishes as well as other large oceanic animals. Without the marine microcrustaceans the existence of advanced forms of life in the ocean in any significant amount would be impossible.

Lobsters, crabs, and shrimp (Fig. 30-8) are highly esteemed as food for man, and the "shellfish" fishery is an economically important one. Unfortunately it is especially vulnerable to man's polluting activities in coastal areas, since the sensitive larval stages live in the coastal estuaries.

As with virtually all animal groups, some representatives conflict with man's interests and consequently are labeled pests. Barnacles foul the bottom of ships. Certain boring crustaceans are damaging to wharves and pilings. Some carry disease in tropical countries, some are parasitic on man's food, and some destroy man's crops, especially rice.

CENTIPEDES (CLASS CHILOPODA)

The centipedes are active land forms with a preference for moist places, such as under logs or stones, where they feed upon earthworms, insects, etc. Their bodies are somewhat flattened dorsoventrally and may contain from a few up to 177 somites (Fig. 30-9). Each somite, except the one behind the head and the last two, bears one pair of jointed appendages. Those of the first body segment are modified to form poison claws, which they use to kill their prey. Most species are harmless to man. The common house centipede *Cermatia*, with fifteen pairs of legs, is often seen around bathrooms and damp cellars, where they catch insects.

FIG. 30-9. Centipede *Scolopendra*. Class Chilopoda. Most segments have one pair of appendages each. First segment bears pair of poison claws, which in some species can inflict serious wounds. Some tropical forms are nearly a foot long. Centipedes are carnivorous and prey upon earthworms, insect larvae, and even larger prey.

MILLIPEDES (CLASS DIPLOPODA)

The name millipede literally means thousand-legged (Fig. 30-10). Although they do not have that many legs, they do have many appendages in proportion to their length. Their cylindric bodies are made up of from 25 to 100 somites. Each of the four thoracic somites have one pair of appendages, but the remaining somites have two pairs each.

Millipedes are less active than centipedes; they are herbivorous, rather than carnivorous, living on decayed plant and animal matter, and sometimes living plants. They prefer dark moist places under stones or logs. When disturbed they often roll up into a ball. Common examples are *Spirobolus* and *Julus*.

INSECTS (CLASS INSECTA)

The insects are the most successful biologically of all the groups of arthropods. Although they comprise only one class out of more than sixty classes of animals, there are more species of insects than all the others combined. It is estimated that the recorded number of insect species is about 800,000, with thousands of other species yet to be discovered and classified. Examples of the orders of insects are shown in Fig. 30-11.

The science of entomology occupies the time and resources of many skilled men all over the world.

The struggle between man and his insect competitors seems to be an endless one, yet insects are interwoven into our economy in so many useful roles that man would have a difficult time without them.

Distribution. Insects have spread into practically all habitats that will support life, except the sea. They are found in fresh water, soils, forests, deserts, and wastelands, on mountain tops, and as parasites in and on the bodies of plants and animals.

The wide distribution of insects is made possible by their powers of flight and their highly adaptable nature. Their small size allows them to be carried afar by currents of wind and water. Their well-protected eggs can be carried by birds and other animals. Their agility and agressiveness enable them to fight for every possible niche in a location.

Insects are well adapted to dry and desert regions because the chitinous exoskeleton prevents evaporation and because insects extract fluid from food and fecal material, as well as utilize moisture from the water by-products of metabolism.

Food habits. The majority of insects feed upon plant juices and plant tissues (**phytophagous**). Some insects will restrict their feeding to certain varieties of plants; others, such as the grasshoppers, will eat almost any plant. The caterpillars of many moths and butterflies will eat the foliage of only certain plants. Monarch butterflies are poisonous to birds because their caterpillars assimilate cardiac

FIG. 30-10. Millipede *Spirobolus*. Class Diplopoda. They have two pairs of jointed appendages on each of their segments except the four thoracic segments and sometimes one or two tail segments, which have one pair each. Long abdomen is comprised of double segments, which accounts for two pairs of appendages on each apparent somite. In contrast to centipedes, millipedes are usually vegetarian animals. (Courtesy Carolina Biological Supply Co., Burlington, N. C.)

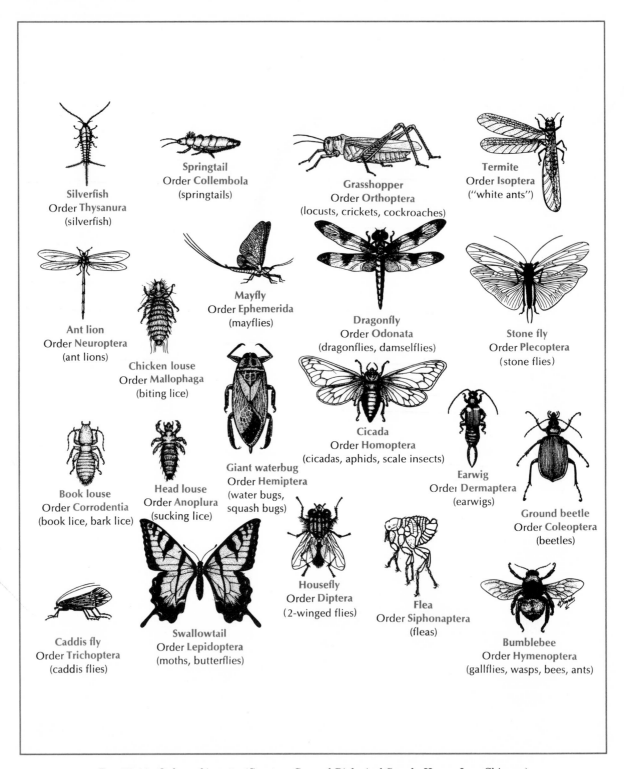

FIG. 30-11. Orders of insects. (Courtesy General Biological Supply House, Inc., Chicago.)

glycosides from a certain species of milkweed. Certain ants are known to maintain fungus gardens as a source of food. Many beetles and insect larvae live upon dead animals (**saprophagous**). Others are **predaceous.**

Many insects are parasitic. Fleas, for instance, live on the blood of mammals, and the larvae of many varieties of wasps live upon spiders and caterpillars. In turn, many are parasitized by other insects.

Protection and coloration. Even in the same species the color pattern may vary seasonally, or there may be sexual color differences. Color patterns are probably highly adaptive, such as protective coloration, warning coloration, and mimicry (Fig. 30-12).

The chitinous exoskeleton affords protection for many; others, such as stink bugs, have repulsive odors and taste; and others protect themselves by aggressiveness or by swiftness in running from danger. Bats can detect their prey (often moths) by echolocation; certain moths have evolved special ultrasonic ears (tympanic membranes) by which they can pick up the bat chirps and thus evade capture.

Sound production and reception. Some sounds are the result of the rubbing together of rough surfaces. Grasshoppers rub the femur of the last pair of legs over the ridges on the forewings. Male crickets scrape their wing covers to produce their characteristic chirping. The hum of mosquitoes is due mainly to the rapid vibration of their wings. The long, drawn-out sound of the cicada is produced in a special chamber between the thorax and abdomen in which the rapid contractions of a muscle cause a membrane to vibrate at different pitches. Sound may be a means of communication by which insects are able to warn of danger, call to their mates, etc. Sound waves may be picked up by the tympanic membrane or the antennae.

Reproduction. The sexes are always separate in insects, and fertilization is internal. Most insects are **oviparous,** but a few are **viviparous** and bring forth their young alive. **Parthenogenesis** occurs in aphids, gall wasps, and others. The queen honeybee may lay more than a million eggs during her lifetime; on the other hand, some viviparous flies bring forth a single young at a time. Forms that make no provision for their young produce more eggs than those that have to provide for their larvae.

Metamorphosis and growth. Insects illustrate metamorphosis more dramatically than any other group. The transformation of the hickory horned devil into the beautiful royal walnut moth is indeed an astonishing change.

Most insects (about 88%) undergo a **complete metamorphosis** that separates the physical processes of growth (**larva**) from those of differentiation (**pupa**) and reproduction (**adult**) (Fig. 30-13). This type of metamorphosis is often called **holometabo-**

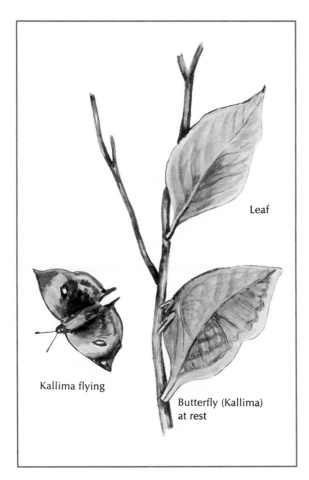

Leaf

Kallima flying

Butterfly (Kallima) at rest

FIG. 30-12. Striking case of protective resemblance in butterfly *Kallima,* which resembles leaf when perched on twig. This butterfly is native of East Indies and was first described by famous English naturalist Alfred Russell Wallace.

Fig. 30-13. Stages in the metamorphosis of the angulifera moth *Callosamia angulifera.* **A,** Caterpillar (larva). **B,** Pupa. **C,** Adult.

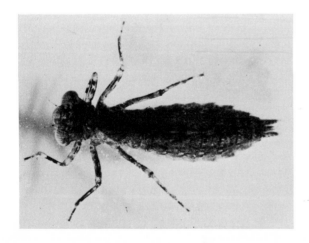

Fig. 30-14. Dragonfly naiad. Order Odonata. Found in bottom of pools and streams.

lous. Each stage functions efficiently without competition with the other stages, for the larvae often live in entirely different surroundings and eat different food from the adults. The wormlike larvae are known as caterpillars, maggots, bagworms, fuzzy worms, grubs, etc. After a series of growth or **instar** stages the larva forms a case or cocoon by spinning silk threads about itself, and within the cocoon goes into a resting period, the pupa (chrysalis). The final metamorphosis occurs within the cocoon and the winged adult emerges full grown. The stages, then, are egg, larva, pupa, and adult. The adult undergoes no further molting. The hormonal control of molting of a butterfly was described on p. 191 and illustrated in Fig. 14-10.

Some insects, such as mayflies, dragonflies, and stoneflies, undergo a type of partial metamorphosis (**hemimetabolous**). The eggs are laid in water

Fig. 30-15. **A,** Young praying mantids (nymphs) emerging from their egg capsule. Egg capsules (oothecae) are glued to shrubbery and other objects in late summer and fall. When eggs hatch in spring, enormous swarm of wingless nymphs emerges from single capsule. **B,** Praying mantis gets its name from the way it often holds its forelimbs but it is far more interested in preying on other insects than in pious devotions. Both views are about life size. Order Orthoptera.

and develop into aquatic **naiads** (Fig. 30-14), which have tracheal gills or other modifications for an aquatic life. They grow by successive molts, crawl out of the water, and, after the last molt, become winged adults. The stages are egg, naiad, and adult.

Some other insects, such as grasshoppers and true bugs, have a form of gradual metamorphosis (**paurometabolous**) in which immature forms called nymphs are terrestrial and resemble the adult in appearance, except that they are wingless and of a different body proportion (Fig. 30-15). At each molt the nymph looks more like the adult, the wings developing in later stages. The stages here are egg, nymph, and adult.

A few insects, such as silverfish and springtails, undergo no metamorphosis at all (**ametabolous**). The young, or **juveniles,** are like the adult except in size. The stages are egg, juvenile, and adult.

Social instincts. Some insects, such as bees, ants, and termites, exhibit complicated patterns of social instincts and have worked out complex societies involving division of labor. In these societies

FIG. 30-16. A, Bald-faced hornet *Vespula maculata.* **B,** Paper nest of bald-faced hornet *Vespula maculata,* cut open to show one of the horizontal combs. These nests are attached to bushes or trees and are comprised of fibers of weather-worn wood. Larvae are reared in comb cells. Order Hymenoptera.

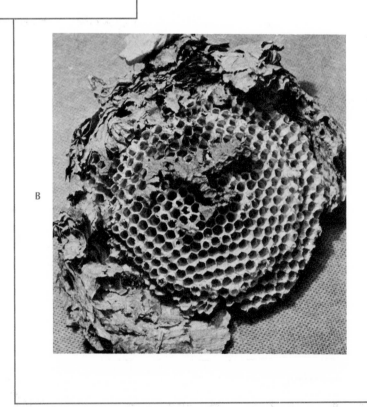

the adults of one or both sexes live together with the young in a cooperative manner. Among the bumblebees the groups are small and the groupings last only a season, but in honeybees as many as 60,000 to 70,000 bees may be found in a single hive. Of these, there are a single **queen,** a few hundred **drones** (males), and the rest are **workers** (infertile females). The workers gather nectar from flowers, manufacture honey, collect pollen, secrete wax, take care of the young, and ventilate and guard the hive. Each worker appears to do a specific task in all this multiplicity of duties. Their life span is only a few weeks. One drone (sometimes more) fertilizes the queen; from this single mating enough sperm are stored in her spermatheca to last a lifetime. Drones are usually driven from the hive or killed by the workers at the end of the summer. A queen may live as long as five seasons, during which time she may lay a million eggs. She is responsible for keeping the hive going during the winter, and only one reigning queen will be tolerated in a hive at one time.

Honeybees have evolved an efficient system of communication by which their scouts inform the workers of the location and quantity of food sources. (See Animal communication, p. 230.)

The social vespids, or paper wasps (Fig. 30-16), also have a caste system of queens, workers, and drones. They construct a nest of papery material consisting of wood or foliage chewed up and elaborated by the wasps.

Termite colonies contain two main castes—the fertile individuals, both males and females, and the infertile individuals. Some of the reproductive individuals may have wings and may leave the colony, mate, lose their wings, and start a new colony. Reproductive individuals without wings may, under certain conditions, substitute for the king and queen. The sterile members are wingless and become workers and soldiers. Within the castes, there are also different types. These caste differentiations are not due to genetic differences but to extrinsic factors. Reproductive individuals and soldiers secrete **ectohormones** containing inhibiting substances that are passed to the nymphs through a mutual feeding process called **trophallaxis.** This causes them to become sterile workers. In large populations some of the nymphs are not so inhibited and these

differentiate into the female individuals. The phenomenon of trophallaxis, or exchange of nutrients, appears to be common among all social insects because it serves to integrate the colony. The process involves the feeding of the young by the queen and workers, which in turn may receive a drop of saliva from the young. It may also involve mutual licking, shampooing, etc.

Relation to man's welfare. Insects are both beneficial and harmful to man's interests. Some **beneficial insects** (Fig. 30-17) produce useful products, such as honey and beeswax from bees and silk from the silkworms. Some 25,000 cocoons are necessary to make a single pound of silk, and 50 million pounds of silk are produced annually. Shellac is made from a wax secreted by the lac insects of the family Coccidae.

Insects are necessary to cross-fertilize fruits and crops. Bees, for example, are indispensible in raising fruits, clover, and other crops. Orchard owners in Washington and Oregon are having to import millions of honeybees from California to pollinate the fruit crops because the heavy use of pesticides has killed nearly all the native bees.

Insects and higher plants formed an early relationship of mutual adaptations that have been to each other's advantage. Insects exploit flowers for food, and flowers exploit insects for pollination. Each floral development of petal and sepal arrangement is correlated with the sensory adjustment of certain pollinating insects. Among these mutual adaptations are amazing devices of allurements, traps, specialized structures, precise timing, etc.

Many predaceous insects, such as tiger beetles, aphid lions, ant lions, praying mantes, and ladybird beetles, destroy other insects harmful to man. Some insects control harmful ones by parasitizing them and laying their eggs where their own young, when hatched, may devour the host. Dead animals are quickly taken care of by maggots hatched from eggs laid in the carcasses. Finally, insects and their larvae serve as an important source of food for birds, fish, and other animals.

Harmful insects (Fig. 30-18) include those that eat and destroy plants and fruits, such as grasshoppers, chinch bugs, corn borers, cotton-boll weevils, grain weevils, San Jose scales, and scores of others. Some insects are harmful to domestic ani-

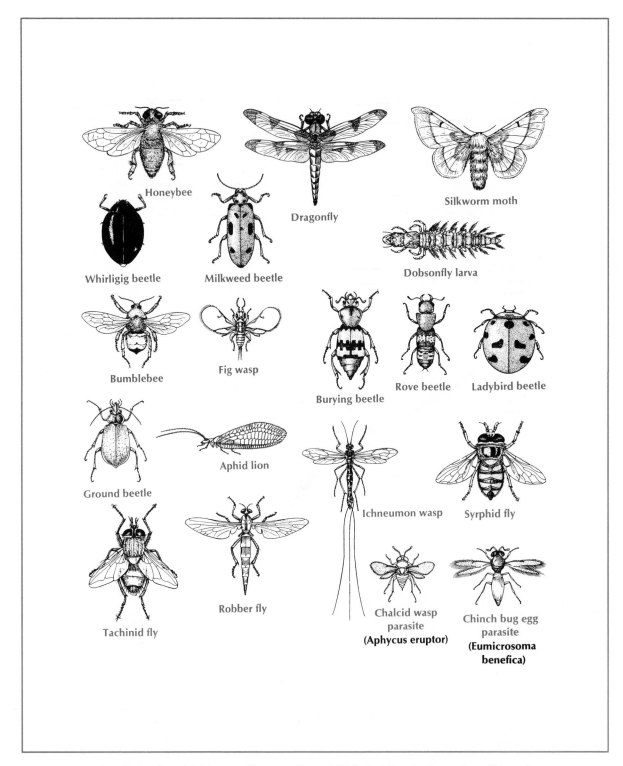

FIG. 30-17. Beneficial insects. (Courtesy General Biological Supply House, Inc., Chicago.)

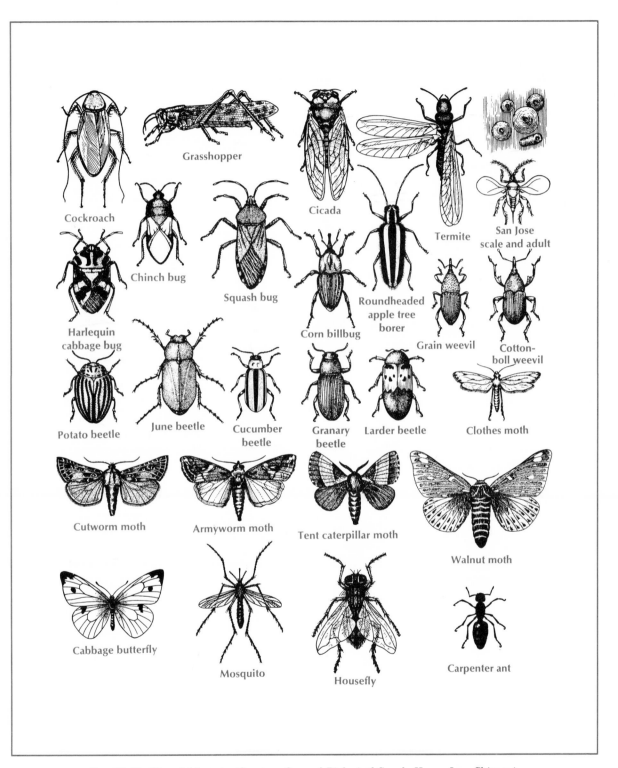

FIG. 30-18. Harmful insects. (Courtesy General Biological Supply House, Inc., Chicago.)

mals. This large group includes lice, blood-sucking flies, warble flies, and botflies.

Insects also carry diseases. Malaria, carried by the *Anopheles* mosquito, is still one of the world's greatest killers; yellow fever is carried by another mosquito, *Aedes,* and elephantiasis is also transmitted by mosquitoes. Fleas carry the plague, which at many times in history has almost wiped out whole populations. The housefly is the vector for typhoid and the louse for typhus fever, the tsetse fly carries African sleeping sickness, and a blood sucking bug, *Rhodnius,* is a carrier of Chagas fever.

In addition to all this, there is tremendous destruction of food, clothing, and property by weevils, cockroaches, ants, clothes moths, carpet beetles, and termites. Not the least of insect pests is the bedbug *(Cimex),* a blood-sucking hemipterous insect of cosmopolitan distribution. Man probably acquired this pest from bats in caves where men once lived in his early evolution.

Control of insects. Insecticides have been devised for controlling insects, but the wide and unwise use of chemicals has brought harm both to man and to the other animals of his environment. Many insecticides, such as the chlorinated hydrocarbons, persist in the environment and accumulate as residues in the bodies of higher animals. In recent years man has tried to be more selective in choosing pesticides that are specific for insect control and are less harmful to the forms he desires to protect. Many insects have developed strains resistant to insecticides. This constant fighting back by insects keeps scientists continually on the hunt for new and less persistent pesticides.

One method of control is to develop insect-resistant crops. So many factors (yield, quality, etc.) of resistant crops are involved that such a process requires the teamwork of specialists from many related fields. Some progress, however, has been made as in a corn hybrid that resists the corn borer and corn ear worm.

Three types of biologic controls are being worked on by the United States Department of Agriculture and others. One of these is the use of pathogens, such as *Bacillus thuringiensis,* used to control the leaf-eating lepidopteran that takes a heavy toll of California's lettuce crops. This spore-forming bacterium forms a protein crystal that is poisonous to lepidopteran larvae. However, it attacks all lepidopterans (butterflies and moths), not just the specific pest. A second type of control is the use of various viruses that are natural enemies of insects, but which could be cultivated in large numbers and applied at the most opportune time. A number of viruses have been isolated that seem to have potential as insecticides. However, specific viruses are difficult to raise and would be expensive to put into commercial production. A third method is to interfere with the metabolism or reproduction of the insect pests by introducing natural predators or by the sterile male approach. There have already been some successes in this area, such as the vedalia beetle brought from Australia to counteract the work of the cottony-cushion scale on citrus plants, and parasites introduced from Europe for control of the alfalfa weevil. The sterile male approach has been effective in eradicating screwworm flies, a livestock pest. Large numbers of male insects, sterilized by irradiation, are introduced into the natural population; females that mate with the sterile males lay infertile eggs. Insect sex attractants (pheromones) have also been utilized to trap pests. Recently, the sex attractant of the gypsy moth, a serious pest of forest and shade trees, has been identified and synthesized. In Australia it has been found that dung beetle larvae that develop from eggs laid in dung have been highly effective in controlling the buffalo fly that develops in the same place.

31

Echinoderms, hemichordates, and arrowworms

These three groups, the echinoderms, the hemichordates, and the arrowworms, belong to the Deuterostomia division of the bilateral animals, that is, those animals in which the mouth is formed from a new embryonic opening, the stomadeum, rather than from the embryonic blastopore as was the case with the Protostomia (annelids, mollusks, and arthropods). In the Deuterostomia, the blastopore becomes the anus. This distinction may seem to be a peculiar basis for separating large groups of animals, but it actually underlies a very fundamental difference in embryonic development in these two great evolutionary branches of animals (Fig. 24-1, p. 355). And as we have already seen, embryonic development is the best clue we have of invertebrate evolution, representing as it does a telescoped evolutionary history.

The chordates also belong to the Deuterostomia. So we would suspect that the echinoderms and the chordates originated from a common ancestor—and indeed this is what zoologists who have studied the evidence believe. Another feature the Deuterostomia have in common is that all form their coelom by pinching off coelomic sacs (called enterocoelous) rather than by the prostostomian method of creating a coelom by splitting apart an embryonic sheet of cells (called schizocoelous).

ECHINODERMS (PHYLUM ECHINODERMATA*)

The echinoderms are marine forms and include the sea stars, brittle stars, sea urchins, sea cucumbers, and sea lilies. They represent a bizarre group sharply distinguished from all other members of the animal kingdom. Their name is derived from the spiny characteristic of their integument.

Echinoderms are abundant along the seashore and extend to depths of many thousand feet. None can move rapidly and some are sessile. None of them are parasitic. There are about 6,000 species, which include some of the most fascinating, conspicuous, and brightly colored animals in the ocean.

Because of their spiny nature, echinoderms, except for their eggs and young, have limited use as food for other animals. Sea stars are, however, a pest to the shellfish industries. Feeding mainly on mollusks, crustaceans, and other invertebrates, a single sea star can eat as many as a dozen clams or oysters a day.

The primitive pattern of the echinoderms seems to have included radial symmetry, radiating grooves (ambulacra), and a tendency for the body openings (oral side) to face upward, as seen in sessile forms.

*E-ki″no-der′ma-ta (Gr. *echinos*, sea urchin, hedgehog, + *derma*, skin, + *ata*, characterized by).

431

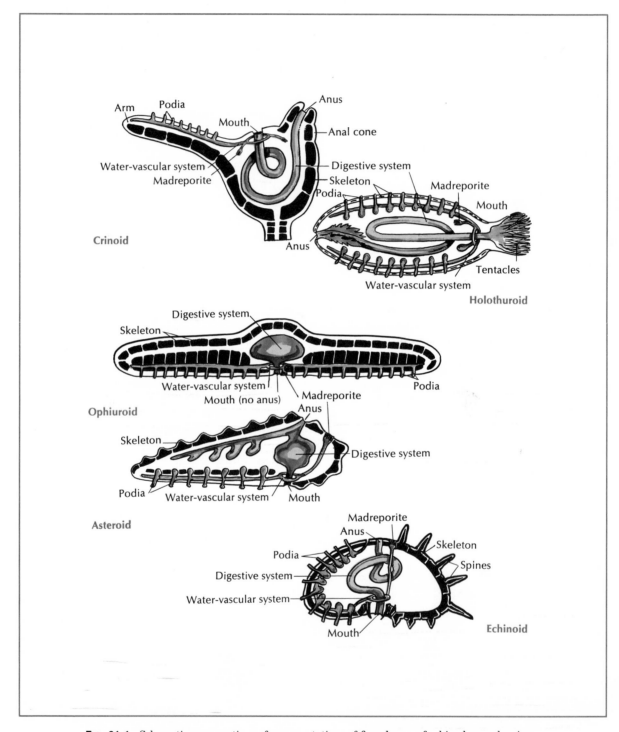

Fig. 31-1. Schematic cross sections of representatives of five classes of echinoderms showing comparatively the skeletal, water-vascular, and digestive systems. Skeleton is shown in black.

Radial symmetry with arm extensions would be an advantage to a sessile animal; it could get its food from all directions. Living crinoids follow this primitive pattern.

Probably the most striking characteristics of the echinoderms are the spiny endoskeleton, the water-vascular system, the pedicellariae, and the radial symmetry—which is not limited to echinoderms, but is unusual in a group with such complex organ systems.

CLASSIFICATION

There are about 6,000 living and 20,000 extinct or fossil species. Five classes of existing echinoderms are recognized (Fig. 31-1), and about that many extinct classes are known to invertebrate paleontologists.

Class Asteroidea (as″ter-oi′de-a)—**sea stars.** Star-shaped echinoderms, with the arms not sharply marked off from the central disk; tube feet with suckers.

Class Ophiuroidea (o″fi-u-roi′de-a)—**brittle stars, basket stars.** Star shaped, with the arms sharply marked off from the central disk; tube feet without suckers.

Class Echinoidea (ek″i-noi′de-a)—**sea urchins, sea biscuits, sand dollars.** More or less globular echinoderms with no arms; compact skeleton or test; movable spines; tube feet with suckers.

Class Holothuroidea (hol″o-thu-roi′de-a)—**sea cucumbers.** Cucumber-shaped echinoderms with no arms; spines absent; tube feet with suckers.

Class Crinoidea (kri-noi′de-a)—**sea lilies.** Body attached during part or all of life by an aboral stalk of dermal ossicles; five arms branching at base and bearing pinnules; tentacle-like tube feet for food collecting.

CHARACTERISTICS

1. Enterocoelous animals
2. Body typically **pentamerous,** usually with five radiating areas **(ambulacra)** containing podia **(tube feet)** alternating with areas without podia (interambulacra)
3. **Endoskeleton** or calcareous ossicles with spines; covered by epidermis, usually ciliated; **pedicellariae** (in some)
4. A unique hydraulic system, the **water-vascular system,** of coelomic origin, equipped with podia or tube feet that function in locomotion, respiration, and sensory perception
5. Coelom extensive, giving rise during development to the hemal and water-vascular systems
6. Respiration by **dermal branchiae** (skin gills), by tube feet, by **respiratory tree** (holothuroids), or by **bursae** (ophiuroids)
7. Reduced circulatory system—a hemal or lacunar system enclosed in coelomic channels; excretory organs absent
8. Digestive system usually complete; intestine and anus absent in ophiuroids
9. Nervous system diffuse and uncentralized; usually two or three nerve rings around the digestive tract, each with radial nerves and nerve nets; sensory system poorly developed
10. Usually dioecious, with simple gonads and ducts; external fertilization
11. Development with holoblastic radial cleavage; several types of bilateral free-swimming larvae which usually metamorphose to radially symmetric adults

SEA STARS (CLASS ASTEROIDEA)

Sea stars are found in all oceans, usually on rocky seashores but also at great depths in the ocean far from the shoreline. More than a thousand species are known. Although five rays is the usual number, there are species that have as many as forty or fifty; the number, however, is not constant if in excess of five or six (Fig. 31-2). At low tide the rocks along the shore may have large numbers of sea stars upon them. Sea stars are often brightly colored.

One of the most unique characteristics of echinoderms is the **water-vascular system.** This system consists of water-filled canals and specialized **tube feet** that, in most sea stars, are used in locomotion.

Sea stars use this hydraulic water-vascular system to control the movements of the many tube feet, which can extend and twist about by muscles in the wall. When the tube foot touches a substratum, it becomes attached by its sucker and contracts to draw the animal forward. On a soft surface, such as muck or sand, the suckers are ineffective, and so the tube feet are employed as legs in a stepping process. When inverted, the sea star twists its rays until some of its tube feet attach to the substratum, then slowly rolls over.

The tube feet are also used in opening bivalves, which are the chief source of food. The sea star arches its body over a clam and grips the opposite valves with its tube feet. A force of some 1,300 grams can be exerted by the star in this manner. When the bivalve's muscles tire and its valves gape even slightly, the star everts its cardiac stomach

433

FIG. 31-2. Group of common west coast echinoderms. **A,** Pacific purple stars *(Pisaster);* **B,** sun star *(Solaster);* **C,** common sea cucumber *(Stichopus);* **D,** sea urchin *(Strongylocentrotus);* **E,** 21-rayed star *(Pycnopodia);* **F,** 5-rayed star *(Dermasterias).*

and inserts it between the valves. Secretions from the stomach and digestive glands begin digesting the prey in its shell. The partly digested food is then sucked up into the stomach and digestive glands where digestion is completed. Sea stars are most active at night, when they search for food. Their reactions are mainly to touch, light, temperature, and chemicals.

Since most sea stars breathe through **skin gills,** small fingerlike projections of the coelom through pores of the body wall, it is important that the body surface be kept free of debris. This is the function of the **pedicellariae,** minute pincers located around the base of the spines, that snap shut when touched.

Sea stars can regenerate lost parts. A single ray attached to a portion of the central disk can regen-

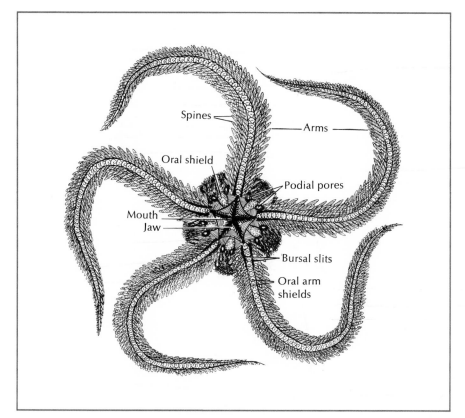

Spines

Arms

Oral shield

Podial pores

Mouth
Jaw

Bursal slits

Oral arm
shields

FIG. 31-3. Spiny brittle star
(Ophiothrix) oral view.

erate an entire animal. Many of them also have the power of autotomy, or voluntarily discarding a part.

Recent depradations on coral reefs by the sea star known as the crown-of-thorns, *Acanthaster planci,* has caused much concern. This sea star apparently eats the live coral polyps that secrete the limestone coral reefs, leaving only the skeletons of the coral. Regeneration of the reefs appears to be slow, although the places from which the polyps are stripped are replaced by algae that may play a part in the economy of the coral reefs. The shrimp *Hymenocera elegans,* which feeds on the sea stars by sucking out the inner tissues of the star, may help some in controlling the problem. But the natural enemy of the crown-of-thorns star is the Pacific giant triton, a marine snail. It is thought that the star has been allowed to multiply because shell collectors have taken too many of the tritons (100,000 were collected in a ten-year period), thus upsetting the ecologic balance between predator and prey.

BRITTLE STARS AND BASKET STARS (CLASS OPHIUROIDEA)

Brittle stars, like the sea stars, have a central disk with five or more rays (Fig. 31-3). The arms, however, are long, slender, and highly mobile, and are sharply marked off from the disk. There are more species of ophiuroids than of any other class of echinoderms. They are found in both shallow and deep water and have a wide distribution. The basket stars have complexly branched rays.

Brittle stars can often be found under stones and in seaweed at low tide. At high tide they are active, searching for small animals to capture with their rays. They move with a writhing, serpantlike motion of the arms.

SEA URCHINS AND SAND DOLLARS (CLASS ECHINOIDEA)

Echinoids lack rays and have rounded bodies with movable spines (Fig. 31-2, *D*). Sand dollars are disk shaped and have very short spines. Some of them have perforated tests to facilitate burrowing. Sea urchins are hemispherical or ovoid and have longer spines. A common east-coast urchin is *Arbacia punctulata,* found in both deep water and shallow tide pools; a common west-coast urchin is *Strongylocentrotus* (Fig. 31-2, *D*). Echinoids are omnivorous and scavengers.

SEA CUCUMBERS (CLASS HOLOTHUROIDEA)

Sea cucumbers are soft bodied, bottom-dwelling forms. Their bodies are elongated, with mouth and retractile **tentacles** at one end, and the anus at the other (Figs. 31-1 and 31-2, *C*). They are often found buried in the mud or sand of low-tide pools, with their tentacles extending up into clearer water. Common species are *Cucumaria* and *Thyone* along the eastern seacoast and *Stichopus* (Fig. 31-2, *C*) and several species of *Cucumaria* on the west coast.

Sea cucumbers feed upon small organisms that they entangle in the sticky mucus of their tentacles and suck into the mouth.

When irritated, sea cucumbers may, by muscular contractions, cast out some of the viscera through the anus or by rupture of the body wall, then regenerate the lost parts.

SEA LILIES AND FEATHER STARS (CLASS CRINOIDEA)

The crinoids are the most primitive of the echinoderms. They were once far more common than now. They are mostly deepwater forms, but some species live near shore. They differ from other echinoderms by being attached to a substratum for part or all of their lives. The beautifully colored feather stars have long many-branched arms and are free swimming. Sea lilies are pallid and are attached by stalks (Fig. 31-4). Most stalked crinoids are from 6 to 12 inches long; most of the free-swimming forms average a foot or so across the arms.

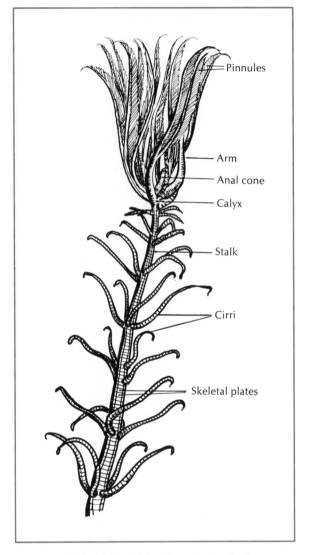

FIG. 31-4. Crinoid (sea lily) with part of stalk. Some crinoids have stalks 2 feet long.

HEMICHORDATES (PHYLUM HEMICHORDATA*)

The taxonomic status and phylogenetic relationship of the hemichordates has been somewhat of a puzzle among zoologists for years. Some authorities consider them a subphylum under the chordates and lump them with the Protochordata. This classifica-

*Hem"i-kor-da'ta (Gr. *hemi*, half, + L. *chorda*, cord, + *ata*, characterized by).

tion was based mainly upon certain characteristics of the hemichordates, such as gill slits and a supposed rudimentary notochord (the stomachord), which members of the phylum Chordata possess. Some hemichordates have the beginning of a dorsal hollow nerve cord, a characteristic of the chordates. The group also is somewhat unique in having both invertebrate and chordate characteristics. Their affinity with the echinoderms is marked because their larva is remarkably like certain echinoderm larvae. There are also other resemblances to echinoderms, such as the coelomic pouches that act like a water-vascular system and the general plan of the nervous system, characteristic of both groups. Both the hemichordates and the echinoderms also share many habits in common, such as feeding methods and ecologic niches. Because of these and other reasons, the hemichordates are placed in a phylum of their own.

CLASSES OF HEMICHORDATA

Class Enteropneusta (en″ter-op-neus′ta)—**acorn, or tongue, worms.** Body vermiform with no stalk; active, free living, and burrowing; many gill slits in a row on each side of anterior region. Examples: *Balanoglossus, Saccoglossus.*

Class Pterobranchia (ter″o-bran′ke-a)—**pterobranchs.** Compact body with stalk; sessile in chitinous tubes with lophophore bearing ciliated arms; a single pair of gill slits in pharynx or none. Example: *Cephalodiscus.*

GENERAL CHARACTERISTICS

The **tongue worms** are vermiform hemichords that are common in sand or mud flats in rather shallow water, although specimens have been collected at great depths. They vary in length from an inch or two to several feet. The genus *Balanoglossus* has a worldwide distribution. Several species of *Saccoglossus* (Fig. 31-5) are found along both the east and west coasts. Some are bright orange and red.

Tongue worms are delicate, sluggish animals that use their proboscis and collar for burrowing through sandy or muddy sea bottoms. By taking in water through the pores into the coelomic sacs, the proboscis and collar are stiffened for burrowing. Then, by contracting the body-wall musculature, the excess water is driven through the gill slits while the silt and mud, containing usable organic food, is

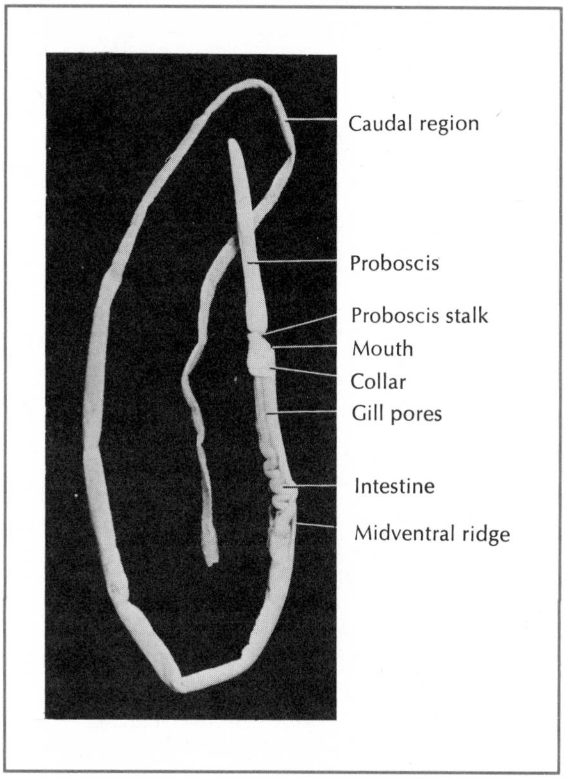

FIG. 31-5. Tongue, or acorn, worm *Saccoglossus.* Proboscis appears dense and rigid, with collar folded just behind it. Note coils of intestinal tract. Class Enteropneusta.

passed along the ventral esophagus into the intestine, where digestion occurs. Like earthworms, hemichordates pass a great deal of indigestible material through the alimentary canal. Investigations in recent years have shown that those that do not burrow are suspension feeders who collect food particles by means of the mucus on ciliary tracts of the proboscis and anterior part of the collar. Food (plankton and other small organisms) is collected in mucous ropes that pass to the mouth. Unwanted particles are carried posteriorly on ciliary paths of the trunk and are rejected.

The **pterobranchs** are similar to the tongue worms, but certain structural differences are correlated with their sedentary mode of life. They are

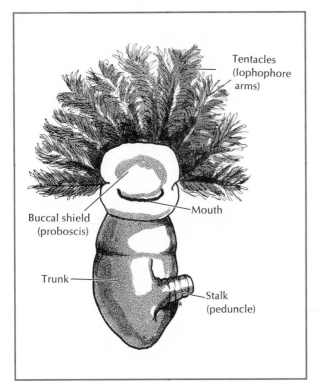

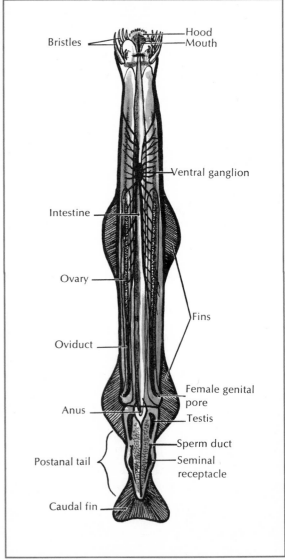

Fig. 31-6. *Cephalodiscus,* a pterobranch. These tiny sessile forms do not exceed 2 mm. in length and live in coenecium tubes in which they can move about. Ciliated bands on arms carry currents of food toward mouth. Patches of stiff epithelial cells in pharyngeal region are called a notochord (?) by some. Pair of so-called gill slits serve merely as outlet for water. It is thought that these deep-sea organisms are close to ancestral stock of both echinoderms and chordates. (Modified from Lang and others.)

Fig. 31-7. Arrowworm *Sagitta,* ventral view. These worms, rarely more than 2 or 3 inches long, form important part of marine plankton in both littoral and open sea waters. They have many resemblances to certain pseudocoelomates, and some authorities hesitate to call them coelomate animals, although they are enterocoelous and are placed in Deuterostomia. Among their features are postanal tail, hood, fins, and stratified epidermis.

small animals, usually 1 to 7 mm. in length, although the stalk may be longer. *Cephalodiscus* (Fig. 31-6) are free living, but many individuals live together in gelatinous tubes, which often form an anastomosing system. Through apertures in these tubes, they extend their crown of tentacles. By means of this ciliated crown they collect their food.

THE ARROWWORMS (PHYLUM CHAETOGNATHA*)

These arrowworms are marine animals and are considered by some to be related to the nematodes and by others to be related to the annelids. How-

*Ke-tog'na-tha (Gr. *chaeton*, bristle, + *gnathos*, jaw).

ever, they actually show no distinct relations to other groups and probably represent an early offshoot from the main line leading to other deuterostomes. There are fewer than fifty known species. Their small, straight bodies, which resemble miniature torpedoes, are from 1 to 3 inches long and are transparent and difficult to see.

Most of these forms swim near the surface, although sometimes they may be found at a depth of several hundred feet. They have the habit of coming to the surface at night and descending during the day. They are rapid swimmers. Arrowworms are predaceous and feed on small plants and animals. Most of them form an important part of plankton populations.

The most common representative is *Sagitta*, the common arrowworm (Fig. 31-7).

32

Protochordates and fishes

CHORDATES

The phylum Chordata derives its name from one of the few common characteristics of this group—the **notochord** (Gr. *noton*, back, + L. *chorda*, cord). This structure is possessed by all members of the phylum, either in the larval or embryonic stages or throughout life. The notochord is a rodlike, semi-rigid body of vacuolated cells, which extends, in most cases, the length of the body between the enteric canal and the central nervous system. Its primary purpose is to support and to stiffen the body, that is, to act as a skeletal axis.

The structural plan of chordates retains many of the features of invertebrate animals, such as bilateral symmetry, anterior-posterior axis, coelom, tube-within-a-tube arrangement, metamerism, and cephalization.

One distinctive feature of the chordates, of great significance to the evolution of this group, is the **endoskeleton.** This structure has the advantage of allowing continuous growth, without the necessity of shedding a restricting outer shell. For this reason vertebrate animals can attain great size; some of them are the most massive in the animal kingdom. Endoskeletons provide much surface for muscle attachment, and size differences between animals result mainly from the amount of muscle tissue they possess. More muscle tissue necessitates greater development of body systems, such as circulatory, digestive, respiratory, and excretory. Thus it is seen that the endoskeleton is a crucial factor in the development and specialization of the higher animals. Its function has shifted more from a protective to a supportive one. However, endoskeletons still retain some protective functions as, for example, the cranium for the brain and the thorax for important visceral organs.

Ecologically the chordates are among the most adaptable of organic forms and are able to occupy most kinds of habitat. Chordates are of primary interest because they illustrate so well the broad biologic principles of evolution, development, and relationship. They represent as a group the background of man himself.

ANCESTRY AND EVOLUTION

That vertebrates have evolved from invertebrates is not questioned by most biologists. The best evidence available is that which involves the protochordates and the echinoderms (p. 431). The great chasm, however, in the ancestral lineage between the invertebrates and the chordates has never been bridged despite many attempts of biologists to do so. We simply do not know where the chordate ancestors originated. There are no fossil records. The primitive chordates, or **protochordates,** are so different from the invertebrates that they throw little light on the problem. Apparently the early chordates were soft bodied and only the vertebrates were sufficiently hard and durable to be laid down as fossil forms. It has been possible to trace with considerable success the evolutionary patterns of many vertebrates, for the sequence of their fossil records is convincing.

The ancestor of the chordates probably was sessile and quite simple; it must have had a filter device for collecting food, an alimentary canal, and a reproductive system—all enclosed in a soft body.

CHARACTERISTICS

The three distinctive characteristics of chordates are the **notochord, dorsal tubular nerve cord,** and **pharyngeal gill slits.** These characteristics are always found in the early embryo, although they may be altered or may disappear in later stages of the life cycle.

Notochord. This rigid, yet flexible, rodlike structure, extending the length of the body, is the first part of the endoskeleton to appear in the embryo. As a rigid axis for muscle attachment, it permits undulatory movements of the body. In most of the protochordates and in primitive vertebrates, the notochord persists throughout life. In all vertebrates a series of cartilaginous or bony vertebrae are formed from the connective tissue sheath around the notochord and replace it as the chief mechanical axis of the body.

Dorsal tubular nerve cord. In the invertebrate phyla the nerve cord (often paired) is ventral to the alimentary canal and is solid, but in the chordates the cord is dorsal to the alimentary canal and is a tube (although the hollow center may be nearly obliterated during growth). The anterior end becomes enlarged to form the brain. The hollow cord is produced in the embryo by the infolding of ectodermal cells on the dorsal side of the body above the notochord. Among the vertebrates the nerve cord lies in the neural arches of the vertebrae,

and the anterior brain is surrounded by a bony or cartilaginous cranium.

Pharyngeal gill slits. Pharyngeal gill slits are perforated slitlike openings that lead from the pharyngeal cavity to the outside. They are formed by the invagination of the outside ectoderm and the evagination of the endodermal lining of the pharynx. The two pockets break through when they meet, to form the slit. In higher vertebrates these pockets may not break through and only grooves are formed instead of slits; most traces of them usually disappear. The slits have in their walls supporting frameworks of gill bars, which, in aquatic vertebrates, develop into gills.

SUBPHYLA

There are three subphyla under phylum Chordata. Two of these subphyla are small, lack a vertebral column, and are of interest primarily as borderline or first chordates (protochordates). Since members of these subphyla lack a cranium (braincase), they are also referred to as Acrania. The third subphylum is provided with a vertebral column and is called Vertebrata. Since members of this phylum have a cranium, it is also called Craniata.

Protochordata (Acrania)

Subphylum Urochordata (u″ro-kor-da′ta)—**Tunicata.** Notochord and nerve cord only in free-swimming larva; adults sessile and encased in tunic.

Subphylum Cephalochordata (sef″a-lo-kor-da′ta). Notochord and nerve cord found along entire length of body and persist throughout life; fishlike in form.

Craniata

Subphylum Vertebrata (ver″te-bra′ta). Bony or cartilaginous vertebrae surround spinal cord; notochord in all embryonic stages and persists in some of the fish. This subphylum may also be divided into two groups (superclasses) according to whether they have jaws.

Superclass Agnatha (ag′na-tha). Without true jaws or appendages.

Superclass Gnathostomata (na″tho-sto′ma-ta). With jaws and (usually) paired appendages.

THE PROTOCHORDATES: FORERUNNERS OF VERTEBRATES

Two of the three subphyla of the chordates are often referred to collectively as the **protochordates,** that is, the first or early chordates. These primitive borderline forms have little economic importance but are of great interest to biologists because they exhibit the characteristics of chordates in simple form.

TUNICATES (SUBPHYLUM UROCHORDATA)

The tunicates (Fig. 32-1) are found in all seas from near the shoreline to great depths. Most of them are sessile as adults, although some are free living. The name tunicate is suggested by the nonliving tunic that surrounds them and contains cellulose. They vary in size from microscopic forms to several inches in length. They may be considered as degenerative or specialized members of the chordates, for they lack many of the common characteristics of chordates.

Of the three recognized classes of Urochordata, the **ascidians** (also called "sea squirts") are the most common and best known. Most ascidians are attached to the substrate as adults but have evolved from free-moving ancestors. Each of the solitary and colonial forms has its own outer covering (test), but among the compound forms many individuals may share the same test. In some of these compound ascidians each member has its own incurrent siphon, but the excurrent siphon is common to the group. The structure of a sea squirt is shown in Fig. 32-1.

Of the three chief characteristics of the chordates, adult tunicates have only one, the pharyngeal gill slits. However, the larva form gives away the secret of their true relationship, for it possesses all three of the chordate characteristics—a notochord, a dorsal tubular nerve cord, and gill slits.

CEPHALOCHORDATES (SUBPHYLUM CEPHALOCHORDATA)

The cephalochordates are the marine lancelets. These include *Amphioxus (Branchiostoma),* one of the classic animals in zoology. The amphioxus has the three distinctive characteristics of chordates

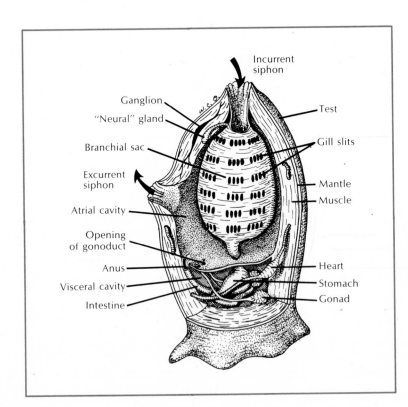

Incurrent
siphon

Ganglion
"Neural" gland
Branchial sac
Excurrent
siphon
Atrial cavity
Opening
of gonoduct
Anus
Visceral cavity
Intestine

Test
Gill slits
Mantle
Muscle

Heart
Stomach
Gonad

FIG. 32-1. Structure of adult solitary, simple ascidian. Arrows indicate direction of water currents.

in simple form, and in other ways it may be considered a blueprint of the phylum. It has a long, slender, laterally compressed body 2 to 3 inches long, with both ends pointed (Fig. 32-2).

Not only are the three chief characters of the chordates—dorsal nerve cord, notochord, and pharyngeal gill slits—well represented in amphioxus, but also are the secondary characteristics, such as postanal tail, liver diverticulum, hepatic portal system, and the beginning of a ventral heart. The separation of the dorsal and ventral roots of the spinal nerves may indicate the early condition in the vertebrate ancestors. *Amphioxus* is often placed close in affinity to the higher chordates, the vertebrates. Many authorities place it near the primitive fish, ostracoderms, but whether it comes before or after these fish in the evolutionary line is not settled. Many regard the amphioxus as a highly specialized or degenerate member of the early chordates and believe that the overdeveloped notochord was developed in them as a correlation to their burrowing habits. The extension of the notochord into the tip of the snout may be one of the reasons for the small development of the brain of the amphioxus. There are other objections to considering amphioxus as a generalized ancestral type of chordate, and some authorities therefore think it is a divergent side branch of some stage intermediate between the early filter-feeding prevertebrates and the vertebrates.

VERTEBRATES (SUBPHYLUM VERTEBRATA)

The third subphylum of the chordates, Vertebrata, has the same characteristics that distinguish the other two subphyla, but in addition it has a braincase, or **cranium,** and a spinal column of **vertebrae** that forms the chief skeletal axis of the body.

AMMOCOETE LARVA AS VERTEBRATE ARCHETYPE

The oldest known group of vertebrates is the Agnatha, which includes the extinct ostracoderms and the existing cyclostomes (lampreys and hagfish). It is logical, therefore, to look for a vertebrate ancestor among these primitive forms. Adult cyclostomes are too specialized and too degenerative in many respects for meeting the requirements of such a generalized type. The ammocoete larva of lampreys, however, possesses many of the basic structures one would expect to find in a chordate archetype (Fig. 32-3). It has a heart, ear, eye, thyroid gland, and pituitary gland, which are charac-

FIG. 32-2. Structure of *Amphioxus.* Photomicrograph of juvenile; gonads not present.

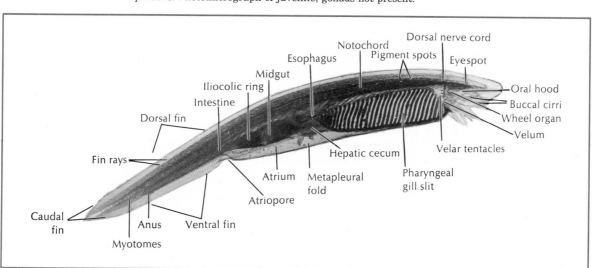

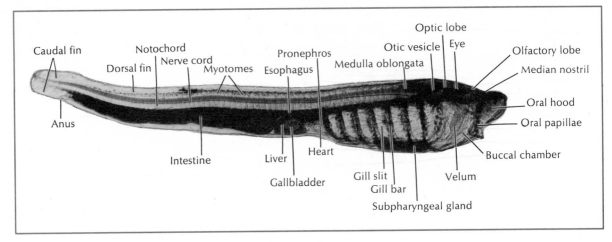

FIG. 32-3. Structure of ammocoete larva. Photograph of stained slide.

teristic of vertebrates but are lacking in the amphioxus. This larva is so different from the adult
lamprey that it was for a long time considered to be
a separate species; not until it was shown to metamorphose into the adult lamprey was the exact
relationship explained. This eellike larva spends
several years buried in the sand and mud of shallow streams, until it finally emerges as an adult
that may continue to live in fresh water (freshwater
lampreys) or else may migrate to the sea (marine
lampreys).

LAMPREYS AND HAGFISH (CLASS CYCLOSTOMATA [AGNATHA])

The primitive cyclostomes are the only existing
vertebrates without jaws (superclass Agnatha). This
distinguishes them from the remaining vertebrates,
which have jaws (superclass Gnathostomata). Cyclostomata are represented by about fifty species, almost equally divided between two orders.

Classification

Order 1. Petromyzontia (pet"ro-mi-zon' te-a) — **lampreys.** Mouth suctorial with horny teeth; gill pouches,
seven pairs. Examples: *Entosphenus, Petromyzon.*

Order 2. Myxinoidea (mik"si-noi' de-a) — **hagfish and
slime eels.** Mouth terminal with four pairs of tentacles;
gill pouches, ten to fourteen pairs. Examples: *Myxine,
Bdellostoma.*

Lampreys (order Petromyzontia)

All the lampreys of the northern hemisphere belong to the family Petromyzontidae. The destructive
marine lamprey *Petromyzon marinus* is found on
both sides of the Atlantic Ocean (America and
Europe) and may attain a length of 3 feet. Other
genera also have a wide distribution in North America and Eurasia and are usually from 6 to 24 inches
long. Of the 19 species of lampreys in North America, about half are of the nonparasitic brook type;
the others are parasitic.

All lampreys spawn in the spring on shallow
gravel beds in streams of fresh water. The males
clear away the pebbles from a sandy bottom and form
a type of pit (Fig. 32-4). When a female anchors her-
self to a pebble over one of these pits, a male winds
his tail around her and discharges his sperm over
the eggs as they are extruded from her body into the
depression. The adults (brook lampreys) soon die
after their spawning act.

The eggs hatch in about 2 weeks into small larvae
(ammocoetes), which stay in the nest until they are
about 1/2 inch long; they then burrow into the mud
and sand and emerge at night to feed. The ammocoete period lasts from three to seven years, before
the larva rapidly metamorphoses into an adult.

Parasitic lampreys either migrate to the sea, if
marine, or else remain in fresh water, where they
attach themselves by their suckerlike mouth to fish
and, with their sharp horny teeth, rasp away the

FIG. 32-4. Lamprey nest with four individuals hard at work removing pebbles. About 1/3 natural size. (Courtesy J. W. Jordan, Jr.)

FIG. 32-5. Whitefish with sea lamprey *Petromyzon marinus* attached. When lamprey attaches itself with its sucking mouth, it proceeds to rasp off small bits of flesh with its chitinous teeth. Lamprey then injects an anticoagulant and sucks blood of host fish. Such wounds often prove fatal to fish, especially if point of attachment is in abdominal region of body. In many Great Lakes regions commercial fishing has been greatly reduced by the lampreys. Many devices have been used to control lampreys, such as poison and electrical barriers, but none have been wholly effective. (U. S. Fish and Wildlife Service.)

flesh and suck out the blood (Fig. 32-5). To promote the flow of blood, the lamprey injects an anticoagulant into the wound. When gorged, the lamprey releases its hold but leaves the fish with a large gaping wound that may prove fatal. The parasitic freshwater adults live a year or more before spawning and then die; the marine forms may live longer.

The nonparasitic lampreys do not feed after emerging as adults, for their alimentary canal degenerates, and within a few months they die.

The invasion of the upper Great Lakes by the sea lamprey has destroyed nearly all of the lake trout and whitefish there. Control methods of trapping to prevent spawning or of poisoning the larvae have been only partially successful.

Hagfish and slime eels (order Myxinoidea)

The members of this order, hagfish and slime eels, are often called "borers" because they burrow into fish for flesh consumption. They characteristically attach to the gills of fishes, then bore inside with a tonguelike boring apparatus. They devour the muscle and viscera, leaving only a bag of bones. They are all marine and spawn on the ocean floor. Some may reach a length of 36 inches. The hagfish *Myxine* on the Atlantic coast and the hagfish *Polistotrema* on the Pacific coast are the most common species in North America.

Lampreys and hagfish are probably not closely related according to protein differences in their blood. Because of the similarities between the hemoglobin of lampreys and that of higher vertebrates, the lamprey is considered to be nearer the vertebrate ancestral line, whereas the hagfish evolved into a side branch, with no close affinities to other groups. The external appearance of a hagfish and a lamprey is compared in Fig. 32-6.

SHARKS, SKATES, AND RAYS (CLASS CHONDRICHTHYES*)

The group that includes sharks, skates, and rays is an ancient one and has left many fossil forms. One of their distinctive features is their degenerate cartilaginous skeleton. Although there is some calcification, bone is entirely absent throughout the class.

Classification

Subclass Elasmobranchii (e-las″mo-bran′ ke-i). Gills in separate clefts along pharynx.
    Order Selachii (se-la′ ke-i) — **modern sharks.** Body spindle shaped; five to seven pairs of lateral gills not covered by operculum. Example: *Squalus.*
    Order Batoidei (ba-toi′ de-i) — **skates and rays.** Body spread out; pectoral fins enlarged and attached to head and body; five pairs of gill slits on ventral side; spiracles large. Example: *Raja* (common skate).
Subclass Holocephali (hol″o-cef′ a-li) — **chimaeras.** Gill slits covered with operculum; aberrant shape; jaws with tooth plates; without scales. Example: *Chimaera.*

*Kon-drik′ thi-es (Gr. *chondros*, cartilage, + *ichthys*, fish).

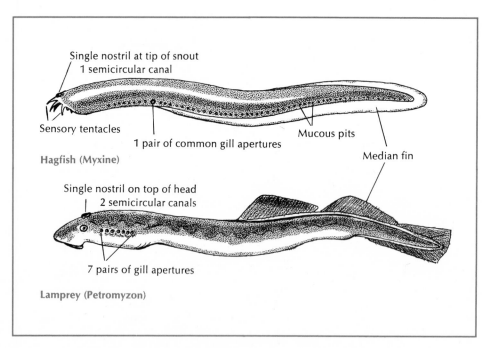

Single nostril at tip of snout
1 semicircular canal

Sensory tentacles

1 pair of common gill apertures

Mucous pits

Median fin

Hagfish (Myxine)

Single nostril on top of head
2 semicircular canals

7 pairs of gill apertures

Lamprey (Petromyzon)

FIG. 32-6. Comparison of hagfish and lamprey (class Cyclostomata). Mucous pits indicate enormous amount of mucin produced by hagfish.

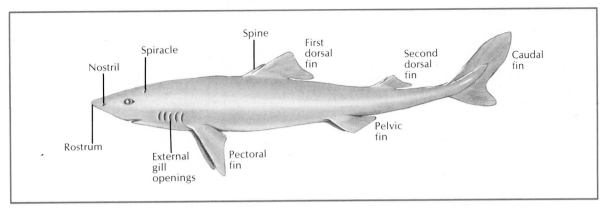

FIG. 32-7. Dogfish shark *Squalus acanthias*. Order Selachii.

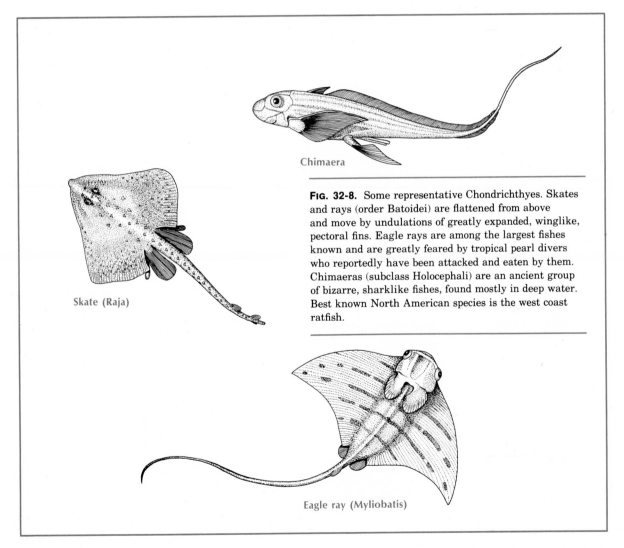

Chimaera

Skate (Raja)

Eagle ray (Myliobatis)

FIG. 32-8. Some representative Chondrichthyes. Skates and rays (order Batoidei) are flattened from above and move by undulations of greatly expanded, winglike, pectoral fins. Eagle rays are among the largest fishes known and are greatly feared by tropical pearl divers who reportedly have been attacked and eaten by them. Chimaeras (subclass Holocephali) are an ancient group of bizarre, sharklike fishes, found mostly in deep water. Best known North American species is the west coast ratfish.

Characteristics

1. **Body fusiform** or **spindle shaped**
2. **Mouth ventral; two olfactory sacs which do not break into the mouth cavity;** jaws present
3. Skin with **placoid** scales and **mucous glands;** teeth modified placoid scales
4. **Endoskeleton entirely cartilaginous**
5. Digestive system with a J-shaped stomach and intestine with a spiral valve
6. Circulatory system of several pairs of aortic arches; two-chambered heart
7. Respiration by means of five to seven pairs of gills with separate and exposed gill slits; **no operculum**
8. **No swim bladder**
9. Brain of two olfactory lobes, two cerebral hemispheres, two optic lobes, a cerebellum, and a medulla oblongata; ten pairs of cranial nerves
10. Sexes separate; oviparous or ovoviviparous; direct development; fertilization internal

With the exception of the whale, the sharks are the largest living vertebrates. The larger sharks may reach 40 to 50 feet in length, although the common dogfish shark seldom exceeds 3 feet (Fig. 32-7).

Skates and rays are specialized for bottom dwelling. The pectoral fins are used like wings in swimming (Fig. 32-8). The gill openings are on the underside of the head, and the spiracles (on top of the head) are unusually large. Respiratory water is taken in through these spiracles to prevent clogging the gills, for their mouth is often buried in sand. The teeth are adapted for crushing the prey — mainly mollusks, crustaceans, and an occasional small fish. In the stingrays and eagle rays (Fig. 32-8), the caudal and dorsal fins have disappeared and the tail is slender and whiplike. The stingray tail is armed with one or more saw-toothed spines that can inflict dangerous wounds. Electric rays have certain dorsal muscles modified into powerful electric organs, which can give severe shocks and stun their prey. Stingrays also have electric organs.

BONY FISHES (CLASS OSTEICHTHYES*)

The bony fish are the most familiar of all the fishlike animals. Their skeletons are partly bony, and most fish are covered by dermal scales. The skull and pectoral girdles are covered by investing bony plates

*Os"te-ik' thi-es (Gr. *osteon*, bone, + *ichthys*, fish).

in the dermal skin. The bony fish have the gills covered by opercular folds provided with bony supports. The class as a whole (especially the Teleostei) has reached at the present time a climax of success. Their 30,000 species indicate that they are the most numerous of all vertebrates, at least in number of species and probably also in number of individuals. One fifth are strictly freshwater dwellers. Most of the rest are marine, although some divide their time between the two kinds of aquatic habitats.

Classification

Subclass Actinopterygii (ak"ti-nop"te-ryj'e-i) – **ray-finned fish.** Paired fins supported by dermal rays and without basal lobed portions; one dorsal fin (may be divided); nasal sacs open only to outside.
Superorder Chondrostei (kon-dros'te-i) – **primitive ray-finned fish.** Includes sturgeons, paddlefish, and bichirs. Scaleless or with longitudinal rows of bony scutes; skeleton mostly cartilaginous.
Superorder Holostei (ho-los'te-i) – **intermediate ray-finned fish,** gars, and bowfins. Well-developed bony structure, approaching the teleosts in structure (Fig. 32-9).
Superorder Teleostei (tel"e-os'te-i) – **climax bony fish.** Body covered with thin scales without bony layer (cycloid or ctenoid) or scaleless; endoskeleton mostly bony; more than 30 different orders and some 350 families recognized. These orders may be placed in two basic groups: Isospondyli, or soft-rayed fish, such as the tarpon, herring, salmon, and trout, and Acanthopterygii, or spiny-rayed fish which includes most of the teleosts, such as perch, codfish, sole, crappie, and seahorse.
Subclass Choanichthyes (ko"a-nik'thi-es) – **lobe-finned or air-breathing fish.** Body primitively fusiform, but slender to thick in existing forms; paired fins lobed or axial; two dorsal fins; primitive cosmoid scale modified to thin cycloid type. These include the Crossopterygii, or lobe-finned fish, and the Dipnoi, or lungfish.

Characteristics

1. **Skeleton bony;** vertebrae numerous
2. Skin with mucous glands and with embedded dermal scales of three types: **ganoid, cycloid,** or **ctenoid;** some without scales; no placoid scales
3. Fins both median and paired, with **fin rays of cartilage or bone**
4. **Mouth terminal** with many teeth (some toothless); jaws present; olfactory sacs paired and may or may not open into mouth
5. Respiration by gills supported by bony gill arches and covered by a **common operculum**

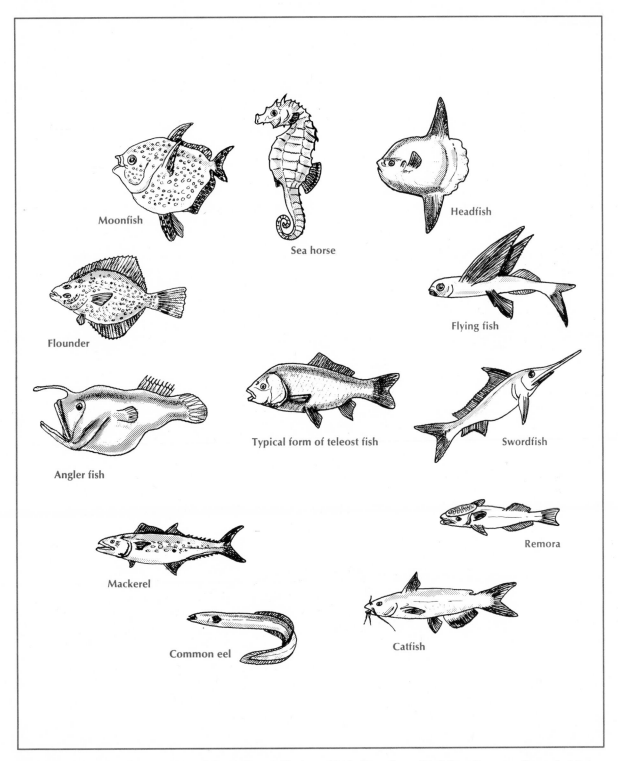

FIG. 32-9. Adaptive radiation of bony fishes (teleosts). Variety of body forms have fitted them for many diverse habitats and conditions of existence. It is not always possible, however, to explain all their adaptive shapes and structures.

6. **Swim bladder** often present with or without duct connected to pharynx
7. Circulation consisting of a two-chambered heart and four pairs of aortic arches
8. Nervous system of a brain with small olfactory lobes and cerebrum and large optic lobes and cerebellum; ten pairs of cranial nerves
9. Sexes separate; fertilization usually external; larval forms may differ greatly from adults

Evolutionary patterns of bony fish

In no other vertebrate group do we see better examples of adaptive radiation, where from certain generalized types, fish species have evolved whose adaptations fit them for virtually every kind of aquatic ecologic niche (Fig. 32-9). Some have fusiform or streamlined bodies for reducing friction and for rapid swimming. Predaceous fish have trim, elongated bodies and powerful tail fins for swift pursuit. Sluggish bottom-feeding forms have flattened bodies for movement and concealment on the ocean floor. Many fish have striking protective coloration; some are fitted for deep-sea existence.

Structural adaptations of fish

Swim bladder. The swim bladder (air bladder), found in many bony fish (Fig. 32-10), is a development from the paired lungs of primitive Osteichthyes, or their ancestors the placoderms, that lived in alternate wet and dry regions where lungs were necessary for survival. Functional lungs are present in existing lungfish. In all other bony fish the lungs have mostly lost their original function and have become swim bladders.

The swim bladder may serve as a hydrostatic organ, as a sense organ, or as an organ of sound production. The swim bladder can alter the specific gravity of the fish by filling with gas, thus lessening the buoyancy or by emptying and increasing buoyancy. In those fish in which the swim bladder is not

FIG. 32-10. Anatomy of a bony fish (yellow perch *Perca flavescens*).

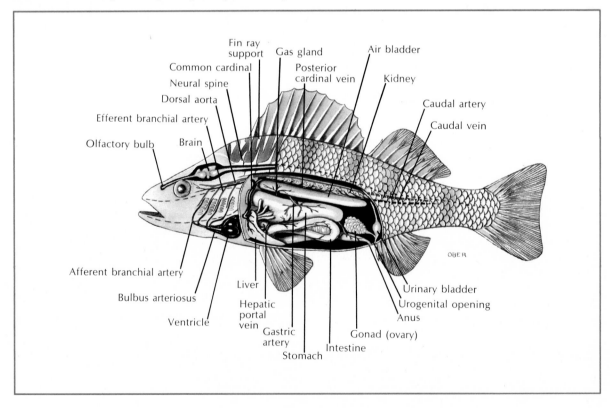

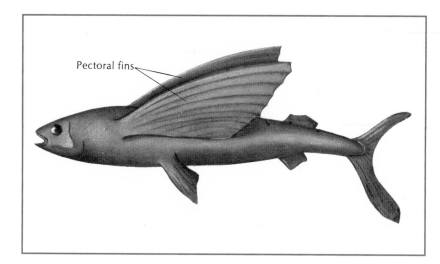

Pectoral fins

FIG. 32-11. Flying fish *Cypselurus*, example of highly specialized structural and functional adaptation among fish. Flying fish do not fly but glide, with their modified pectoral fins held in rigid position. By vigorous side-to-side swimming movements of their tails, they taxi for several feet on surface and gain sufficient momentum for their takeoff. Their flight usually lasts only a few seconds and distance of their glide varies from a few feet to several hundred feet.

connected to the pharynx by a duct, gases (oxygen, carbon dioxide, nitrogen) are secreted into the bladder by a special anterior gland, the **red gland,** with its network of blood vessels, the **rete mirabile.** A posterior **oval gland** of the swim bladder has the power to absorb these gases to lessen its size. In those fish with a swim bladder connected to the pharynx, the bladder may be filled by gulping air.

Fins. Fins in fish are always of two kinds: (1) **paired,** which include the **pectoral** and **pelvic** fins (Fig. 32-10); and (2) **unpaired,** which consist of the **dorsal, caudal,** and **anal** fins. Fish swim mainly by lateral movements of the tail and tail fin, while the paired fins are held closely against the side and the other unpaired fins are spread out to keep the animal in a vertical position. The body as a whole is thrown into a sinuous curve, not merely the tail (see p. 94). When swimming quietly they may use their paired lateral fins. The only vertebrate parasite of man, the candiru of the South American catfish, has a slender body adapted for entering the sex organ orifices of man. One species is especially feared for this reason because it may mistake the flow of urine during micturition for the respiratory current of water from fish gill cavities in which it normally lives.

The flying fish use their large and extended pectoral fins for gliding (Fig. 32-11), and the climbing perch of India use their gill cover and anal spines for ascending the branches of shrubs and trees found

in their habitats. The "walking" catfish *(Clarias batrachus),* an exotic species introduced into Florida in recent years, can also breathe on land as well as in fresh and brackish water. If threatened by severe drought or deliberate poisoning of water, it migrates overland by means of special adaptations of its modified posterior gills and pectoral-fin spines. Its ecologic status has not yet been appraised, although its rapid spread has caused some concern.

FIG. 32-12. Starry flounder *Platichthys.* In common with other flatfish, this one has both eyes on same side of head. Note twisted, distorted mouth. Flounders have remarkable ability to change colors in imitation of their background. Superorder Teleostei.

"Bloodless fish." Although fish have the characteristic vascular pattern of vertebrates, such as red blood, nucleated corpuscles, and specialized hemoglobin for transporting oxygen, certain fish collected from the Antarctic regions are exceptions to this rule. The icefish (*Chaenocephalus*) has transparent blood with little iron and no red blood corpuscles; white corpuscles, however, are present. Oxygen transportation appears to take place only by physical solution in the plasma with low oxygen capacity. Such fish apparently are able to survive only in very cold water where metabolic demand for oxygen is low.

Color. The males of some species, such as the horned dace and darters, have beautiful colors during the breeding season, but the most striking colors are found in tropical fish, especially those that live in and around coral reefs. Color is chiefly due to the presence of pigment cells (**chromatophores**) found in the dermal layer of the skin.

Many fish can change their color patterns to harmonize with their suroundings by contracting or expanding the pigment in the chromatophores. The flounder is one of these (Fig. 32-12). Other fish can alter their color to a lesser extent.

Scales. Scales grow throughout the life of a fish. The age of many fish, such as salmon and trout, can be determined on the scales by the interruptions (winter marks, or annuli) between regular groups of circuli. During winter slower growth results in fewer lines of growth spaced close together; in summer there is more growth, with the growth lines farther apart.

Fish scales (and also the skin) serve as reflectors for camouflage or display. Certain fish secrete nitrogenous compounds that form tiny crystals, arranged in stacks of alternating layers of crystals and in cytoplasmic spacing to reflect particular colors. Since the platelets are arranged in overlapping layers, all wavelengths are included and a sheen silvery color results because each array of crystals reflects light like small mirrors.

33

Amphibians

CLASS AMPHIBIA*

Members of the class Amphibia were the first vertebrates to attempt the transition from water to land. Strictly speaking, their crossopterygian ancestors were the animals that made the first attempt with any success—a feat that would be impossible for a poorly adapted transitional form. Amphibians are not completely land adapted and hover between aquatic and land environments. Structurally they are between the fish on the one hand and the reptiles on the other. Few amphibians can stray far from moist conditions, but some have developed devices for keeping their eggs out of water, where the larvae would be exposed to enemies.

Amphibians have developed limbs in place of fins, lungs in place of gills, and some skin changes. All larval forms retain a link with the aquatic life by having gills, and some retain gills throughout life. There are also other structural differences from fish that are mainly correlated with their mode of life, such as skeletal and muscular differences. There are about 2,000 species of amphibians.

*Am-fib' e-a (Gr. *amphi*, both, double, + *bios*, life).

HOW VERTEBRATES HAVE MET THE TERRESTRIAL MODE OF LIFE

The transition from water to land was gradual, and certain modifications were more striking than others. Amphibians represent a transitional stage between the purely aquatic and the purely terrestrial walks of life. Some structural modifications are summarized below.

Dry skin for protection against desiccation. Terrestrial forms are protected from drying out by having hard, dry skins. Instead of the soft epidermis of aquatic forms, the outer layers of land forms are cornified and are comprised of dead cells. Toads are an example. Most amphibians, however, retain the smooth, moist skin of typically aquatic forms.

Special egg for laying on land. Life on land required an egg protected against drying out and mechanical injury. The **amniotic egg** thus is provided with a tough but porous shell, large yolk, and special sacs and membranes (amnion, chorion, allantois). Since it requires internal fertilization, many modifications of mating habits have occurred. This type of egg is first found in reptiles. **The lack of it has kept amphibians close to water.**

Replacement of gills with lungs. Lungs are ideal for breathing on land because they are situated deep in the body to protect their delicate structure from the drying action of air. Along with this change in position has come the development of special air passageways, such as the trachea and bronchi, which have no counterparts in gill-breathing animals.

New pulmonary circulation. In fish the gill circulation (aortic arches) is placed directly in the path of the blood from the ventral aorta. Terrestrial forms have modified this aortic arch plan into a double circulation—a systemic circulation over the body and a pulmonary circulation to the lungs. The heart must take on additional chambers so that part of the double heart receives blood from the body and the other part receives blood from the lungs. The amphibians have developed a three-chambered heart, a transitional stage between the two-chambered heart of fish, and the fully divided four-chambered heart of higher vertebrates.

Jointed appendages for locomotion on land. The paddlelike fins of aquatic animals are replaced on land by jointed appendages, which became specialized for walking, running, climbing, flying, etc.

Changing priorities in the special senses. In water, olfactory organs carry a greater burden of sensory impression because of the poor development of other sense organs. Accordingly, the olfactory lobes in many fish are exceptionally large. In most terrestrial animals the lens is accommodated for distant vision, whereas fish are mostly near-sighted. The sense of hearing is also more keenly developed in fully terrestrial vertebrates. It is doubtful whether fish can hear in the ordinary sense of the word, although they are sensitive to vibrations through their lateral line organs.

Ammonia replaced with urea or uric acid. Freshwater forms excrete their nitrogenous waste chiefly in the form of ammonia, which, although highly toxic, can be easily washed away in the abundance of water that surrounds aquatic animals. Terrestrial animals must conserve their water and so have converted toxic ammonia to relatively nontoxic urea, or into uric acid, as the final product of nitrogenous waste excretion.

BRIEF CLASSIFICATION

Order 1. Gymnophiona (jim″no-fi′o-na) **(Apoda)—caecilians.** Body wormlike; limbs and limb girdle absent.

Order 2. Urodela (u″ro-de′la) **(Caudata)—salamanders, newts.** Body with head, trunk, and tail; usually two pairs of equal limbs.
Order 3. Salientia (sa″li-ench′e-a) **(Anura)—frogs, toads.** Head and trunk fused; no tail; two pairs of limbs.

CHARACTERISTICS

1. Skeleton mostly bony; ribs present in some, absent in others
2. Body forms vary greatly from an elongated trunk with distinct head, neck, and tail to a compact, depressed body with fused head and trunk and no intervening neck
3. **Limbs usually four (tetrapod); webbed feet often present**
4. **Skin smooth and moist with many glands,** some of which may be poisonous; **pigment cells (chromatophores)** common; **no scales,** except concealed dermal ones in some
5. Mouth usually large with small teeth in upper or both jaws; **two nostrils open into anterior part of mouth cavity**
6. Respiration by gills, lungs, skin, and pharyngeal region either separately or in combination
7. **Circulation with three-chambered heart,** two auricles and one ventricle, and a double circulation through the heart
8. Separate sexes; metamorphosis usually present; **eggs with jelly-like membrane coverings**

NATURAL HISTORY OF AMPHIBIANS

Caecilians. There are about fifty species of this little-known order Gymnophiona (Fig. 33-1). They are mostly found in Africa, Asia, and South America. In North America no species is found north of Mexico. They are strictly burrowing forms and are rarely seen above the earth's surface.

Salamanders and newts. Order Urodela has about one hundred and fifty species. Although found to a limited extent in other parts of the world, the temperate part of North America is the chief home of the tailed amphibians. The largest caudate known is the Japanese salamander, more than 5 feet long. Most of those in North America are from 3 to 6 inches long, although a few are longer (*Necturus*). Some common eastern North American species are shown in Fig. 33-2.

These forms have primitive limbs set at right angles to the body with the forelimbs and hind limbs about the same size. Many never leave the

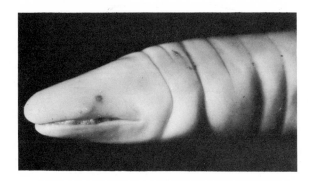

FIG. 33-1. Head and anterior region of caecilian. Order Gymnophiona (Apoda). These legless and wormlike amphibians may reach length of 18 inches and diameter of 3/4 inch. Their body folds give them appearance of segmented worm. They have many sharply pointed teeth, a pair of tiny eyes mostly hidden beneath the skin, a small tentacle between eye and nostril, and some forms have embedded mesodermal scales. (Courtesy General Biological Supply House, Inc., Chicago.)

FIG. 33-2. A, Red-backed salamander *Plethodon cinereus* has dorsal reddish stripe with gray to black sides. **B,** Two-lined salamander *Eurycea bislineata* is yellow to brown with two dorsolateral black stripes. **C,** Long-tailed salamander *Eurycea longicauda* is yellow to orange with black spots that form vertical stripes on sides of tail. **D,** Slimy salamander *Plethodon glutinosus* is black with white spots.

water in their entire life cycle, although others assume a terrestrial life, living in moist places under stones and rotten logs, usually not far from the water. All have gills at some stage of their lives; some lose their gills when they become adults.

Some species have neither lungs nor gills, but breathe through the skin and pharyngeal region. This is true with the family Plethodontidae, a common group in North America assumed to have originated in swift mountain streams of the Appalachian mountains. Mountain brook water, which is cool and well oxygenated, is excellent for cutaneous breathing. The cold temperature of the water slows down metabolism and thus less oxygen is required.

The salamanders and newts show less diversity in breeding habits than do the frogs and toads. Both external and internal fertilization are found in the group, but some males deposit their sperm in capsules called **spermatophores.** These are placed on leaves, sticks, and other objects, and the female picks them up with her cloacal lips and thus fertilizes her eggs. Aquatic species lay their eggs in clusters or stringy masses in the water; terrestrial forms may also lay their eggs in the water or in moist places.

Blind salamanders are found in limestone caves in certain parts of the United States and Austria. One of these species is *Typhlotriton spelaeus,* which has functional eyes in the larval form and lives near the mouth of caves. As an adult, it withdraws deeper into the caves and the eyes degenerate. If the larval forms are kept in the light, they retain functional eyes when they mature; if they metamorphose in the dark, they lose their eyesight.

Some urodeles exhibit **neoteny,** that is, sexual reproduction during larval periods. This is strik-

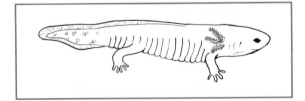

FIG. 33-3. An axolotl, larval form of tiger salamander *Ambystoma tigrinum.* In Mexico and southwestern states, larva does not metamorphose but breeds in this form.

ingly shown by the tiger salamander, *Ambystoma tigrinum.* This species is widely distributed over the United States into Central Mexico. The larvae, known as **axolotls** (Fig. 33-3), breed in Mexico and in the southwestern part of the United States, where they apparently never acquire the adult form. When transferred to the eastern states, they lose their gills, assume lungs, and become native tiger salamanders.

Salamanders and newts live on worms, small arthropods, and small mollusks. Most of them will eat only things that are moving.

Frogs and toads. Toads usually have dry warty skins, and frogs tend to have smooth, slimy skins. Whenever they live in or near the water, their skins are more or less smooth and slimy; in deserts or dry regions, however, their skins are rough and warty. The skin is usually well supplied with glands. One type of gland secretes mucus used mainly as a lubricant; the other type secretes a granular poison that may be highly irritating. In the giant South American toad *Bufo marinus* the poison is concentrated in the parotid glands (Fig. 33-4). The poison of *Dendrobates,* another South American frog, is used by Indian tribes to poison the points of their arrows.

A skin modification occurs in the so-called "hairy frog," *Astylosternus,* found in the Cameroons of Africa (Fig. 33-5, *B*). The males have fine cutaneous filaments on the thighs, groins, and sides, which have a resemblance to hair. These structures may have a respiratory function. The female surinam toad *Pipa* has a modified skin on her back for carrying eggs and young.

Anurans are usually defenseless, but in the tropics and subtropics many frogs and toads are aggressive, jumping and biting at their potential enemies. Some defend themselves by feigning death. Most anurans can blow up their lungs so that they are difficult to swallow. Bullfrogs in captivity will not hesitate to snap at tormenters and are capable of inflicting painful bites.

Unlike the salamanders, which mostly have internal fertilization, nearly all frogs and toads fertilize their eggs externally. The fertilized egg develops into a larva that hatches and changes during a period of metamorphosis into a juvenile frog (Fig. 33-6). When the female lays her eggs, the male em-

FIG. 33-4. Giant South American toad *Bufo marinus*. Some of these are more than 6 inches long. Their large parotid glands produce most poisonous secretion to be found among amphibians.

FIG. 33-5. Three frogs with unusual features. **A,** "Flying" frog of Borneo, *Polypedates nigropalmatus*. Large webs between digits of feet aid it in gliding from higher elevation to lower one. **B,** "Hairy frog" of Africa, *Astylosternus*. Hairlike filaments on groins and sides of male are actually cutaneous papillae, probably used for respiration. These filaments are unusually well developed during breeding season. **C,** Bell toad of northwestern Pacific coast, *Ascaphus truei*. Male's cloacal appendage serves as copulatory organ for fertilizing female's eggs. This frog and certain ovoviviparous frogs of Africa are only known salientians that have internal fertilization.

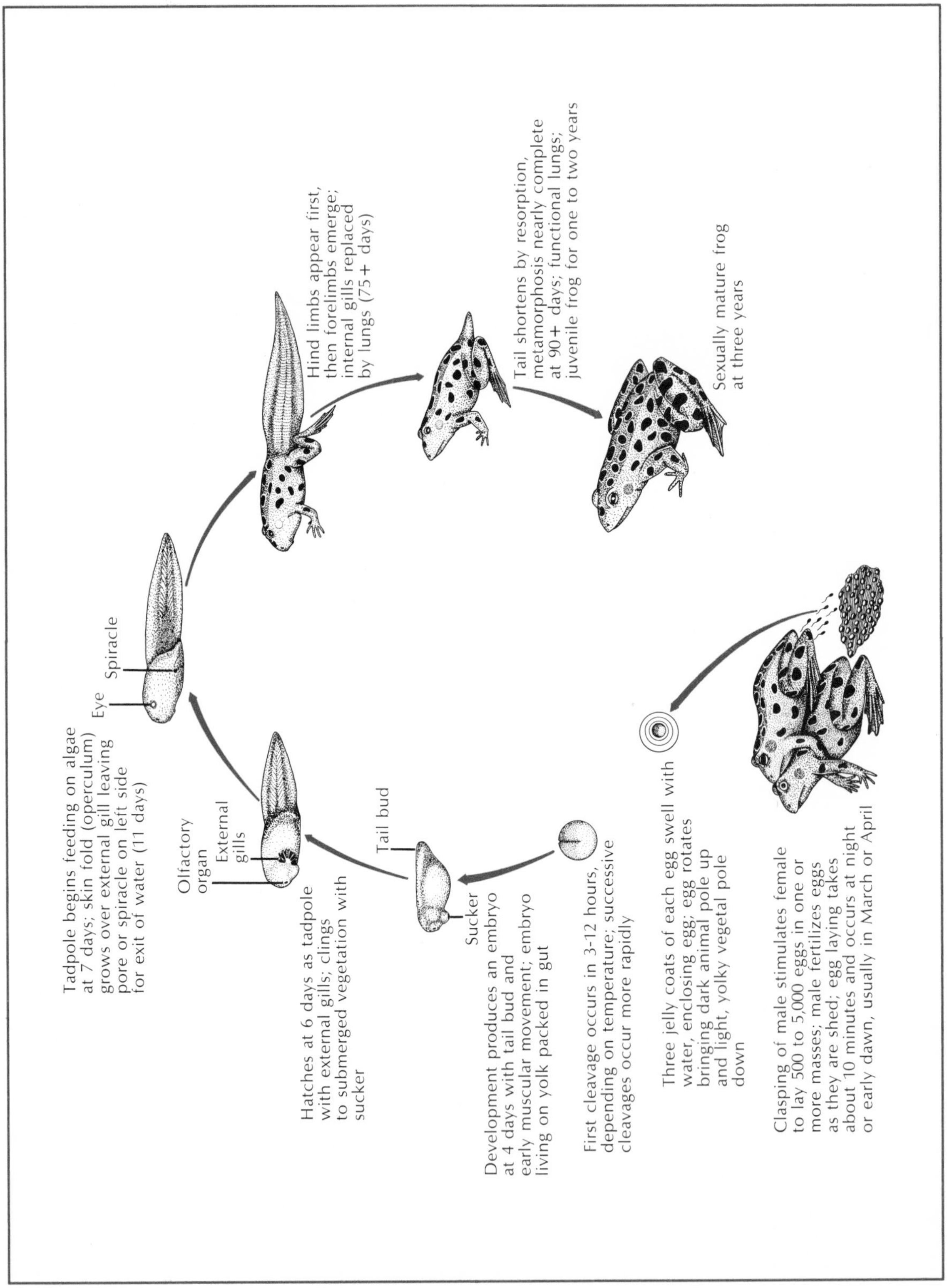

Hind limbs appear first, then forelimbs emerge; internal gills replaced by lungs (75+ days)

Tail shortens by resorption, metamorphosis nearly complete at 90+ days; functional lungs; juvenile frog for one to two years

Sexually mature frog at three years

Tadpole begins feeding on algae at 7 days; skin fold (operculum) grows over external gill leaving pore or spiracle on left side for exit of water (11 days)

Eye   Spiracle

Olfactory organ

External gills

Tail bud

Sucker

Hatches at 6 days as tadpole with external gills; clings to submerged vegetation with sucker

Development produces an embryo at 4 days with tail bud and early muscular movement; embryo living on yolk packed in gut

First cleavage occurs in 3–12 hours, depending on temperature; successive cleavages occur more rapidly

Three jelly coats of each egg swell with water, enclosing egg; egg rotates bringing dark animal pole up and light, yolky vegetal pole down

Clasping of male stimulates female to lay 500 to 5,000 eggs in one or more masses; male fertilizes eggs as they are shed; egg laying takes about 10 minutes and occurs at night or early dawn, usually in March or April

FIG. 33-6. Life cycle of frog.

A

B

C

Fig. 33-7. A, Bullfrog *Rana catesbeiana* is largest of all American frogs. **B,** Green frog *Rana clamitans* is next to bullfrog in size. Body is usually green, especially around jaws; has dark bars on sides of legs. **C,** Leopard frog *Rana pipiens* has light-colored dorsolateral ridges and irregular spots.

Fig. 33-8. Spring peeper *Hyla crucifer* with its resonating vocal sac enlarged just before it gives its high note, which can be heard a long distance. It will be seen that the whole body as well as vocal sac is inflated. Spring peepers are small (1 to 1½ inches), light brown, and marked with characteristic "X" on the back. (Courtesy R. Fuson.)

braces her and discharges his sperm over the eggs as they extrude from her body. A notable exception is the famed bell toad *(Ascaphus)* of the Pacific coast region (Fig. 33-5, *C*). This small toad (family Ascaphidae), which is only about 2 inches long, is found in swift mountain streams of low temperature from British Columbia to northern California and has a conspicuous extension of cloaca, which serves as an intromittent or copulatory organ for fertilizing internally the eggs of the female.

The largest anuran is *Rana goliath,* which is more than 1 foot long from tip of nose to anus; it is found in west Africa. This giant will eat animals as big as rats and ducks. The smallest frog recorded is *Phyllobates limbatus,* which is only about 1/2 inch long. This tiny frog, which is more than covered by a dime, is found in Cuba. Our largest American frog is the bull frog *(Rana catesbeiana),* which reaches a length of 8 or 9 inches (Fig. 33-7, *A*). The bullfrog is native east of the Rocky Mountains but has been introduced into most of the western states. The leopard frog *(R. pipiens)* (Fig. 33-7, *C*) has a wider variety of habitats and, with all its subspecies and phases, is perhaps the most widespread of all the North American frogs in its distribution. It has been found in some form in nearly every state and extends far into northern Canada and as far south as Panama. *Rana clamitans* (green frog, Fig. 33-7, *B*) is confined mainly to the eastern half of the United States, although it has been introduced elsewhere. Within the range of any species of frogs, they are often restricted to certain habitats (for instance, to certain streams or pools) and may be absent or scarce in similar habitats of the range. A common American tree frog, the spring peeper, is shown in Fig. 33-8.

34

Reptiles and birds

Reptiles and birds are so similar structurally that birds have often been referred to as "glorified reptiles." Reptiles arose from the amphibians through a primitive reptilelike amphibian (called a stem reptile) before the Permian period, perhaps 300 million years ago (Fig. 34-1). *Seymouria,* a small partly aquatic tetrapod fossil found in Texas, is a connecting link between the amphibians and reptiles, for it has characteristics of both groups.

The adaptive radiation of reptiles was especially pronounced in the Triassic period, but the Jurassic (150 to 200 million years ago) was the true age of reptiles. Yet most of the great reptilian orders died out completely by late Cretaceous time (perhaps 100 million years ago). No one factor can be singled out for blame. A combination of climatic and ecologic factors, excessive specialization, low reproduction rate, etc., may have been responsible, but all these are speculative. Many believe that fierce competition of the mammals was especially lethal. Why have some survived? Turtles had their protective shells, snakes and lizards evolved in habitats of dense forests and rocks where they could meet the competition of any tetrapod, and crocodiles, because of their size and natural defense and offense, had few enemies in their aquatic habitats.

Since the bones of birds are light and disintegrate quickly, it is only under the most favorable conditions that their remains are preserved as fossils. The earliest known bird is *Archaeopteryx,* a land form about the size of a crow with both reptile and bird characteristics. It had feathers, which the reptiles did not have, but it also had many reptilian characteristics such as teeth and an elongated tail.

The evidence, therefore, is strikingly in favor of a reptilian ancestor of birds. The ancestors of birds no doubt came from a branch of the stem-ruling reptiles (thecodonts) (Fig. 34-1). By 1950 over 700 different fossil species had been recorded. The modernization of birds took place chiefly during the Cretaceous period, and they were thoroughly modern by early Cenozoic times (80 million years ago).

REPTILES (CLASS REPTILIA*)

Reptiles are the first completely terrestrial group. They include snakes, lizards, turtles, tortoises, alligators, and crocodiles. There are more than 6,000 species of reptiles in the world, and of these more than 300 species are found in the United States. Reptiles were the first vertebrate class to break away from breeding in the water. Although many

*Rep-til′ e-a (L. *repere,* to creep).

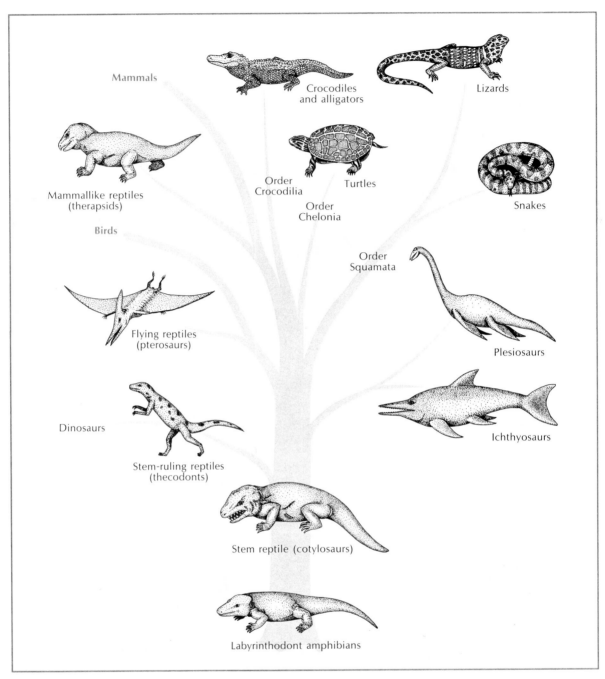

FIG. 34-1. Adaptive radiation of reptiles. First vertebrates to possess land were reptiles, amazing in their variety. Transition from certain labyrinthodont amphibians to reptiles occurred in Carboniferous period to Mesozoic times. This transition was effected by development of amniote egg, which made land existence possible, although this egg may well have developed before oldest reptiles had ventured far on land. Explosive adaptation by reptiles may have been due partly to variety of ecologic niches into which they could move. Fossil record shows that lines arising from stem reptiles led to ichthyosaurs, plesiosaurs, and stem-ruling reptiles. Some of these returned to the sea. Later radiations led to mammallike reptiles, turtles, flying reptiles, birds, dinosaurs, etc. Of this great assemblage, the only reptiles now in existence belong to four orders (Chelonia, Crocodilia, Squamata, and Rhynchocephalia). The Rhynchocephalia, not shown in this diagram, is represented by only one living species, the tuatara (Sphenodon) of New Zealand. How the mighty have fallen!

of them, such as alligators, snakes, and turtles, live in or near the water, they always return to the land to lay their eggs.

Orders

Order 1. Squamata (squa-ma′ta). Skin of horny epidermal scales or plates, which is shed; teeth attached to jaw. Examples: snakes (3,000 species), lizards (3,800 species), chameleons.

Order 2. Testudinata (tes-tu″di-na′ta) **(Chelonia).** Body in a bony case of dermal plates with dorsal carapace and ventral plastron; jaws without teeth but with horny sheaths. Examples: turtles and tortoises (400 species).

Order 3. Crocodilia (krok″o-dil′e-a) **(Loricata).** Four-chambered heart; forelimbs usually with five digits, hind limbs with four digits. Examples: crocodiles and alligators (25 species).

Order 4. Rhynchocephalia (ring″ko-se-fa′le-a). Example: *Sphenodon*—only species existing.

CHARACTERISTICS

1. **Body covered with an exoskeleton of horny epidermal scales; integument with few glands**
2. **Limbs paired, usually with five toes;** absent in snakes
3. Respiration by lungs; **no gills**
4. **Three-chambered heart; crocodiles with four-chambered heart**

5. **Twelve pairs of cranial nerves**
6. Sexes separate; **amniote eggs, which are covered with leathery shells**

LIZARDS AND SNAKES (ORDER SQUAMATA)

Lizards (Fig. 34-2) vary in length from 1 or 2 inches to several feet. They are found from the hottest desert to forested regions and the water.

The skin of lizards is flexible, with the scales arranged in rows. On the ventral surface they have small overlapping scales instead of transverse scutes (as in snakes). Some lizards have thin scales (osteoderms) in the skin in contrast to the thick, bony (dermal) scales of crocodiles and alligators.

Some lizards (although cold blooded) have evolved methods for regulating their body temperature by orienting their bodies either at right angles to the sun for maximum effect or parallel to the rays for minimum effect. Thus basking becomes a fine art with these animals. They can also regulate the absorption of heat by expanding their skin chromatophores when they are cold or by contracting them when the body becomes too warm.

The only poisonous lizard in the United States is the Gila monster found in the southwestern states. With poison fangs in the lower jaw, it works its poison into its prey by chewing movements. Living

Fig. 34-2. Carolina anole *Anolis.* Family Iguanidae. These lizards are popularly called chameleons because they can change their color, like true chameleons of the Old World.

in the same region as the Gila monster is *Phrynosoma,* the "horned toad," which is really a lizard.

Snakes represent one of the most specialized groups of animals in the world. Their loss of limbs is not restricted to the appendages but also applies to the pectoral and pelvic girdles (the latter being found as vestiges in pythons). Snakes also differ from lizards in having no sternum, eyelids, external ear openings, or bladder. The jaws of snakes are loosely connected to the skull so that the mouth may be greatly expanded, enabling them to swallow food much greater in diameter than their own bodies. They can breathe while swallowing a large object because the glottis is located far anterior in the floor of the mouth, just behind the teeth. Snakes conserve water through the reabsorption of water from their fecal content by the walls of the cloaca and large intestine.

Like lizards, snakes bear rows of scales that overlap as do shingles on a roof. In moving, the snakes project the margins of the scutes and use them for clinging against a surface while the body is driven forward by lateral undulations. Snakes also move by alternately throwing the body into coils and then straightening it out. Most species can swim by lateral convolutions of the body.

The forked tongue of the serpent is merely a sensory organ for detecting chemical stimuli. The tongue can be withdrawn into a sheath in the floor of the mouth. When the mouth is closed, the tongue can be thrust out of a groove between the two jaws. In poisonous snakes a pair of teeth on the maxillary bones are modified as fangs. These are grooved or tubular for conducting the poison from the poison sac (a modified salivary gland) into the prey that is bitten. This mechanism is therefore on the order of a hypodermic needle.

There are two types of snake venom. The neuro-

FIG. 34-3. Rat snake *Elaphe obsoleta*—a harmless snake and one of the four largest snakes in the United States. This living specimen is more than 6 feet long, but some are known to grow longer. Kills its prey by constriction.

toxic type acts mainly on the nervous system, affecting the optic nerves (causing blindness) or the phrenic nerve of the diaphragm (causing paralysis of respiration). The other type is hemolytic, that is, it breaks down the red blood corpuscles and blood vessels and produces extensive extravasation of blood into the tissue spaces. Many venoms have both neurotoxic and hemolytic properties.

There are more than 100 species of snakes in the United States (Figs. 34-3 to 34-5), of which fewer than thirty species are venomous. The latter include rattlesnakes (Fig. 34-4), copperheads, water moccasins, and coral snakes (Fig. 34-5). All these are pit vipers (except the coral snakes) and derive their name from a small cavity just anterior to the eye. This pit is a specialized infrared heat detector that is extremely sensitive to temperature and enables the snake to detect a warm-blooded animal. When striking, their fangs are erected to a position at right angles to the mouth surface. Just how many people are bitten each year by venomous snakes in the United States is not accurately known but probably not more than a few hundred. Of those bitten, only a small percentage die. More people are bitten by the copperhead *Agkistrodon mokasen* than by any other species, although more deaths are caused by the Texas diamondback rattlesnake *Crotalus atrox* (Fig. 34-4). Rattlesnakes are characterized by the rattle, which consists of horny, ringlike segments held loosely together on the end of their tails. Whenever the skin is shed, the posterior part remains behind to form another ring of the rattle. The number of rattles is not an accurate indication of age, for the snake often sheds more than once a year, and rings of the rattle are frequently lost. Our largest poisonous snake in the United States is the eastern diamondback rattlesnake *(Crotalus adamanteus),* which may reach a length of 8 feet. Most

FIG. 34-4. Texas diamondback rattlesnake *Crotalus atrox.* More deaths are caused by this snake than by any other species of poisonous snakes in United States. (Courtesy E. P. Haddon, U. S. Fish and Wildlife Service.)

FIG. 34-5. Coral snake *Micrurus fulvius,* a very poisonous snake that inhabits southern United States and tropical countries. Only representative of cobra family in North America. Body bands black, red, and yellow. (Courtesy F. M. Uhler, U. S. Fish and Wildlife Service.)

FIG. 34-6. Boa constrictor *(Boa constrictor)*. This specimen was kept at DePauw University for twenty-three years, until its death (1961) at nearly 30 years of age. During its vigorous years it ate from 15 to 30 rats or pigeons each summer (by first constricting them to death) but refused to eat in winter. During its last years it ate only 4 or 5 rats a year. It grew 2 to 3 feet in captivity; was nearly 9 feet long.

of the large constrictor snakes are found in the tropics (Fig. 34-6).

TORTOISES AND TURTLES
(ORDER TESTUDINATA)

Tortoises and turtles are enclosed in shells consisting of a doral carapace and a ventral plastron. The shell is built in with the thoracic vertebrae and ribs. Into this shell the head and appendages can be retracted for protection (in many). The term "tortoise" is usually given to the land forms, whereas the term "turtle" is reserved for the aquatic forms.

Turtles range from a few inches in diameter to the great marine ones that may weigh a thousand pounds. Sea turtles (Fig. 34-7) usually grow larger than land forms, although some of the latter in the Galápagos and Seychelles Islands may weigh several hundred pounds. Most turtles are rather sluggish in their movements, which may account for their longevity; some are believed to live more than 100 years.

Turtles eat both vegetable and animal products. Many marine forms capture fish and other vertebrates. Land tortoises live on insects, plants, and berries. The common box turtle *(Terrapene)* (Fig. 34-8) grows fat during the wild strawberry season.

CROCODILES AND ALLIGATORS
(ORDER CROCODILIA)

Crocodiles are the largest members of Reptilia; some have been captured that were more than 25 feet long. This order is divided into crocodiles (Fig. 34-9) and alligators. Crocodiles have relatively long slender snouts; alligators have shorter and broader snouts. Although all are carnivorous, many will not attack man. The "man-eating" members of the group are found mainly in Africa and Asia. The estuarine crocodile *(Crocodylus porosus)* found in southern Asia grows to a great size and is very much feared. Alligators are usually less aggressive than crocodiles. Alligators are almost unique among reptiles in being able to make definite sounds. The male alligator can give loud bellows in the mating season. Vocal sacs are found on each side of the throat and are inflated when he calls. In the United States, *Alligator mississipiensis* is the only species of alligator; *Crocodylus americanus* is the only species of crocodiles. The latter is confined to Florida and is almost extinct. The American alligator is found from North Carolina to Florida and west to Texas. Few specimens now caught are more than 12 feet long. Although alligators can put up a severe fight when cornered, they are timid toward man and will avoid him.

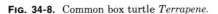
FIG. 34-8. Common box turtle *Terrapene*.

FIG. 34-7. Green sea turtle *Chelonia*. Note that limbs are modified into flippers. Such turtles are strictly aquatic except when they lay their eggs on sandy shore. Some of these turtles weigh as much as 400 to 500 pounds.

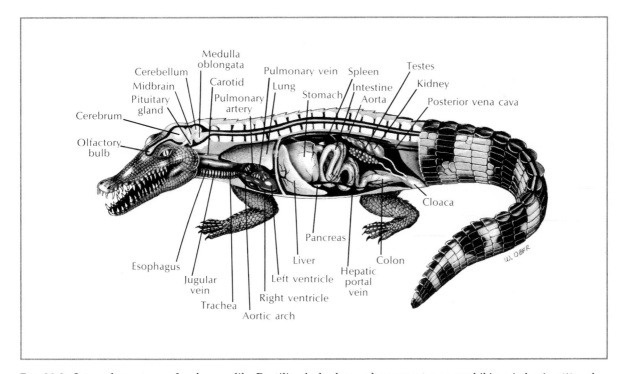

FIG. 34-9. Internal structures of male crocodile. Reptilian body shows advancement over amphibians in having (1) scaly skin for dry land, (2) amniotic egg for land existence, (3) better ossified skeleton and more efficient limbs, (4) heart partly or completely separated into four chambers, and (5) beginning regulation of body temperature by behavior patterns.

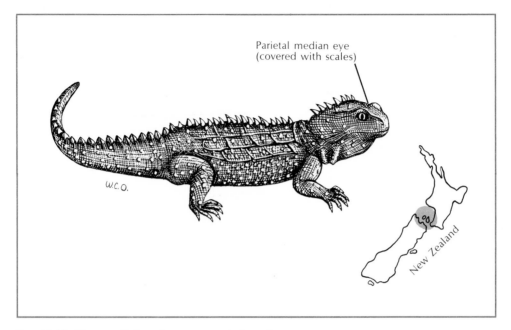

Parietal median eye
(covered with scales)

W.C.O.

New Zealand

FIG. 34-10. Tuatara *(Sphenodon punctatum)*, the only living representative of order Rhynchocephalia. This "living fossil" reptile has well-developed parietal "eye" with retina and lens on top of head. Eye is covered with scales and is considered nonfunctional but may have been important sense organ in early reptiles. The tuatara is found only on islands of Cook Strait, New Zealand.

RHYNCHOCEPHALIA

This order, which has many fossil forms, is represented by only one living species, the tuatara *(Sphenodon punctatum)* of New Zealand (Fig. 34-10). This "living fossil" is found on one or two islands in Cook Strait and is protected by the New Zealand government. It is a lizardlike form about 2 feet or less in length and has a number of primitive characteristics, such as unique skull peculiarities and a well-marked **parietal eye** that is less degenerate than those in other animals and is one of its most distinguishing structures. This form may not be an ancestor from which modern lizards have descended but may represent an independent specialized type. The members of this order appeared in the Triassic period and all became extinct during the Mesozoic, except for *Sphenodon.*

BIRDS (CLASS AVES*)

There are more than 8,600 species of birds distributed all over the world. Their taxonomy has been thoroughly worked out, and most authorities do not expect to find many new species in the future.

Despite their gift of flight, many birds have adapted themselves to certain climatic zones and do not stray from them. Thus the arctic and tropic regions have extensive bird life that is unique for those areas.

Colombia in South America has reported the largest list of bird species in any one region, over 1,700 species. In contrast, Texas in the United States reports 545 species, which probably exceeds the number seen in most other states. The most numerous bird in the world appears to be the domestic fowl; the most numerous wild species are the starling and house sparrow. The total bird population in the world is estimated to be about 100 billion, or about thirty times man's population. At least 2% of all bird species are on the en-

*A′vez (L. *avis*, bird).

Fig. 34-11. Group of ratite birds.

Cassowary (Casuarius)
(Australia and adjacent
islands)

Kiwi (Apteryx)
(New Zealand)

These flightless birds (ratites) have evolved from flying
ancestors but live either in localities where there are
few carnivorous enemies, such as in Australia and New Zealand,
or on plains where they can see and outdistance their
potential predators

Ostrich (Struthio)
(Africa)

Rhea (Rhea)
(South America)

dangered list of rare birds, the ivory-billed wood-pecker being the rarest in the United States. Many birds have become extinct in the past few hundred years, including our own passenger pigeon in 1914.

RÉSUMÉ OF MORE IMPORTANT ORDERS

Class Aves (birds) is made up of about twenty-seven orders of living birds and a few fossil orders. Of these 20 are represented by North American spe-cies. Existing birds are divided into two groups: (1) **ratite** (Ratitae—ra-ti′te) (Fig. 34-11), or those that have a flat sternum with poorly developed pec-toral muscles and are flightless, and (2) **carinate** (Carinatae—kar″i-na′te), or those that have a keeled sternum with large pectoral muscles and can fly. Most paleontologists think that the flightless forms were derived from those that could fly. Most flightless forms are found where there are few car-nivorous enemies, or else where they can outrun predators.

FIG. 34-12. European white stork *(Ciconia ciconia)*. This bird is woven into folklore of many European countries such as Holland and Denmark, where it often nests on tops of chimneys. Order Ciconiiformes. (Courtesy Chicago Natural History Museum.)

FIG. 34-13. Pair of trumpeter swans *(Cygnus buccinator)*. These rare birds are almost extinct and are rigidly protected in their few nesting sites. Order Anseriformes. (Courtesy W. E. Banko, U. S. Fish and Wildlife Service.)

In the list below are ten of the more important of the twenty orders of North American carinate birds.

Order Pelecaniformes (pel″e-kan″i-for′mes) — **pelicans, cormorants, gannets, boobies**

Order Ciconiiformes (si-ko″ne-i-for′mes) — **herons, bitterns, storks** (Fig. 34-12), **ibises, spoonbills, flamingos**

Order Anseriformes (an″ser-i-for′mes) — **swans** (Fig. 34-13), **geese, ducks**

Order Falconiformes (fal″ko-ni-for′mes) — **eagles** (Fig. 34-14), **hawks, vultures, falcons, condors** (Fig. 34-15), **buzzards**

Order Galliformes (gal″li-for′mes) — **quail, grouse, pheasants, ptarmigan, turkeys, domestic fowl** *(Gallus domesticus)*

Order Charadriiformes (ka-red″re-i-for′mes) — **shore birds, such as gulls, oyster catchers, plovers, sandpipers, terns, woodcocks**

Order Columbiformes (ko-lum″bi-for′mes) — **pigeons, doves**

Order Strigiformes (stri″ji-for′mes) — **owls** (Fig. 34-16)

Order Apodiformes (a-pod″i-for′mes) — **swifts, hummingbirds** (Fig. 34-17)

Order Passeriformes (pas″er-i-for′mes) — **perching birds** (Fig. 34-18)

Fig. 34-14. Bald eagle *(Haliaeetus leucocephalus).* Although this eagle is national emblem of our country, it is found in only a few restricted areas. Order Falconiformes. (Courtesy Chicago Natural History Museum.)

Fig. 34-15. South American or Andean condor *Vultur,* largest flying bird (by weight) in the world. It may have a wingspread of more than 12 feet, lives chiefly on carrion. Order Falconiformes. (Courtesy Smithsonian Institution, Washington, D. C.)

471

FIG. 34-16.
Screech owl *(Otus asio).*
Order Strigiformes.

FIG. 34-17. Ruby-throated hummingbird *(Archilochus colubris).* Of fourteen species of hummingbirds in United States, ruby-throated bird is the only one found east of the Mississippi. Order Apodiformes.

FIG. 34-18. Mockingbird *(Mimus polyglottos).* One of the most famous songsters of North America. It sings at any season at any time of day and is great imitator of other birds' songs. Order Passeriformes. (Courtesy Natural Science Museum, Cleveland.)

CHARACTERISTICS

1. Body usually spindle shaped, with four divisions: head, neck, trunk, and tail; **neck disproportionately** long for balancing and food gathering
2. Limbs paired, with the **forelimbs usually adapted for flying;** posterior pair variously adapted for perching, walking, and swimming; foot with four toes (chiefly)
3. Epidermal **exoskeleton of feathers** and **leg scales;** thin integument of epidermis and dermis; **pinna of ear rudimentary**
4. **Skeleton fully ossified with air cavities or sacs;** jaws covered with **horny beaks;** small ribs; sternum well developed with keel or reduced with no keel; **no teeth**
5. Nervous system well developed, with brain and twelve pairs of cranial nerves
6. Circulatory system of **four-chambered heart,** with the **right aortic arch persisting**
7. Respiration by slightly expansible lungs, with thin air sacs among the visceral organs and skeleton; **syrinx (voice box) near junction of trachea and bronchi**
8. Excretory system by metanephric kidney; **no bladder; urine of urates, semisolid**
9. Sexes separate; **females with left ovary** and **oviduct**
10. Fertilization internal; **egg with much yolk** and **hard calcareous shells;** embryonic membranes in egg during development; **incubation external;** young active at hatching **(precocial)** or helpless and naked **(altricial)**

ADAPTATIONS AND NATURAL HISTORY

Flight. Before birds acquired the power of flight, they may have passed through a sequence pattern of swift running, flying leaps, tree climbing, parachute gliding from tree to tree, and a general arboreal existence. Many of their adaptations, for example, the perching mechanism and active climbing habits, indicate such an apprenticeship. Feathers were an absolute requirement for true flight, for there is no reason to suppose that they passed through a stage of wings composed of skin membranes, such as the winged reptiles (pterodactyls) had. In ordinary flying, birds have to make use of the same principles and solve the same problems that confront human heavier-than-air aircraft.

Wings are required to function both as propellers and planes. In flying, birds elevate the wings, and then pull them forward, downward, and backward. When the wing is moved downward, the air is dis-placed and the bird is kept up or raised; the backward movement gives horizontal velocity to the body; and the air resistance, which retards the forward movements, also has a lifting force because the bird's wing is convex above and offers less resistance than the lower concave surface. Between strokes the loss of altitude may be slight in swift flying, but it is quite evident in the up-and-down flight of woodpeckers and goldfinches. Rapidity of flight within limits lessens the energy required to fly. It is estimated that the energy expended by a pigeon in taking off is five times as great as that needed to maintain its regular speed.

There are three types of bird flight—gliding, flapping, and soaring.

Gliding flight. In this type a bird attains a certain velocity and then planes without moving the wings, or, having reached a certain altitude, it descends without wing stroke. This is probably the most primitive method of flying.

Flapping flight. This type (already described) refers to the up-and-down movement of the wings and represents the most complicated mechanism of flight. It is ordinary bird flight and is often described as a screwlike wing motion, with the primary feathers acting as propellers and the secondary feathers furnishing the plane lift.

Soaring flight. In soaring flight, birds usually take advantage of updrafts and air currents so that flapping is dispensed with for considerable periods of time. The long, narrow wings of gulls and albatrosses and the short, broad wings of hawks are equally effective for this type of flight. Hawks often make use of the ascending warm air that arises from warmed areas of the earth to mount in circles to great heights.

Speed of flight. How fast birds fly depends to a great extent on the conditions of the air. Against a strong headwind their speed is naturally cut down. Under favorable conditions most small songbirds fly from 20 to 40 miles per hour. Larger birds such as ducks usually fly from 40 to 80 miles per hour. Some swifts are known to fly more than 100 miles per hour, and some birds of prey probably exceed this in diving after their prey. Many estimates of speed, however, are purely guesswork and lack accurate measurement.

Feathers. Feathers are modified from the epi-

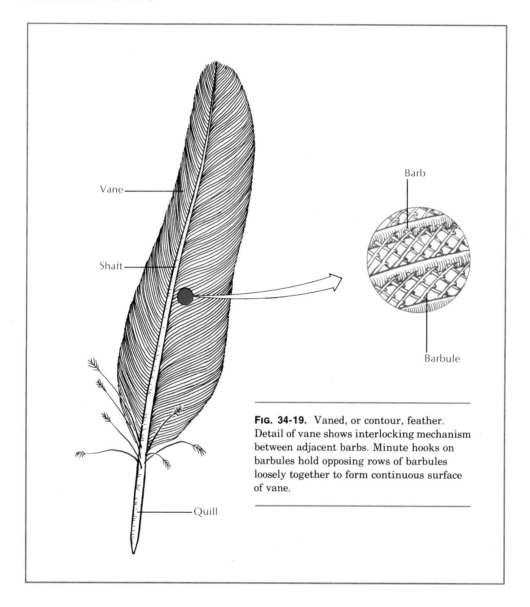

Vane

Shaft

Quill

Barb

Barbule

Fig. 34-19. Vaned, or contour, feather. Detail of vane shows interlocking mechanism between adjacent barbs. Minute hooks on barbules hold opposing rows of barbules loosely together to form continuous surface of vane.

dermis and serve for insulation, support of the body in flight, protection of the skin, and regulation of body heat. A typical feather consists of a hollow **quill,** or calamus, thrust into the skin and a **shaft,** or rhachis, which is a continuation of the quill and bears the **barbs** (Fig. 34-19). On the sides of each barb are the smaller **barbules,** whose opposing rows are held together by small **barbicels** with hooks (hamuli). Some flightless birds (ostrich) lack this interlocking mechanism and have a fluffy plum-age. If the barbs and barbicels form a flat expansive surface, the structure is called a **vane;** when the barbs form only a fluffy mass, it is called **down.**

Feathers are divided into three types:

1. The **contour feather** consists of a central shaft and parallel barbs, arranged to form a vane. There are several types of contour feathers, such as the wing feathers, which include primaries on the hand, secondaries on the forearm, tertiaries on the upper arm (humerus), and tail feathers. Each feather is

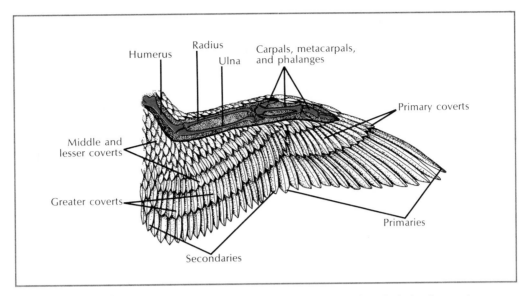

FIG. 34-20. Spread right wing of typical bird. Forelimb of bird modified for flight by changes in carpus and hand. Humerus, ulna, and radius are little changed, but there are only two free carpals (ulnare and radiale); other carpal bones fuse with metacarpals to form carpometacarpus. Of the three digits (II, III, and IV), second has short metacarpal and one phalanx and bears alula; third has long metacarpal and two phalanges; fourth has one long metacarpal and one phalanx. Primary flight feathers are supported by "hand" (digits III and IV and metacarpus), secondary flight feathers by ulna and radius, and tertiary feathers (if present) by humerus.

usually covered above and below by rows of other feathers known as **coverts** (Fig. 34-20).

2. **Down feathers** are entirely of down and are found interspersed among the contour feathers. They have short quills and short barbules. They are especially abundant on the breasts and abdomens of certain birds such as water birds and birds of prey.

3. **Filoplume feathers** are a kind of hair feathers in which the shaft is greatly reduced with few or no barbs. They are the pinfeathers that remain on a plucked bird. Modified filoplumes form rictal bristles about the mouth of flycatchers and whippoorwills.

The term **plumage** refers to all the feathers collectively. **Molting** is the shedding and replacement of feathers. At certain seasons, usually spring and fall, the molting continues until it is complete. Usually a bird has four moltings during its first year of life, but birds in poor health may omit a molt entirely. A molt usually requires about 6 weeks.

Feather counts vary greatly with the different species. A hummingbird may have less than 1,000 feathers, whereas a whistling swan may have more than 25,000. The same species may show a seasonal variation, having the greatest number in the winter. The large wing and tail feathers are remarkably constant in number. The bare parts on vultures are exposed to ultraviolet rays to keep away bacteria that might be encouraged by matted feathers.

Coloration. Coloration is most striking in tropical birds, but many that dwell in the temperate zones are also renowned for their colors. The color is partly due to pigments and partly to interference colors produced by light reflection and refraction. Some of the pigments (melanins) are granules of yellow, black, red, and a few others. Other pigments are the lipochromes, such as zooxanthin (yellow) and zooerythrin (red). Colors produced this way always appear the same, but those due to reflection and refraction may appear different, depending on the conditions under which they are seen. White is produced by reflection. The surface markings as well as the internal structure play a part in this interference coloration. Whenever there is an absense of pigment, an albino may result.

Molting may be accompanied by a change in the color of the plumage. The ptarmigan is white in winter and mottled brown at other seasons. The juvenile plumage in young birds may resemble that of the adult female. In **sexual dimorphism** the male is usually the brighter. Examples are the cardinal and scarlet tanager.

Songs. When air passes through the syrinx, the semilunar membrane vibrates to produce sounds. In the song sparrow more than 800 different song variations have been recorded. Nearly all songbirds studied have different songs for different purposes, many of which convey precise information.

In addition to songs, which are more common in the males during the breeding season, birds have distinctive call notes, which are uttered when they are alarmed or are used for attracting mates or young, as well as for territory rights. Many birds have great powers of mimicry; among these are the brown thrasher, catbird, mocking bird, myna, and, of course, the parrot.

Food. Because birds are very active their food requirement is large. Their body metabolism and body temperature are the highest known among animals. Many birds are strictly vegetarian, consuming great quantities of foliage, seeds, and fruits, because of the less concentrated nature of the food. Others such as birds of prey are entirely carnivorous. Many small songbirds such as flycatchers and woodpeckers live exclusively upon insects. Still other birds are omnivorous. During the nesting period many songbirds feed their nestlings insects, although the adults themselves are vegetarians. Some birds, such as the hummingbird that lives upon the nectar of flowers and the kingfisher that lives upon fish, have rather restricted food habits. The hummingbird requires so much food that it would starve to death at night except that it goes into a state of torpor, with low breathing rate, low heart rate, and low body temperature.

Care of young. To produce offspring, all birds lay eggs that must be incubated by one or both parents. Cowbird eggs require only 9 to 10 days for hatching; most songbirds, about 14 days; the hen, 21 days; ducks and geese, at least 28 days; and the wandering albatross, 73 days. Most of the duties of incubation fall upon the female, although in many instances both parents share in the task, and occasionally, as with the phalarope, only the male performs this work.

Nests vary from depressions on the ground to huge and elaborate affairs. Some birds simply lay their eggs on the bare ground or rocks. Some of the most striking nests are the pendant nests constructed by orioles (Fig. 34-21), the neat lichen-covered nests of hummingbirds (Fig. 34-22) and flycatchers, the chimney-shaped mud nests of cliff swallows, and the huge brush pile nests of the Australian brush turkey.

When birds hatch, they are of two types: **precocial** or **altricial.** The precocial young, such as quail, fowl, ducks, and most water birds, are covered with down when hatched and can run or swim as soon as their plumage is dry. The altricial ones, on the other hand, are naked and helpless at birth and remain in the nest for a week or more. Nesting success is very low with many birds, especially in altricial species.

Many birds, such as the eagle and some of the songbirds, are known to mate for life. Others mate only for the rearing of a single brood. In most bird populations there are usually many sexually mature individuals that have no mates at all. Elaborate courtship rituals are found in many birds, such as the prairie chicken, sage grouse, bower birds, great crested grebes, and others.

Bird populations. The National Audubon Society and the Federal Fish and Wildlife Service have sponsored many censuses. Some are concerned with game birds and species on the verge of extinction. Emphasis is often placed upon breeding birds, making use of the territorial singing of the mates. In 1914 a bird census made in the northeastern United States revealed about 125 pairs of birds per 100 acres (open farms) and 199 pairs per 100 acres of woodland. In 1949 a survey made in a spruce-fir forest in Maine gave a count of 370 pairs of breeding birds per 100 acres. Another count in the same region in 1950 showed 385 pairs of breeding birds per 100 acres. In areas where pesticides are used extensively, bird populations have declined, sometimes severely.

Selection of territories. A territory is selected in the spring by the male, who jealously guards it against all other males of the same species. The male sings a great deal to help him establish priority on his domain. The female apparently wanders from one territory to another until she settles down with a male. When members of another species trespass, the pair usually pay little attention. Competition is greatest among the members of the same species. Song sparrows, however, try to keep off members of other species as well as their own. A recent study indicates that birds may also defend their territories against other species because of environmental limitations and changes or because competition for food or other factors between different species (usually closely related) may occur.

Behavior and intelligence. Among their amazing behavior reactions are the elaborate and skillfully built nests characteristic of many species, the migratory instinct that enables birds to travel thousands of miles without deviating from their courses, the dexterity they show in food capture, the courtship and mating rituals of some, and their power to produce music.

It is not always possible to distinguish between strictly inborn reflexes and acquired associations. Many of the former are found in newly hatched birds, for they are able to perform them without previous experience, but the parents are undoubtedly responsible for furnishing the stimuli in the development of the instinctive behavior of the young. The power birds have of radically changing their food and nesting habits indicates that they can establish simple associations. For example, when their ground nests were threatened by mongooses, certain birds of the West Indies took to building in trees.

FIG. 34-21. A, Pendant nest of Baltimore oriole. **B,** Nest of red-eyed vireo.

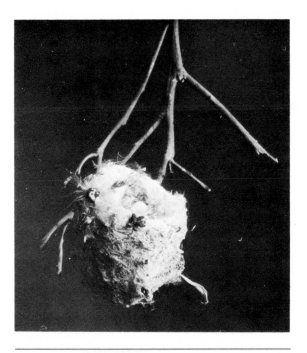

FIG. 34-22. Nest of hummingbird with two eggs. Collected in Arizona.

35

Mammals

CLASS MAMMALIA*

The term "mammalia" was given by Linnaeus (1758) to that group of animals which are nourished by milk from the breasts of the mother. There are only between 4,000 and 5,000 named species of mammals, together with many thousands of subspecies. The evolutionary transition from reptiles to mammals was characterized by metabolic advancements that have made this group the dominant one in the animal kingdom. In them are carried to a high level of efficiency those potentialities that were more or less latent in other vertebrates. Mammals have stressed the nervous system as their chief contribution to animal evolution. As a climax to the evolutionary development of the nervous system, man himself has been the outcome.

ORIGIN AND RELATIONSHIPS

The ancestor of all mammals appears to have been an early type of reptile that lived when the dinosaurs flourished during the Jurassic period. When the dinosaurs vanished near the beginning of the Cenozoic era, mammals suddenly erupted into the

*Mam-ma′ le-a (L. *mamma*, breast).

numerous ecologic niches vacated by reptiles. During the Eocene and Oligocene periods most of the orders of mammals as we know them originated.

RÉSUMÉ OF LIVING ORDERS

Many authorities think that when all mammals are classified there will be some 20,000 species and subspecies. Fifteen thousand or more species and subspecies have been named, divided among seventeen or eighteen living orders. According to Simpson's classification of mammals, there are eighteen living and fourteen extinct orders of mammals. The eighteen living orders are listed below.

Order Monotremata (mo″no-tre′ mah-tah) – **egg-laying mammals, e.g., duck-billed platypus, spiny anteater** (Fig. 35-1)

Order Marsupialia (mar-su″pe-a′ le-a) – **pouched mammals, e.g., opossums** (Fig. 35-2), **kangaroos, kaola**

Order Insectivora (in″sek-tiv′ o-ra) – **insect-eating mammals, e.g., shrews, hedgehogs, moles**

Order Chiroptera (ki-rop′ ter-a) – **flying mammals** (Fig. 35-3)

Order Dermoptera (der-mop′ ter-a) – **flying lemurs**

Order Carnivora (kar-niv′ o-ra) – **flesh-eating mammals, e.g., dogs, wolves** (Fig. 35-4), **cats, bears, weasels**

FIG. 35-1. Mammal that lay eggs—duck-billed platypus *(Ornithorhynchus anatinus)*. Order Monotremata. These monotremes are found only in restricted regions of Australia, Tasmania, and New Guinea, where they are found in rivers and ponds; they feed on insects and shellfish. They make complicated burrows in banks of their water habitats. Eggs are about 3/4 inch in size and are incubated by mother for about 2 weeks before hatching. From her fur, young suck milk that seeps from her milk glands. (Courtesy New York Zoological Society.)

FIG. 35-2. Mother opossum *(Didelphis virginiana)* with four young. Young are first carried in her abdominal pouch and later on her back. Order Marsupialia. (Courtesy R. M. Blake, U. S. Fish and Wildlife Service.)

FIG. 35-3. Large brown bat *(Eptesicus fuscus)* often found in old barns and church steeples. They frequently hang head down as this one is doing. Single young is usually born in June. Order Chiroptera.

FIG. 35-4. Coyote *(Canis latrans)*. His cunning in avoiding capture and his destruction of livestock has made him a serious pest in certain localities. Order Carnivora, family Canidae. (Courtesy E. P. Haddon, U. S. Fish and Wildlife Service.)

FIG. 35-5. Prairie dog *(Cynomys ludovicianus)* feeding. Where they are numerous, these rodents do a great deal of damage by burrowing and by destroying crops. Order Rodentia. (Courtesy D. A. Spencer, U. S. Fish and Wildlife Service.)

FIG. 35-6. Muskrat *(Ondatra zibethica),* most common fur-bearing animal in America. Order Rodentia, family Muridae.

Order Tubulidentata (tu"bu-li-den-ta'ta) — **aardvarks**
Order Rodentia (ro-den'te-a) — **gnawing mammals, e.g., squirrels, rats, woodchucks** (Figs. 35-5 and 35-6)
Order Pholidota (fol"i-do'ta) — **pangolins**
Order Lagomorpha (lag"o-mor'fa) — **rabbits, hares, pikas**
Order Edentata (e"den-ta'ta) — **toothless mammals, e.g., sloths, anteaters, armadillos** (Fig. 35-7)
Order Cetacea (se-ta'she-a) — **fishlike mammals, e.g., whales, dolphins, porpoises**
Order Proboscidea (pro"bo-sid'e-a) — **proboscis mammals, e.g., elephants**
Order Hyracoidea (hy"ra-koi'de-a) — **hyraxes, e.g., conies**
Order Sirenia (si-re'ne-a) — **sea cows, e.g., manatees** (Fig. 35-8)
Order Perissodactyla (pe-ris"so-dak'ty-la) — **odd-toed hoofed mammals** (Fig. 35-9)
Order Artiodactyla (ar"te-o-dak'ty-la) — **even-toed hoofed mammals** (Figs. 35-10 and 35-11)
Order Primates (pri-ma'tez) — **highest mammals, e.g., lemurs** (Fig. 35-12), **monkeys, apes** (Fig. 35-13), **man**

FIG. 35-7. Armadillo *(Dasypus novemcinctus).* Although timid and inoffensive, armadillo has survived mainly because of its protective coat of armor. Order Edentata. (Courtesy Smithsonian Institution, Washington, D. C.)

FIG. 35-8. Manatees *(Trichechus mamatus),* or sea cows, are aquatic mammals that sometimes reach length of 10 feet and live in estuaries of tropical and subtropical America. Order Sirenia. (Courtesy Chicago Natural History Museum.)

FIG. 35-9. Malay tapir *(Tapirus indicus).* Order Perissodactyla. (Courtesy Smithsonian Institution, Washington, D. C.)

FIG. 35-10. Group of Alaskan caribou *(Rangifer arcticus)*. Order Artiodactyla. (Courtesy Chicago Natural History Museum.)

FIG. 35-11. Hippopotamus *(Hippopotamus amphibius)* and young. Order Artiodactyla. (Courtesy Smithsonian Institution, Washington, D. C.)

FIG. 35-12. Ring-tailed lemur *(Lemur catta).* Order Primates, suborder Lemuroidea. (Courtesy Chicago Natural History Museum.)

CHARACTERISTICS

1. **Body covered with hair,** which may be reduced in some
2. **Integument with sweat, sebaceous,** and **mammary glands**
3. Mouth with teeth on both jaws
4. **Movable eyelids** and **fleshy external ears**
5. Four limbs (reduced or absent in some) adapted for many forms of locomotion
6. Circulatory system of a four-chambered heart, **persistent left aorta,** and **nonnucleated red blood corpuscles**
7. Respiratory system of lungs and a voice box
8. **Muscular partition between thorax and abdomen**
9. Excretory system of metanephros kidneys and ureters that usually open into the bladder
10. Nervous systems of a well-developed brain and twelve pairs of cranial nerves
11. **Warm blooded**
12. Separate sexes
13. Internal fertilization; **eggs develop in a uterus** with **placental attachment** (except monotremes); **fetal membranes (amnion, chorion, allantois)**
14. Young nourished by **milk from mammary glands**

484

FIG. 35-13. Chimpanzee *(Pan satyrus).* Order Primates, superfamily Hominoidea. (Courtesy Smithsonian Institution, Washington, D. C.)

ADAPTATIONS AND NATURAL HISTORY

Mammals, with their highly developed nervous system, alertness, and adaptations, display a remarkable range of activity. They demonstrate the highest range of intelligence patterns in the animal kingdom and have worked out ingenious devices for survival.

Size ranges. The smallest mammal known is the pigmy shrew *(Sorex),* with a body length of less than 1½ inches and weighing only a fraction of an ounce. The largest is the whale, certain species of which may reach a length of 103 feet and may weigh more than 100 tons. The African elephant, which may reach a height of 11 feet at the shoulders and may weigh 7 to 8 tons, is the largest living terrestrial mammal. In no other group of animals do we find such a great size range, which indicates the wide adaptability of mammals.

Hair: a unique mammalian characteristic. Hair probably originated from the tactile sensory pits of fish and amphibians and from apical bristles that at first were of sensory function. The distribution was the same as that of scales. Although in most mammals the scales are replaced with hair, they do persist to some extent in odd forms such as the scaly anteater and on the scaly tails of the beaver, rat, opossum, lemur, and shrew. In the whale and some other aquatic mammals hair is reduced to only a few bristles on the upper lip. The structure and development of hair was described on p. 85.

Mammals usually have two kinds of hair: (1) the thick and soft underhair next to the skin for insulation and (2) the coarser and longer guard hair for protection against wear. Around the nose and eyes of many are the long tactile **vibrissae.** The hair covering, or **pelage,** varies greatly with different mammals, depending on the climate.

Hair is usually shed once or twice a year. The hairs making up the fur of most animals cease to grow after reaching a certain length, but those on the scalp of man continue to grow unless they are shed. Among the varieties of hair are the spines of the porcupine and hedgehog, the stiff bristles of hogs, and the wool of sheep.

Body temperature. Mammals, like birds, are warm blooded (homoiothermic or endothermal). Many of the larger members have temperatures from 100° to 103° F. Man's temperature is 98.6° F. Primitive mammals (monotremes) have very imperfect temperature control; their body temperature may vary 10° F. or more daily. Such mammals are called heterothermal and their temperature is regulated chiefly by variations of heat production (increased by activity) and by the surrounding temperature. Many newly born mammals are naked, helpless, and lack any capacity to regulate their body temperature (poikilothermal).

Smaller desert mammals are mostly fossorial (fitted for digging burrows) and nocturnal; larger ones make physiologic adjustments of their homeostatic mechanisms of thermoregulation or osmoregulation to avoid undue loss of water by evaporation. A steep temperature gradient exists between the surface of the pelage and the skin surface when mammals are exposed to high solar radiation. Heat from the pelage is returned to the environment by reradiation and convection. Sweating and panting are important avenues of heat loss by evaporative cooling. The camel can store heat during the day, allowing its body temperature to rise several degrees. The heat is lost during the cool night. It can also endure a much greater loss of body water (dehydration tolerance) than other mammals.

Arctic animals as foxes, Eskimo dogs, caribou, and Arctic sea gulls make use not only of thick fur and feathers for better insulation but also of special vasomotor controls. The arteries and veins run close together and intertwine. When the warm arterial blood enters a limb, it is cooled by the returning cold venous blood by a countercurrent exchange. In this way, valuable heat is returned to the body core instead of lost to the environment.

Hibernation. Many mammals solve the problems of winter scarcity of food and low climatic temperatures by undergoing (1) a state of **lethargy** or (2) **true hibernation.** In lethargy the mammal becomes dormant for several weeks, during which time it has intermittent periods of sleep and wakefulness. When asleep it responds quickly to stimulation and its metabolic processes are only a little lower than those of an active animal. There is little to no lowering of temperature. Such animals may remain in a dormant condition for considerable lengths of time, but the light sleepers frequently interrupt their winter sleep by stirring around.

Good examples of this type of winter sleeper are skunks, bears, opossums, badgers, and raccoons. True hibernation, on the other hand, involves an inactive state in which bodily processes are vastly lowered, due to a marked fall in temperature to a new level which may be only slightly higher than that of the surroundings (woodchucks, bats, ground squirrels, and marmots). The hibernator always monitors its body temperature, and if freezing threatens, it will awaken and warm up by violent shivering.

Summer torpor, or **aestivation,** is practiced by some mammals, especially by ground squirrels in the western parts of our country.

Reproduction and secondary sex characteristics. Mammals are all viviparous except the monotremes, which lay eggs (Fig. 35-1). Most mammals have definite mating seasons, usually in the winter or spring. The female mating function is restricted to a periodic cycle, known as the **estrus cycle,** or "heat." At other times, they will not allow the male to approach them. The estrus cycle is marked by certain characteristic changes in the vagina and uterus. The cycle is divided into the **anestrum,** or resting period; **proestrum,** or preparation for mating; **estrum,** or period of accepting the male; and **metestrum,** or period of regressive changes in the uterine and vaginal walls. Unless fruitfully mated, the female rat comes in estrus about every 4 days; the female dog, about every 6 months; the female house mouse, every 4 to 6 days; and the cow, every 21 days. Those animals that have only a single estrus during the breeding season are called **monoestrus;** those that have a recurrence of estrus during the breeding season are called **polyestrus.** Dogs, foxes, and bats belong to the first group; field mice and squirrels are all polyestrus.

Gestation, or period of pregnancy, varies greatly among the mammals. Mice and rats have about 21 days; rabbits and hares, 30 to 36 days; cats and dogs, 60 days; cows, 280 days; and elephants, 22 months. The marsupials (opossum) have a very short gestation period of 13 days (Fig. 35-2). Small rodents that serve as prey for so many carnivores produce, as a rule, more than one litter of several young each season. Field mice are known to produce seventeen litters of four to nine each in a year.

Most carnivores have but one litter of three to five young a year. Large mammals, elephants and horses, have only one young. Those young born with hair and open eyes and ability to move around are called **precocial** (ungulates and jackrabbits); those that are naked, blind, and helpless (carnivores and rodents) are known as **altricial.**

Population surveys. Surveys of population are made by trapping, observation, tracks, signs, and other devices. In some favorable situations actual counts of all the individuals of a species population can be made such as Darling did in his famous study of various herds of red deer. In other instances indirect methods of marking captured specimens and releasing them must be used. From the ratio of recaptures to the marked numbers one can estimate the number of the whole population of the species in the area being studied. These surveys show that the population of any species of mammal tends to vary from year to year. The population of most mammals is greatest just after the breeding season because of the additional young members. Whenever a species declines rapidly, an epidemic disease is suspected.

A cycle of animal population may take several years, during which time a species may build up to a peak and then may decline. Meadow mice, for instance, usually have a cycle of about three to four years; the snowshoe hare, ten years; and squirrels, five years. Usually smaller mammals have a shorter cycle than do larger species.

The cycle of the Canadian lynx is closely correlated with the cycle of the rabbit or hare, for the lynx preys on these animals.

Population densities of the various species of mammals depend on shelters, food, enemies, and kind of species. Small mammals are usually far more common per unit area than are larger ones. Mice and shrews under favorable conditions may number 100 to 200 or even more per acre; jackrabbits, about 10 to 20 per square mile; and deer, about 12 to 15 per square mile.

Ecologic density refers to the habitable parts of any total area being studied. Rarely do we find an area as big as a township that has a uniformity of ecologic conditions all over it. Such an area may include grassland, forest, marsh, and streams, and any particular species population will be far more concentrated in those habitats for which it is best adapted.

Homes and shelters. Nearly all mammals have places where they rear their young or shelter themselves when they are not actively engaged in feeding or other pursuits. Many have permanent quarters from which they seldom stray far and to which they return to rest. Others such as deer, which range far and wide to forage, are usually not restricted to a permanent home but take advantage of whatever is at hand. Arboreal species, squirrels, raccoons, opossums, and mice, make use of holes in trees. Rabbits, woodchucks, prairie dogs, coyotes, and skunks excavate burrows in the ground.

Food and storage. Animals may be divided into herbivorous, carnivorous, omnivorous, and insectivorous. **Herbivorous** animals feed upon grasses and vegetation and include most domestic animals, deer, elephants, rabbits, squirrels, and hosts of others. The **carnivorous** forms feed mainly upon the herbivorous animals and include foxes, weasels, cats, dogs, fishes, wolverines, lions, tigers, and others. **Omnivorous** mammals live on both plants and animals. Examples of these are man, raccoons, rats, and bears. Those that subsist chiefly on insects are called **insectivorous** and include bats, moles, shrews, etc.

Migration and emigration. Migration refers to the periodic passing from and the return to a region. Many mammals have the practice, although their ranges of migration are less striking than that of birds. The breeding grounds of the fur seal are on the Pribilof Islands off the coast of Alaska, but the females and young winter far to the south of the islands. Toward spring all of them migrate back to the islands for breeding and for rearing their young.

Emigration means the movement away from a territory with no intention of returning. The lemming (*Lemmus*) of Norway and Sweden is one of the best known cases of emigration. These small rodents are found on the high plateaus where they live on the moss and vegetation of that rough terrain. But every four or five years they suffer from overpopulation and overflow into the lower land masses, into the rivers and fiords, and even into the sea, where they perish.

Immigration is the coming into a new region. At first introduced here and there in small numbers, the gray squirrels have spread in the past few decades over more than 40% of the total land surface of England, Scotland, and Wales and have also invaded several counties in Ireland – a good example of immigration.

Economic importance. Man's welfare has been closely related to that of other mammals. He has domesticated the horse, cow, sheep, pig, goat, dog, and many others. He uses these for beasts of burden, for food, and for clothing and many other products.

Still basically a hunter, man hunts mammals for pleasure as well as for profit. Of fur-bearing mammals, the muskrat at present seems to be the most valuable. Many mink and fox farms have proved profitable. A number of wild fur-bearing animals have been almost exterminated.

Many mammals, including the small ones, destroy insects and are thus helpful. Skunks feed on grubs and cutworms. Moles and shrews consume enormous numbers of insect larvae. On the other hand, rabbits, field, mice, and woodchucks can damage crops and fruit trees. Gophers and prairie dogs (Fig. 35-5) do damage with their burrows and destroy valuable crops. A dozen jackrabbits will eat as much as a sheep. Predatory mammals, wolves, coyotes (Fig. 35-4), and panthers destroy livestock. Probably the most destructive mammal is the house rat.

Mammals also carry disease. Bubonic plague and typhus are carried by house rats. Tularemia, or rabbit fever, is transmitted to man by the wood tick carried by rabbits, woodchucks, muskrats, and other rodents. Rocky Mountain spotted fever is carried by ticks on ground squirrels. Trichina worms and tapeworms are acquired by man through hogs, cattle, and other mammals.

REFERENCES TO PART VI

Barrington, E. J. W. 1967. Invertebrate structure and function. Boston, Houghton Mifflin Co.

Barrington, E. J. W. 1965. The biology of hemichordata and protochordata. University reviews in biology, no. 2. London, Oliver & Boyd, Ltd.

Bellairs, A. 1970. The life of reptiles. New York, Universe Books. 2 vols.

Bent, A. C. 1919-1953. Life history of North American birds (many volumes). Washington, D. C., United States National Museum.

Borror, D. J., and D. M. DeLong. 1971. An introduction to the study of insects. New York, Holt, Rinehart & Winston, Inc.

Chapman, R. F. 1969. The insects—structure and function. New York, American Elsevier Publishing Co., Inc.

Cousteau, J. Y., and J. Dugan. 1964. The living sea. New York, Pocket Books. (Paperback.)

Cousteau, J. Y., and F. Dumas. 1953. The silent world. New York, Harper and Row, Publishers (Perennial Library). (Paperback.)

Crowell, S., et al. (editors). 1965. Behavioral physiology of coelenterates. Amer. Zool. **5**:335.

Dales, R. P. 1963. Annelids. London, Hutchinson & Co. (Publishers) Ltd.

Denton, E. 1971. Reflectors in fishes. Sci. Amer. **224**:64-72 (Jan.).

Hadzi, J. 1963. The evolution of the metazoa. Elmsford, N. Y., Pergamon Press, Inc.

Hinton, S. 1969. Seashore life of Southern California. Berkeley, University of California Press.

Johnson, C. W. G. 1963. The aerial migration of insects. Sci. Amer. **209**:132.

Laverack, M. S. 1963. The physiology of earthworms. Elmsford, N. Y., Pergamon Press, Inc.

Mayr, E. 1968. Principles of systematic zoology. New York, McGraw-Hill Book Co., Inc.

Minton, S. A., and M. R. Minton. 1969. Venomous reptiles. New York, Charles Scribner's Sons.

Nelson, B. 1968. Galápagos: islands of birds. New York, William Morrow & Co.

Olsen, O. W. 1962. Animal parasites; their biology and life cycles. Minneapolis, Burgess Publishing Co.

Russell-Hunter, W. D. 1968. A biology of the lower invertebrates. New York, The Macmillan Co.

Russell-Hunter, W. D. 1969. A biology of higher invertebrates. New York, The Macmillan Co.

Scheffer, V. B. 1969. The year of the whale. New York, Charles Scribner's Sons. (Paperback.)

Wood, D. L., R. M. Silverstein, and M. Nakajima (editors). 1970. Control of insect behavior by natural products. New York, Academic Press, Inc.

Young, J. Z. 1962. The life of vertebrates, ed. 2. New York, Oxford University Press, Inc.

glossary

aboral (ab-o′ral) (L. *ab,* from, + *os,* mouth). Pertaining to a region opposite the mouth.

Acanthocephala (a-kan″tho-sef′a-la) (Gr. *akantha,* spine, thorn, + *kephale,* head). A phylum composed of spinyheaded worms that are pseudocoelomate parasites.

acanthor (a-kan′thor) (Gr. *akantha,* spine, thorn). First larval form of acanthocephs in the intermediate host.

acoelomate (a-se′lo-mate) (Gr. *a,* not, + *koilos,* cavity). Without a coelom, e.g., flatworms and proboscis worms.

Actinopterygii (ak″ti-nop″ter-yj′e-i) (Gr. *aktino,* ray, + *pterygion,* fin or small wing). One of the two main groups of bony fish: the ray-finned fish.

adenine (ad′e-nen) (Gr. *aden,* gland, + *ine,* suffix). A component of nucleotides and nucleic acids.

adenosine (a-den′o-sen) **(di-, tri-) phosphate** (ADP and ATP). Certain phosphorylated compounds that function in the energy cycle of cells.

adipose (ad′i-pos) (L. *adipis,* fat). Fatty (tissue).

adrenaline (ad-ren′al-in) (L. *ad,* to, + *renis,* kidney). A hormone produced by the adrenal, or suprarenal, gland.

aerobic (a″er-o′bik) (Gr. *aer,* air, + *bios,* life). Oxygen-dependent form of respiration.

afferent (af′er-ent) (L. *ad,* to, + *ferre,* to bear). Pertaining to a structure (blood vessel, nerve, etc.) leading toward some point.

Agnatha (ag′na-tha) (Gr. *a,* not, + *gnathos,* jaw). A class of vertebrates that includes the modern lampreys and hagfish and the extinct ostracoderms.

allantois (a-lan′to-is) (Gr. *allas,* sausage, + *eidos,* form). One of the extraembryonic membranes of the amniote egg.

allele (al-lel′) (Gr. *allelon,* of one another). One of a pair, or series, of genes that are alternative to each other in heredity and are situated at the same locus in homologous chromosomes. Allele genes may consist of a dominant and its correlated recessive, or two correlated dominants, or two correlated recessives.

alula (al′u-la) (L. dim. of *ala,* wing). The first digit or thumb of a bird's wing, much reduced in size.

alveolus (al-ve′o-lus) (L. dim. of *alveus,* hollow). A small cavity or pit, such as a microscopic air sac of the lungs, terminal part of an alveolar gland, or bony socket of a tooth.

ambulacra (am″bu-la′kra) (L. *ambulare,* to walk). Radiating grooves where podia of water-vascular system project to outside.

amictic (a-mik′tik) (Gr. *a,* without, + *miktos,* mixed or blended). Pertaining to diploid eggs of rotifers or the females that produce such eggs.

amino acid (a-me′no) (amine, an organic compound). An organic acid with an amino radical (NH_2). Makes up the structure of proteins.

amitosis (am″i-to′sis) (Gr. *a,* not, + *mitos,* thread). A form of cell division in which mitotic nuclear changes do not occur; cleavage without separation of daughter chromosomes.

amnion (am′ni-on) (Gr. caul, probably from dim. of *amnos,* lamb). One of the extraembryonic membranes forming a sac around the embryo in amniotes.

amphiblastula (am″fi-blas′tu-la) (Gr. *amphi,* both, on both sides + *blastos,* bud). A blastula in the larval stage of sponges in which one hemisphere bears flagella cells and the other does not.

amylase (am′i-las) (L. *amylum,* starch, + *ase,* suffix meaning enzyme). An enzyme that breaks down starches into smaller units.

anadromous (a-nad'ro-mus) (Gr. *ana*, up, + *dromos*, running). Referring to those fish that migrate upstream to spawn.

anaerobic (an"a-er-o'bik) (Gr. *an*, not, + *aer*, air, + *bios*, life). Not dependent on oxygen for respiration.

androgen (an'dro-jen) (Gr. *andros*, man, + *genes*, born). Any of a group of male sex hormones, e.g., testosterone.

angstrom (Å) (ang'strum) (after Angstrom, physicist). A minute unit of length equal to 1/10,000 of a micron, or 1/100,000,000 of a centimeter.

anhydrase (an-hi'dras) (Gr. *an*, not, + *hydor*, water, + *ase*, enzyme suffix). An enzyme involved in the removal of water from a compound. Carbonic anhydrase promotes the conversion of carbonic acid into water and carbon dioxide.

anlage (ahn'lah-ge) (pl., anlagen) (Ger. *an*, on + *lagen*, to put). The basis of subsequent development; first beginning.

aperture (ap'er-tur) (L. *aperire*, to uncover). An opening; the slight entrance and exit of certain mollusk shells; longer passages are called siphons.

apopyle (ap'o-pil) (Gr. *apo*, away from, + *pyle*, gate). Opening of the radial canal into the spongocoel.

arboreal (ar-bor'e-al) (L. *arbor*, tree). Living in trees.

archenteron (ar-ken'ter-on) (Gr. *archein*, to begin, + *enteron*, gut). The central cavity of a gastrula that is lined with endoderm, representing the future digestive cavity.

archeocyte (ar'ke-o-sit) (Gr. *arche*, beginning, + *kytos*, cell). Ameboid cells of varied function in sponges.

ascon (as'kon) (Gr. *askos*, bladder). Simplest form of canal in sponges, having incurrent canals leading directly into central cavity.

autosome (aw'to-som) (Gr. *autos*, self, + *soma*, body). Any chromosome that is not a sex chromosome.

autotomy (aw-tot'o-my) (Gr. *autos*, self, + *tomos*, cutting). The automatic breaking off of a part of the body.

autotroph (aw"to-trof') (Gr. *autos*, self, + *trophos*, feeder). An organism that makes its organic nutrients from inorganic raw materials.

avicularium (a-vik"u-la'ri-um) (L. *avicula*, dim. of *avis*, bird + *aria*, like). Modified zooid attached to the surface of the major zooid in ectoprocta.

axolotl (ak'so-lot"l) (Nahuatl, *atl*, water, + *xolotl*, doll). Larval stage of *Ambystoma tigrinum mexicanum* exhibiting neotenic reproduction.

axopodia (ak"so-po'di-a) (L. *axis*, an axis, + Gr. *podos*, foot). Stiff pseudopodia with axial filaments, but not for locomotion.

benthos (ben'thos) (Gr. depth of the sea). Those organisms that live along the bottom of seas and lakes.

biogenesis (bi"o-gen'e-sis) (Gr. *bios*, life, + *genesis*, birth). The doctrine that life originates only from preexisting life.

biomass (bi'o-mas) (Gr. *bios*, life, + *maza*, lump or mass). The weight of a species population per unit of area.

biome (bi'om) (Gr. *bios*, life, + *ome*, group). Complex of communities characterized by climatic and soil conditions; the largest ecologic unit.

bipinnaria (bi"pin-nar'i-a) (L. *bi*, double, + *pinna*, wing, + *aria*, like). A free-swimming larval stage of certain sea stars.

blastopore (blas'to-por) (Gr. *blastos*, germ, + *poros*, passage). Opening into archenteron of the gastrula; future mouth in some, future anus in others.

blepharoplast (blef'ah-ro-plast") (Gr. *blepharon*, eyelid). A granule at the base of a flagellum or cilium; a kinetosome.

brachiolaria (bra"ki-o-la'ri-a) (L. *brachiolum*, a small arm + *aria*, like). An asteroid larva with preoral processes.

buffer (buf'er). Any substance or chemical compound that tends to keep pH constant when acids or bases are added.

caenogenesis (se"no-jen'i-sis) (Gr. *kainos*, new, + *genesis*, birth). In the development of an organism, the new stages that have arisen in adaptive response to the embryonic mode of life, such as the fetal membranes of amniotes.

carboxyl (kar-bok'sil) (carbon + oxygen + yl, chemical radical). The acid group of organic molecules—COOH.

carotene (kar'o-ten) (L. *carota*, carrot). A red, orange, or yellow pigment belonging to the group of carotenoids; precursor of vitamin A.

catadromous (ka-tad'ro-mus) (Gr. *kata*, down, + *dromos*, running). Referring to those fish that migrate from fresh water to the ocean to spawn.

catalyst (kat'a-list) (Gr. *kata*, down, + *lysis*, a loosening). A substance that accelerates a chemical reaction but does not become a part of the end product.

cecum (se'kum) (L. *caecus*, blind). A blind pouch at the beginning of the large intestine, or any similar pouch.

Cenozoic (se"no-zo'ik) (Gr. *kainos*, recent, + *zoe*, life). The geologic era from the Mesozoic to the present (about 75 million years).

centriole (sen'tre-ol) (Gr. dim. of *kentron*, center of a circle). A minute granule, usually found in the centrosome and considered to be the active division center of the cell.

centromere (sen'tro-mer) (Gr. *kentron*, center, + *meros*, part). A small body or constriction on the chromosome where it is attached to a spindle fiber.

Chaetognatha (ke-tog'nath-a) (Gr. *chaite*, bristle, hair, + *gnathos*, jaw). Small marine worms, often called arrowworms, with curved bristles on each side of mouth; an enterocoelomate phylum.

chelicera (ke-lis'e-ra) (Gr. *chele*, claw). Pincerlike head appendage on the members of the subphylum Chelicerata.

chloragogue (klo′ra-gog) (Gr. *chloros*, green, + *agogos*, leader). Excretory tissue in the outer surface of the enteron of annelids and some other invertebrates.

choanocyte (ko″a-no-sit′) (Gr. *choane*, funnel, + *kytos*, cell). Flagellate collar cells that line cavities and canals in sponges.

cholinergic (ko″li-ner′jik). Indicating the type of nerve fiber that releases acetylcholine from axon terminal.

chorion (ko′re-on) (Gr. membrane). The outer of the double membrane that surrounds the embryo of the amniotes; in mammals it helps form the placenta.

chromatid (kro′ma-tid) (Gr. *chroma*, color, + *id*, daughter). A half chromosome between early prophase and metaphase in mitosis; a half chromosome between synapsis and second metaphase in meiosis; at the anaphase stage each chromatid is known as a daughter chromosome.

chromomere (kro′mo-mer) (Gr. *chroma*, color, + *meros*, part). The chromatin granules of characteristic size on the chromosome; may be identical with genes or clusters of genes.

cilium (sil′i-um) (L. eyelash). Threadlike organ capable of lashing movements.

circadian (sir″ka-de′an) (L. *circa*, around, + *dies*, day). Recurring in about 24 hours.

climax (kli′maks) (Gr. *klimax*, ladder). A state of dynamic equilibrium; a culmination of the succession in the biota of a community.

clitellum (kli-tel′um) (L. *clitellae*, packsaddle). A thickened, glandlike body on certain portions of midbody segments of earthworms and leeches.

clone (Gr. *klon*, twig). A group of animals produced by asexual reproduction from a single individual.

coelogastrula (se″lo-gas′tru-la) (Gr. *koilos*, hollow, + *gaster*, stomach). The typical gastrula derived from a coeloblastula; a two- or three-layered stage in embryology.

coelom (se′lom) (Gr. *koilos*, hollow). The body cavity in triploblastic animals, lined with mesoderm.

coelomocyte (se″lo-mo-sit′) (Gr. *koiloma*, a hollow, + *kytos*, cell). Primitive or undifferentiated cell of the coelom and water-vascular system of echinoderms.

coenzyme (ko-en′zim) (L. *cum*, with, + Gr. *en*, in, + *zyme*, leaven). A required substance in the activation of an enzyme.

collenchyme (ko-leng′kim) (Gr. *kolle*, glue, + *enchyma*, infusion). Loose mesenchyme cells scattered through noncellular material.

commensalism (ko-men′sal-iz″m) (L. *cum*, together, + *mensa*, table). A symbiotic relationship in which one benefits and the other is unharmed.

community (kom-mu′ni-ti) (L. *communitas*, community). An assemblage of organisms that are associated together in a common environment and interact with each other in a self-sustaining and self-regulating relation.

cotylosaur (kot′i-lo-sor″) (Gr. *kotyle*, cup or vessel, + *sarous*, lizard). A primitive group of fossil reptiles that arose from the labyrinthodont amphibians and became the ancestral stem of all other reptiles.

Ctenophora (te-nof′o-ra) (Gr. *ktenos*, comb, + *phoros*, bearing). A small phylum of marine animals consisting of three germ layers and eight rows of comb plates by which they move.

cytochrome (si′to-krom″) (Gr. *kytos*, cell, + *chroma*, color). One of the hydrogen carriers in aerobic respiration.

cytosome (si′to-som) (Gr. *kytos*, cell + *soma*, body). The cell body as opposed to the nucleus.

cytostome (si′to-stom) (Gr. *kytos*, cell, + *stoma*, mouth). Mouth of unicellular animal.

deoxyribose (de-ok′se-ri″bos) (*deoxy*, loss of oxygen, + *ribose*, pentose sugar). A 5-carbon sugar having 1 oxygen atom less than ribose; a component of deoxyribose nucleic acid (DNA).

dioecious (di-e′shus) (Gr. *dis*, twice, + *oikos*, house). Referring to a condition in which male and female organs occur in different individuals.

dipleurula (di″ploor′u-la) (Gr. *dis*, twice, + *pleura*, side). A hypothetical, bilateral larval form of echinoderms and hemichordates; ancestral form of most deuterostomial animals.

diploid (dip′loid) (Gr. *diploos*, double, + *eidos*, form). Having the somatic number of chromosomes, or twice the number characteristic of a gamete of a given species.

DPN Abbreviation of diphosphopyridine nucleotide, a hydrogen carrier in respiration. Now called NAD, nicotinamide adenine dinucleotide.

Echiuroidea (ek″i-u-roi′de-a) (Gr. *echis*, adder, + *oura*, tail, + *eidos*, form, + *ea*, pl. suffix). A phylum of wormlike animals that inhabit marine coastal mud flats.

ecologic equivalence Ecologic type of the same requirements, which are in similar but geographically separated environments.

ecologic niche The status of an organism in a community with reference to its responses and behavior patterns.

ecosystem (ek′o-sys″tem) (Gr. *oikos*, home, + system). An ecologic unit consisting of both biotic communities and the nonliving (abiotic) environment that interact to produce a stable system.

ecotone (ek′o-ton) (Gr. *oikos*, home, + *tonos*, stress). The transition zone between two adjacent communities.

ectohormone (ek″to-hor′mone) (Gr. *ektos*, outside, + *hormon*, exciting). A pheromone; a substance secreted externally by an organism to influence the behavior of other organisms; an ectocrine.

ectoplasm (ek′to-plaz″m) (Gr. *ektos*, outside, + *plasma*, form). The cortex of a cell or that part of cytoplasm just under the cell surface; contrasts with endoplasm.

emulsion (e-mul′shun) (L. *emulsus*, milked out). A colloidal system in which both phases are liquids.

endergonic (end″er-go′nik) (Gr. *endon*, within, + *ergon*, work). Pertaining to a chemical reaction that requires energy.

endocrine (en'do-krin) (Gr. *endon*, within, + *krinein*, to separate). Referring to a gland without a duct that releases its product directly into the blood or lymph.

endoplasm (en'do-plaz"m) (Gr. *endon*, within, + *plasma*, form). That portion of cytoplasm that immediately surrounds the nucleus.

endoplasmic reticulum (en'do-plaz"mik) (Gr. *endon*, within, + *plasma*, mold or form). The cytoplasmic double membrane with ribosomes (rough) or without ribosomes (smooth).

endostyle (en'do-stil") (Gr. *endon*, within, + *stylos*, column). A ciliated groove in the floor of the pharynx of tunicates, amphioxus, and ammocoetes, used for getting food; may be homologous to the thyroid gland of higher forms.

enterocoel (en'ter-o-sel") (Gr. *enteron*, gut, + *koilos*, hollow). A type of coelom that is formed by the outpouching of a mesodermal sac from the endoderm of the primitive gut.

enterocoelomate (en"ter-o-sel'o-mat) (Gr. *enteron*, gut, + *koilos*, hollow, + *ate*, state of). Those that have an enterocoel, such as the echinoderms and the vertebrates.

Entoprocta (en"to-prok'ta) (Gr. *entos*, within, + *proktos*, anus). A phylum of sessile animals that have the anus enclosed in the ring of ciliated tentacles.

enzyme (en'zim) (Gr. *en*, in, + *zyme*, leaven). A protein substance produced by living cells that is capable of speeding up specific chemical transformations, such as hydrolysis, oxidation, or reduction, but is unaltered itself in the process; a biologic catalyst.

epididymis (ep"i-did'i-mis) (Gr. *epi*, over, + *didymos*, testicle). That part of the sperm duct that is coiled and lying near the testis.

epigenesis (ep"i-jen'e-sis) (Gr. *epi*, over, + *genesis*, birth). The embryologic view that an embryo is a new creation that develops and differentiates step by step from an initial stage; the progressive production of new parts that were nonexistent as such in the original zygote.

epigenetics (ep"i-je-net'iks) (Gr. *epi*, over, + *genesis*, birth). That study of the mechanisms by which the genes produce phenotypic effects.

estrogen (es'tro-jen) (Gr. *oistros*, frenzy, + *genes*, born). An estrus-producing hormone; one of a group of female sex hormones.

euryhaline (yu"re-ha'lin) (Gr. *eurys*, broad, + *halinos*, made of salt). Describing aquatic animals that are able to live in water of a wide range of salinity.

eurytopic (yu"re-top'ic) (Gr. *eurys*, broad, + *topos*, place). Referring to an organism with a wide range of distribution.

exergonic (ek"ser-go'nik) (Gr. *exo*, outside of, + *ergon*, work). Describing an energy-yielding reaction.

exocrine (ek'so-krin) (Gr. *exo*, outside, + *krinein*, to separate). Referring to the type of gland that releases its secretion through a duct.

exteroceptor (ek'ster-o-sep"ter) (L. *exterus*, outward, + *capere*, to take). A sense organ near the skin or mucous membrane that receives stimuli from the external world.

FAD Abbreviation for flavine adenine dinucleotide, a hydrogen acceptor in the respiratory chain.

fermentation (fur"men-ta'shun) (L. *fermentum*, ferment). The conversion of organic substances into simpler substances under the influence of enzymes, with little or no oxygen involved (anaerobic respiration).

fiber, fibril (L. *fibra*, thread). These two terms are often confused. Fiber is a strand of protoplasmic material produced or secreted by a cell and lying outside the cell, or a fiberlike cell. Fibril is a strand of protoplasm produced by a cell and lying within the cell.

Foraminifera (fo-ram"i-nif'er-a) (L. *foramen*, hole, + *ferre*, to bear). An order of sarcodine protozoans with slender branched pseudopodia (myxopodia) that are extruded through holes in their calcareous shells.

gamete (gam'eet) (Gr. *gamos*, marriage). A mature germ cell, either male or female.

Gastropoda (gas-trop'o-da) (Gr. *gaster*, stomach, + *podos*, foot). A class of mollusks consisting of slugs and snails.

Gastrotricha (gas-trot'ri-ka) (Gr. *gaster*, stomach, + *trichos*, hair). A class of aquatic pseudocoelomate animals with cilia or bristles on the body.

gel (jel) (L. *gelare*, to freeze). That state of a colloidal system in which the solid particles form the continuous phase and the fluid medium the discontinuous phase.

gene (jeen) (Gr. *genes*, born). That part of a chromosome that is the hereditary determiner and is transmitted from one generation to another. It occupies a fixed chromosomal locus and can best be defined only in a physiologic or operational sense.

genome (jeen'om) (Gr. *genos*, race). The total number of genes in a haploid set of chromosomes.

genotype (jen'o-tip) (Gr. *genos*, race, + *typos*, form). The genetic constitution, expressed and latent, of an organism; the particular set of genes present in the cells of an organism; opposed to phenotype.

genus (pl., genera) (jee'nus) (L. race). A taxonomic rank between family and species.

germ layer In the animal embryo, one of three basic layers (ectoderm, endoderm, mesoderm) from which the various organs and tissues arise in the multicellular animal.

germ plasm The germ cells of an organism, as seen from the somatoplasm.

gestation (jes-ta'shun) (L. *gestare*, to bear). The period in which offspring are carried in the uterus.

glochidium (glo-kid'i-um) (Gr. dim. of *glochis*, projecting point). A bivalve larva of freshwater clams.

Golgi body (gol'je) (after Golgi, Italian histologist). A cytoplasmic component that may play a role in certain cell secretions or may represent a region where high-energy compounds from the mitchondria collect.

habitat (hab′i-tat) (L. *habitare,* to dwell). The place where an organism normally lives or where individuals of a population live.

haploid (hap′loid) (Gr. *haploos,* single). Having the reduced number of chromosomes typical of gametes, as opposed to the diploid number of somatic cells.

hectocotylus (hek″to-kot′i-lus) (Gr. *hekaton,* hundred, + *kotyle,* cup). Transformed, and sometimes autonomous, arm for transfer of sperm in male cephalopods.

Hemichordata (hem″i-kor-da′ta) (Gr. *hemi,* half, + L. *chorda,* cord). A phylum of wormlike animals with close affinities to the chordates; body of proboscis, collar, and trunk, with stomochord or rudimentary notochord.

hermaphrodite (hur-maf′ro-dit) (Gr. *Hermaphroditos,* containing both sexes; from Greek mythology, son of Hermes and Aphrodite). An organism with both male and female organs. Hermaphroditism commonly refers to an abnormal condition in which male and female organs are found in the same animal; monoecious is a normal condition for the species.

heterotroph (het′er-o-trof″) (Gr. *heteros,* another, + *trophos,* feeder). An organism that obtains both organic and inorganic raw materials from the environment in order to live.

heterozygote (het″er-o-zi′got) (Gr. *heteros,* another, + *zygotos,* yoked together). An organism in which the pair of alleles for a trait is composed of different genes (usually dominant and recessive); derived from a zygote formed by the union of gametes of dissimilar genetic constitution.

holozoic (hol″o-zo′ik) (Gr. *holos,* whole, + *zoikos,* of animals). Referring to the type of nutrition that involves ingestion of solid organic food.

homology (ho-mol′o-ji) (Gr. *homologia,* similarity). Similarity in embryonic origin and adult structure, based on descent from a common ancestor.

homozygote (ho″mo-zi′got) (Gr. *homos,* same, + *zygotos,* yoked together). An organism in which the pair of alleles for a trait is composed of the same genes (either dominant or recessive but not both).

humoral (hu′mer-al) (L. *humor,* a fluid). Pertaining to a body fluid such as blood or lymph.

hydrolysis (hi-drol′i-sis) (Gr. *hydor,* water, + *lysis,* a loosening). The decomposition of a chemical compound by the addition of water; the splitting of a molecule into its groupings so that the split products acquire hydrogen and hydroxyl groups.

hydroxyl (hi-drok′sil) (Gr. *hydor,* water, + oxygen, + yl). Containing an OH— group, a negatively charged ion formed by alkalies in water.

hypertonic (hi″per-ton′ik) (Gr. *hyper,* over, + *tonos,* tension). Referring to a solution whose osmotic pressure is greater than that of another solution with which it is compared; contains a greater concentration of particles and gains water through a semipermeable membrane from a solution containing fewer particles.

hypothalamus (hi″po-thal′a-mus) (Gr. *hypo,* under, + *thalamos,* inner chamber). A ventral part of the forebrain beneath the thalamus; one of the centers of the autonomic nervous system.

hypotonic (hi″po-ton′ik) (Gr. *hypo,* under, + *tonos,* tension). Referring to a solution whose osmotic pressure is less than that of another solution with which it is compared or taken as standard; contains a lesser concentration of particles and loses water during osmosis.

inductor (in-duk′tor) (L. *inducere,* to introduce). In embryology a tissue or organ that causes the differentiation of another tissue or organ.

infusoriform (in″fu-sor′i-form) (L. Infusoria, ciliates, + *forma,* form). A larval stage in dicyemids of phylum Mesozoa.

introvert (in′tro-vurt″) (L. *intro,* inward, + *vertere,* to turn). In the sipunculid, the anterior narrow portion that can be withdrawn into the trunk.

invagination (in-vaj″i-na′shun) (L. *in,* in, + *vagina,* sheath). An infolding of a layer of tissue to form a saclike structure.

isotope (i′so-top) (Gr. *isos,* equal, + *topos,* place). One of several different forms of a chemical element, differing from each other physically but not chemically.

keratin (ker′a-tin) (Gr. *keratos,* horn). A protein found in epidermal tissues and modified into hard structures such as horns, hair, and nails.

kinetosome (ki-neet′o-som) (Gr. *kinetos,* moving, + *soma,* body). The granule at the base of the flagellum or cilium; similar to centriole.

kinin (kin′in) (Gr. *kinetos,* moving). A type of local hormone that is released near its site of origin.

Kinorhyncha (kin″o-ring′ka) (Gr. *kineo,* move, + *rhynchos,* beak). A class of pseudocoelomate animals belonging to the phylum Aschelminthes (Hyman). Same as Echinodera.

labyrinthodont (lab″i-rin′tho-dont) (Gr. *labyrinthos,* labyrinth, + *odontos,* tooth). A group of fossil stem amphibians from which most amphibians later arose. They date from the late Paleozoic.

lacteal (lak′te-al) (L. *lactis,* milk). Referring to one of the lymph vessels in the villus of the intestine.

lacunar (la-ku′nar) (L. *lacuna,* channel). Pertaining to epidermal canal system peculiar to acanthocephs.

lagena (la-je′na) (L. large flask). Portion of the primitive ear in which sound is translated into nerve impulses; evolutionary beginning of cochlea.

lemniscus (lem-nis′kus) (L. ribbon). One of a pair of internal projections of the epidermis from the neck region of Acanthocephala that function in fluid control of the protrusion and invagination of the proboscis.

leukocyte (lu′ko-sit) (Gr. *leukos,* white, + *kytos,* cell). A common type of white blood cell with beaded nucleus.

lipase (li′pas) (Gr. *lipos,* fat, + *ase,* enzyme suffix). An enzyme that converts fats into fatty acids and glycerin; it may also promote the reverse reaction.

lipid, lipoid (lip'id) (Gr. *lipos,* fat). Pertaining to certain fattylike substances that often contain other groups such as phosphoric acid.

lithosphere (lith'o-sfer) (Gr. *lithos,* rock, + *sphaira,* a ball). The rocky component of the earth's surface layers.

littoral (lit'o-ral) (L. *litus,* seashore). The floor of the sea from the shore to the edge of the continental shelf.

lophophore (lo'fo-for) (Gr. *lophos,* crest, + *phoros,* bearing). Tentacle-bearing ridge or arm that is an extension of the coelomic cavity in lophophorate animals.

luciferase (lu-sif'er-as) (L. *lux,* light, + *ferre,* to bear, + *ase,* enzyme suffix). An enzyme involved in light production in organisms.

lunule (lu'nul) (L. dim. of *luna,* moon). Slitlike openings in sand dollar test.

macronucleus (mak"ro-nu'kle-us) (Gr. *makros,* large, + L. *nucleus,* nut). The larger of the two kinds of nuclei in ciliate protozoa; controls all cell functions except reproduction.

madreporite (mad're-po"rit) (It. *madre,* mother, + *poro,* pore + *ite,* suffix for part of body). A sievelike structure on the surface (or interior) of echinoderms, connected to the water-vascular system for the intake of water.

marsupial (mar-su'pi-al) (Gr. *marsypion,* pouch). One of the pouched mammals of the subclass Metatheria.

Mastigophora (mas"ti-gof'o-ra) (Gr. *mastix,* whip, + *phoros,* bearing). A protozoan class whose members have flagella for locomotion; sometimes called flagellates.

matrix (ma'triks) (L. *mater,* mother). The intercellular substance of a tissue, or that part of a tissue into which an organ or process is set.

maxilla (mak-sil'a) (L. jaw). One of the upper jawbones in vertebrates; one of the head appendages in arthropods.

maxilliped (mak-sil'i-ped) (L. *maxilla,* jaw, + *pedis,* foot). One of the three pairs of head appendages located just posterior to the maxilla in crustaceans.

medulla (me-dul'a) (L. marrow). The inner portion of an organ in contrast to the cortex or outer portion; hindbrain.

medusa (me-du'sa) (Greek mythology, female monster with snake-entwined hair). A jellyfish, or the free-swimming stage in the life cycle of coelenterates.

meiosis (mi-o'sis) (Gr. *meion,* to make small). That nuclear change by which the chromosomes are reduced from the diploid to the haploid number.

menopause (men'o-poz) (Gr. *menos,* month, + *pauein,* to cease). In the human female that time when reproduction ceases; cessation of the menstrual cycle.

menstruation (men"stru-a'shun) (L. *mensis,* month). The discharge of blood and uterine tissue from the vagina at the end of a menstrual cycle.

mesenchyme (mes'eng-kim) (Gr. *mesos,* middle, + *enchyma,* infusion). A generalized embryonic tissue; the most primitive connective tissue.

mesoglea (mes"o-gle'a) (Gr. *mesos,* middle, + *gloia,* glue). The jellylike gelatinous filling between the ectoderm and endoderm of certain coelenterates and comb jellies.

mesosoma (mes"o-so'ma) (Gr. *mesos,* middle, + *soma,* body). The middle region of the body of certain invertebrates.

metabolism (me-tab'o-liz"m) (Gr. *metabole,* change). A group of processes that includes nutrition, production of energy (respiration), and synthesis of more protoplasm; the sum of the constructive (anabolism) and destructive (catabolism) processes.

metamerism (me-tam'er-iz-m) (Gr. *meta,* after, + *mere,* part). Condition of segmentation.

metamorphosis (met"a-mor'fo-sis) (Gr. *meta,* beyond, + *morphe,* form). A sudden change in body form after the completion of embryonic development.

metasoma (met"a-so'ma) (Gr. *meta,* after, + *soma,* body). Posterior part of the body of certain invertebrates.

micron (μ) (mi'kron) (Gr. *mikros,* small). One one-thousandth of a millimeter; about 1/25,000 of an inch.

micronucleus (mi"kro-nu'kle-us) (Gr. *mikros,* small, + L. *nucleus,* nut). A small nucleus found in ciliate protozoa; controls the reproductive functions of these organisms.

microsome (mi'kro-som) Gr. *mikros,* small, + *soma,* body). A constituent of cytoplasm that contains RNA and is the site of protein synthesis. Now called a ribosome.

miocyte (mi'o-sit) (Gr. *myos,* muscle, + *kytos,* cell). A contractile cell of the sponges.

miracidium (mi"ra-sid'e-um) (Gr. *meirakidion,* youthful person). A minute ciliated larval stage in the life of flukes.

mitochondria (mit"o-kon'dre-a) (Gr. *mitos,* a thread, + *chondros,* a small roundish mass). Minute granules, rods, or threads in the cytoplasm and the seat of important cellular enzymes.

Mollusca, mollusk (mol-lus'ka, mol'usk) (L. *molluscus,* soft). Mollusca is a major phylum of schizocoelomate animals; body typically of visceral mass, foot, and shell; comprises snails, clams, squids, and others.

monoecious (mo-ne'shus) (Gr. *monos,* one, + *oikos,* house). Having male and female organs in same individual.

monosaccharide (mo"no-sak'a-rid) (Gr. *monos,* one, + *sakcharon,* sugar). A simple sugar that cannot be decomposed into smaller sugar molecules.

morphogenesis (mor"fo-jen'e-sis) (Gr. *morphe,* form, + *genesis,* birth). Development of the architectural features of organisms.

morphology (mor-fol'o-ji) (Gr. *morphe,* form, + *logos,* study). The science of structure. Includes cytology, or the study of cell structure; histology, or the study of tissue structure; and anatomy, or the study of gross structure.

mutation (mu-ta'shun) (L. *mutare,* to change). A stable and abrupt change of a gene; the heritable modification of a character.

myofibril (mi″o-fi′bril) (Gr. *myos,* muscle, + L. *fibra,* thread). A contractile filament within muscle or muscular fiber.

myosin (mi′o-sin) (Gr. *myos,* muscle). A protein found in muscle; important component in the contraction of muscle.

myxedema (mik″se-de′ma) (Gr. *myxa,* slime, + *oidema,* a swelling). A disease that results from thyroid deficiency in the adult.

nacre (na′ker) (F. mother-of-pearl). The iridescent inner layer of mollusk shell.

nekton (nek′ton) (Gr. *nektos,* swimming). Term for the actively swimming organisms in the ocean.

nematogen (nem′a-to-jen) (Gr. *nema,* thread, + *genes,* born). One of the dimorphic forms of dicyemids, which produces vermiform embryos.

Nematomorpha (nem″a-to-mor′fa) (Gr. *nematos,* thread, + *morphe,* form). Hairworms, a pseudocoelomate class of phylum Aschelminthes.

neoteny (ne-ot′e-ni) (Gr. *neos,* new, + *teinein,* to stretch). Condition of permanent, sexually mature, larval state.

notochord (no′to-kord) (Gr. *noton,* back, + L. *chorda,* cord). A rod-shaped cellular body along the median plane and ventral to the central nervous system in chordates.

nucleic acid (nu-kle′ik) (L. *nucleus,* nut). One of a class of molecules composed of joined nucleotides; chief types are deoxyribonucleic acid (DNA), found only in cell nuclei (chromosomes), and ribonucleic acid (RNA), found both in cell nuclei (chromosomes and nucleoli) and in cytoplasm (microsomes).

nucleolus (nu-kle′o-lus) (dim. of nucleus). A deeply staining body within the nucleus of a cell and containing RNA.

nucleoprotein (nu″kle-o-pro′tein). A molecule composed of nucleic acid and protein; occurs in two types, depending on whether the nucleic acid portion is DNA or RNA.

nucleotide (nu′kle-o-tid″). A molecule consisting of phosphate, 5-carbon sugar (ribose or deoxyribose), and a purine or a pyrimidine; the purines are adenine and guanine, and the pyrimidines are cytosine, thymine, and uracil.

nymph (nimf) (L. *nympha,* a young woman). The immature form of an insect that undergoes a gradual metamorphosis.

ontogeny (on-toj′e-ni) (Gr. *ontos,* being, + *gennao,* bring forth). The development of an individual from egg to senescence.

operculum (o-pur′ku-lum) (L. cover). The gill cover in bony fish.

organism (or′gan-iz″m). An individual plant or animal, either unicellular or multicellular.

osmosis (os-mo′sis) (Gr. *osmos,* impulse). The process in which water migrates through a semipermeable membrane, from a side containing a lesser concentration to the side containing a greater concentration of parti-

cles (solute). The diffusion of a solvent (usually water) through a semipermeable membrane.

osphradium (os-fra′di-um) (Gr. dim. of *osphra,* strong scent). A sense organ for testing incoming water in mollusks.

osteichthyes (os″te-ik′thi-ez) (Gr. *osteon,* bone, + *ichthys,* fish). A class of vertebrates comprising the bony fish.

ostium (os′te-um) (L. mouth). A mouthlike opening.

oviparity, oviparous (o″vi-par′i-ti, o-vip′a-rus) (L. *ovum,* egg, + *parere,* to bring forth). Reproduction in which eggs are released by the female; development of offspring occurs outside the maternal body.

ovoviviparity, ovoviviparous (o″vo-viv″i-par′i-ti, o″vo-vi-vip′a-rus) (L. *ovum,* egg, + *vivere,* to live, + *parere,* to bring forth). Reproduction in which eggs develop within the maternal body without nutrition by the female parent.

oxidation (ok″si-da′shun) (Gr. *oxys,* sharp, keen). Rearrangement of a molecule to create a high-energy bond; a chemical change in which a molecule loses one or more electrons.

Paleozoic (pa″le-o-zo′ik) (Gr. *palaios,* old, + *zoe,* life). The geologic era between the Precambrian and the Mesozoic, approximately from 550 to 200 million years ago.

palingenesis (pal″in-jen′e-sis) (Gr. *palin,* backward, + *genesis,* birth). The stages in the development or ontogeny of an animal that are inherited from ancestral species, such as gill slits in the unborn of mammals.

papilla (pa-pil′a) (L. nipple). A small nipplelike projection.

parapodia (par″a-po′di-a) (Gr. *para,* beside + *podos,* foot). The segmental appendages in polychaete worms that serve in breathing, locomotion, and creation of water currents.

parasympathetic (par″a-sim″pa-thet′ik) (Gr. *para,* beside, + *sympathes,* sympathetic). One of the subdivisions of the autonomic nervous system, whose centers are located in the brain, anterior part of the spinal cord, and posterior part of the spinal cord.

parenchyma (pa-reng′ki-ma) (Gr. *para,* beside, + *enchyma,* infusion). A type of mesenchyme with closely packed cells filling in spaces between organs; the term is also used for the chief functional cells or tissues of an organ in contrast to the supporting tissues (stroma).

parthenogenesis (par″the-no-jen′e-sis) (Gr. *parthenos,* virgin, + *genesis,* birth). The development of an unfertilized egg; a type of sexual reproduction.

pathogenic (path″o-jen′ic) (Gr. *pathos,* disease, + *gennao,* produce). Producing a disease.

pedicellaria (ped″i-sel-a′ri-a) (NL. *pedicellus,* little foot, + *aria,* like). Minute pincerlike organ on surface of certain echinoderms.

pedogenesis (pe″do-jen′e-sis) (Gr. *apis,* child, + *genesis,* birth). Reproduction by young or larval forms, especially parthenogenesis.

peduncle (pe-dung'kl) (L. dim. of *pedis*, foot). A stalk; a band of white matter joining different parts of the brain.

pelagic (pe-laj'ik) (Gr. *pelagos*, the open sea). Pertaining to the open ocean.

Pelecypoda (pel″e-sip'o-da) (Gr. *pelekus*, hatchet, + *podos*, foot). A class of the phylum Mollusca comprising clams, mussels, and oysters.

pentadactyl (pen″ta-dak'til) (Gr. *pente*, five, + *daktylos*, finger). With five digits.

peptidase (pep'ti-das) (Gr. *peptein*, to digest, + *ase*, enzyme suffix). An enzyme that splits di- or polypeptides into amino acids.

peristalsis (per″i-stal'sis) (Gr. *peri*, around, + *stalsis*, contraction). The series of alternate relaxations and contractions by which food is forced through the alimentary canal.

peritoneum (per″i-to-ne'um) (Gr. *peri*, around, + *teinein*, to stretch). The membrane that lines the abdominal cavity and covers the viscera.

petrifaction (pet″ri-fak'shun) (Gr. *petra*, stone, + L. *facere*, to make). The changing of organic matter into stone.

pH A symbol of the relative concentration of hydrogen ions in a solution; pH values are from 0 to 14, and the lower the value, the more acid or hydrogen ions in the solution.

phagocyte (fag'o-siit) (Gr. *phagein*, to eat, + *kytos*, cell). A white blood cell of the body, that devours and destroys microorganisms or other harmful substances.

phenotype (fe'no-tiip) (Gr. *phainein*, to show). The visible characters; opposed to genotype of the hereditary constitution.

Phoronida (fo-ron'i-da) (Gr. *phoros*, bearing, + L. *nidus*, nest). A phylum of wormlike, marine, tube-dwelling, schizocoelomate animals.

phosphagen (fos'fa-jen) (phosphate + glycogen). A term for creatine-phosphate and arginine-phosphate, which store and may be sources of high-energy phosphates.

phosphorylation (fos″fo-ri-la'shun). The addition of a phosphate group, such as H_2PO_3, to a compound.

phylogeny (fi-loj'e-ni) (Gr. *phylon*, tribe, + *gennao*, bring forth). The evolutionary history of a group of organisms.

phylum (pl., phyla) (fi'lum) (Gr. *phylon*, race, tribe). A chief category of taxonomic classification into which living things are divided.

pinacocyte (pin'a-ko-sit) (Gr. *pinaks*, plank, + *kytos*, cell). Flat cell found on surface and lining of sponge cavities.

pinocytosis (pi″no-si-to'sis) (Gr. *pinein*, to drink, + *kytos*, cell, + *osis*, condition). A process of cell drinking.

placenta (pla-sen'ta) (L. flat cake; from Gr. *plax, plakos*, anything flat and broad). The vascular structure, embryonic and maternal, through which the embryo and fetus are nourished while in the uterus.

plankton (plangk'ton) (Gr. *planktos*, wandering). The floating animal and plant life of a body of water.

plasma membrane (plaz'ma) (Gr. *plasma*, formed). The thin membrane that surrounds the cytosome; considered a part of the cytoplasm.

plastid (plas'tid) (Gr. *plastes*, one who forms, + *id*, daughter). A small body in the cytoplasm that often contains pigment.

Platyhelminthes (plat″y-hel-min'thes) (Gr. *platys*, flat, + *helmins*, worm). Flatworms; a phylum of acoelomate animals; consists of planarians, flukes, and tapeworms.

pleopod (ple'o-pod) (Gr. *plein*, to swim, + *podos*, foot). One of the swimming feet on the abdomen of a crustacean.

plesiosaur (ple'si-o-sor″) (Gr. *plesios*, near, + *sauros*, lizard). A long-necked, marine reptile of Mesozoic times.

pleura (ploor'a) (Gr. side). The membrane that lines each half of the thorax and covers the lungs.

plexus (plek'sus) (L. braid). A network, especially of nerves or of blood vessels.

pluteus (ploot'e-us) (L. a shed or painter's easel). The free-swimming bilateral larva of echinoids and ophiurans.

polarization (po″ler-i-za'shun) (L. *polaris*, pole). The arrangement of positive electric charges on one side of a surface membrane and negative electric charges on the other side (in nerves and muscles).

polymorphism (pol″i-mor'fizm) (Gr. *polys*, many, + *morphe*, form). The presence in a species of more than one type of individual.

polyp (pol'ip) (L. *polypus*, many-footed). The sessile stage in the life cycle of coelenterates.

polypeptide (pol″i-pep'tid) (Gr. *polys*, many, + *peptein*, to digest). A molecule consisting of many joined amino acids, not as complex as a protein.

polyphyletic (pol″i-fi-let'ik) (Gr. *polys*, many, + *phylon*, tribe). Derived from more than one ancestral type; contrasts with monophyletic, or from one ancestor.

polysaccharide (pol″i-sak'a-rid) (Gr. *polys*, many, + *sakcharon*, sugar). A carbohydrate composed of many monosaccharide units, such as glycogen, starch, and cellulose.

Porifera (po-rif'e-ra) (L. *porus*, pore, + *ferre*, to bear). The phylum of sponges.

Priapulida (pri″a-pu'li-da) (Gr. *priapos*, phallus, + *ida*, pl. suffix). A small phylum of pseudocoelomate animals.

progesterone (pro-jes'ter-on) (L. *pro*, before, + *gestare*, to carry). Hormone secreted by the corpus luteum and the placenta; prepares the uterus for the fertilized egg and maintains the capacity of the uterus to hold the embryo and fetus.

prosopyle (pro'so-pil) (Gr. *pros*, forward, + *pyle*, gate). Connection between the incurrent and radial canals in certain sponges.

prothrombin (pro-throm'bin) (L. *pro*, before, + *thrombus*, clot). A constituent of blood plasma that is changed to thrombin by thrombokinase in the presence of calcium ions; involved in blood clotting.

pterosaur (ter'o-sor) (Gr. *pteron*, feather, + *sauros*, lizard). An extinct flying reptile that flourished during the Mesozoic.

puff The pattern of swelling of specific bands or gene loci on giant chromosomes during the larval and imaginal stages of flies.

pygidium (pi-jid'i-um) (Gr. dim. of *pyge,* rump). The terminal segment in annelids.

pylorus (pi-lo'rus) (Gr. *pyle,* gate, + *ouros,* watcher). The opening between the stomach and duodenum, which is guarded by a valve.

pyrenoid (pi're-noid) (Gr. *pyren,* fruit stone, + *eidos,* form). A protein body in the chloroplasts of certain organisms that serves as a center for starch formation.

Radiolaria (ra"de-o-la're-a) (L. dim. of radius). A group of sarcodine protozoans, characterized by silicon-containing shells.

radula (rad'u-la) (L. *radere,* to scrape). Rasping tongue of certain mollusks.

redia (re'de-a) (from Redi, Italian biologist). A larval stage in the life cycle of flukes; it is produced by a sporocyst larva, and in turn gives rise to many cercariae.

rete mirabile (re'te mi-rab'i-le) (L. wonderful net). A network of small blood vessels so arranged that the incoming blood runs parallel to the outgoing, with the result that a counter-exchange is possible between the two bloodstreams.

retina (ret'i-na) (L. dim. of *rete,* a net). The sensitive, nervous layer of the eye.

rhabdocoel (rab'do-sel) (Gr. *rhabdos,* rod, + *koilos,* a hollow). A member of a group of free-living flatworms possessing a straight, unbranched digestive cavity.

rostrum (ros'trum) (L. ship's beak). A snoutlike projection on the head.

Rotifera (ro-tif'er-a) (L. *rota,* wheel, + *ferre,* to bear). A class of microscopic pseudocoelomate animals belonging to the phylum Aschelminthes.

saprophytic (sap"ro-fit'ic) (Gr. *sapros,* rotten, + *phyton,* plant). Referring to an organism living on decayed organic matter.

Sarcodina (sar"ko-di'na) (Gr. *sarkos,* flesh). A class of Protozoa; includes *Amoeba,* Foraminifera, Radiolaria, etc.; characterized by pseudopodia.

sarcolemma (sar"ko-lem'a) (Gr. *sarkos,* flesh, + *lemma,* rind). The thin noncellular membrane of striated muscle fiber or cell.

schizocoel, schizocoelomate (skiz'o-seel) (Gr. *schizein,* to split). Schizocoel is a coelum formed by a splitting of embryonic mesoderm. Schizocoelomate is an animal with a schizocoel, such as an arthropod or mollusk.

scleroblast (skle'ro-blast") (*skleros,* hard, + *blastos,* bud). A mesenchyme cell that secretes spicules in sponges.

sclerotic (skle-rot'ik) (Gr. *skleros,* hard). Pertaining to the tough outer coat of the eyeball.

scrotum (skro"tum) (L. bag). The pouch that contains the testes and accessory organs in most mammals.

seminiferous (sem"i-nif'er-us) (L. *semen,* semen, + *ferre,* to bear). Pertaining to the tubules that produce or carry semen in the testes.

semipermeable (sem"i-pur'me-a-bl") (L. *semi,* half, + *permeabilis,* capable of being passed through). Permeable to small particles, such as water and certain inorganic ions, but not to colloids, etc.

septum (sep'tum) (L. fence). A wall between two cavities.

sere (ser) (L. *serere,* to join). The sequence or series of communities that develop in a given situation from pioneer to terminal climax communities during ecologic succession.

serum (ser'um) (L. whey). The plasma of blood that separates on clotting; the liquid that separates from the blood when a clot is formed.

simian (sim'e-an) (L. *simia,* ape). Referring to monkeys.

Sipunculida (si-pun-kyu'li-da) (L. *sipunculus,* small siphon). A phylum of wormlike schizocoelomate animals.

soma (so'ma) (Gr. body). The body of an organism in contrast to the germ cells (germ plasm).

somatic (so-mat'ik) (Gr. *soma,* body). Referring to the body, such as somatic cells in contrast to germ cells.

speciation (spe'shi-a'shun) (L. *species,* kind). The evolving of two or more species by the splitting of one ancestral species.

spermatheca (spurm"a-the'ka) (Gr. *sperma,* seed, + *theke,* a case). A sac in the female reproductive organs for the storage of sperm.

sphincter (sfingk'ter) (Gr. *sphingein,* to bind tight). A ring-shaped muscle capable of closing a tubular opening by constriction.

sporocyst (spo'ro-sist) (Gr. *sporos,* seed, + *kystis,* pouch). A larval stage in the life cycle of flukes; it originates from a miracidium.

Sporozoa (spo"ro-zo'a) (Gr. *sporos,* seed, + *zoon,* animal). A class of parasitic Protozoa.

sporozoite (spo"ro-zo'it) (Gr. *sporos,* seed, + *zoon,* animal, + *ite,* offspring). A motile spore formed from the zygote in many Sporozoa.

stenohaline (sten"o-hal'in) (Gr. *stenos,* narrow, + *halinos,* made of salt). Adapted only to narrow changes in salinity of water.

stenotopic (sten"o-top'ik) (Gr. *stenos,* narrow, + *topos,* place). Referring to an organism with restricted range.

stereogastrula (ste"re-o-gas'tru-la) (Gr. *stereos,* solid + *gaster,* stomach). A solid type of gastrula, such as the planula of coelenterates.

sterol, steroid (ste'rol, ste'roid) (Gr. *stereos,* solid, + *ol* [L. *oleum,* oil]). One of a class of organic compounds containing a molecular skeleton of four fused carbon rings; it contains cholesterol, sex hormones, adrenocortical hormones, and vitamin D.

stoma (sto'ma) (Gr. mouth). A mouthlike opening.

stratum (L. covering). A horizontal layer or division of a biologic community that exhibits stratification of habitats (ecologic).

substrate (sub'strat) (L. *substratus,* strewn under). A substance that is acted upon by an enzyme.

symbiosis (sim"be-o'sis) (Gr. *sym,* with, + *bios,* life). The living together of two different species in an intimate relationship; includes mutualism, commensalism, and parasitism.

synapse (si-naps′) (Gr. *synapsis*, union). The place at which a nerve impulse passes from an axon of one nerve cell to a dendrite of another nerve cell.

syncytium (sin-sish′i-um) (Gr. *syn*, together, + *kytos*, cell). A mass of protoplasm containing many nuclei and not divided into cells.

tagma (tag′ma) (Gr. arrangement). Body division of an arthropod, containing two or more segments.

taiga (ti′ga) (Russ.). Habitat zone characterized by large tracts of coniferous forests, long, cold winters, and short summers; most typical in Canada and Siberia.

telencephalon (tel″en-sef′a-lon″) (Gr. *telos*, end, + *encephalon*, brain). The most anterior vesicle of the brain.

teleology (te″le-ol′o-ji) (Gr. *telos*, end, + *logos*, study). The philosophic view that natural events are goal directed and are preordained; contrasts with scientific view of causalism.

template (tem′plet). A pattern or mold guiding the formation of a duplicate; often used with reference to gene duplication.

tentaculocyst (ten-tak′u-lo-syst″) (L. *tentaculum*, feeler, + Gr. *kystis*, pouch). A sense organ of several parts along the margin of medusae and derived from a modified tentacle: sometimes called rhopalium.

tetrapoda (te″tra-po′da) (Gr. four-footed ones). Four-legged vertebrates; the group includes amphibians, reptiles, birds, and mammals.

therapsid (the-rap′sid) (Gr. *theraps*, an attendant). Extinct Mesozoic mammal-like reptile, from which true mammals evolved.

thrombokinase (throm″bo-kin′as) (Gr. *thrombos*, lump, + *kinein*, to move, + *ase*, enzyme suffix). Enzyme released from blood platelets that initiates the process of clotting; transforms prothrombin into thrombin in presence of calcium ions; thromboplastin.

tornaria (tor-na′ri-a) (L. *tornare*, to turn, to round off). Free-swimming larva of certain Enteropneusta; similar to the auricular or bipinnaria larva of certain echinoderms.

trachea (tra′ke-a) (ML. windpipe, trachea). The windpipe; any of the air tubes of insects.

transduction (trans-duk′shun) (L. *trans*, across, + *ducere*, to lead). Transfer of genetic material from one bacterium to another through the agency of virus.

trochophore (trok′o-for″) (Gr. *trochos*, wheel, + *phoros*, bearing). A free-swimming ciliated marine larva characteristic of schizocoelomate animals; common to many phyla.

trophallaxis (trof″al-lak′sis) (Gr. *trophos*, feeder, + *allaxis*, exchange). Exchange of food between young and adults, especially among those of certain social insects.

trophozoite (trof″o-zo′it) (Gr. *trephein*, to nourish, + *zoon*, animal, + *ite*, offspring). That stage in the life cycle of a sporozoan in which it is actively absorbing nourishment from the host.

tundra (ton′dra) (Russ.). Terrestrial habitat zone, between taiga in south and polar region in north; characterized by absence of trees, short growing season, and mostly frozen soil during much of the year.

typhlosole (tif′lo-sol) (Gr. *typhlos*, blind, + *solen*, channel). A longitudinal fold projecting into the intestine in certain invertebrates such as the earthworm.

ungulate (ung′gu-lat) (L. *ungula*, hoof). Hoofed mammals.

urethra (u-re′thra) (Gr. *ourethra*, urethra). The tube from the urinary bladder to the exterior in both sexes.

uriniferous tubule (u″ri-nif′er-us) (L. *urina*, urine, + *ferre*, to bear). One of the tubules in the kidney extending from a malpighian body to the collecting tubule.

Urochordata (u″ro-kor-da′ta) (Gr. *oura*, tail, + L. *chorda*, cord). A subphylum of chordates; often called the Tunicata.

utricle (u′tri-kl) (L. *utriculus*, little bag). That part of the inner ear containing the receptors for dynamic body balance; the semicircular canals lead from and to the utricle.

vacuole (vak′u-ol) (L. *vacuus*, empty, + *ole*, dim.). A fluid-filled space in a cell.

vagility (va-jil′i-ty) (L. *vagus*, wandering). Ability to tolerate environmental variation or the ability to cross ecologic barriers. Example: Birds have high and mollusks very low vagility.

veliger (vel′i-jer) (L. *velum*, sail, awning). Posttrochophoral larva stage in certain mollusks.

vestige (ves′tij) (L. *vestigium*, footprint). A rudimentary structure that is well developed in some other species or in the embryo.

villus (vil′us) (L. tuft of hair). A small fingerlike process on the wall of the small intestine and on the embryonic portion of the placenta.

virus (vi′rus) (L. slimy liquid poison). A submicroscopic noncellular particle, composed of a nucleoprotein core and a protein shell; parasitic and will grow and reproduce in a host cell.

viscera (vis′er-a) (L., pl. of *viscus*, internal organ). Internal organs in the body cavity.

vitalism (vi′tal-iz″m) (L. *vita*, life). The view that natural processes are controlled by supernatural forces and cannot be explained through the laws of physics and chemistry alone; contrasts with mechanism.

vitamin (vi′ta-min) (L. *vita*, life). An organic substance contributing to the formation or action of cellular enzymes; essential for the maintenance of life.

vitelline membrane (vi-tel′in) (L. *vitellus*, yolk of an egg). The noncellular membrane that encloses the egg cell.

viviparity, viviparous (viv″i-par′i-ti, vi-vip′a-rus) (L. *vivus*, alive, + *parere*, to bring forth). Reproduction in which eggs develop within the female body, with nutritional aid of maternal parent; offspring are born as juveniles.

xanthophyll (zan′tho-fil) (Gr. *xanthos*, yellow, + *phyllon*, leaf). One of a group of yellow pigments found widely among plants and animals; the xanthophylls are members of the carotenoid group of pigments.

zoochlorella (zo′o-klo-rel″la) (Gr. *zoon*, animal, + *chloros*, green, + dim. ending). One of the green symbiotic algae found in certain invertebrates.

zygote (zi′got) (Gr. *zygotos*, yoked). The cell formed by the union of a male and a female gamete; the fertilized egg.

index*

*Bold face numbers refer to illustrations.